Origins and Natures of Inflation, Dark Matter and Dark Energy

Origins and Natures of Inflation, Dark Matter and Dark Energy

Editor

Kazuharu Bamba

Basel • Beijing • Wuhan • Barcelona • Belgrade • Novi Sad • Cluj • Manchester

Editor
Kazuharu Bamba
Faculty of Symbiotic
Systems Science
Fukushima University
Fukushima
Japan

Editorial Office
MDPI
St. Alban-Anlage 66
4052 Basel, Switzerland

This is a reprint of articles from the Special Issue published online in the open access journal *Universe* (ISSN 2218-1997) (available at: https://www.mdpi.com/journal/universe/special_issues/origins).

For citation purposes, cite each article independently as indicated on the article page online and as indicated below:

Lastname, A.A.; Lastname, B.B. Article Title. *Journal Name* **Year**, *Volume Number*, Page Range.

ISBN 978-3-7258-0755-0 (Hbk)
ISBN 978-3-7258-0756-7 (PDF)
doi.org/10.3390/books978-3-7258-0756-7

Contents

About the Editor

Kazuharu Bamba

Kazuharu Bamba is an Associate Professor at Fukushima University in Japan. He has been working on inflationary cosmology, dark energy problems, and modified gravity theories. He is an Editor of *Universe, Symmetry and Entropy* (MDPI).

Preface

According to recent, very precise cosmological observations, including type Ia supernovae (NaIa), cosmic microwave background (CMB) radiation, the large-scale structure of the universe, baryon acoustic oscillations (BAOs), and weak lensing, it has been established that not only inflation in the early universe but also expansion of the current universe is accelerating. If the current universe is spatially flat, it is well known that the energy components of the current universe are classified into three components: dark energy (about 70%), dark matter (about 25%), and baryon (about 5%).

Various studies have been proposed to explain the origins of dark energy, dark matter, and baryon, as well as the physical mechanism of inflation from theoretical physics and astronomical points of view based on fundamental physics theories such as general relativity, quantum field theories, particle physics, and astrophysics. In addition to gravitational waves from colliding neutron stars or binary black holes, if the primordial gravitational waves from inflation are detected through the B-mode polarization of CMB radiation by future detectors, the information of the energy scale of inflation can be obtained.

There are two possibilities for the origin of dark matter, namely, new particles in particle-theory models beyond the standard model, and astrophysical objects. In addition, there are two representative approaches that can be used to investigate the true properties of dark energy, with its negative pressure, to lead to the late-time cosmic acceleration. One introduces some unknown matter, called "dark energy", within the framework of general relativity. The other extends general relativity to larger scales. The latter approach is recognized as geometrical dark energy.

The main aim of this Special Issue is to understand the origins and true natures of inflation, dark matter, and dark energy. The organization of this reprint is as follows. The first part (containing one article) details inflation based on holographic space–time. The second part (containing six articles) concerns the origin and nature of dark matter including magnetized quark nugget dark matter and axion-like particles. The third part (comprising four articles) covers the origin of dark energy and the extended theories of gravitation.

I would like to sincerely acknowledge MDPI. I am also greatly appreciative of the Managing Editor, Ms. Athena Li, for her kindness, support, and assistance throughout this project. Moreover, I am extremely grateful to the Editor-in-Chief, Professor Dr. Lorenzo Iorio, for giving me the opportunity to serve as Guest Editor of this Special Issue. I would also like to thank all of the authors for submitting their articles to this Special Issue of *Universe*.

Kazuharu Bamba
Editor

Editorial

Origins and Natures of Inflation, Dark Matter and Dark Energy

Kazuharu Bamba

Faculty of Symbiotic Systems Science, Fukushima University, Fukushima 960-1296, Japan; bamba@sss.fukushima-u.ac.jp

check for **updates**

Citation: Bamba, K. Origins and Natures of Inflation, Dark Matter and Dark Energy. *Universe* **2024**, *10*, 144. https://doi.org/10.3390/universe10030144

Received: 13 September 2023
Accepted: 12 March 2024
Published: 15 March 2024

Various precise cosmological observations, e.g., Supernovae Ia (SNe Ia) [1,2], the cosmic microwave background (CMB) radiation [3–8], the large-scale structure (LSS) of the universe [9,10], the baryon acoustic oscillations (BAO) [11,12], and the effect of weak lensing [13–15], have strongly suggested that the accelerated expansion of the present universe is realized in addition to inflation in the early universe [16–19]. In particular, according to the recent Planck results [7,8], for the spatially flat universe, the energy of the current universe is composed of the following three components: (i) Dark energy (around 70%), an unknown type of energy with negative pressure; (ii) Dark matter (around 25%), which does not shine and has only its gravitational interaction; (iii) Baryon (around 5%), i.e., basically protons and nucleons.

It is expected that more detailed and precise observational data in terms of modern cosmology will be obtained by the Euclid satellite [20] of the European Space Agency (ESA) [21–27]. Moreover, the events of gravitational waves have been detected [28,29], along with cosmology, through further future observations of the gravitational waves not only from astrophysical compact objects, but also the origins in the early universe, including inflation and cosmological phase Electro-Weak (EW) and QCD transitions [30–33].

Two representative approaches have been explored so that the mechanism of the accelerated expansion of the late-time universe can be understood. The first approach is to assume the existence of dark energy such as the cosmological constant within general relativity. The second is to extend a gravity theory from general relativity at large scale. The latter is interpreted as a kind of geometrical dark energy. Various extended theories of gravity have been studied (for detailed reviews of the physics of the cosmic acceleration, dark energy, alternative theories of gravity, and their cosmological and astrophysical applications and investigations, see, e.g., Refs. [34–63] and references therein). This Special Issue of *Universe*, "Origins and Natures of Inflation, Dark Matter and Dark Energy", collects eleven original research manuscripts on the topics of inflation, dark matter, and dark energy. The Special Issue is organized as follows. Firstly, the topic of inflation [64] related to the origin of dark matter is discussed. Secondly, both theoretical and experimental studies of dark matter are described. In particular, quark-nugget dark matter [65–67] and axion-like particles [68–70] are investigated. After that, the subjects of dark energy [71,72] and modified gravity theories [73] are explored. In the end, acting as a summary of sorts, a recent review on modern cosmology in terms of dark energy, dark matter, as well as inflation [74] is included in this Special Issue. See below for a brief overview of the ten research articles and one review included in this Special Issue.

In Ref. [64], as a candidate of small primordial black holes, the discretely charged dark matter is studied in inflationary cosmology with the holographic spacetime. A new model of black holes created by inflation is proposed. The Big Bang universe is realized by the decay of the black holes, and the charge of a discrete symmetry has the smallest value. The fraction of the inflationary black holes carrying this charge is determined for the case in which the universe enters the matter-dominated stage from the radiation-dominated stage at a cosmic temperature of approximately 1 eV.

In Ref. [65], as a candidate of dark matter, the limits on magnetized quark nugget constructed by up, down, and strange quarks, whose numbers are approximately equal, are analyzed based on episodic natural events. This is an application of the implication that the center part of magnetar is composed of quark nuggets, which are a liquid state with a ferromagnetic nature. Magnetized quark nuggets are known to form and aggregate before decaying, and their mass distribution is broad and stabilized by magnetic fields. Through the magnetopause, magnetized quark nuggets can interact with ordinary matter. During this process, their translational velocity decreases and their rotational velocity increases, and the energy of electromagnetic fields radiates. In this work, rare events compatible with the property of magnetized quark nuggets are explored. The strength of magnetic fields covering quark nuggets is constrained and a proposal to test whether magnetized quark nuggets can be a candidate of dark matter is supported.

In Ref. [66], for a possible candidate of dark matter, magnetized quark nuggets are evaluated based on their radial impacts on the earth. At the early stage of the universe, magnetized quark nuggets formed and aggregated before decaying through the weak force with a wide distribution of mass. An event has been reported which may support the presence of magnetized quark nuggets. The parameter of magnetic fields on the surface determines the distribution of the mass of magnetized quark nuggets and the cross section of the interaction. Sufficient energy may be transferred to create craters that do not originate from meteorites. In the present work, the computer simulations for the energy deposition of magnetized quark nuggets are performed for an environment containing peat saturated by water, soft sediments, and granite. Moreover, the report of the excavation of the crater is shown. Five agreement points of the observations with the computer simulations support the second event, which suggests magnetized quark nuggets. Furthermore, the potential qualification of more events for magnetized quark nuggets is discussed.

In Ref. [67], the possibility that the multi-modal events of the Horizon-10T are related to quark nuggets of axion fields is discussed. Multi-modal events with several peaks, implying they originated from clustering, were reported by the Horizon-10T collaboration. It is proposed that the events of the annihilation of dark matter would lead to these multi-modal events in a dark matter model of quark nuggets of axion fields. This is because it is too difficult to understand these events based on an ordinary interpretation with cosmic rays. It is demonstrated that various observational results such as the frequency of their appearance, the intensity, the distribution of space, the duration of time, and the property of the clustering may be compatible with the nature of the emission from the events of the atmosphere for the annihilation of quark nuggets of axion fields. In addition, in light of the ordinary air showers of the cosmic rays, many properties relating to the events of quark nuggets of axion fields are discussed.

In Ref. [68], a photon collider of the resonance of the stimulation with the fields of the focused lasers is investigated. The present collider with three beams is used for the direct production of particles like axions. Two beams are used to create axion-like particles and the other beam is used to simulate the decay of such particles. This research explores how suitable the photon collider is for examining particles like axions whose mass range is about eV. It is shown that the particles like axions with a mass around the eV range may be probed. In addition, the sensitivity of the coupling between particles like axions and photons is analyzed.

In Ref. [69], a pilot survey of particles like axions is performed using a photon collider. The photon collider is used for the resonance of the stimulation, and it has three beams with the lasers emitting short pulses. In the case of the present photon collider of the resonance of the stimulation, three laser beams with short pulses are focused into a vacuum so that the particles like axions, with a mass range around eV, can be detected systematically. In order to realize such a collider, a proof-of-principle experiment is described. The incident angles of these three beams are made large to solve the problem in that the overlap of the spacetime of the lasers with short pulses must be maintained. Moreover, a way of evaluating the bias of the states of the polarization is investigated. This method is important in a system for a collision with variable incident angles in the future. This paper describes the consequences of this pilot survey, as well as this method by using the exploited system. The result of this survey is compatible with the null state. The largest possible value for the minimum of the coupling between particles like axions and photons is also derived.

In Ref. [70], the plan and composition of a photon collider are presented. The photon collider that provides the resonance for the stimulation has three very strong laser beams with variable angles. The purpose of this collider is to survey particles like axion whose mass scale is around eV. The angle of the emission of these three laser beams can be changed, and therefore the energy of the collision for the system of the center of mass may vary. As a result, the mass range around eV can be surveyed continuously. Furthermore, through the calibration of laser beams, the mechanism of the variation of the angle is verified. The realistic value of the sensitivity of the photon collider is also projected for a future survey.

In Ref. [71], a scenario in which dark energy is unified with dark matter is proposed as a novel version of a dark energy model of generalized Chaplygin gas. The evolutions of the Hubble parameter and distance modulus for the present scenario under considerations and the ΛCDM model are explored. The theoretical consequences are verified using cosmological observations. In addition, two geometric diagnostics are analyzed to distinguish the new model from ΛCDM. Furthermore, with different observational data points, the trajectories of the evolution for the planes of the diagnostic are explicitly depicted to investigate the geometric property of the proposed new model.

In Ref. [72], the solutions of the homogeneous and anisotropic spacetime of the Bianchi type I are derived for a quintom theory with multifield chirality. In such an extended chiral model, the energy density of one or two scalar fields is negative. When a degree of freedom of this theory is removed, the original quintom theory appears. The Kasner type analytic solutions and an exponential form with anisotropy are found in terms of the potential of the scalar field with its specialized functional expression. Moreover, based on the Noether symmetry, the theories are classified by their symmetries and the laws of conservation are also demonstrated.

In Ref. [73], a solution of charged, nonlinear black holes with is explored as part of the Rastall gravity theory. The model parameter in the theory does not influence the solution of the linear gravitational field equation for a charged black hole with spherical symmetry. On the other hand, if a nonlinear electrodynamic source exists, a new spherically symmetric black hole solution involved with the Rastall parameter, mainly originated from the non-vanishing trace part of the nonlinear electrodynamic source, is derived. In addition, it is demonstrated that the new black hole solution is regarded as the Reissner–Nordström one for the anti-de Sitter spacetime, where the cosmological constant includes the model parameter of the Rastall gravity. When the case is limited to general relativity, in which the Rastall parameter vanishes, the new solution corresponds to the solution of Reissner–Nordström spacetime. Furthermore, by analyzing the geodesic deviation of gravitational field equations and thermodynamic properties, including the first law of thermodynamics, it is shown that this black hole solution is stable, differing from the charged case with the linearity, in which the second-order phase transition occurs.

In Ref. [74], with recent various cosmological observational data, the constrains on dark energy models in which a dynamical scalar field plays the role of dark energy are overviewed in detail. Such scalar fields are classified into two types: a canonical scalar field called quintessence, whose value of the equation of state is larger than -1 and less than $-1/3$; and a kind of non-canonical scalar field called the phantom field, whose value of the equation of state can be less than -1. The value of the equation of state of the cosmological constant is -1. The energy density of such a scalar field can lead to the late-time accelerated expansion of the universe. The background and theoretical motivations of these models are presented. A scenario in which dark energy interacts with dark matter is also described. The recent observational constraints on the theoretical model parameters are explained. It is demonstrated that the ΛCDM model with spatial flatness is favored by the observations, and that the dark energy models consisting of such a scalar field may still be compatible with the cosmological observations.

It is considered that the eleven papers that comprise this Special Issue will provide useful references for future works investigating the origins and natures of the mechanism of inflation, dark matter and dark energy in modern physics and cosmology.

Funding: This work was supported, in part, by the JSPS KAKENHI Grant Number JP21K03547 and 23KF0008.

Acknowledgments: Contributions of all the authors are highly appreciated by the Guest Editor (Kazuharu Bamba) of this Special Issue.

Conflicts of Interest: The author declares no conflicts of interest.

References

1. Perlmutter, S.; Aldering, G.; Goldhaber, G.; Knop, R.A.; Nugent, P.; Castro, P.G.; Deustua, S.; Fabbro, S.; Goobar, A.; Groom, D.E.; et al. Measurements of Omega and Lambda from 42 High-Redshift Supernovae. *Astrophys. J.* **1999**, *517*, 565. [CrossRef]
2. Riess, A.G.; Filippenko, A.V.; Challis, P.; Clocchiatti, A.; Diercks, A.; Garnavich, P.M.; Gillil, R.L.; Hogan, C.J.; Jha, S.; Kirshner, R.P.; et al. Observational Evidence from Supernovae for an Accelerating Universe and a Cosmological Constant. *Astron. J.* **1998**, *116*, 1009. [CrossRef]
3. Spergel, D.N.; Verde, L.; Peiris, H.V.; Komatsu, E.; Nolta, M.R.; Bennett, C.L.; Halpern, M.; Hinshaw, G.; Jarosik, N.; Kogut, A.; et al. First Year Wilkinson Microwave Anisotropy Probe (WMAP) Observations: Determination of Cosmological Parameters. *Astrophys. J. Suppl.* **2003**, *148*, 175. [CrossRef]
4. Spergel, D.N.; Bean, R.; Doré, O.; Nolta, M.R.; Bennett, C.L.; Dunkley, J.; Hinshaw, G.; Jarosik, N.; Komatsu, E.; Page, L.; et al. Wilkinson Microwave Anisotropy Probe (WMAP) three year results: Implications for cosmology. *Astrophys. J. Suppl.* **2007**, *170*, 377. [CrossRef]
5. Komatsu, E.; Dunkley, J.; Nolta, M.R.; Bennett, C.L.; Gold, B.; Hinshaw, G.; Jarosik, N.; Larson, D.; Limon, M.; Page, L.E.A.; et al. Five-Year Wilkinson Microwave Anisotropy Probe (WMAP) Observations: Cosmological Interpretation. *Astrophys. J. Suppl.* **2009**, *180*, 330. [CrossRef]
6. Bennett, C.L.; Hill, R.S.; Hinshaw, G.; Larson, D.; Smith, K.M.; Dunkley, J.; Gold, B.; Halpern, M.; Jarosik, N.; Kogut, A.; et al. Seven-Year Wilkinson Microwave Anisotropy Probe (WMAP) Observations: Cosmological Interpretation. *Astrophys. J. Suppl.* **2011**, *192*, 18. [CrossRef]
7. Aghanim, N. et al. [Planck Collaboration] Planck 2018 results. VI. Cosmological parameters. *Astron. Astrophys.* **2020**, *641*, A6; Erratum in *Astron. Astrophys.* **2021**, *652*, C4.
8. Akrami, Y. et al. [Planck Collaboration] Planck 2018 results. X. Constraints on inflation. *Astron. Astrophys.* **2020**, *641*, A10.
9. Tegmark, M.; Strauss, M.A.; Blanton, M.R.; Abazajian, K.; Dodelson, S.; Sandvik, H.; Wang, X.; Weinberg, D.H.; Zehavi, I.; Bahcall, N.A.; et al. Cosmological parameters from SDSS and WMAP. *Phys. Rev. D* **2004**, *69*, 103501. [CrossRef]
10. Seljak, U.; Makarov, A.; McDonald, P.; Anderson, S.F.; Bahcall, N.A.; Brinkmann, J.; Burles, S.; Cen, R.; Doi, M.; Gunn, J.E.; et al. Cosmological parameter analysis including SDSS Ly-alpha forest and galaxy bias: Constraints on the primordial spectrum of fluctuations, neutrino mass, and dark energy. *Phys. Rev. D* **2005**, *71*, 103515. [CrossRef]
11. Eisenstein, D.J.; Zehavi, I.; Hogg, D.W.; Scoccimarro, R.; Blanton, M.R.; Nichol, R.C.; Scranton, R.; Seo, H.J.; Tegmark, M.; Zheng, Z.; et al. Detection of the Baryon Acoustic Peak in the Large-Scale Correlation Function of SDSS Luminous Red Galaxies. *Astrophys. J.* **2005**, *633*, 560. [CrossRef]
12. Alam, S.; Aubert, M.; Avila, S.; Ball, C.; Bautista, J.E.; Bershady, M.A.; Bizyaev, D.; Blanton, M.R.; Bolton, A.S.; Bovy, J.; et al. Completed SDSS-IV extended Baryon Oscillation Spectroscopic Survey: Cosmological implications from two decades of spectroscopic surveys at the Apache Point Observatory. *Phys. Rev. D* **2021**, *103*, 083533. [CrossRef]
13. Jain, B.; Taylor, A. Cross-correlation Tomography: Measuring Dark Energy Evolution with Weak Lensing. *Phys. Rev. Lett.* **2003**, *91*, 141302. [CrossRef] [PubMed]
14. Abbott, T.M.; Abdalla, F.B.; Alarcon, A.; Aleksić, J.; Allam, S.; Allen, S.; Amara, A.; Annis, J.; Asorey, J.; Avila, S.; et al. Dark Energy Survey year 1 results: Cosmological constraints from galaxy clustering and weak lensing. *Phys. Rev. D* **2018**, *98*, 043526. [CrossRef]
15. Abbott, T.M. et al. [DES Collaboration] Dark Energy Survey Year 3 results: Cosmological constraints from galaxy clustering and weak lensing. *Phys. Rev. D* **2022**, *105*, 023520. [CrossRef]
16. Guth, A.H. The Inflationary Universe: A Possible Solution to the Horizon and Flatness Problems. *Phys. Rev. D* **1981**, *23*, 347. [CrossRef]
17. Sato, K. First Order Phase Transition of a Vacuum and Expansion of the Universe. *Mon. Not. Roy. Astron. Soc.* **1981**, *195*, 467–479. [CrossRef]
18. Starobinsky, A.A. A New Type of Isotropic Cosmological Models Without Singularity. *Phys. Lett.* **1980**, *91B*, 99. [CrossRef]
19. Linde, A.D. A New Inflationary Universe Scenario: A Possible Solution of the Horizon, Flatness, Homogeneity, Isotropy and Primordial Monopole Problems. *Phys. Lett.* **1982**, *108B*, 389. [CrossRef]
20. Available online: https://www.esa.int/Science_Exploration/Space_Science/Euclid (accessed on 12 September 2023).

21. Laureijs, R.; Amiaux, J.; Arduini, S.; Augueres, J.L.; Brinchmann, J.; Cole, R.; Cropper, M.; Dabin, C.; Duvet, L.; Ealet, A.; et al. Euclid Definition Study Report. *arXiv* **2011**, arXiv:1110.3193.
22. Amendola, L.; Appleby, S.; Avgoustidis, A.; Bacon, D.; Baker, T.; Baldi, M.; Bartolo, N.; Blanchard, A.; Bonvin, C.; Borgani, S.; et al. Cosmology and fundamental physics with the Euclid satellite. *Living Rev. Rel.* **2013**, *16*, 6. [CrossRef]
23. Amendola, L.; Appleby, S.; Avgoustidis, A.; Bacon, D.; Baker, T.; Baldi, M.; Bartolo, N.; Blanchard, A.; Bonvin, C.; Borgani, S.; et al. Cosmology and fundamental physics with the Euclid satellite. *Living Rev. Rel.* **2018**, *21*, 2. [CrossRef]
24. Blanchard, A.; Camera, S.; Carbone, C.; Cardone, V.F.; Casas, S.; Clesse, S.; Ilić, S.; Kilbinger, M.; Kitching, T.; Kunz, M.; et al. Euclid preparation. VII. Forecast validation for Euclid cosmological probes. *Astron. Astrophys.* **2020**, *642*, A191.
25. Nesseris, S.; Sapone, D.; Martinelli, M.; Camarena, D.; Marra, V.; Sakr, Z.; Garcia-Bellido, J.; Martins, C.J.A.P.; Clarkson, C.; Silva, A.D.; et al. Euclid: Forecast constraints on consistency tests of the ΛCDM model. *Astron. Astrophys.* **2022**, *660*, A67. [CrossRef]
26. Casas, S.; Cardone, V.F.; Sapone, D.; Frusciante, N.; Pace, F.; Parimbelli, G.; Archidiacono, M.; Koyama, K.; Tutusaus, I.; Camera, S.; et al. Euclid: Constraints on f(R) cosmologies from the spectroscopic and photometric primary probes. *arXiv* **2023**, arXiv:2306.11053.
27. Ballardini, M.; Akrami, Y.; Finelli, F.; Karagiannis, D.; Li, B.; Li, Y.; Sakr, Z.; Sapone, D.; Achúcarro, A.; Baldi, M.; et al. *Euclid*: The search for primordial features. *arXiv* **2023**, arXiv:2309.17287.
28. Abbott, B.P. et al. [LIGO Scientific Collaboration and Virgo Collaboration] Observation of Gravitational Waves from a Binary Black Hole Merger. *Phys. Rev. Lett.* **2016**, *116*, 061102. [CrossRef] [PubMed]
29. Abbott, B.P. et al. [LIGO Scientific Collaboration and Virgo Collaboration] GW170817: Observation of Gravitational Waves from a Binary Neutron Star Inspiral. *Phys. Rev. Lett.* **2017**, *119*, 161101. [CrossRef]
30. Abbott, B.P.; Abbott, R.; Abbott, T.D.; Acernese, F.; Ackley, K.; Adams, C.; Adams, T.; Addesso, P.; Adhikari, R.X.; Adya, V.B.; et al. Multi-messenger Observations of a Binary Neutron Star Merger. *Astrophys. J. Lett.* **2017**, *848*, L12. [CrossRef]
31. Barack, L.; Cardoso, V.; Nissanke, S.; Sotiriou, T.P.; Askar, A.; Belczynski, C.; Bertone, G.; Bon, E.; Blas, D.; Brito, R.; et al. Black holes, gravitational waves and fundamental physics: A roadmap. *Class. Quant. Grav.* **2019**, *36*, 143001. [CrossRef]
32. Abdalla, E.; Abellán, G.F.; Aboubrahim, A.; Agnello, A.; Akarsu, O.; Akrami, Y.; Alestas, G.; Aloni, D.; Amendola, L.; Anchordoqui, L.A.; et al. Cosmology intertwined: A review of the particle physics, astrophysics, and cosmology associated with the cosmological tensions and anomalies. *JHEAp* **2022**, *34*, 49–211. [CrossRef]
33. Agazie, G.; Anumarlapudi, A.; Archibald, A.M.; Arzoumanian, Z.; Baker, P.T.; Bécsy, B.; Blecha, L.; Brazier, A.; Brook, P.R.; Burke-Spolaor, S.; et al. The NANOGrav 15 yr Data Set: Evidence for a Gravitational-wave Background. *Astrophys. J. Lett.* **2023**, *951*, L8. [CrossRef]
34. Nojiri, S.; Odintsov, S.D. Introduction to modified gravity and gravitational alternative for dark energy. *Int. J. Geom. Meth. Mod. Phys.* **2007**, *4*, 115. [CrossRef]
35. Copeland, E.J.; Sami, M.; Tsujikawa, S. Dynamics of dark energy. *Int. J. Mod. Phys. D* **2006**, *15*, 1753–1936. [CrossRef]
36. Padmanabhan, T. Dark energy and gravity. *Gen. Rel. Grav.* **2008**, *40*, 529–564. [CrossRef]
37. Durrer, R.; Maartens, R. Dark Energy and Dark Gravity. *Gen. Rel. Grav.* **2008**, *40*, 301–328. [CrossRef]
38. Sotiriou, T.P.; Faraoni, V. $f(R)$ Theories Of Gravity. *Rev. Mod. Phys.* **2010**, *82*, 451–497. [CrossRef]
39. Cai, Y.F.; Saridakis, E.N.; Setare, M.R.; Xia, J.Q. Quintom Cosmology: Theoretical implications and observations. *Phys. Rept.* **2010**, *493*, 1–60. [CrossRef]
40. Felice, A.D.; Tsujikawa, S. $f(R)$ theories. *Living Rev. Rel.* **2010**, *13*, 3. [CrossRef]
41. Amendola, L.; Tsujikawa, S. *Dark Energy*; Cambridge University Press: Cambridge, UK, 2010.
42. Faraoni, V.; Capozziello, S. *Beyond Einstein Gravity*; Springer: Dordrecht, The Netherlands, 2010.
43. Nojiri, S.; Odintsov, S.D. Unified cosmic history in modified gravity: From $F(R)$ theory to Lorentz non-invariant models. *Phys. Rept.* **2011**, *505*, 59–144. [CrossRef]
44. Capozziello, S.; Laurentis, M.D. Extended Theories of Gravity. *Phys. Rept.* **2011**, *509*, 167–321. [CrossRef]
45. Clifton, T.; Ferreira, P.G.; Padilla, A.; Skordis, C. Modified Gravity and Cosmology. *Phys. Rept.* **2012**, *513*, 1–189.
46. Weinberg, D.H.; Mortonson, M.J.; Eisenstein, D.J.; Hirata, C.; Riess, A.G.; Rozo, E. Observational Probes of Cosmic Acceleration. *Phys. Rept.* **2013**, *530*, 87–255. [CrossRef]
47. Bamba, K.; Capozziello, S.; Nojiri, S.; Odintsov, S.D. Dark energy cosmology: The equivalent description via different theoretical models and cosmography tests. *Astrophys. Space Sci.* **2012**, *342*, 155–228. [CrossRef]
48. Will, C.M. The Confrontation between General Relativity and Experiment. *Living Rev. Rel.* **2014**, *17*, 4. [CrossRef] [PubMed]
49. Joyce, A.; Jain, B.; Khoury, J.; Trodden, M. Beyond the Cosmological Standard Model. *Phys. Rept.* **2015**, *568*, 1–98. [CrossRef]
50. Bamba, K.; Odintsov, S.D. Inflationary cosmology in modified gravity theories. *Symmetry* **2015**, *7*, 220–240. [CrossRef]
51. Cai, Y.F.; Capozziello, S.; Laurentis, M.D.; Saridakis, E.N. $f(T)$ teleparallel gravity and cosmology. *Rept. Prog. Phys.* **2016**, *79*, 106901. [CrossRef]
52. Nojiri, S.; Odintsov, S.D.; Oikonomou, V.K. Modified Gravity Theories on a Nutshell: Inflation, Bounce and Late-time Evolution. *Phys. Rept.* **2017**, *692*, 1–104. [CrossRef]
53. Langlois, D. Dark energy and modified gravity in degenerate higher-order scalar–tensor (DHOST) theories: A review. *Int. J. Mod. Phys. D* **2019**, *28*, 1942006. [CrossRef]
54. Frusciante, N.; Perenon, L. Effective field theory of dark energy: A review. *Phys. Rept.* **2020**, *857*, 1–63. [CrossRef]
55. Olmo, G.J.; Rubiera-Garcia, D.; Wojnar, A. Stellar structure models in modified theories of gravity: Lessons and challenges. *Phys. Rept.* **2020**, *876*, 1–75. [CrossRef]

56. Saridakis, E.N.; Lazkoz, R.; Salzano, V.; Moniz, P.V.; Capozziello, S.; Jimenez, J.B.; Laurentis, M.D.; Olmo, G.J. *Modified Gravity and Cosmology*; Springer: Cham, Switzerland, 2021.
57. Bamba, K. Review on Dark Energy Problem and Modified Gravity Theories. *LHEP* **2022**, *2022*, 352. [CrossRef]
58. Bahamonde, S.; Dialektopoulos, K.F.; Escamilla-Rivera, C.; Farrugia, G.; Gakis, V.; Hendry, M.; Hohmann, M.; Said, J.L.; Mifsud, J.; Valentino, E.D. Teleparallel gravity: From theory to cosmology. *Rept. Prog. Phys.* **2023**, *86*, 026901. [CrossRef]
59. Arai, S.; Aoki, K.; Chinone, Y.; Kimura, R.; Kobayashi, T.; Miyatake, H.; Yamauchi, D.; Yokoyama, S.; Akitsu, K.; Hiramatsu, T.; et al. Cosmological gravity probes: Connecting recent theoretical developments to forthcoming observations. *PTEP* **2023**, *2023*, 072E01. [CrossRef]
60. Odintsov, S.D.; Oikonomou, V.K.; Giannakoudi, I.; Fronimos, F.P.; Lymperiadou, E.C. Recent Advances in Inflation. *Symmetry* **2023**, *15*, 1701. [CrossRef]
61. de Haro, J.; Nojiri, S.; Odintsov, S.D.; Oikonomou, V.K.; Pan, S. Finite-time cosmological singularities and the possible fate of the Universe. *Phys. Rept.* **2023**, *1034*, 1–114. [CrossRef]
62. Heisenberg, L. Review on f(Q) gravity. *Phys. Rept.* **2024**, *1066*, 1–78. [CrossRef]
63. Yousaf, Z.; Bamba, K.; Bhatti, M.Z.; Farwa, U. Quasi-static evolution of axially and reflection symmetric large-scale configuration. *arXiv* **2023**, arXiv:2311.10369.
64. Banks, T.; Fischler, W. Discretely Charged Dark Matter in Inflation Models Based on Holographic Space-Time. *Universe* **2022**, *8*, 600. [CrossRef]
65. Vandevender, J.P.; VanDevender, A.P.; Wilson, P.; Hammel, B.F.; McGinley, N. Limits on Magnetized Quark-Nugget Dark Matter from Episodic Natural Events. *Universe* **2021**, *7*, 35. [CrossRef]
66. Vandevender, J.P.; Schmitt, R.G.; Mcginley, N.; Duggan, D.G.; McGinty, S.; Vandevender, A.P.; Wilson, P.; Dixon, D.; Girard, H.; Mcrae, J. Results of Search for Magnetized Quark-Nugget Dark Matter from Radial Impacts on Earth. *Universe* **2021**, *7*, 116. [CrossRef]
67. Zhitnitsky, A. Multi-Modal Clustering Events Observed by Horizon-10T and Axion Quark Nuggets. *Universe* **2021**, *7*, 384. [CrossRef]
68. Homma, K.; Ishibashi, F.; Kirita, Y.; Hasada, T. Sensitivity to axion-like particles with a three-beam stimulated resonant photon collider around the eV mass range. *Universe* **2023**, *9*, 20. [CrossRef]
69. Ishibashi, F.; Hasada, T.; Homma, K.; Kirita, Y.; Kanai, T.; Masuno, S.; Tokita, S.; Hashida, M. Pilot Search for Axion-Like Particles by a Three-Beam Stimulated Resonant Photon Collider with Short Pulse Lasers. *Universe* **2023**, *9*, 123. [CrossRef]
70. Hasada, T.; Homma, K.; Kirita, Y. Design and Construction of a Variable-Angle Three-Beam Stimulated Resonant Photon Collider toward eV-Scale ALP Search. *Universe* **2023**, *9*, 355. [CrossRef]
71. Al Mamon, A.; Dubey, V.C.; Bamba, K. Statefinder and O_m Diagnostics for New Generalized Chaplygin Gas Model. *Universe* **2021**, *7*, 362. [CrossRef]
72. Paliathanasis, A. Bianchi I Spacetimes in Chiral–Quintom Theory. *Universe* **2022**, *8*, 503. [CrossRef]
73. Nashed, G.G.L. Nonlinear Charged Black Hole Solution in Rastall Gravity. *Universe* **2022**, *8*, 510. [CrossRef]
74. Avsajanishvili, O.; Chitov, G.Y.; Kahniashvili, T.; Mandal, S.; Samushia, L. Observational Constraints on Dynamical Dark Energy Models. *Universe* **2024**, *10*, 122. [CrossRef]

Article

Discretely Charged Dark Matter in Inflation Models Based on Holographic Space-Time

Tom Banks [1],* and Willy Fischler [2]

[1] Department of Physics and NHETC, Rutgers University, Piscataway, NJ 08854, USA
[2] Department of Physics and Texas Cosmology Center, University of Texas, Austin, TX 78712, USA
* Correspondence: tibanks@ucsc.edu

Abstract: The holographic space-time (HST) model of inflation has a potential explanation for dark matter as tiny primordial black holes. Motivated by a recent paper of Barrau, we propose a version of this model where some of the inflationary black holes (IBHs), whose decay gives rise to the Hot Big Bang, carry the smallest value of a discrete symmetry charge. The fraction f of IBHs carrying this charge is difficult to estimate from first principles, but we determine it by requiring that the crossover between radiation and matter domination occurs at the correct temperature $T_{eq} \sim 1\,\text{eV} = 10^{-28} M_P$. The fraction is small, $f \sim 2 \times 10^{-9}$, so we believe this gives an extremely plausible model of dark matter.

Keywords: inflationary cosmology; primordial black holes; dark matter

Citation: Banks, T.; Fischler, W. Discretely Charged Dark Matter in Inflation Models Based on Holographic Space-Time. *Universe* 2022, *8*, 600. https://doi.org/10.3390/universe8110600

Academic Editor: Kazuharu Bamba

Received: 30 September 2022
Accepted: 8 November 2022
Published: 14 November 2022

Publisher's Note: MDPI stays neutral with regard to jurisdictional claims in published maps and institutional affiliations.

1. Introduction

The HST model of inflation [1–9] is a finite quantum mechanical model, which gives a very economical explanation of known facts about the very early universe. HST models are based on Jacobson's Principle: the Einstein equations are the hydrodynamic equations of the area law $S = A/4G$ applied to any causal diamond in a Lorentzian space-time. Therefore one should search for quantum models whose hydrodynamics agree with some particular solution of Einstein's equations. The features of HST inflation models are easily summarized:

- The model consists of a large number of independent quantum systems, describing the universe as viewed from different geodesics in an FRW space-time. The relation between proper time and the area of the holographic screen of a diamond with past tip on the singular beginning of the universe is matched to the relationship between the time in the quantum theory and the entropy of the density matrix assigned to the diamond.

- The Hamiltonian is time-dependent to ensure that the degrees of freedom inside a given causal diamond form an independent subsystem. This also provides a natural resolution of the Big Bang singularity: when the Hilbert space of a diamond is small enough, the hydrodynamic description breaks down, but the quantum mechanics is well defined and finite.

- A particular soluble model, in which, for each proper time t, the modular Hamiltonian of a diamond is the L_0 generator of a cutoff conformal field theory on an interval of length I with a UV cutoff l, such that $I/l \gg 1$ (but t-independent) and central charge scaling like t^2 is "dual" to a flat FRW geometry with scale factor

$$a(t) = \sinh^{1/3}(3t/R_I). \tag{1}$$

These models have no localized excitations and saturate the covariant entropy bound at all times.

- Inflationary models are obtained by insisting that the dynamics follow the soluble model for a large number of e-folds (80 is what seems to fit the data of our universe), after which the diamond Hilbert space slowly expands so that it can fit $>e^{80}$ copies of the original space. What one would have called gauge copies of the causal diamond in a de Sitter space with radius R_I become localized excitations of the expanded diamond, with all of the statistical properties of black holes of radius $\sim R$. See Figure 1 for a cartoon of how this happens. This gives rise to a novel theory of CMB fluctuations, with $\zeta \sim (R_I \epsilon)^{-1}$ and the scalar-to-tensor ratio $r \sim \epsilon^{-2}$. Properties of spinning black holes and the $1+1$ CFT model of the horizon pin down the coefficients in these relations, in which $\epsilon = -\frac{\dot{H}}{H^2}$. One requires a different slow roll metric than conventional inflationary models to fit the data on the CMB.
- Evaporation of the "inflationary black holes" (IBHs) gives rise to the Hot Big Bang and baryogenesis [1].
- The only element of very early universe cosmology that is not explained simply by the model is dark matter. We have speculated [11–13] that mergers of the tiny IBHs might form a collection of primordial black holes (PBHs) consistent with astronomical data [14].

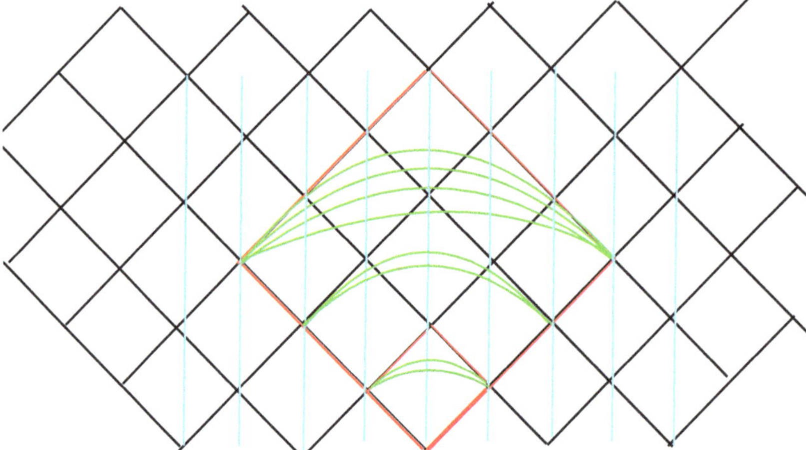

Figure 1. Holographic Inflationary Cosmology in Conformal Time: Equal Time Surfaces are Hyperbolae Interpolating Between Diamonds. There are additional green spacelike surfaces that we have omitted for clarity, interpolating between those shown, such that the proper time between consecutive surfaces along the central geodesic is always the Planck time. When a non-central geodesic penetrates the causal diamond of the central geodesic, consistency with the fact that that geodesic is still experiencing inflation implies that it must behave like an isolated quantum system with finite entropy given by the area law.

Recently, Barrau [15] has argued that the merger scenario cannot work, but that a model in which evaporation of the IBHs' left over Planck scale remnants could explain dark matter. While we remain somewhat skeptical that one can come to Barrau's negative conclusion without dedicated computer simulations, we found the idea of remnants intriguing. GellMann's totalitarian principle is an axiomatization of a known fact about quantum systems. Transition matrix elements exist unless some (approximate) conservation law forbids them. Put simply, Planck mass black hole remnants cannot be ruled out by Hawking's thermodynamic arguments, but they are implausible unless there is some quantum number that prevents them from decaying.

In models of quantum gravity, charges carried by local excitations are always coupled to gauge fields. The most innocuous kind of gauge field, from the point of view of a dark matter model, is associated with a discrete gauge group like Z_N. In gauge theories, charged particles always experience long-range interactions. For discrete gauge theories, these are just Aharonov–Bohm interactions with topological cosmic strings, so Z_N charged particles will behave like neutral dark matter. In order to distinguish between Z_N charged remnants and stable dark matter particles with only gravitational interactions, one would have to locate a cosmic string and observe the AB effect. We note that models of particle dark matter almost always invoke a discrete symmetry to stabilize the dark matter candidate. In gravitational theories, any exact discrete symmetry must be gauged.

The theory of supersymmetry (SUSY) breaking contains a natural discrete gauge symmetry in a universe like our own with a small positive cosmological constant. If one imagines the model of our universe to be part of a (possibly discrete) family of models that converges to zero c.c., then the limiting model is likely to be supersymmetric. This is a "phenomenological" observation. There are no perturbative string models in Minkowski space that violate SUSY. There are no known sequences of AdS/CFT models with tunable c.c. which violate SUSY in the limit [2] of small c.c.

On the other hand generic SUGRA Lagrangians with SUSY-preserving minima have negative c.c. because of a term proportional to the square of the superpotential in the vacuum energy. One of us [16] pointed out long ago that the criterion for a supersymmetric vacuum with vanishing c.c. was preservation of an R symmetry. The R symmetry must be discrete [17,18].

The R symmetry acts chirally on the gravitino and keeps it massless, but, in dS space, there are processes where the R symmetry is broken by absorption and re-emission of gravitinos at the horizon. In [19], it was postulated that R symmetry violating terms in the low-energy effective Lagrangian, induced by this non-local effect, would trigger the super-Higgs effect in a self-consistent manner. This leads to an equation for the gravitino mass in terms of the c.c.,

$$m_{3/2} = K\Lambda^{1/4},\qquad(2)$$

where it has proven difficult to estimate the constant K.

A discrete R symmetry and a light gravitino does not, at first, sound like a recipe for obtaining stable R-charged black hole remnants. The remnants can emit gravitinos and reduce their R charge. However, there are many examples in field theory where the lowest charge under some discrete gauge group is carried by a very heavy particle. This means that the symmetry breaking induced by the effective gravitino mass leaves over a discrete subgroup of the high-energy discrete gauge symmetry. The only instability of the heavy R-charged black holes will be moving through the horizon, or arise through spontaneous nucleation of a black hole of opposite charge, which is a highly improbable process.

We note that discrete charges may not be the only way to stabilize black hole remnants. A referee kindly pointed out [20] which constructs black hole remnants using non-commutative geometry.

2. Phenomenology of Discretely Charged PBH Dark Matter

In the context of the HST model of inflation, it is simple to incorporate discrete charges that stabilize a fraction f of IBHs at the Planck scale. Inflation is followed by an early matter-dominated era in which the matter is composed of IBHs. For comparison, we can calculate the expected fraction of magnetically charged black holes using the black hole entropy formula, according to which the expected fraction of black holes in a random sample is

$$f = e^{-\frac{Q^2 M_P^2}{\alpha M^2}},\qquad(3)$$

where Q is the integer valued magnetic charge and α is the value of the fine structure constant at the scale of the Schwarzschild radius. Taking $M \sim M_P$, $Q = \pm 1$ and α equal to its value at the scale of unification of standard model couplings, $\alpha_U \approx \frac{1}{25}$, this gives $f \sim 10^{-12.5}$. There are many issues with this estimate, the most serious of which is using a formula from statistical mechanics for a low-entropy system, but it gives us a general sense that f should be small, but not doubly exponentially small.

In HST inflation models, the number density of IBHs at the end of inflation is

$$n_{IBH} \sim CR_I^{-3},\tag{4}$$

where $C \sim 1/30$ is the minimal dilution factor necessary to assure that the IBHs do not immediately coalesce to form a maximum entropy state. The inflationary Hubble radius, R_I in Planck units, is determined by matching to the size of CMB fluctuations

$$R_I = \epsilon^{-1}10^5,\tag{5}$$

and ϵ is bounded by the requirement that slow roll expansion is faster than fast scrambling of the black holes. This again follows from the requirement that black holes remain isolated quantum subsystems during the slow roll era. The bound is

$$\epsilon > (\ln R_I)^{-1}.\tag{6}$$

This is satisfied for $\epsilon \sim 0.1$. We insist on being close to the bound because we require the highest probability of initial conditions that lead to a universe with localized excitations. Since the power spectrum of CMB fluctuations in these models scales like ϵ^{-2}, this value is roughly consistent with the data.

The universe remains matter-dominated until a time $t_D = 10240\pi g^{-1}R_I^3$, when most of the IBHs decay into radiation. g is the number of particle species below the Hawking temperature of the IBHs. The resulting energy density of radiation at the beginning of the Hot Big Bang is

$$\rho_\gamma = \frac{C}{R_I^2}t_D^{-2} = Cg^2(10240\pi)^{-2}R_I^{-8} = \frac{\pi^2}{30}gT_{RH}^4.\tag{7}$$

If a fraction f of the IBHs leave over Planck scale remnants, then their energy density at reheating is

$$\rho_{rem} = fCR_I^{-3}t_D^{-2},\tag{8}$$

and

$$\frac{\rho_{rem}}{\rho_\gamma} = fR_I^{-1}.\tag{9}$$

This ratio grows like $\frac{T_{RH}}{T}$ as the radiation gas cools, and hits 1 when

$$T_{eq} = (R_I)^{-1}fT_{RH} = 10^{-28},\tag{10}$$

where the last equality is the observed temperature at which matter radiation crossover occurs. T_{RH} is given by

$$\approx \frac{g^{1/4}}{100\pi R_I^2} \sim 0.5 \times 10^{-13}.\tag{11}$$

Thus

$$f \approx 2 \times 10^{-9}.\tag{12}$$

So, what appears to be a reasonable estimate of the probability of discretely charged black holes being formed in HST models is consistent with the data. We consider this estimate reasonable because of our rough calculation of the probability of a random black hole having an ordinary magnetic charge. It seems unlikely that having a Z_N charge would be less probable. Magnetic energy distorts the geometry of the black hole, reducing the area of the horizon and there is no such effect for discrete charges. Given the uncertainties in these estimates, it is perhaps best to view the value $f \sim 10^{-9}$ as a feature of a phenomenological model, which must be verified by a more detailed model of quantum gravity.

One more issue needs to be addressed. In previous publications [11], we have argued that, if a fraction $f \sim 10^{-24}$ of black holes of mass $\sim 10^{11}$ are formed during the early matter-dominated era, then these could account for the observed value of the matter radiation crossover. In the present scenario, these are unnecessary and could even become an embarrassment. Since these PBHs are not cosmologically stable, their decay could lead to signatures that have been ruled out by observation. It is possible that during the matter-dominated era below $T = 10^{-28}$, the unstable PBHs merge into more stable ones before too much Hawking radiation has been emitted. Ongoing computer simulations will determine whether this is plausible [21]. If it is, we will have two competing models that account for the data. It seems highly unlikely that the probabilities work out so that both contributions to dark matter have comparable densities, but, if they do, one could have a scenario where some of the dark matter decays. Such models have been invoked to explain some of the apparent discrepancies between data and the LCDM model. From the present point of view, the simplest idea is that the fraction of merged IBHs which could survive down to $T = 10^{-28}$ is negligible and that discretely charged dark matter (DCDM) accounts for everything we see.

3. Conclusions

Motivated by a suggestion of Barrau, we propose that HST inflation models incorporate a discrete Z_N gauge symmetry and that a fraction $f \sim 10^{-9}$ of the erstwhile inflationary horizon volumes in the model carry the smallest value of Z_N charge. This discrete symmetry group could be the remnant of a larger discrete R symmetry, broken by gravitino interactions with the horizon, which generate the gravitino mass. The resulting models account, at the order of magnitude level, with everything we know about the cosmology of the very early universe. Inflation ends in an early matter-dominated era, dominated by IBHs with Schwarzschild radii approximately equal to the inflationary horizon size. Most of the IBHs decay, producing the Hot Big Bang and baryogenesis. Those charged under Z_N become the dark matter. And the rest is history.

Author Contributions: Conceptualization, T.B. and W.F.; writing—original draft preparation, T.B. and W.F.; writing—review and editing, T.B. and W.F. All authors have read and agreed to the published version of the manuscript.

Funding: The work of T.B. is supported in part by the DOE under grant DE-SC0010008. The work of W.F. is supported by the NSF under grant PHY-1914679.

Data Availability Statement: Not applicable.

Conflicts of Interest: The authors declare no conflict of interest.

Notes

[1] It is often stated erroneously that black hole evaporation cannot give rise to baryogenesis. This is incorrect. The decrease in mass of the black hole breaks CPT. The full decay process is not thermal because the equilibrium is changing. In a previous paper [9], we argued that for the tiny IBHs, if one postulates order one CP violation in decay matrix elements, one gets close to the required value for the baryon-to-entropy ratio. Other papers on gravitational baryogenesis are [10].

[2] SUSY violating relevant perturbations of SUSic models represent large objects embedded in AdS space. The physics far from the center become exactly supersymmetric.

References

1. Banks, T.; Fischler, W. Holographic Inflation Revised. In *The Philosophy of Cosmology*; Cambridge University Press: Cambridge, UK, 2017. [CrossRef]

2. Banks, T.; Fischler, W. The holographic spacetime model of cosmology. *Int. J. Mod. Phys. D* **2018**, *27*, 1846005. [CrossRef]

3. Banks, T.; Fischler, W. Holographic cosmology 3.0. *Phys. Scr.* **2005**, *T117*, 56–63. doi: 10.1238/Physica.Topical.117a00056. [CrossRef]

4. Banks, T.; Fischler, W. Holographic Cosmology. *arXiv* **2004**, arXiv:hep-th/0405200.

5. Banks, T.; Fischler, W.; Mannelli, L. Microscopic quantum mechanics of the p = rho universe. *Phys. Rev. D* **2005**, *71*, 123514. [CrossRef]

6. Banks, T.; Fischler, W. The Holographic Approach to Cosmology. *arXiv* **2004**, arXiv:hep-th/0412097.
7. Banks, T.; Fischler, W. Holographic Theories of Inflation and Fluctuations. *arXiv* **2011**, arXiv:1111.4948.
8. Banks, T.; Fischler, W.; Torres, T.J.; Wainwright, C.L. Holographic Fluctuations from Unitary de Sitter Invariant Field Theory. *arXiv* **2013**, arXiv:1306.3999.
9. Banks, T.; Fischler, W. CP Violation and Baryogenesis in the Presence of Black Holes. *arXiv* **2015**, arXiv:1505.00472
10. Davoudiasl, H.; Kitano, R.; Kribs, G.D.; Murayama, H.; Steinhardt, P.J. Gravitational baryogenesis. *Phys. Rev. Lett.* **2004**, *93*, 201301. [CrossRef] [PubMed]
11. Banks, T.; Fischler, W. Primordial Black Holes as Dark Matter. *arXiv* **2020**, arXiv:2008.00327.
12. Banks, T.; Fischler, W. Entropy and Black Holes in the Very Early Universe. *arXiv* **2021**, arXiv:2109.05571.
13. Hook, A. Baryogenesis from Hawking Radiation. *Phys. Rev. D* **2014**, *90*, 083535. [CrossRef]
14. Carr, B.; Raidal, M.; Tenkanen, T.; Vaskonen, V.; Veermäe, H. Primordial black hole constraints for extended mass functions. *Phys. Rev. D* **2017**, *96*, 023514. [CrossRef]
15. Barrau, A. The holographic space-time and black hole remnants as dark matter. *Phys. Lett. B* **2022**, *829*, 137061. [CrossRef]
16. Banks, T. Cosmological breaking of supersymmetry? *Int. J. Mod. Phys. A* **2001**, *16*, 910–921. [CrossRef]
17. Witten, E. New Issues In Manifolds Of SU(3) Holonomy. *Nucl. Phys. B* **1986**, *268*, 79. [CrossRef]
18. Komargodski, Z.; Seiberg, N. Comments on the Fayet-Iliopoulos Term in Field Theory and Supergravity. *J. High Energy Phys.* **2009**, *6*, 7. [CrossRef]
19. Banks, T. Breaking SUSY on the Horizon. *arXiv* **2014**, arXiv:hep-th/0206117.
20. Nicolini, P. Noncommutative Black Holes, The Final Appeal To Quantum Gravity: A Review. *Int. J. Mod. Phys. A* **2009**, *24*, 1229–1308. [CrossRef]
21. Banks, T.; Suresh, A. Simulation of black hole mergers in a model of the very early universe. *manuscript in preparation*.

universe

Article

Limits on Magnetized Quark-Nugget Dark Matter from Episodic Natural Events

J. Pace VanDevender [1,*], Aaron P. VanDevender [1], Peter Wilson [2], Benjamin F. Hammel [3] and Niall McGinley [4]

[1] VanDevender Enterprises LLC, 7604 Lamplighter Lane NE, Albuquerque, NM 87109, USA; sig@netdot.net
[2] School of Geography and Environmental Sciences, Ulster University, Cromore Road, Coleraine, County Londonderry BT52 1SA, UK; p.wilson@ulster.ac.uk
[3] Department of Materials Science and Engineering, University of Pennsylvania, 3231 Walnut Street, Philadelphia, PA 19104, USA; bfhammel@seas.upenn.edu
[4] Ardaturr, Churchill PO, Letterkenny, Co. F928982 Donegal, Ireland; niallmacfhionnghaile@gmail.com
* Correspondence: pace@vandevender.com

Abstract: A quark nugget is a hypothetical dark-matter candidate composed of approximately equal numbers of up, down, and strange quarks. Most models of quark nuggets do not include effects of their intrinsic magnetic field. However, Tatsumi used a mathematically tractable approximation of the Standard Model of Particle Physics and found that the cores of magnetar pulsars may be quark nuggets in a ferromagnetic liquid state with surface magnetic field $B_0 = 10^{12\pm1}$ T. We have applied that result to quark-nugget dark matter. Previous work addressed the formation and aggregation of magnetized quark nuggets (MQNs) into a broad and magnetically stabilized mass distribution before they could decay and addressed their interaction with normal matter through their magnetopause, losing translational velocity while gaining rotational velocity and radiating electromagnetic energy. The two orders of magnitude uncertainty in Tatsumi's estimate for B_0 precludes the practical design of systematic experiments to detect MQNs through their predicted interaction with matter. In this paper, we examine episodic events consistent with a unique signature of MQNs. If they are indeed caused by MQNs, they constrain the most likely values of B_0 to 1.65×10^{12} T $+/- 21\%$ and support the design of definitive tests of the MQN dark-matter hypothesis.

Keywords: dark matter; quark nugget; magnetized quark nugget; MQN; nuclearite; magnetar; strangelet; slet; Macro

Citation: VanDevender, J.P.; VanDevender, A.P.; Wilson, P.; Hammel, B.F.; McGinley, N. Limits on Magnetized Quark-Nugget Dark Matter from Episodic Natural Events. *Universe* **2021**, 7, 35. https://doi.org/10.3390/universe7020035

Received: 23 December 2020
Accepted: 24 January 2021
Published: 4 February 2021

Publisher's Note: MDPI stays neutral with regard to jurisdictional claims in published maps and institutional affiliations.

1. Introduction

The great majority of mass in the Universe is non-luminous material called dark matter [1]. Gravity from dark matter literally holds galaxies together [2]. The nature of dark matter has been studied for decades but remains one of the most puzzling mysteries in science [3]. Most dark-matter candidates are assumed to interact with normal matter only through gravity, but stronger interactions are consistent with requirements for dark matter if their effective interaction time is billions of years [3]; Magnetized Quark Nuggets (MQNs) are one such candidate for dark matter. Previously published work [4] shows the primordial origin of MQNs and their compatibility with requirements [5] of dark matter. Their origin in the very early Universe was in the magnetic aggregation of Λ^0 particles (consisting of one up, one down, and one strange quark) into a broad mass distribution of stable ferromagnetic MQNs before they could decay. After t ≈ 66 μs after the big bang, mean MQN mass is between ~10^{-6} kg and ~10^4 kg, depending on the surface magnetic field B_0. The corresponding mass distribution is sufficient for MQNs to meet the requirements of dark matter in the subsequent processes, including those that determine the Large Scale Structure (LSS) of the Universe and the Cosmic Microwave Background (CMB).

For the last four decades, searches for dark-matter candidates have focused on particles beyond the Standard Model of Particle Physics [6]. MQNs are composed of Standard Model quarks. However, mathematical techniques for applying the Model in the ~90-MeV-energy scale of MQNs rely on approximations. In addition, the key result that leads to MQNs requires a Quantum Chromo Dynamics (QCD) fine-structure constant $\alpha_c \approx 4$ at this energy scale; the value of α_c at this energy scale is not known. To the extent that the calculations can be performed, MQNs are consistent with the Standard Model and do not require a new particle Beyond the Standard Model (BSM). Therefore, MQNs have been somewhat controversial as dark-matter candidates.

In the case of this paper, the controversy may be mitigated because we are not claiming discovery. The paper does find that the theory of MQNs is consistent with a reported observation. However, the event is very rare, is not reproducible, was not recorded by multiple observers, and cannot be quantitatively validated after the fact by anyone else. Consequently, the evidence does not meet today's standard for a discovery.

Quarks are components of many particles in the Standard Model of Particle Physics. Ensembles of strange, up, and down quarks (in approximately equal numbers) are called quark nuggets and are thought to be stable at sufficiently large masses [7] and qualify as candidates for dark matter. Quark nuggets are also called strangelets [8], nuclearites [9], Axion Quark Nuggets (AQNs) [10], slets [11], Macros [5], and MQNs [4]. A brief summary of four decades of research [4–35] on quark-nugget charge-to-mass ratio, formation, stability, and detection is provided for convenience and completeness as Appendix A: Quark-Nugget Research Summary.

Most models of quark nuggets do not include effects of their intrinsic magnetic field. However, Xu [26] has shown that the low electron density, permitted in stable quark nuggets, limits surface magnetic fields from ordinary electron ferromagnetism to ~2×10^7 T. Tatsumi [27] examined ferromagnetism from a One Gluon Exchange interaction and concluded that the surface magnetic field could be sufficient to explain the ~10^{12} T magnetic fields inferred for magnetar cores. Since the result depends on the currently unknown value of α_c at the 90 MeV energy scale, the result needs to be confirmed with relevant observations and/or advances in QCD calculations.

Tatsumi's result has recently been applied to quark-nugget dark matter. By definition of ferromagnetic, the lowest energy state in Tatsumi's ferromagnetic liquid is with magnetic dipoles aligned. The individual quark nuggets are formed with baryon number A = 1, as are neutrons and protons, and may have spin and a corresponding surface magnetic field similar to that of protons and neutrons. However, unlike protons and neutrons, ferromagnetic dipoles of quark nuggets (MQNs) align upon aggregation and maintain the surface magnetic field. Their magnetic field at substantial distances is large because their aggregated size is large, not because their intrinsic magnetization (and corresponding surface magnetic field) is necessarily larger than that of other baryons.

Previous work [4] addressed the formation and aggregation of magnetized quark nuggets (MQNs) in the early Universe into a broad and magnetically stabilized mass distribution with baryon number A between ~10^3 and 10^{37} before they could decay by the weak interaction; addressed the compatibility of MQNs with the requirements of dark matter; and addressed their interaction with normal matter through their magnetopause [28], while losing translational velocity, gaining rotational velocity, and radiating electromagnetic energy [36].

Electromagnetic energy accumulated in the Universe from MQNs is unfortunately not detectable because the plasma in most of the Universe is too low density to establish the magnetopause effect and the electromagnetic radiation from the rest of interstellar space is too low frequency to propagate through the solar-wind plasma and reach Earth. However, MQNs transiting through the plasma and gas around Earth spin up to MHz frequencies and should be detectable as they exit the magnetosphere [36]. Predicted event rates are strongly dependent on the MQN magnetic field B_0.

Uncertainty in Tatsumi's estimate of $B_0 = 10^{12 \pm 1}$ T is too large to design a system to systematically look for MQNs. In this paper, we examine one type of episodic event consistent with a unique signature of MQNs, i.e., an MQN impacting Earth on a nearly tangential trajectory, penetrating the ground for a portion of its path, and emerging where it can be observed. We calculate what would be observed and compare the results with extant observations. Such episodic events are impossible to predict or reproduce and fall short of the standard for evidence of discovery in physics. Therefore, we are not asserting the discovery of MQNs but are examining consistency with MQNs and the resulting constraints on B_0. The results are useful for designing systematic tests of the MQN dark-matter hypothesis.

Without a reasonably small uncertainty in B_0, success in fielding systematic experiments is unlikely. We attempted such an experiment by instrumenting a 30 sq-km area of the Great Salt Lake in Utah, USA, and looked for acoustic signals from MQN impacts. No impacts were observed in 2200 hours of recording. Subsequent theory [4] explained the null result and showed that the predicted mass distribution of MQNs means that impacts are very rare. Even a planet-sized detector is marginal. Consequently, we have turned to observations of episodic, naturally occurring events to narrow the uncertainty in B_0.

Reference 4 shows the most important unknown in the theory of MQNs is the surface magnetic field parameter B_0, which quantifies the uncertainty in the distribution of MQN mass and the event rate. The mean of the surface magnetic field $<B_S>$ is related to B_0 in reference [4], through

$$< B_S >= \left(\frac{\rho_{QN}}{10^{18}(\text{kg/m}^3)} \right) \left(\frac{\rho_{DM\_T=100\text{Mev}}}{1.6 \times 10^8 (\text{kg/m}^3)} \right) B_0 \tag{1}$$

In Equation (1), ρ_{QN} is the MQN mass density, and ρ_{DM} is the density of dark matter at time $t \sim 65~\mu$s, when the temperature T in the early Universe was ~ 100 MeV [37]. If better values of ρ_{QN}, ρ_{DM}, and B_0 are determined by observations, then a more accurate value of $<B_S>$ can be calculated with Equation (1).

In this paper, we show that a nearly tangential impact and transit through a chord of Earth by an MQN provides a unique signature: a magnetically levitated mass of greater than nuclear density that ionizes and excites the atmosphere for many minutes. We compare the results of analytic calculations and computer simulations of such an event to the observations on 6 August 1868, published by M. Fitzgerald [38] in the *Proceedings of the Royal Society*, the premier scientific publication at that time. The event's unusual characteristics are consistent with an extremely rare, nearly tangential, MQN impact. Similar events have been reported elsewhere [39], yet Fitzgerald's report is the best documented and only scientifically published event we have found, making it the most suitable for comparison with theory.

As noted above, Tatsumi [27] estimated that magnetar cores have $B_0 = 10^{12 \pm 1}$ T. The results reported in this paper and reference 4 constrain the most likely values of B_0 to 1.65×10^{12} T $+/- 21\%$ and will permit the design of a systematic experiment to test the MQN hypothesis.

2. Materials and Methods

This study used:

1. Analytic methods presented in detail in the results section.
2. Computational simulations coupling the Rotating Magnetic Machinery module and the Nonlinear Plasticity Solid Mechanics module of the 3D, finite-element, COMSOL Multiphysics code [40]. Details are included in Appendix B: COMSOL Simulation of Rotating Magnetized Sphere Interaction with Plastically Deformable Conductor.
3. Original field work at the location reported by Fitzgerald [38] in County Donegal, Ireland, is documented in Appendix C: Field Investigation of Fitzgerald's Report to Royal Society. The GPS locations are included to facilitate replication, subject to acquiring permission from the property owners listed in Acknowledgements. Radiocarbon dating was conducted by Beta Analytic Inc. 4985 SW 74th Court, Miami, FL 33155, USA.

3. Results

3.1. Nearly Tangential Impact and Transit of MQNs through Earth

Figure 1 illustrates three MQN impacts on an idealized portion of a spherical Earth.

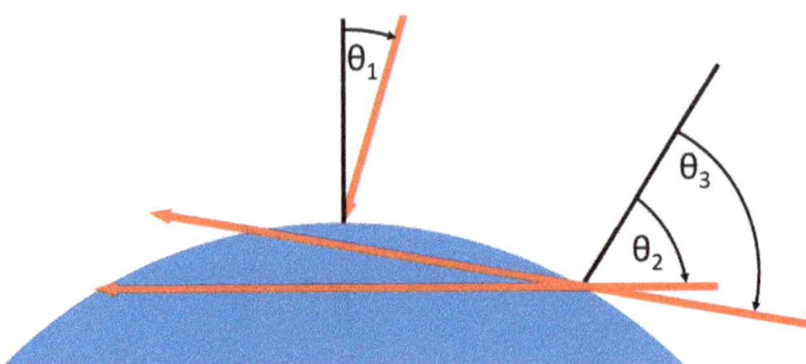

Figure 1. Three MQN trajectories are shown as red arrows, along with their angle of impact θ with respect to the normal surface of an idealized, not to scale, Earth (blue). The trajectory with impact angel $\theta_1 \ll 90°$ is a more common radial impact. Nearly tangential trajectories that transit through Earth are represented by trajectories with impact angles between θ_2 and θ_3. MQNs on a θ_2 trajectory emerge from Earth with negligible velocity after transiting a distance x_{max}, the maximum range of an MQN. MQNs on a θ_3 trajectory emerge from Earth with considerable velocity.

Only massive MQNs have enough momentum to penetrate significant distances through water or rock. Therefore, we focus on MQNs between 10^5 kg (maximum mass associated with B_o ~1.3×10^{12} T) and 10^{10} kg (maximum mass associated with B_o ~3×10^{12} T), and use these extremes to illustrate each calculation.

3.2. Slowing Down in Passage through a Portion of Earth

Hypervelocity MQNs ionize surrounding matter through a shock wave and interact with that matter through a magnetopause in the same way that Earth interacts with the solar wind through its magnetopause. The equations governing the interaction are derived in reference [28] and are summarized in Equations (2) through (6) for convenience.

The force equation for a high-velocity body with instantaneous radius r_m, and velocity v, moving through a fluid of density ρ_p with a drag coefficient K ~1, is

$$F_e \approx K\pi r_m^2 \rho_p v^2 \tag{2}$$

The geometric radius r_{QN} of the MQN of mass m and mass density ρ_{QN} is

$$r_{QN} = \left(\frac{3m}{4\pi\rho_{QN}}\right)^{\frac{1}{3}} \tag{3}$$

However, MQNs have a velocity-dependent interaction radius [28] equal to the radius of their magnetopause.

$$r_m \approx \left(\frac{2B_o^2}{\mu_0 K\rho_p v^2}\right)^{\frac{1}{6}} r_{QN} \tag{4}$$

MQNs with mass 10^5 kg have r_{QN} ~4×10^{-5} m. For B_o ~1.3×10^{12} T and $v = 250$ km/s, an MQN passing through water of density 1000 kg/m³ has the magnetopause radius r_m ~0.025 m. The corresponding values for mass $m = 10^9$ kg with B_o ~3×10^{12} T are r_{QN} ~9×10^{-4} m and r_m ~0.71 m. Although their nuclear density makes these massive MQNs physically small, their large magnetic fields and high velocities make their interaction radius and cross section very large, even in solid-density matter.

The interaction radius of an MQN varies as velocity $v^{-1/3}$ in Equation (4). Including that velocity dependence in the calculation with initial velocity v_0 gives velocity v as a function of distance x:

$$v = v_0 \left(1 - \frac{x}{x_{\max}} \right)^{\frac{3}{2}} \tag{5}$$

in which x_{max} is the stopping distance for an MQN:

$$x_{\max} = \left(\frac{3m}{2\pi r_{QN}^2} \right) \left(\frac{\mu_0 v_0^2}{2K^2 \rho_p^2 B_0^2} \right)^{\frac{1}{3}} \tag{6}$$

MQNs with mass 10^5 kg, B_o ~1.3 $\times$ 10^{12} T, and velocity v_0 ~250 km/s have x_{max} ~241 km passing through water. The corresponding value for mass of 10^9 kg with B_o ~3 $\times$ 10^{12} T is x_{max} ~3000 km.

From the geometry in Figure 1, the angle of incidence θ_2 for trajectories that emerge from Earth with negligible velocity is given by

$$\theta_2 = \cos^{-1} \left(\frac{x_{\max}}{2 r_{Earth}} \right) \tag{7}$$

in which r_{Earth} is the radius of the Earth, or ~6.38 $\times$ 10^6 m. For $\pi/2 > \theta > \theta_2$, MQNs emerge from Earth with velocity

$$v_{exit} = v_0 \left(1 - \frac{x_{exit}}{x_{\max}} \right)^{\frac{3}{2}} \text{ for } x_{exit} = 2 r_{Earth} \cos(\theta) \tag{8}$$

Integrating Equation (5) gives the transit time t_{exit}

$$t_{exit} = \left(\frac{2 x_{\max}}{v_0} \right) \left(\left(\frac{x_{\max}}{x_{\max} - x_{exit}} \right)^{\frac{1}{2}} - 1 \right) \tag{9}$$

Transit time t_{exit} is strongly dependent on $(x_{max} - x_{exit})$ and is infinite at $x_{max} = x_{exit}$. A 10^5 kg MQN with $B_o = 1.3 \times 10^{12}$ T, initial velocity $v_0 = 250$ km/s, and incidence angle $\theta_2 = 88.91817°$ penetrates a distance $x_{exit} = 240.09$ km of water and emerges in t_{exit} ~54 s with $v_{exit} = 10$ m/s. If the incidence angle $\theta_2 = 88.92278°$, then $x_{exit} = 239.89$ km, transit time t_{exit} ~24 s, and $v_{exit} = 100$ m/s.

During the transit, the MQN is falling towards the center of Earth with acceleration $g = 9.8$ m/s^2, which is not considered in Equations (7) through (9). The deviation from the straight-line approximation is $\delta r \sim \frac{1}{2} g t_{exit}^2$ and the corresponding fractional error in path length introduced by neglecting gravity is

$$\delta \sim \frac{\delta r}{\sqrt{x_{exit}^2 + (\delta r)^2}} \tag{10}$$

For the example in the previous paragraph, fractional error in path length because of gravity is δ ~0.06 for $v_{exit} = 10$ m/s and is δ ~0.012 for v_{exit} ~100 m/s. In general, fractional error decreases with increasing v_{exit}, decreasing B_o, decreasing MQN mass, and increasing mass density of material transited (granite with $\rho_p = 2300$ kg/m^3 or water $\rho_p = 1000$ kg/m^3).

If the MQN slows to <<10 m/s, gravity dominates its motion, and it does not emerge from the ground or water. Fast objects are difficult to perceive, especially if an observer is not primed to expect an event. A human observer requires ~0.25 s to perceive an object as a thing [41]. If the object is about 1 m in diameter and moving at > 100 m/s, the object will have moved > 25 m before cognitive acquisition and may be out of range before the observer can process the image well enough to be confident of what was seen. Therefore, we suggest that only MQNs emerging with 10 m/s $\leq v_{exit} \leq$ 100 m/s are likely to be reported by human observers.

Rearranging Equation (8) gives the incidence angle θ_{vexit} for a given v_{exit}:

$$\theta_{vexit} = \cos^{-1}\left(\left(\frac{x_{max}}{2r_{Earth}}\right)\left(1 - \left(\frac{v_{exit}}{v_0}\right)^{\frac{2}{3}}\right)\right) \tag{11}$$

For v_{exit} = 10 m/s and 100 m/s, the corresponding impact angles are, respectively, θ_{10} and θ_{100}, which will be used in estimating the event rates for directly observable MQN events.

3.3. Estimated Event Rates

The number of events per year on Earth is estimated as follows:

1. Earth is moving about the galactic center, in the direction of the star Vega, and through the dark-matter halo with a velocity of ~230 km/s [42]. Therefore, dark matter streams into the Earth frame of reference with mean streaming velocity ~230 km/s.
2. Dark matter in the halo also has a nearly Maxwellian velocity distribution with mean velocity of ~230 km/s, so the ratio of streaming velocity to Maxwellian velocity is approximately 1 [42].
3. Approximating the velocity of dark matter streaming from the direction of Vega as ~230 km/s, we calculate the cross section $A_{10\text{-}100}$ for transiting a chord through Earth and emerging with velocity between v_{10} = 10 m/s and v_{100} = 100 m/s:

$$A_{10-100} = 2\pi r_{Earth}^2 (\sin\theta_{100} - \sin\theta_{10}) \tag{12}$$

More generally, the cross section for MQNs impacting between θ_{min} and θ_{max} is

$$A = 2\pi r_{Earth}^2 (\sin\theta_{max} - \sin\theta_{min}) \tag{13}$$

4. MQNs can have masses between 10^{-23} kg and 10^{10} kg [4]. We approximate such a large range by (1) associating the flux of all MQNs that have mass between 10^i kg and 10^{i+1} kg with a representative mass $10^{i+0.5}$ kg (which we call the representative decadal mass) for $-23 \leq i \leq 10$; (2) calculating the behavior of each decadal-mass MQN; (3) assuming all the MQNs in that decadal range behave the same way. The associated number flux is called the decadal flux F_{m_decade} (number N/y/m^2/sr) and was computed [4] as a function of B_0 from simulations of the aggregation of quark nuggets from their formation in the early Universe and evolution to the present era.
5. For $A_{10\text{-}100\_m\_decade}$, defined as the $A_{10\text{-}100}$ appropriate to a decadal mass m, the number of events per year per steradian for MQNs streaming from the direction of Vega and emerging with velocity between 10 and 100 m/s is $F_{m\_decade} A_{10\text{-}100\_m\_decade}$, summed over all decadal masses m.
6. For random velocity, approximately equal to streaming velocity, reference [36] shows that 5.56 sr is the effective solid angle that generalizes the streaming result to include MQNs from all directions.
7. Therefore, the total number of events per year somewhere on Earth with v_{exit} between 10 and 100 m/s is

$$N = 5.56 \sum_{m\_decade} \left(F_{m\_decade} A_{10-100\_m\_decade}\right) \tag{14}$$

In Section 3.6, we will show that MQNs transiting Earth along a chord spin up to MHz frequencies and emit substantial radio frequency (RF) power. If and only if the RF is not absorbed by the surrounding plasma, these MQNs can be detected by their RF emissions propagating around Earth in the waveguide between the ground and the ionosphere. Their cross section is given by Equation (13) with $\theta_{max} = \pi/2$ and $\theta_{min} = \theta_2$ from Equation (7).

Figure 2 shows the estimated number of events per year somewhere on Earth as a function of B_0. Two modes of transit (through granite or water) and both potential modes of detection (human or radiofrequency) are considered.

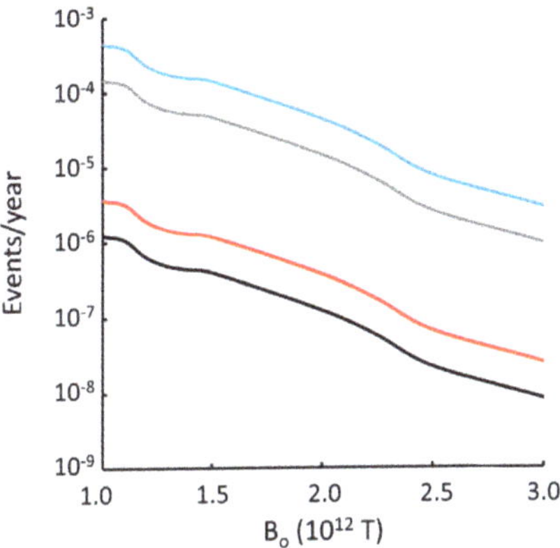

Figure 2. Estimated number of events per year somewhere on Earth as a function of B_0 for MQNs with 10^5 kg $\leq m \leq 10^{10}$ kg. The four solid-line curves correspond to event rates based on interstellar dark-matter density [3,42] of ~7 $\times$ 10^{-22} kg/m^3: MQNs transiting through water and emerging with any velocity v_{exit} (blue); MQNs transiting through water and emerging with velocity 10 m/s $\leq v_{exit} \leq$ 100 m/s (gray); MQNs transiting through granite and emerging with any velocity v_{exit} (red); MQNs transiting through granite and emerging with velocity 10 m/s $\leq v_{exit} \leq$ 100 m/s (black).

Figure 2 shows that detection on or near water is much more likely than detection deep within continents and shows the event rate is strongly dependent on parameter B_0. Unless the RF is absorbed by the surrounding plasma, the detection of MQNs by RF instruments (sensitive to any velocity) is much more likely than the detection of MQNs at 10 to 100 m/s, observable by human observers; however, records of human observations span centuries. Even one reliable report of an MQN event with the characteristics of a nearly tangential transit would suggest low values of B_0 and a mechanism that enhances the density of dark matter inside the solar system compared to that of interstellar space, as briefly described in Section 4.5.

3.4. Rotation at Megahertz Frequencies

MQNs interact with matter through its magnetopause. The shape of the magnetopause depends on the angle between the MQN velocity and the MQN magnetic moment, as illustrated in Figure 3.

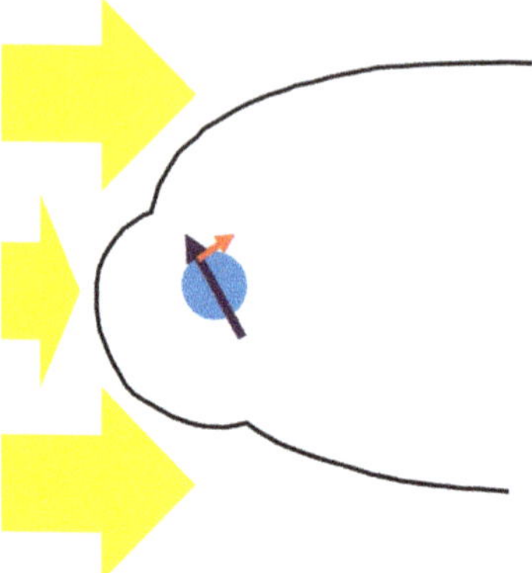

Figure 3. Cross sectional view of the magnetopause is shown (black line) between an MQN (blue circle) with magnetic moment (purple vector) at an angle of 60° to the velocity of the plasma (yellow arrows) flowing into the rest frame of the MQN. The plasma flow produces a net force (red arrow) centered at the top of the MQN and a corresponding torque vector into the page. The magnetopause is the locus of points at which the plasma pressure (on the left in Figure 3) is balanced by the magnetic pressure of the compressed magnetic field on the right. The complex shape of the magnetopause and resulting torque have been computed by Papagiannis [43] for Earth, illustrated in Figure 3, and extended to the case of MQNs. The effect can be understood by considering that the mean distance between the magnetopause and the MQN on the top half of Figure 3 is less than on the bottom half, which means that the magnetic field is compressed more on the top than on the bottom. Since force is transmitted by the compressed magnetic field, the net force is a push on the top, as shown by the red arrow.

Time-dependent asymmetry of the magnetopause produces a velocity-dependent and angle-dependent torque on the MQN and causes the MQN to oscillate initially about an equilibrium [36], as shown in Figure 4. Since the quark nugget slows down as it passes through ionized matter, the decreasing forward velocity reduces the torque with time, so the time-averaged torque in one half-cycle is greater than the opposing time-averaged torque in the next half-cycle. The amplitude of the oscillation necessarily grows. Once the angular momentum is sufficient to give continuous rotation, the net torque continually accelerates the angular motion to produce a rapidly rotating quark nugget. As shown in Figure 4, MHz frequencies are quickly achieved, even with a 10^6 kg quark nugget moving through granite (2300 kg/m^3 density matter) by the time v has slowed to 220 km/s.

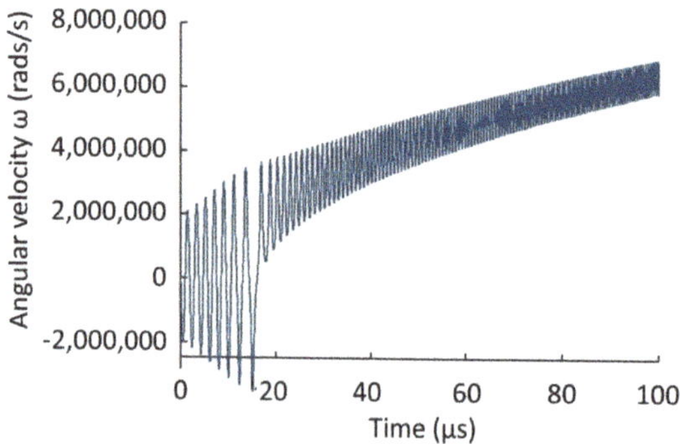

Figure 4. Estimated angular velocity in the first 10^{-4} s for 10^6 kg quark nugget with velocity $v = 220$ km/s, initial angle $\chi = 0.61$ rad, initial angular velocity $\omega = 0$, and passing through matter with mass density of 2300 kg/m^3. Note the initial oscillation about 0 until a full rotation occurs, after which the angular velocity increases rapidly.

As developed in reference [36] and summarized in Equations (15) through (19) for convenience, hypervelocity MQNs transiting through matter experience a torque that causes them to rotate with a frequency that depends on B_o, MQN mass m, MQN velocity v, and density ρ_p of the surrounding material. Rotating magnetic dipoles radiate at power P, where

$$P = \frac{Z_o}{12\pi}\left(\frac{\omega}{c}\right)^4 m_m^2 \tag{15}$$

in SI units, with $Z_o = 377\ \Omega$, ω = angular frequency, and c = the speed of light in vacuum. Magnetic dipole moment $m_m = 4\pi\, B_o\, r_{QN}^3/\mu_o$.

Assuming the energy loss per cycle from electromagnetic radiation to the surrounding plasma is just balanced by the energy gain per cycle from the torque T,

$$\int_0^{\frac{2\pi}{\omega}} T\,dt = \frac{2\pi P}{\omega^2} \tag{16}$$

in which

$$T = C_2 \rho_p^{0.5} v B_o r_{QN}^3 F_\chi$$
$$F_\chi = MIN(ABS(\tan\chi), ABS(\cot\chi))\frac{\tan\chi}{ABS(\tan\chi)} \tag{17}$$

The constant $C_2 = 1400$ with units of N s kg$^{-0.5}$ m$^{-1.5}$ T^{-1}, and the angle of rotation χ is the angle between the velocity of the incoming plasma and the magnetic moment.

The rate of change of angular velocity ω for MQN of mass m, with moment of inertia $I_{mom} = 0.4\, m\, r_{QN}^2$, and experiencing torque T, is

$$\frac{d\omega}{dt} = \frac{T}{I_{mom}} \tag{18}$$

Combining Equations (15) through (18) gives

$$\int_0^{\frac{2\pi}{\omega}} \frac{v(t)F_\chi(\chi(\omega t))}{\omega^2}\,dt = \frac{5.54\times10^{-22}B_o r_{QN}^3}{\rho_x^{0.5}} \tag{19}$$

which is solved numerically for the equilibrium angular velocity ω or frequency $f = \omega/(2\pi)$.

After emerging from dense matter, MQN rotational frequency decreases as rotational energy is radiated away with power P. Since P varies as ω^4 and the rotational energy varies as ω^2, the frequency as a function of time is not exponential. Solving for $\omega(t)$ gives

$$C_{RF} = \frac{1.08\times10^{43}}{B_o^2 m^{\frac{1}{3}}}$$
$$\omega(t) = \omega_0\sqrt{\frac{C_{RF}}{\omega_0^2 t + C_{RF}}} \tag{20}$$

Representative results for slowing down during transit through water and granite and for parameters of greatest interest are shown in Appendix D: Tables of MQN Interactions with Water and Granite.

The tables show RF frequencies at emergence from water or granite are weakly dependent on B_o but strongly dependent on mass and range from 7 MHz to 0.3 MHz for mass m between 10^5 kg and 10^{10} kg, respectively. Rotational energy ranges from ~0.1 MJ to ~1000 MJ and equals ~10^{-11} times the translational energy at impact. RF power emission at emergence ranges from ~4 GW to 22 TW.

The tables also show that the RF power, calculated with Equations (14) and (19) at $t = 1200$ seconds after emergence, ranges from ~6 MW to ~200 GW for the most massive MQNs with B_o between 1.3×10^{12} T and 3.0×10^{12} T, respectively.

Event rates in Figure 2 assume MQNs have the same flux as interstellar dark matter. However, the MQN flux may exceed that of interstellar space. Some MQNs passing through a portion of the solar photosphere are slowed to less than escape velocity from the solar system. Some of these are subsequently deflected by the net gravity of the planets so that they cannot return to the Sun and be absorbed. They accumulate. Therefore, this aerocapture process can enhance the local flux of MQNs and make Figure 2 a worst-case scenario. Enhancement factor, multiplied by observation time, would have to be >> 1000 for MQNs with nearly tangential trajectories to be observed.

Since RF detection occurs in real time, Figure 2 shows that the predicted event rate is too low to use RF to observe MQNs with nearly tangential trajectories, even if the RF is not absorbed by the plasma surrounding the MQN. Therefore, RF detection is best done in space [36] where the cross section is much larger and RF cutoff can be avoided.

In contrast, direct observations by human observers can cover centuries and may be recorded for us to analyze. The next section explores the observables in such an event to compare with observations published in the *Proceedings of the Royal Society* regarding an event on 6 August 1868 [38].

3.5. Simulations of a Rotating MQN with Plastically Deformable Conducting Witness Plate

The force between a rotating magnetized sphere and a plastically deformable conducting material was simulated by coupling the Rotating Magnetic Machinery module and the Nonlinear Plasticity Solid Mechanics module of the 3D, finite-element, COMSOL Multiphysics code [40]. The geometry is illustrated in Figure 5.

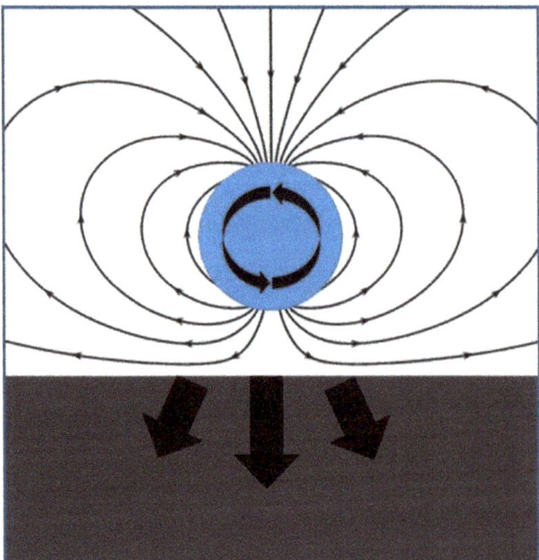

Figure 5. Geometry of simulation of rotating magnetized sphere above a highly conducting material. Magnetized, rotating sphere (blue) is shown above conducting material (gray). Arrows inside sphere indicate rotation and arrows in conducting material indicate force on the material. Arrows in air (white) indicate magnetic field lines at one moment in time. The axis of rotation is the y-axis, out of the plane of the figure. The magnetic axis of the magnetized sphere is initially in the x-direction and remains in the xz-plane.

Details of the simulation are provided in Appendix B: COMSOL simulation of rotating magnetized sphere interaction with plastically deformable conductor.

A small-scale experiment validated the COMSOL force calculation. A 3 mm or 6 mm thick copper plate was placed below a rotating spherical magnet and suspended with a calibrated spring to measure the force on the plate as a function of 1) separation between the center of the sphere and the front of the plate and 2) rotational frequency of the sphere. The measured force agreed with the force computed by the COMSOL simulation to within 10%. The agreement validates the computational method with the rotating coordinate system in the COMSOL module.

The computational mesh cannot resolve the microscopic diameter of an MQN. Therefore, we approximate the MQN with a 0.1 m radius, magnetized sphere with surface magnetic induction $B = 17{,}000$ T, which corresponds to an MQN with mass $m = 7 \times 10^7$ kg and $B_0 = 1 \times 10^{12}$ T.

Simulations with different values of electrical conductivity σ and frequencies f showed that the dynamics of the problem scales with the electromagnetic skin depth λ:

$$\lambda = \frac{1}{\sqrt{\pi \mu_0 \sigma f}} \tag{21}$$

Simulations converged best with low frequency. Consequently, we used frequency $f = 10$ Hz and varied the conductivity σ to explore the time-averaged force as a function of λ. The scaling with λ let us apply the results to higher frequencies and more realistic values of σ. Distance between the center of the sphere at $z = 0$ and the surface of the simulated peat was $z_p = -0.3$ m. Results are shown in Figure 6.

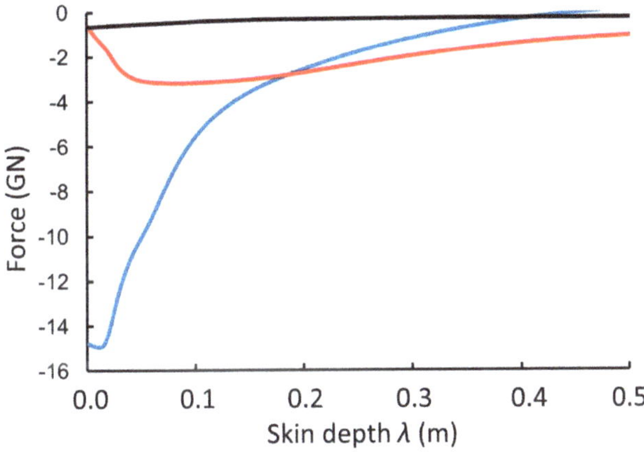

Figure 6. Components of the time-averaged force between the simulated quark nugget and conducting slab as a function of the electromagnetic skin depth λ. The negative (<0) force F_z (blue) opposes gravity and levitates the rotating magnetized sphere for $\lambda < 0.5$ m, with the most negative value for $\lambda < 0.03$ m. The force generated by the magnetic field traveling through the deformable conductor in the x-direction, as the magnetized sphere rotates about the y-axis, generates a propulsive force F_x (red). F_x is much less than F_z for small skin depths. The much smaller F_y (black) illustrates $\pm 5\%$ error in the calculation, since symmetry requires $F_y = 0$.

The levitating force is strongly dependent on the skin depth and decreases rapidly with increasing skin depth. A skin depth of 0.05 m, corresponding to $\sigma = 10^7$ S/m and $f = 10$ Hz for $\sigma f = 10^8$ SHz/m, was chosen for the simulation of plastic deformation.

The radius of the magnetized sphere was set at 0.1 m and its magnetic induction field was set at 2085 T, corresponding to an MQN with mass ~9×10^7 kg and $B_0 = 1 \times 10^{12}$ T. The radius of the rotating volume in the COMSOL mesh was set at $r = 0.2$, and the front surface of the $4 \times 4 \times 2$ m deformable conductor was at $r = 0.3$ m.

For a time-averaged force of 10^7 N in the z-direction (the direction opposing gravity), the maximum magnetic induction in the peat is 18.5 T and the maximum induced current density is 3×10^8 A/m^2. The time-averaged forces were -1.1×10^7 N, -0.3×10^7 N, and -0.05×10^7 N in the z-, x-, and y-directions, respectively.

Deformation of the material as a function of time from the integrated simulation is shown in Figure 7 as contour plots for four times and two orthogonal directions.

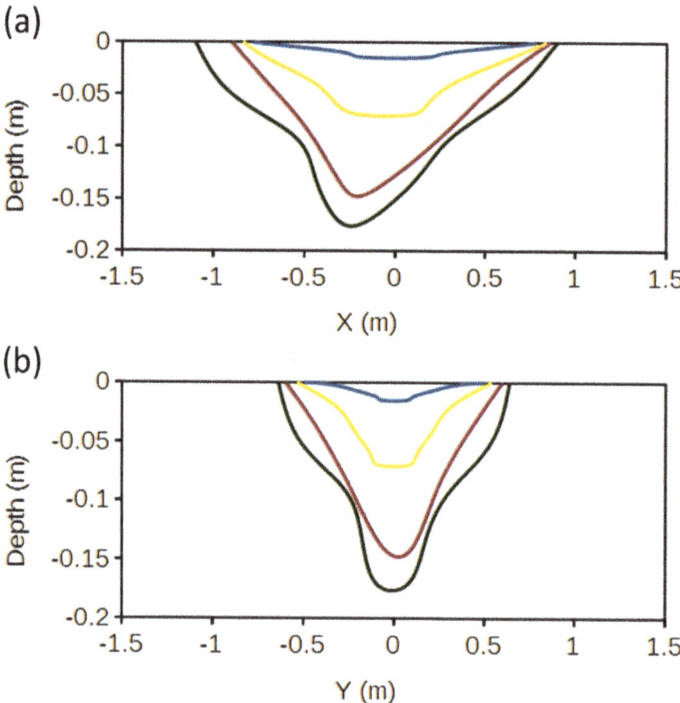

Figure 7. Contours of the hole formed by the rotating the magnetic field of the magnetized sphere in (**a**) the x-direction and (**b**) the y-direction for times 2.5 ms (blue), 10 ms (gold), 20 ms (red), and 30 ms (black). The magnetic field sweeps through plastically deformable conducting material, and the displacement of the bottom of the hole is approximately -0.25 m in the x-direction. The same contours for the y-direction, which is along the axis of rotation, show the deformation is symmetric about $y = 0$, as expected. In both cases, the vertical axis has a different scale from the horizontal axis.

A rotating magnetic dipole is equivalent to two oscillating current loops oriented at 90° and with currents 90° out of phase. The linear superpositions of the two magnetic fields induce currents that are almost independent of orientation but produce a net force in the direction (x) perpendicular to the axis of rotation (y). That explains the shape of the deformations in Figure 7a,b.

The simulation has many limitations. The MQN cannot be realistically resolved, so the gradient in the B field is not exact. The rotating sphere cannot move downward as material is displaced, so the deformation stops when the applied stress reaches equilibrium with the strain specified in the stress–strain curve. The yield strength of the material is a constant, independent of the degree of compression and of flow of liquid driven by the $\mathbf{J} \times \mathbf{B}$ force in the material, so the deformation rate is only qualitative. Consequently, these simulations provide only semi-quantitative results to compare with observations.

Despite these limitations, these simulations clearly show that a rapidly rotating magnetized sphere with sufficient mass and sufficient magnetic dipole field, such as a massive MQN, will create a hole in a plastically deformable conducting material by currents induced in its interior. They also show the sphere will experience a force that moves it through the material to create a trench in conducting, deformable material.

In addition, the results in Figure 6 clearly indicate that the levitating force decreases rapidly with increasing skin depth, which varies inversely with the square root of frequency. Therefore, the MQN height above conducting material decreases as the rotational energy is depleted by RF emissions and by the resistive dissipation of current induced in the peat.

The RF power emissions are predicted to be megawatts to many gigawatts and are certainly sufficient to ionize and excite the surrounding air. Thermal and magnetohydrodynamic motion and mixing of the air around an MQN complicates simple estimates of the shape of the luminosity; however, Equation (15) shows that the radiated power varies as the fourth power of the frequency, so the local electric field varies as the square of the frequency. Therefore, the ionization and excitation of air will diminish faster than the height above the peat.

These simulations assume induced current flows in the conducting medium. If the surface electric field, from the rate of change of magnetic induction, is sufficient to form a plasma at the air–ground interface, then magnetic pressure will deform the material as if $\lambda \sim 0$.

3.6. Comparison with M. Fitzgerald's Report to the Royal Society

M. Fitzgerald's report [37] to the Royal Society describes an event that occurred on 6 August 1868. The reported observations are consistent with the characteristics of a nearly tangential MQN impact as developed in this paper: a luminous orb with clear skies overhead, persisting for much longer than weather-related ball lightning [44], moving slowly across and into a plastically deformable conducting medium (peatland), decreasing in diameter with time, and creating trenches in the ground. Since it is difficult to obtain copies of proceedings more than a century old, Fitzgerald's description is quoted from the published report in Appendix C: Field Investigation of M. Fitzgerald's Report to Royal Society. In summary, a ~0.60 m diameter, light-emitting orb was observed travelling at ~1 m/s over and into a peat bog in County Donegal, Ireland, for ~20 min. During that time, its diameter decreased to ~0.08 m; it displaced over 10^5 kg of water-saturated peat; it produced approximately 1 m wide trenches in the peat.

Although we initially dismissed his report as incredible, we eventually realized that peat only grows a few centimeters per century [45], so the features he described must still be visible if the reported event actually occurred. His report was sufficiently detailed to enable an investigation, which we conducted from 2004 through 2006.

Fitzgerald's reported features and our corresponding findings are summarized by the following:

- An approximately 6.4 m square hole described by Fitzgerald on the course from the crown of the ridge to the south of Meenawilligan, towards the town of Church Hill. We found a 6.4 m square hole 0.7 m deep along that course.
- An approximately 180 m distance reported to the next deformation. We found the deformation had been partially destroyed by draining of the field for sheep grazing. If this deformation were still the reported 100 m length, the southern end would be 175 ± 2 m from the hole.
- An approximately 100 m long, 1.2 m deep, and 1 m wide trench. As stated above, this deformation has been truncated by the owner having drained the field. The remaining trench is currently 63 ± 1 m long, 0.2 ± 0.05 m deep (soft to 0.8 ± 0.05 m), and 1.2 ± 0.1 m wide. Carbon dating of peat inside and outside the trench confirms a disturbance occurred, consistent with the report.
- Unspecified distance to the third excavation. We found the distance to be 5 ± 0.3 m.
- Curved trench formed when the stream bank was "torn away" for 25 m and dumped into the stream. We found the remaining curved trench to be 25 ± 1 m long and 1.4 ± 0.1 m deep. The 1863 Ordnance Survey map does not show the stream diversion that Fitzgerald reported as occurring on 6 August 1868. Therefore, the event happened after 1863. Fitzgerald's submission to the Royal Society is dated 20 March 1878, so the

event occurred before 1868. Therefore, the event is independently dated between 1863 and 1878.

- Cave in the stream bank directly opposite the end of the "torn away" bank. We found the cave at that position. It is currently 0.45 ± 0.08 m wide, 0.3 ± 0.06 m high, and 0.5 ± 0.1 m deep. However, its proximity to the water line raises the possibility that its origin was flowing water and not the event Fitzgerald reports.

The extant features support Fitzgerald's account. This part of Ireland was, and still is, sparsely populated and did not have local newspapers that might have recorded an unusual sound or tremor. Therefore, no contemporaneous and independent eyewitness report confirms his account.

The vegetation is dominated by members of the heath (Ericaceae) and sedge (Cyperaceae) families. Surface features would become obscured to some degree by the growth of vegetation and debris carried in by wind, reducing the time a depression could be observed from a distance. A nearby crater, attributed [46] to a vertical impact of a 10 kg MQN in 1985, has changed little in the past 35 years, which is consistent with the current state of the features Fitzgerald attributed to the 1868 event.

The size of the trenches and the yield strength of peat gives a downward force of $>10^7$ N, which implies a rotating, magnetically levitated mass of $\sim 10^6$ kg. That mass and the volume of the ~ 0.08 m diameter luminosity implies a mass density $>10^9$ kg/m^3, which is inconsistent with normal matter. Its levitation implies an extremely large magnetic dipole rotating at >1 MHz to levitate the large mass.

The repulsive force of electromagnetic induction by an MQN, as derived in this paper, are consistent with the levitation of an MQN core and the displacement of the peat in its path. The yield strength of peat was measured and found to be 530 kN/m$^2 \pm 23\%$, which is the value used in the simulations of plastic deformation above. Electrical conductivity σ within 0.2 m of the surface was measured to be 22 mS/m $\pm 30\%$, which is consistent with published values for peatlands [47], as follows: 25 mS/m near the surface and ~ 380 mS/m at up to 2 m depth. Table A2 in Appendix D (Tables of MQN Interactions with Water and Granite) gives 3 to 9 MHz for the frequency of a massive MQN when it first emerges from the ground. The corresponding skin depth $\lambda \sim 1.1$ to 2.0 m for $\sigma = 22$ mS/m and $\lambda \sim 0.27$ to 0.47 m for $\sigma = 385$ mS/m. As shown in Figure 6, these values of λ are too large to produce the reported levitation and deformation if the induced current is flowing through the peat, as assumed in the simulations.

However, the large electric field from the rate of change of the magnetic induction should cause the surface to flash over and form a plasma on top of the peat, similar to how the air-water interface in pulsed power devices flash over. (Montoya, R. and Danneskiold, J. Five seconds at F/16, with a broken camera. *Sandia Lab News* (7 June 2018). https://www.sandia.gov/news/publications/labnews/articles/2018/08-06/Randy.html. (accessed on 24 July 2020).) If so, then the effective skin depth $\lambda \sim 0$ and the deformation are consistent with the frequencies calculated for the MQN. Computationally or experimentally simulating such a process would be extremely difficult, so surface flashover is consistent with relevant experience but has not been confirmed in the field.

Since the field with the second deformation (the remaining 63 m of trench) has not been drained, its slope (10%) is almost the same as it was in 1868. The simplicity of this particular deformation allows us to estimate the minimum mass of the core of this object from the yield strength and the volume. The product of the minimum pressure P, which we equate to the measured yield strength 530 ± 120 kN/m^2, times the volume change ΔV, is the minimum work required to compress the trench (1.4 m wide, 1.2 m deep, and 100 m long) and is $\sim 10^8$ joules. The energy for this work came purely from gravitational energy as the "globe of fire" descended the slope from the beginning of the trench to its terminus. The corresponding mass is $m \sim 10^8/(gh) \sim 10^6$ kg, where g is the acceleration of gravity in m/s^2 and h is the change in distance toward Earth's center in meters and agrees with the mass estimated from the yield strength and trench diameter.

If the induced currents are flowing in a plasma on the surface, the pressure equals the magnetic pressure P_m produced by the magnetic field B:

$$P_m = \frac{B^2}{2\mu_0} \tag{22}$$

Since the pressure supports a mass m against the acceleration g of gravity, $P_m = mg/(\pi r^2)$ for r = the half-width of the trench and $\mu_0 = 4\,\pi \times 10^{-7}$ H/m. The value of B spatially and temporarily averaged over the effective area πr^2 is ~5.4 T and is consistent with the spatial and temporal maximum value of $B = 18.5$ T from the simulation.

Fitzgerald reports that the diameter of the luminous ball diminished from ~0.6 m diameter at the beginning to ~0.08 m at the end of the event; however, our survey shows that the width and depth of the depressions in the peat were about the same at the beginning and end of the event. Therefore, the core of the object remained unchanged, with diameter <0.08 m, volume of $<3 \times 10^{-4}$ m^3, and density of $>10^6/(3 \times 10^{-4})$ or $>3 \times 10^9$ kg/m^3—at least 200,000 times the density of normal matter. Matter does not exist with density between 3×10^9 kg/m^3 and nuclear density. Such a large mass density implies nuclear density matter and is consistent with the ~10^{18} kg/m^3 mass density of MQNs.

Based on Table A1 in Appendix D (Tables of MQN Interactions with Water and Granite) and on the MQN's motion to the northeast, the impact must have been >125 km southwest of the 1868 observations for impact velocity ~250 km/s. That location is deep in the Atlantic Ocean, >80 km from land, and too far away to be heard by Fitzgerald. Therefore, Table A1 in Appendix D (Tables of MQN Interactions with Water and Granite) applies and indicates the impact was necessarily >200 km from land and is unlikely to be found.

Uncertainties in the current distribution and in applying the quasi-static and uncompressed yield strength to such a dynamic process leads us to assign $+/-$ an order of magnitude error bar to the mass estimate. The resulting $10^{6+/-1}$ kg mass corresponds to the maximum mass in the mass distributions [4] for $B_o = 1.65 \times 10^{12}$ T $+/- 21\%$. For comparison, the magnetic moments and mass densities of protons and neutrons, which are also baryons, correspond to magnetic fields $B_o = 1.5 \times 10^{12}$ T and 2.5×10^{12} T, respectively, in reasonable agreement with our value for MQNs. The smaller range is a considerable improvement over Tatsumi's $10^{12+/-1}$ T estimate and permits the design of a systematic test of the MQN dark-matter hypothesis.

4. Discussion

The principal result of this paper is reducing uncertainty in the key parameter of the MQN theory to $B_o = 1.65 \times 10^{12}$ T $+/- 21\%$. The result depends on (1) Fitzgerald having accurately reported what he observed, (2) the event being caused by a nearly tangential MQN impact as we have calculated, and (3) the absence of a more likely explanation.

4.1. Fitzgerald's Accuracy

We have done what due diligence is possible on Fitzgerald's qualifications as a reliable observer. The County records indicate he was the assistant surveyor for County Donegal during this period. That was a responsible position and is consistent with the detailed report to the Royal Society. We also know that the Royal Society had sufficient confidence in his report to have the president of the Society read it into the proceedings. We have no reason to doubt his integrity.

4.2. Consistency with MQN Impact

The details allowed us to find all the secondary observations, i.e., the reported deformations in the peat bog. Carbon 14 analysis, the ordnance survey map of 1860 (prior to the reported event), and the quantitative agreement between our measurements and Fitzgerald's reported measurements, all support consistency with a nearly tangential MQN impact, as we have calculated.

4.3. Alternative Explanations

We have attempted to identify other possible causes of the secondary observations. Surface features produced by water erosion of peat and settling under gravity are well documented in the geomorphological literature. Sinuous water-cut gullies dissect peat bogs into complex patterns of haggs (residual masses) and groughs (the gullies) [48]. Mass movement processes involving the flow or slide of water-saturated peat cause major disruptions to bog surfaces and can extend for several hundreds of meters downslope [49]. The features described by Fitzgerald are unlike any of the peat erosional features previously reported and cannot be ascribed to conventional geomorphological processes.

One of us (Professor Peter Wilson) is a geomorphologist specializing in peat bogs. He has investigated the four structures Fitzgerald's eyewitness account connected to the event and concludes that the literature on the geomorphology of peat bogs and his forty years of field work in peat bogs have not suggested any other possible causes for three of the four structures. We conclude that the fourth (the cave) was too close to the stream to preclude its being formed by water flow.

Neutron-stars have the right mass density. However, the gravitational force that holds them together is too weak to sustain the required $\sim 10^6$ kg mass. In addition, pulsar magnetic fields [50] are two orders of magnitude less than those of magnetars [51], upon which the MQN model has been constructed.

Primordial Mini-Black Holes (MBHs) [52] have been proposed to explain luminous and levitating events attributed to anomalous non-weather-related ball lightning [53]. Although an MBH does not have a magnetic field that could levitate it, the net shape of the event horizon from the combined gravitational fields of Earth and a black hole can, in principle, direct evaporating particles downward to provide thrust and levitate the mass [53]. However, the lifetime of a 10^6 kg MBH is not consistent with the 20-min event reported by Fitzgerald, and the explosion equivalent to 10 million one-megaton hydrogen bombs characteristic of the final evaporation [52] of an MBH was not observed.

We have also sought alternative explanations from others. Given the eyewitness account and given that our simulations show the MQN hypothesis is consistent with Fitzgerald's eyewitness report, the question becomes can something else reside above and in the peat for 20 min, leave meter-scale (depth and width) structures in the peat, and have a glowing light associated with it. We have presented these results to about a thousand people in university colloquia and in contributed and invited talks at meetings of the American Physical Society and Institute of Electrical and Electronic Engineers in the USA, UK, Russia, and India. None of the audience members have suggested an alternative explanation.

4.4. Limitations to the Evidence

Even though the three criteria identified in the first paragraph of the Discussion are arguably satisfied, the primary event is the "globe of fire" itself and was seen only by Fitzgerald. The Fitzgerald event is not quite singular. Two other similar primary events qualitatively consistent (i.e., meter-scale, luminous, long-lasting, quasi-spherical, and rotating, as observed by the angular momentum imparted to the surrounding water) with nearly tangential MQN impacts were reported by Soviet Navy Captains at sea in 1962 and 1966 [39]. However, the Fitzgerald event is the only one contemporaneously published in the scientific journal of its day and the only one with detailed secondary observables that can be verified by anyone after the event.

We conclude that the primary event is very rare, is not reproducible, was not recorded by multiple observers, and cannot be quantitatively validated after the fact by anyone else. Consequently, the evidence does not meet today's standard for a discovery.

4.5. Significance

However, we also conclude that we can tentatively accept his report and use it to narrow the uncertainty in B_0 since it is consistent with an MQN event, and a more likely explanation has not been found. The resulting uncertainty in B_0 can be used to design an experiment to systematically test the MQN hypothesis within the constrained range of B_0. If MQNs are found as predicted, then the acceptance of the Fitzgerald event as an MQN event will have been validated. If nothing is found, then the experiment will have been another null experiment placing a limit of the mass distribution of MQNs characterized by the tested values of B_0, just as all the single-mass quark-nugget experiments to date have been null experiments that placed a flux limit on that single mass.

The results have one additional consequence. Predicted event rates in Figure 2 are so small that even one observed event is consistent with the major portion of dark matter being composed of MQNs. Observing more than one per century suggests a substantial enhancement factor for dark-matter density inside the solar system compared to that of interstellar space. Nuclear-density MQNs are indestructible and can survive passage through the solar photosphere. The magnetopause interaction [28] with matter in the solar photosphere and subsequent deflection by the combined gravity of the planets offer the possibility of enhancing the flux of MQN dark matter within the solar system. Computing an accurate enhancement factor for MQN impacts on Earth requires detailed Monte Carlo simulations beyond the scope of this paper and strongly depends on B_0, the radial profile of mass density in the solar photosphere, the velocity distribution of dark matter in interstellar space, and scattering of MQNs by planetary gravity.

In contrast to other candidates for dark matter, MQNs are baryons and, therefore, are consistent [4] with the Standard Model for particles and fields. Well known physics can guide additional experiments and observations [36] to test the MQN hypothesis for dark matter, invent ways for collecting useful MQNs, and develop applications for an indestructible source of $\sim 10^{12}$ tesla magnetic fields.

5. Conclusions

The two orders of magnitude uncertainty in Tatsumi's estimate for B_0 precludes the practical design of systematic experiments to detect MQNs through their predicted interaction with matter. In this paper, we theoretically examined the signature of a new class of episodic events consistent with a unique signature of MQNs and reported the results of field investigations of one published event consistent with that signature. Tentatively accepting that the event was indeed caused by MQNs constrains the most likely values of B_0 to 1.65×10^{12} T $+/- 21\%$, which can be used to design a systematic test of the MQN dark-matter hypothesis.

Author Contributions: Conceptualization, J.P.V. and A.P.V.; data curation, J.P.V. and N.M.; formal analysis, J.P.V.; Funding acquisition, J.P.V.; investigation, J.P.V., A.P.V., P.W., B.F.H. and N.M.; methodology, J.P.V., A.P.V., P.W. and N.M.; project administration, J.P.V.; resources, J.P.V.; software, J.P.V.; supervision, J.P.V.; validation, J.P.V., A.P.V. and B.F.H.; visualization, J.P.V.; Writing—Original draft, J.P.V.; Writing—Review and editing, J.P.V., A.P.V., P.W., B.F.H. and N.M. All authors have read and agreed to the published version of the manuscript.

Funding: This research received no external funding.

Institutional Review Board Statement: Not applicable.

Informed Consent Statement: Not applicable.

Data Availability Statement: All final analyzed data generated during this study are included in this published article with its supplements.

Acknowledgments: We gratefully acknowledge S.V. Greene for first suggesting that quark nuggets might explain the geophysical evidence, and W. F. Brinkman for especially helpful suggestions during this work; Peter van Doorn of the Tornado and Storm Research Organization, UK, for introducing us to the 1868 event; Josie "The Post" Duddy of County Donegal, Ireland, for showing us the 1985 event; Jesse Rosen for editing the manuscript. We appreciate William E. Gallagher of Meenawilligan, Kevin Rose, and David Daniels of Booze Allen Hamilton, Arlington, VA, for their assistance in the field work, and Archivists Neve Brennan and Ciara Joyce for locating survey documents from the 1860s.

Conflicts of Interest: The authors declare no conflict of interest.

Appendix A. Quark-Nugget Research Summary

This summary is an update of one published by us in the open-source article, reference [4]. Therefore, it contains similar information and is included for convenience.

Macroscopic quark nuggets are theoretically predicted objects composed of up, down, and strange quarks in essentially equal numbers. They are also called strangelets [8], nuclearites [9], AQNs [10], and slets [11] and are a subset of Macros [5], a more general term for massive dark matter.

In 1971, Bodmer [12] suggested that a collection of up, down, and strange quarks should be stable. Witten [7] and Farhi and Jaffe [8] showed that quark nuggets should be in the ultra-dense, color-flavor-locked (CFL) phase of quark matter and should be a stable candidate for dark matter. Steiner et al. [13] showed that the ground state of the CFL phase is color neutral and that color neutrality forces electric charge neutrality, which minimizes electromagnetic emissions. However, Xia et al. [11] found that quark depletion causes the ratio Q/A of electric charge Q to baryon number A to be non-zero and varying as $Q/A \sim 0.32\, A^{-1/3}$ for $3 < A < 10^5$. In addition to this core charge, they find that there is a large surface charge and a neutralizing cloud of charge to give a net zero electric charge for sufficiently large A. So, quark nuggets with $A \gg 1$ are both dark and very difficult to detect with astrophysical observations.

Bodmer, Witten, and Xia et al. also showed that quark-nugget density should be somewhat larger than the density of nuclei, and their mass can be very large, even as large as the mass of a star. Large quark nuggets are predicted to be stable [7,12–15] with mass between 10^{-8} and 10^{20} kg within a plausible but uncertain range of assumed parameters of quantum chromodynamics (QCD) and the MIT bag model with its inherent limitations [16].

Although Witten assumed that a first-order phase transition formed quark nuggets, Aoki et al. [17] showed that the finite-temperature QCD transition that formed quark nuggets in the hot early Universe was very likely an analytic crossover, involving a rapid change as the temperature varied but not a real phase transition. Recent simulations by T. Bhattacharya et al. [18] support the crossover process.

A combination of quark nuggets and anti-quark nuggets has also been proposed within constraints imposed by terrestrial observations of the neutrino flux [19]. Zhitnitsky [9] proposed that the collapse of an axion domain-wall network generated Axion Quark Nuggets (AQNs) of both quark and anti-quark varieties. The model relies on the hypothetical axion particle beyond the Standard Model, appears to explain a wide variety of longstanding problems, and leads to AQNs with a narrow mass distribution at ~ 10 kg [20]. Atreya et al. [21] also found that CP-violating quark and anti-quark scatterings from moving Z(3) domain walls should form quark and anti-quark nuggets, regardless of the order of the quark-hadron phase transition.

Experiments by A. Bazavov et al. [22], at the Relativistic Heavy Ion Collider (RHIC), have provided the first indirect evidence of strange baryonic matter. Additional experiments at RHIC may determine whether the process is a first order phase transition or crossover process. In either case, quark nuggets could have theoretically formed in the early Universe.

In 2001, Wandelt et al. [23] showed that quark nuggets meet the theoretical requirements for dark matter and are not excluded by observations when the stopping power for quark nuggets in the materials covering a detector is properly considered and when the average mass is $>10^5$ GeV ($\sim 2 \times 10^{-22}$ kg). In 2014, Tulin [24] surveyed additional simulations of increasing sophistication and updated the results of Wandelt, et al. The combined results help establish the allowed range and velocity dependence of the strength parameter and strengthen the case for quark nuggets. In 2015, Burdin, et al. [25] examined all non-accelerator candidates for stable dark matter and also concluded that quark nuggets meet the requirements for dark matter and have not been excluded experimentally. Jacobs, Starkman, and Lynn [5] found that combined Earth-based, astrophysical, and cosmological observations still allow quark nuggets of mass 0.055 to 10^{14} kg and 2×10^{17} to 4×10^{21} kg to contribute substantially to dark matter. The large mass means the number per unit volume of space is small, so detecting them requires a very large area detector.

These studies did not consider an intrinsic magnetic field within quark nuggets. However, Xu [26] has shown that the surface magnetic field of quark nuggets from electron ferromagnetism is limited to ~2×10^7 T, which is too small for magnetars and MQNs. Tatsumi [27] has shown that, under some special values of the currently unknown QCD coupling constant at the ~90 MeV energy scale, a One Gluon Exchange interaction may allow quark nuggets to be ferromagnetic with a surface magnetic field of $10^{12\pm1}$ T. Such a large magnetic field is sufficient for magnetar cores and MQNs. For a quark nugget of radius r_{QN} and a magnetar of radius r_s, the magnetic field scales as $(r_{QN}/r_s)^3$. Therefore, the surface magnetic field of a magnetar is smaller than 10^{12} T because $r_s > r_{QN}$. Since quark-nugget dark matter is bare, the surface magnetic field of what we wish to detect is $10^{12\pm1}$ T.

Although the cross section for interacting with dense matter is greatly enhanced [28] by the magnetic field, which falls off as radius r_{QN}^{-3}, the collision cross section is still many orders of magnitude too small to violate the collision requirements [10,21–23] for dark matter.

Chakrabarty [29] showed that the stability of quark nuggets increases with increasing external magnetic field $\leq 10^{16}$ T, so the large self-field described by Tatsumi should enhance their stability. Ping et al. [30] showed that magnetized quark nuggets should be absolutely stable with the newly developed equivparticle model, so the large self-field described by Tatsumi should ensure that quark nuggets with sufficiently large baryon number will not decay by weak interaction.

The large magnetic field also alters MQN interaction with ordinary matter through the greatly enhanced stopping power of the magnetopause around high-velocity MQNs moving through a plasma [28]. Searches [30] for quark nuggets with underground detectors would not be sensitive to highly magnetized quark nuggets, which cannot penetrate the material above the detector. For example, the paper by Gorham and Rotter [19] about constraints on anti-quark-nugget dark matter assumes that limits on the flux of magnetic monopoles from analysis by Price et al. [30] of geologic mica buried under 3 km of rock are also applicable to quark nuggets that can reach the mica.

Porter et al. [32] and Piotrowski et al. [33] reported the absence of sufficiently fast meteor-like objects in the lower atmosphere constrains the flux of quark nuggets (nuclearites) to approximately that required to explain dark matter. Bassan et al. [34] looked for quark nuggets (nuclearites) with gravitational wave detectors and found signals much less than expected for the flux of dark matter.

In summary, experimental or observational evidence of quark nuggets has yet to be found [35] after decades of searching. However, all of these analyses assumed (1) quark nuggets can reach the detector volume because the cross section for momentum transfer is the geometric cross section and (2) all quark nuggets have a single specific mass. In contrast, (1) the MQN magnetopause cross section [28] is many orders of magnitude larger and prevents all but the extremely rare, mostly massive (>1000 kg) MQNs from being detected and (2) MQNs have a very broad mass distribution [4] which means these experiments do not exclude MQNs.

Appendix B. COMSOL Simulation of Rotating Magnetized Sphere Interaction with Plastically Deformable Conductor

The coupled electromagnetic and mechanical interaction between (1) a massive, rotating, strongly magnetized sphere and (2) a nearby conductor was simulated to model the interaction between an MQN and a nearby plastically deformable, electrically conducting medium. The results are provided in the main body of this paper. Additional computational details on the method of this calculation are not of general interest and are given in the following paragraphs.

The overall geometry of the simulation is shown in Figure A1 and consists of a 20 m radius conducting boundary, a cylindrical rotating coordinate system on the same axis as the conducting boundary, a simulated magnetized sphere within the rotating coordinate system, and a 4 × 4 × 2 m slab of simulated peat. The boundary condition for the 20 m radius is $\mathbf{n} \times \mathbf{A} = 0$, in which $\mathbf{n}$ is the unit vector normal to the boundary and $\mathbf{A}$ is the vector potential, so the magnetic induction vector $\mathbf{B}$ lies along the boundary surface everywhere. The 20 m radius is sufficiently large to make the force on the peat insensitive to the position of the boundary.

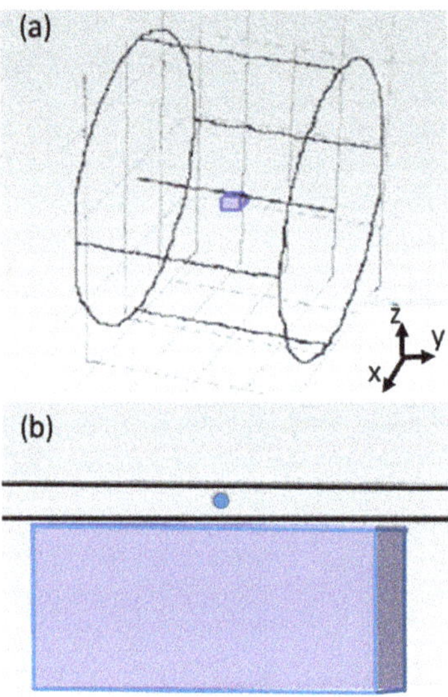

Figure A1. Geometry of simulation of a rotating quark nugget. (**a**) The overall geometry of the simulation shows the 20 m radius conducting boundary with the cylindrical rotating coordinate system centered on the axis and the 4 × 4 × 2 m thick slab of simulated peat. (**b**) Close up of the peat slab with its top surface located 0.3 m below the center of the 0.1 m radius, spherical magnet simulating the quark nugget, which is inside the cylindrical 0.2 m radius rotating coordinate system.

The COMSOL [40] Solid Mechanics module solves for the elastic and plastic deformation of the peat under the volume force $\mathbf{F_v} = \mathbf{J} \times \mathbf{B}$, in which $\mathbf{J}$ is the current density in the peat and $\mathbf{B}$ is the magnetic field.

The results are to be compared to observations of an event in peatland. Peat is 80% to 90% water and is not usually studied with solid mechanics models. Therefore, we estimated the mechanical properties of peat for these first calculations based on the properties of water with the measured electrical conductivity and yield stress of peat. Additional work to refine the properties will affect the results somewhat, but the calculated deformation below illuminates the essential behavior. The properties used in this calculation follow:

Electrical conductivity = 22 mS/m for peat. Relative dielectric constant = 80. Relative permeability = 1. Initial density = 10^3 kg/m^3. Yield stress = 5.30×10^5 Pa. Poisson ratio = 0.4. Young's modulus = 2×10^9 Pa. Isotropic tangent modulus = 1.1×10^8 Pa.

Strain is the fractional change, so it is dimensionless. Therefore, the stress versus strain curve is elastic with Young's modulus until a stress of 5.30×10^5 Pa is exceeded at an elastic strain of 2.0×10^{-4}. Then the ratio of additional stress to additional strain is the isotropic tangent modulus of 1.1×10^8 Pa.

The COMSOL Rotating Magnetic Machinery module solves the rotating and non-rotating parts of the problem in their respective coordinate frames and forces continuity of the scalar magnetic potential V_m in the fixed frame. Since the meshes at the interface of the rotating and non-rotating frames are not identical, the calculation interpolates the scalar magnetic potential between the non-conforming meshes. If the magnetic vector potential **A** has to be interpolated across the boundary, then current is not conserved. In principle, applying Ampère's law only inside the peat slab avoids this problem. Nevertheless, we varied the radius of the rotating coordinate system to assess the degree to which the numerical interpolation technique affects the results. The choice of the radius of the rotating coordinate system affected the magnetic field at the surface of the peat by approximately $\pm 50\%$, so this solution is not ideal. However, choosing 0.2 m for the radius of the rotating coordinate system mitigates the problem. This choice gives an air gap of 0.1 m between the interface and the magnetized sphere and between the interface and the surface of the peat.

We used the elastic–plastic deformation module to investigate the deformation of the peat and its dependence on skin depth λ with its included frequency f, as defined by Equation (21) in the main text. A simulation with the measured conductivity of 22 mS/m and a frequency of 4.5×10^8 Hz took 1500 s of computer time for one 2.2 ns period. Calculating the full deformation at simulation time of 0.03 s would take 6×10^6 hours of computer time, which is prohibitive. Comparison of simulations at 1, 10, and 100 Hz showed that the force on the peat scales with the product of electrical conductivity σ (S/m) and frequency f (Hz), and, therefore, scales with the electromagnetic skin depth λ.

An intermediate skin depth of 0.05 m, corresponding to $\sigma = 10^7$ S/m and $f = 10$ Hz for $\sigma f = 10^8$ SHz/m, was chosen as the baseline case for these exploratory simulations. The radius of the magnetized sphere was set at 0.1 m and its magnetic induction field was set at 2085 T. The radius of the rotating coordinate system was set at 0.2 m and the front surface of the $4 \times 4 \times 2$ m peat slab was at 0.3 m, as shown in Figure A1.

The simulations showed that the total force on the peat scaled as the square of the magnetic field in the magnetized sphere, as it does in Equation (22). In addition, the magnetic field in the 1868 event, which required a force of 10^7 N to support the estimated 10^6 kg mass in Earth's gravitational field, varied as the cube of the radius of the quark nugget, as shown in Figure A2.

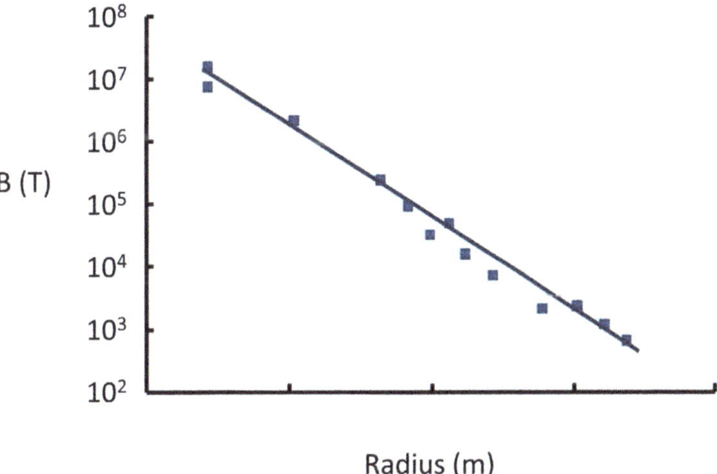

Figure A2. Amplitude of the magnetic induction B in the rotating and magnetized sphere required to produce a time-averaged force of 10^7 N on the peat as a function of the radius of the magnetized sphere. The radius of the magnetized sphere was varied between 5 and 150 mm. The mesh size was too large for the calculation to converge for radii less than 5 mm.

The force required to levitate the sphere for a constant distance from the conducting plane is proportional to the mass of the sphere. The levitating force for the same geometry scales as the square of the surface magnetic field. Therefore, the results are generalized to give the strength of the magnetic field B_S at the surface of the sphere of radius R_S and mass M_S:

$$B_S = \frac{C_1 M_S^{0.5}}{R_S^{3 \pm 0.05}} \tag{A1}$$

in which $C_1 = 1.4 \times 10^{-3 \pm 0.3}$ T m^3 kg$^{-0.5}$. The coefficient of determination (R^2) value is 0.98.

Appendix C. Field Investigation of M. Fitzgerald's Report to Royal Society

A "globe of fire" event [38] in County Donegal, Ireland, on 6 August 1868, was reported to the Royal Society by M. Fitzgerald, Assistant Surveyor for County Donegal. His report gave sufficient detail to let us find the deformations he described and investigated the physics that would cause those deformations.

The original publication in the *Proceedings of the Royal Society* [38] describes a series of depressions in peat caused by a "globe of fire". The report is reproduced below, since the only known copy of the original publication is in London, UK, and is only available to researchers approved by the British Library for access.

"On the 6 August 1868, this neighbourhood being free from the dense black clouds that hung over the mountains of Glenswilly and Glendoan, I went up the latter glen to note anything worthy of observation. On arriving at Meenawilligan, the sky was so black over Bintwilly [or Bin Tuile - 'the mountain of floods'], where lightning and thunder were following each other in rapid succession, that I turned homewards. When I reached Folbane, on looking behind, I noticed a globe of fire in the air floating leisurely along in the direction of Church Hill. After passing the crown of the ridge, where I first noticed it, it descended gradually into the valley, keeping all the way about the same distance from the surface of the land, until it reached the stream between Folbane and Derrora, about 300 yards from where I stood. It then struck the land and reappeared in about a minute, drifted along the surface for about 200 yards, and again disappeared into the boggy soil, reappearing about 20 perches (1 perch = 5.03 m) further down the stream; again it moved along the surface, and again sunk, this time into the brow of the stream, which it flew across and finally lodged in the opposite brow, leaving a hole in the peat bank, where it buried itself.

If it had left no marks behind, I confess that, as I had never seen anything of the kind before, I should hesitate to describe its movements, which surprised me much at the time, but the marks which it left behind of its course and power surprised me more.

I at once examined its course, and found a hole about 20 feet square, where it first touched the land, with the pure peat turned out on the lea as if it had been cut out with a huge knife. This was only one minutes work, and, as well as I could judge, it did not occupy fully that time. It next made a drain about 20 perches in length and 4 feet deep, afterwards ploughing up the surface about 1 foot deep, and again tearing away the bank of the stream about 5 perches in length and 5 feet deep, and then hurling the immense mass into the bed of the stream, it flew into the opposite peaty brink. From its first appearance till it buried itself could not have been more than 20 min, during which it traveled leisurely, as if floating, with an undulating motion through the air and land over one mile. It appeared at first to be a bright red globular ball of fire, about 2 feet in diameter, but its bulk became rapidly less, particularly after each dip in the soil, so that it appeared not more than 3 inches in diameter when it finally disappeared. The sky overhead was clear at the time but about an hour afterwards it became as dark as midnight."

Fitzgerald's report provided enough information to locate the site of his observations. Since the growth rate of peat in the British Isles during the last few thousand years has generally been in the range of 1–6 cm per century [45] and undisturbed peat readily holds its form, the holes and trenches would still be extant after 137 years. During six separate expeditions to the site in 2004–2006, we found the hole, trench, stream diversion, and cave on privately owned land between 54°58.321′ N, 7°54.668′ W and 54°58.294′ N, 7° 54.576′ W. We also made a final visit in 2014 to see how much the site changed over a 10-year period.

Ordnance Survey (OS) maps of 1863 and 1870 were found in the local archive. The 1870 map is a minor revision of the original survey of 1863 rather than a thorough resurvey. A small section of the 1863 map (five years prior to the reported "globe of fire" event of 6 August 1868) is shown in Figure A3. The wavy black lines are the original drawing. Our survey of the area found Fitzgerald's reported features, which are also shown.

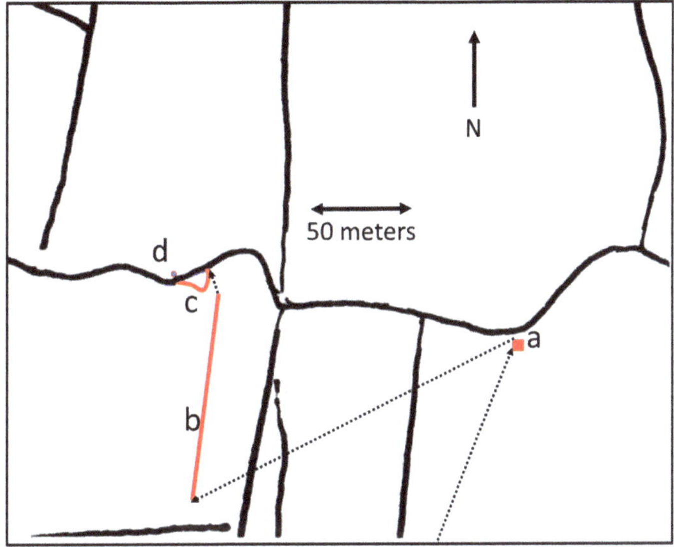

Figure A3. Extract from the 1863 Ordnance Survey map. The locations of the "square hole" (a), the most prominent trench (b), triangular channel (c), and cave (d), and the reported path of the "globe of fire" (dotted line) between features are shown. The field with the "20-foot-square" hole (a) has been drained and is lower than the field with the trench (b) and the triangular channel (c). Therefore, we do not know the relative elevation of the hole (a) and the trench (b) in 1868.

The dominant features of the reported depressions are as follows, with letters referring to locations marked in Figure A3:

- Hole (a): ~6.4 m square depression on the course from the crown of the ridge to the south of Meenawilligan towards the town of Churchill.
- Approximately 180 m to the next depression.
- Straight trench (b): ~100 m long, 1.2 m deep, and 1 m wide.
- Unspecified distance to the third depression.
- Curved trench (c): formed when stream bank was "torn away" for 25 m and dumped into the stream.
- Cave (d): a hole in the stream bank directly opposite the end of the "torn away" bank.

Fitzgerald reported that the "globe of fire" first went into the peat bog near the intersection of (1) the line between the "crown of the ridge" and the town of Church Hill and (2) the stream between Derrora and Falabane. At that location, we found a hole of about 6.0 m square with about 0.6 m of open water at 54°58.294′ N and 7°54.576′ W. The hole is located in a marsh at the intersection of a slight west-to-east flow of surface water and two lesser drainage lines coming from the south.

We determined the contours of the hole at 0.3 and 0.5 m below the top of the peat, which are shown in Figure A4. At the 0.5 m depth, the hole is composed of three trenches intersecting at 90-degree angles to form a "square hole". Each trench is approximately

1.2 m wide and 0.8 m to a hard bottom. The length of the south, east, and north sides are, respectively, 2.85, 6.4, and 8.4 m.

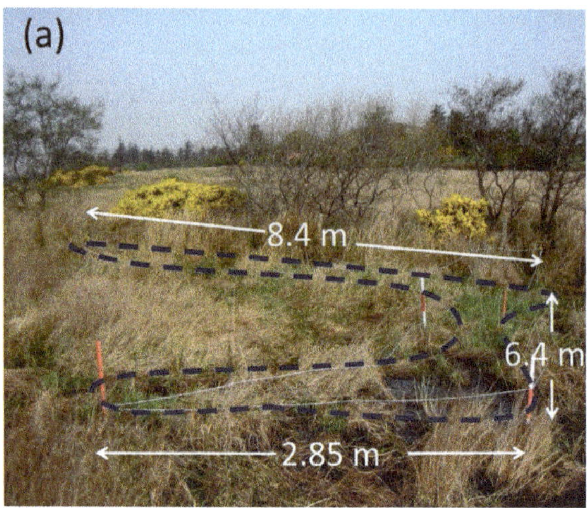

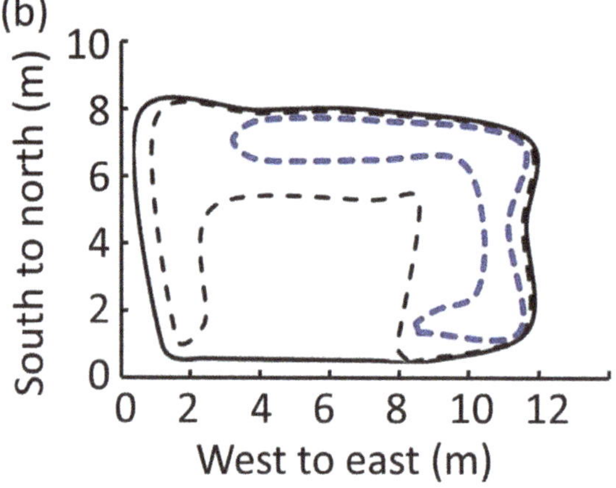

Figure A4. The location consistent with Fitzgerald's "20-foot-square hole," at which the ball of light first disappeared into the peat. (**a**) Photo of the "square hole" with dimensions and 0.5 m deep contour (dashed blue line). (**b**) Contours at three depths are shown: 0 m (solid black line), 0.3 m (dashed black line), and 0.5 m (dashed blue line). The natural feeder drainage flows approximately from the lower right to the upper left. Orientation to north is approximate.

Fitzgerald reported that the "globe of fire" floated over the bog for about 180 m and then cut a 1.2 m deep trench into the bog for about 100 m. A formerly cultivated field lies approximately 150 m west of the hole and has many north–south-aligned trenches. The seventh trench from the eastern edge of the field is the most prominent and is 63 m long. It lies 0.2 m below the adjacent surface and is approximately 1.2 m wide; these parameters are, respectively, 2.2 and 2.3 standard deviations greater than the mean values of the 26 trenches. The trench is further differentiated from the surrounding terrain by greater penetrability; a ski pole readily penetrates ~0.8 m into the trench but penetrates only ~0.3 m into the surrounding peat with the same force of ~130 N.

Probing the rest of the seventh trench revealed that the firm walls of the trench fall off abruptly (in approximately 0.2 m), indicating a well-defined and deep trench containing low-density peat. However, similar measurements of all 26 trenches show that the penetrability into the peat is not as strong a differentiator between trenches as are depth and width. The probe depth of the seventh trench is only 1.1 standard deviations more than the mean depth of all 26 trenches, and three of the other 25 trenches are deeper. Twelve measurements of the variation in the results give a standard deviation of 0.06 m for this measurement.

This trench is located between 54°58.318′ N, 7°54.651′ W and 54°58.282′ N, 7°54.663′ W and is shown in Figure A5.

Figure A5. Photograph looking along the seventh trench.

Two peat samples from 0.8 ± 0.04 m depth (relative to the top of the adjacent ridges) of the third and seventh trenches and one peat sample at the same depth from the ridge west of the seventh trench were carbon-14 dated by a commercial laboratory. The peat from the most prominent (seventh) trench is 620 ± 60 years old at 85 cm depth and adjacent material is 1330 ± 70 years old at same depth—which is consistent with Fitzgerald's trench, having been filled by erosion with a mixture of peat formed at various times between 0 to 1330 years ago. The next most prominent (third) trench, parallel to the seventh trench, was dated at the same depth as a second control and was found to be 2040 ± 50 years old. This second control is downslope from the most prominent trench and is expected to be older. These data support the uniqueness of the seventh candidate trench. However, measurements in peat are complicated because naturally occurring humic acids can circulate through peat and cause carbon dating to give a later date than isolated material would produce.

The field with these trenches was divided into two parts by the owner in 2000 and separated by a new drainage ditch. The southern portion is used for grazing sheep and is about 30 cm lower than the portion discussed above. The seventh trench ends at the new boundary between the fields at a distance of 63 m from the northern end and 158 m from the hole. If this trench had extended into the newly divided southern portion of the field for another 37 m (to make the ~100 m reported by Fitzgerald), then that end would have been 175 m from the hole and consistent with the ~180 m reported by Fitzgerald. From all evidence above, we can conclude that the seventh trench is the most likely candidate for Fitzgerald's "100-m long trench".

At its northern (downhill) end, this trench terminates in a mound of peat that prevents it from draining into the stream, and there is no evidence of subsurface piping draining the trench. The mound may have been created when the stream was realigned about 20 years ago, or the trench may have terminated at the mound when it was formed; there is no way to know for certain.

We found a triangular channel to the south of the existing stream and starting at 5 m west of the trench. It is separated from the stream by a mound of mixed mineral and peat debris at 54°58.319′ N, 7°54.676′ W. The length of the channel is 25 m and its depth is approximately 1.5 m. The landowner said that the water flowed through this channel until the Council redirected the stream in approximately 1989, which agrees with the tree rings observed in one of the trees now growing in the channel bed and with the composition of the mound, which is clearly the material removed from the recently excavated northern channel. The western extremity of this 25 m trench and the stream, as it was recut by the Council, are shown in Figure A6. The photo was taken in 2004 after we cut out the vegetation that had overgrown the trench. In our last trip to the site in 2014, we found the vegetation had again grown throughout the 25 m channel.

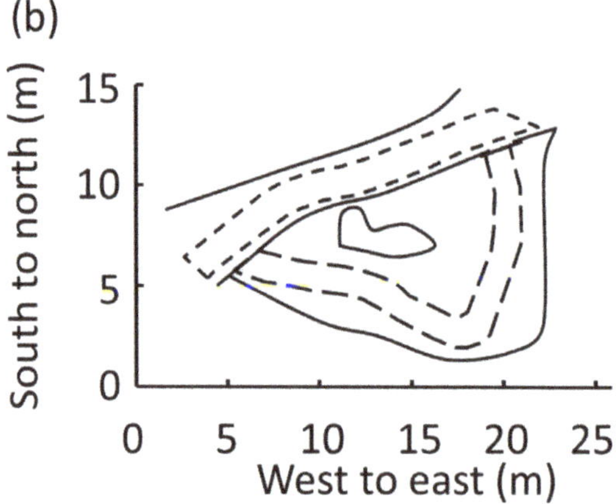

Figure A6. Fitzgerald's "25 m long diversion of the stream" and the current path of the stream, which continues to the left and right of the contour map. (**a**) Photo of the site as seen from the western end; the channel made by the "globe of fire" is on the right and the stream that was recut by the Council is on the left. (**b**) Contour map of the site constructed from the survey and field notes. Solid lines: surface level. Long dashes: the bottom of the channel at −1.2 ± 0.25 m level. Short dashes: the bottom of the stream, as cut by the County Council in the 1980s, at the −1.9 ± 0.2 m level. Orientation to the north is approximate.

Immediately to the south of the western end of the triangular channel, there is a shallow cave in the stream bank at 54°58.321′ N, 7°54.673′ W, as shown in Figure A7.

Figure A7. Cave at the end of the semi-circular channel.

The depth of the cave is 0.5 m and it is located in the north bank of the stream. The photograph was taken when the water depth was only 10 cm; however, we observed flood debris in stream-bank trees downstream from the cave, indicating that the water rises to at least 1.1 m depth at times. The cave appeared the same in 2014 as it did in 2004.

The electrical resistivity was measured at many places and in several seasons. It was consistently 30 to 60 Ω-m. The compressive yield strength was consistently measured to be 530 ± 120 kN/m^2 for uncompressed peat, and the strength increased with increasing compression. Radiation measurements were taken. Only background radiation was detected.

Fitzgerald's reported deformations are compared to our findings in Section 3.6 of the main article.

Appendix D. Tables of MQN Interactions with Water and Granite

Table A1. Representative examples are given for MQNs with impact velocity of 250 km/s transiting through water as a function of B_o and MQN mass m_{qn}.

B_o (T)	1.3×10^{12}	1.5×10^{12}	2×10^{12}	2.5×10^{12}	3×10^{12}	3×10^{12}
m_{qn} (kg)	3.2×10^{5}	3.2×10^{5}	3.2×10^{6}	3.2×10^{6}	3.2×10^{6}	3.2×10^{9}
r_{QN} (m) for $\rho_{QN} = 10^{18}$ kg/m^3	4.2×10^{-5}	4.2×10^{-5}	9.1×10^{-5}	9.1×10^{-5}	9.1×10^{-5}	9.1×10^{-4}
Magnetopause radius r_m (m)	2.5×10^{-2}	2.6×10^{-2}	6.2×10^{-2}	6.7×10^{-2}	7.1×10^{-2}	7.1×10^{-1}
Flux for MQN decadal mass (N/y/m^2/sr)	1.4×10^{-15}	1.4×10^{-15}	1.5×10^{-16}	4.3×10^{-18}	2.7×10^{-19}	1.3×10^{-19}
x_{max} (m)	241,197	219,252	389,995	336,093	297,630	2,977,672
x_{10} m/s (m)	240,914	218,995	389,539	335,700	297,282	2,974,189
x_{100} m/s (m)	239,887	218,062	387,878	334,268	296,014	2,961,506
θ_2 (°)	88.91690	89.01545	88.24855	88.49068	88.66344	76.50504
θ_{10} (°) for $v_{exit} = 10$ m/s	88.91817	89.01660	88.25059	88.49245	88.66500	76.52112
θ_{100} (°) for $v_{exit} = 100$ m/s	88.92278	89.02080	88.25806	88.49888	88.67070	76.57968
t_{exit} (s) for $v_{exit} = 10$ m/s	5.4×10^{1}	5.0×10^{1}	8.8×10^{1}	7.6×10^{1}	6.7×10^{1}	6.7×10^{2}
t_{exit} (s) for $v_{exit} = 100$ m/s	2.4×10^{1}	2.2×10^{1}	3.9×10^{1}	3.4×10^{1}	3.0×10^{1}	3.0×10^{2}
δ fractional error for $v_{exit} = 10$ m/s	6.0×10^{-2}	5.5×10^{-2}	9.7×10^{-2}	8.4×10^{-2}	7.4×10^{-2}	6.0×10^{-1}
δ fractional error for $v_{exit} = 100$ m/s	1.2×10^{-2}	1.1×10^{-2}	1.9×10^{-2}	1.7×10^{-2}	1.5×10^{-2}	1.5×10^{-1}
Cross section for all v_{exit}	4.6×10^{10}	3.8×10^{10}	1.2×10^{11}	8.9×10^{10}	7.0×10^{10}	7.1×10^{12}
Cross section for $v_{exit} = 10$ to 100 m/s	3.9×10^{8}	3.2×10^{8}	1.0×10^{9}	7.5×10^{8}	5.9×10^{8}	6.1×10^{10}

Table A1. *Cont.*

Total number per year	3.5×10^{-4}	2.9×10^{-4}	9.7×10^{-5}	2.1×10^{-6}	1.0×10^{-7}	5.2×10^{-6}
Number per year for 10 to 100 m/s v_{exit}	3.0×10^{-6}	2.5×10^{-6}	8.2×10^{-7}	1.8×10^{-8}	8.7×10^{-10}	4.5×10^{-8}
Frequency (MHz)	7.0×10^{0}	7.0×10^{0}	3.2×10^{0}	3.1×10^{0}	3.0×10^{0}	3.1×10^{-1}
Rotational energy (J)	1.4×10^{5}	1.3×10^{5}	1.3×10^{6}	1.2×10^{6}	1.2×10^{6}	1.2×10^{9}
RF power (MW)	4.4×10^{3}	5.6×10^{3}	4.3×10^{4}	6.2×10^{4}	8.3×10^{4}	8.4×10^{6}
RF power (MW) after 1200 s	6.5×10^{0}	5.0×10^{0}	6.0×10^{1}	3.9×10^{1}	2.8×10^{1}	2.0×10^{5}

Table A2. Representative examples are given for MQNs with impact velocity of 250 km/s transiting through granite as a function of B_o and MQN mass m_{qn}.

B_o (T)	1.3×10^{12}	1.5×10^{12}	2×10^{12}	2.5×10^{12}	3×10^{12}	3×10^{12}
m_{qn} (kg)	3.2×10^{5}	3.2×10^{5}	3.2×10^{6}	3.2×10^{6}	3.2×10^{6}	3.2×10^{9}
r_{QN} (m) for $\rho_{QN} = 10^{18}$ kg/m^3	4.2×10^{-5}	4.2×10^{-5}	9.1×10^{-5}	9.1×10^{-5}	9.1×10^{-5}	9.1×10^{-4}
Magnetopause radius r_m (m)	2.2×10^{-2}	2.3×10^{-2}	5.4×10^{-2}	5.8×10^{-2}	6.2×10^{-2}	6.2×10^{-1}
Flux for MQN decadal mass (N/y/m2/sr)	1.4×10^{-15}	1.4×10^{-15}	1.5×10^{-16}	4.3×10^{-18}	2.7×10^{-19}	1.3×10^{-19}
x_{max} (m)	138,434	125,839	223,837	192,900	170,824	1,709,027
x_{10} m/s (m)	138,272	125,692	223,575	192,674	170,624	1,707,028
x_{100} m/s (m)	137,683	125,156	222,622	191,852	169,897	1,699,749
θ_2 (°)	89.37838	89.43494	88.99486	89.13380	89.23293	82.30288
θ_{10} (°) for v_{exit} = 10 m/s	89.37911	89.43560	88.99604	89.13481	89.23383	82.31194
θ_{100} (°) for v_{exit} = 100 m/s	89.38176	89.43801	89.00032	89.13850	89.23710	82.34492
t_{exit} (s) for v_{exit} = 10 m/s	3.1×10^{1}	2.8×10^{1}	5.1×10^{1}	4.4×10^{1}	3.9×10^{1}	3.9×10^{2}
t_{exit} (s) for v_{exit} = 100 m/s	1.4×10^{1}	1.3×10^{1}	2.3×10^{1}	1.9×10^{1}	1.7×10^{1}	1.7×10^{2}
δ fractional error for v_{exit} = 10 m/s	3.5×10^{-2}	3.1×10^{-2}	5.6×10^{-2}	4.8×10^{-2}	4.3×10^{-2}	3.9×10^{-1}
δ fractional error for v_{exit} = 100 m/s	6.9×10^{-3}	6.3×10^{-3}	1.1×10^{-2}	9.6×10^{-3}	8.5×10^{-3}	8.5×10^{-2}
Cross section for all v_{exit}	1.5×10^{10}	1.2×10^{10}	3.9×10^{10}	2.9×10^{10}	2.3×10^{10}	2.3×10^{12}
Cross section for v_{exit} = 10 to 100 m/s	1.3×10^{8}	1.1×10^{8}	3.3×10^{8}	2.5×10^{8}	1.9×10^{8}	2.0×10^{10}
Total number per year	1.2×10^{-4}	9.7×10^{-5}	3.2×10^{-5}	6.9×10^{-7}	3.4×10^{-8}	1.7×10^{-6}
Number per year for 10 to 100 m/s v_{exit}	9.9×10^{-7}	8.2×10^{-7}	2.7×10^{-7}	5.9×10^{-9}	2.9×10^{-10}	1.4×10^{-8}
Frequency (MHz)	9.0×10^{0}	8.9×10^{0}	4.0×10^{0}	3.9×10^{0}	3.9×10^{0}	3.9×10^{-1}
Rotational energy (J)	2.2×10^{5}	2.2×10^{5}	2.1×10^{6}	2.0×10^{6}	2.0×10^{6}	1.9×10^{9}
RF power (MW)	1.2×10^{4}	1.5×10^{4}	1.1×10^{5}	1.6×10^{5}	2.2×10^{5}	2.2×10^{7}
RF power (MW) after 1200 s	6.7×10^{0}	5.1×10^{0}	6.1×10^{1}	4.0×10^{1}	2.8×10^{1}	2.3×10^{5}

References

1. Aghanim, N.; Akrami, Y.; Ashdown, M.; Aumont, J.; Baccigalupi, C.; Ballardini, M.; Banday, A.J.; Barreiro, R.B.; Bartolo, N.; Basak, S.; et al. (Planck Collaboration), Planck 2018 results. VI. Cosmological parameters. *Astron. Astrophys.* **2020**, *641*, A6. Available online: https://arxiv.org/pdf/1807.06209.pdf (accessed on 10 May 2020).
2. Peebles, P.J.E. *Cosmology's Century, an Inside Story of Our Modern Understanding of The Universe*; Princeton University Press: Princeton, NJ, USA, 2020; pp. 216–297.
3. Salucci, P. The distribution of dark matter in galaxies. *Astron. Astrophys. Rev.* **2019**, *27*, 2. [CrossRef]
4. VanDevender, J.P.; Shoemaker, I.M.; Sloan, T.; VanDevender, A.P.; Ulmen, B.A. Mass distribution of magnetized quark-nugget dark matter and comparison with requirements and direct measurement. *Sci. Rep.* **2020**, *10*, 1–16. [CrossRef]
5. Jacobs, D.M.; Starkman, G.; Lynn, B.W. Macro dark matter. *Mon. Not. R. Astron. Soc.* **2015**, *450*, 3418–3430. [CrossRef]
6. Oerter, R. *The Theory of Almost Everything: The Standard Model, the Unsung Triumph of Modern Physics*; Penguin Group: New York, NY, USA, 2006; pp. 203–218.
7. Witten, E. Cosmic separation of phases. *Phys. Rev. D* **1984**, *30*, 272–285. [CrossRef]
8. Farhi, E.; Jaffe, R.L. Strange matter. *Phys. Rev. D* **1984**, *30*, 2379–2390. [CrossRef]
9. De Rújula, A.; Glashow, S.L. Nuclearites—a novel form of cosmic radiation. *Nature* **1984**, *312*, 734–737. [CrossRef]
10. Zhitnitsky, A.R. 'Nonbaryonic' dark matter as baryonic colour superconductor. *J. Cosmol. Astropart. Phys.* **2003**, *2003*, 010. [CrossRef]

11. Xia, C.-J.; Peng, G.-X.; Zhao, E.-G.; Zhou, S.-G. From strangelets to strange stars: A unified description. *Sci. Bull.* **2016**, *61*, 172–177. [CrossRef]

12. Bodmer, A.R. Collapsed Nuclei. *Phys. Rev. D* **1971**, *4*, 1601–1606. [CrossRef]

13. Steiner, W.S.; Reddy, S.; Prakash, M. Color-neutral superconducting dark matter. *Phys. Rev. D* **2020**, *66*, 094007. [CrossRef]

14. Lugones, G.; Horvath, J.E. Primordial nuggets survival and QCD pairing. *Phys. Rev. D* **2004**, *69*, 063509. [CrossRef]

15. Bhattacharyya, A.; Alam, J.-E.; Sarkar, S.; Roy, P.; Sinha, B.; Raha, S.; Bhattacharjee, P. Relics of the cosmological QCD phase transition. *Phys. Rev. D* **2000**, *61*, 083509. [CrossRef]

16. Chodos, A.; Jaffe, R.L.; Johnson, K.F.; Thorn, C.B.; Weisskopf, V.F. New extended model of hadrons. *Phys. Rev. D* **1974**, *9*, 3471–3495. [CrossRef]

17. Aoki, Y.; Endrődi, G.; Fodor, Z.; Katz, S.D.; Szabó, K.K. The order of the quantum chromodynamics transition predicted by the standard model of particle physics. *Nature* **2006**, *443*, 675–678. [CrossRef] [PubMed]

18. Bhattacharya, T.; Buchoff, M.I.; Christ, N.H.; Ding, H.-T.; Gupta, R.; Jung, C.; Karsch, F.; Lin, Z.; Mawhinney, R.D.; McGlynn, G.; et al. QCD phase transition with chiral quarks and physical quark masses. *Phys. Rev. Lett.* **2014**, *113*, 082001. [CrossRef] [PubMed]

19. Gorham, P.; Rotter, B.J. Stringent neutrino flux constraints on antiquark nugget dark matter. *Phys. Rev. D* **2017**, *95*, 103002. [CrossRef]

20. Ge, S.; Lawson, K.; Zhitnitsky, A. The axion quark nugget dark matter model: Size distribution and survival pattern. *Phys. Rev. D* **2019**, *99*, 116017. [CrossRef]

21. Atreya, A.; Sarkar, A.; Srivastava, A.M. Reviving quark nuggets as a candidate for dark matter. *Phys. Rev. D* **2014**, *90*, 045010. [CrossRef]

22. Bazavov, A.; Ding, H.-T.; Hegde, P.; Kaczmarek, O.; Karsch, F.; Laermann, E.; Maezawa, Y.; Mukherjee, S.; Ohno, H.; Petreczky, P.; et al. Additional strange hadrons from QCD thermodynamics and strangeness freezeout in heavy ion collisions. *Phys. Rev. Lett.* **2014**, *113*, 072001. [CrossRef]

23. Wandelt, B.D.; Davé, R.; Farrar, G.R.; McGuire, P.C.; Spergel, D.N.; Steinhardt, P.J. Self-interacting dark matter, Chapter 5. In *Sources and Detection of Dark Matter and Dark Energy in the Universe*; Cline, D.B., Ed.; Springer: Berlin, Heidelberg, 2001; pp. 263–274.

24. Tulin, S. Self-Interacting dark matter. *AIP Conf. Proc.* **2014**, *1604*, 121.

25. Burdin, S.; Fairbairn, M.; Mermod, P.; Milstead, D.A.; Pinfold, J.L.; Sloan, T.; Taylor, W. Non-collider searches for stable massive particles. *Phys. Rep.* **2015**, *582*, 1–52. [CrossRef]

26. Xu, R.X. Three flavors in a triangle. *Sci. China Phys. Mech. Astron.* **2020**, *63*. [CrossRef]

27. Tatsumi, T. Ferromagnetism of quark liquid. *Phys. Lett. B* **2000**, *489*, 280–286. [CrossRef]

28. VanDevender, J.P.; VanDevender, A.P.; Sloan, T.; Swaim, C.; Wilson, P.; Schmitt, R.G.; Zakirov, R.; Blum, J., Sr.; Cross, J.L.; McGinley, N. Detection of magnetized quark-nuggets, a candidate for dark matter. *Sci. Rep.* **2017**, *7*, 8758. [CrossRef]

29. Chakrabarty, S. Quark matter in a strong magnetic field. *Phys. Rev. D* **1996**, *54*, 1306–1316. [CrossRef]

30. Peng, C.; Xia, C.-J.; Xu, J.-F.; Zhang, S.-P. Magnetized strange quark matter in the equivparticle model with both confinement and perturbative interactions. *Nucl. Sci. Tech.* **2016**, *27*, 98. [CrossRef]

31. Price, P.B.; Salamon, M.H. Search for supermassive magnetic monopoles using mica crystals. *Phys. Rev. Lett.* **1986**, *56*, 1226–1229. [CrossRef]

32. Porter, N.A.; Cawley, M.F.; Fegan, D.; MacNeill, G.C.; Weekes, T.C. A search for evidence for nuclearites in astrophysical pulse experiments. *Irish Astron. J.* **1988**, *18*, 193–196. [CrossRef]

33. Piotrowski, L.W.; Małek, K.; Mankiewicz, L.; Sokolowski, M.; Wrochna, G.; Zadrożny, A.; Żarnecki, A.F. Limits on the flux of nuclearites and other heavy compact objects from the pi of the sky project. *Phys. Rev. Lett.* **2020**, *125*, 091101. [CrossRef] [PubMed]

34. Bassan, M.; Coccia, E.; D'Antonio, S.; Fafone, V.; Giordano, G.; Marini, A.; Minenkov, Y.; Modena, I.; Pallottino, G.; Pizzella, G.; et al. Dark matter searches using gravitational wave bar detectors: Quark nuggets and newtorites. *Astropart. Phys.* **2016**, *78*, 52–64. [CrossRef]

35. Patrignani, C.; Agashe, K.; Aielli, G.; Amsler, C.; Antonelli, M.; Asner, D.M.; Baer, H.; Barnett, R.M.; Basaglia, T.; Bauer, C.W.; et al. Review of particle properties (Particle Data Group). *Chin. Phys.* **2016**, *C40*, 100001.

36. VanDevender, J.P.; Buchenauer, C.J.; Cai, C.; VanDevender, A.P.; Ulmen, B.A. Radio frequency emissions from dark-matter-candidate magnetized quark nuggets interacting with matter. *Sci. Rep.* **2020**, *10*, 13756. [CrossRef] [PubMed]

37. Scherrer, R.J.; Turner, M.S. On the relic, cosmic abundance of stable, weakly interacting massive particles. *Phys. Rev. D* **1986**, *33*, 1585–1589. [CrossRef]

38. Fitzgerald, M. Notes on the occurrence of globular lightning and of waterspouts in County Donegal, Ireland, Quarterly Proceeding at the March 20th, 1878, Proceedings at the Meetings of the Society. *Q. J. Meteorolog. Soc.* **1878**, *1Q1878*, 160–161.

39. Klimov, A. *Analysis of New Reports on Observation of Plasmoids in Atmosphere and Under Water, Proceedings of the AIS-2008 Conference on Atmosphere, Ionosphere, Safety 1, Kaliningrad, Russia, 7–12 July 2008*; Karpov, I.V., Ed.; Russian Foundation of Basic Research: Moscow, Russia, 2008; pp. 38–39. Available online: http://ais2020.ru/archive (accessed on 15 October 2020).

40. COMSOL. Multiphysics Finite Element Code Is a Commercial Code. Available online: http://www.comsol.com (accessed on 22 June 2020).

41. Koch, C. *A Quest for Consciousness*; Roberts: Greenwood Village, CO, USA, 2004; p. 260.

42. Read, J.I. The local density of dark matter. *J. Phys. G Nucl. Part. Phys.* **2014**, *41*, 063101. [CrossRef]

43. Papagiannis, M.D. The torque applied by the solar wind on the tilted magnetosphere. *J. Geophys. Res.* **1973**, *78*, 7968–7977. [CrossRef]
44. Cen, J.; Yuan, P.; Xue, S. Observation of the optical and spectral characteristics of ball lightning. *Phys. Rev. Lett.* **2014**, *112*, 035001. [CrossRef]
45. Feehan, J.; O'Donovan, G. *The Bogs of Ireland*; The Environmental Institute of University College: Dublin, Ireland, 1996.
46. VanDevender, J.P.; Schmitt, R.G.; McGinley, N.; VanDevender, A.P.; Wilson, P.; Dixon, D.; Auer, H.; McRae, J. Results of search of magnetized quark-nugget dark matter from radial impacts on Earth. *arXiv* **2020**, arXiv:2007.04826. Available online: https://arxiv.org/abs/2007.04826 (accessed on 3 February 2021).
47. Theimer, B.D.; Nobes, D.C.; Warner, B.G. A study of the geoelectrical properties of peatlands and their influence on ground-penetrating radar surveying. *Geophys. Prospect.* **1994**, *42*, 179–209. [CrossRef]
48. Wilson, P. Blanket peat. *Ir. Mt. Log* **2019**, *130*, 41–44.
49. Wilson, P.; Griffiths, D.; Carter, C. Characteristics, impacts and causes of the Carntogher bog-flow, Sperrin Mountains, Northern Ireland. *Scott. Geogr. Mag.* **1996**, *112*, 39–46. [CrossRef]
50. Taylor, J.H.; Manchester, R.N.; Lyne, A.G. Catalog of 558 pulsars. *Astrophys. J. Suppl. Ser.* **1993**, *88*, 529–568. [CrossRef]
51. Olausen, S.A.; Kaspi, V.M. The McGill magnetar catalog. *Astrophys. J. Suppl. Ser.* **2014**, *212*, 6. [CrossRef]
52. Hawking, S.W. The quantum mechanics of black holes. *Sci. Am.* **1977**, *236*, 34–40. [CrossRef]
53. Rabinowitz, M. Little black holes: Dark matter and ball lightning. *Astrophys. Space Sci.* **1999**, *262*, 391–410. [CrossRef]

Article

Results of Search for Magnetized Quark-Nugget Dark Matter from Radial Impacts on Earth

J. Pace VanDevender [1,*], Robert G. Schmitt [2], Niall McGinley [3], David G. Duggan [4], Seamus McGinty [5], Aaron P. VanDevender [1], Peter Wilson [6], Deborah Dixon [7], Helen Girard [1] and Jacquelyn McRae [1]

[1] VanDevender Enterprises LLC, 7604 Lamplighter Lane NE, Albuquerque, NM 87109, USA; aaron@vandevender.com (A.P.V.); kentaha2@yahoo.com (H.G.); jacquechan98@gmail.com (J.M.)
[2] Sandia National Laboratories, Albuquerque, NM 87185, USA; rgschmi@sandia.gov
[3] Ardaturr, Churchill PO, Letterkenny, Co., F928982 Donegal, Ireland; niallmacfhionnghaile@gmail.com
[4] National Park and Wildlife Service, Glenveagh National Park, Church Hill, F92XK02 County Donegal, Ireland; dave.duggan@chg.gov.ie
[5] Church Hill, Letterkenny, F928982 County Donegal, Ireland; pacevan@gmail.com
[6] School of Geography and Environmental Sciences, Ulster University, Cromore Road, Coleraine, Co. Londonderry, Northern Ireland BT52 1SA, UK; p.wilson@ulster.ac.uk
[7] DKD Engineering Inc., 801 El Alhambra Cir. NW, Los Ranchos, NM 87107, USA; ddixon@dkdengineeringinc.com
* Correspondence: pace@vandevender.com

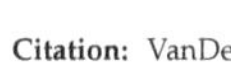
check for
updates

Citation: VanDevender, J.P.; Schmitt, R.G.; McGinley, N.; Duggan, D.G.; McGinty, S.; VanDevender, A.P.; Wilson, P.; Dixon, D.; Girard, H.; McRae, J. Results of Search for Magnetized Quark-Nugget Dark Matter from Radial Impacts on Earth. *Universe* **2021**, *7*, 116. https://doi.org/10.3390/universe7050116

Academic Editors: Kazuharu Bamba and Maxim Yu. Khlopov

Received: 20 February 2021
Accepted: 18 April 2021
Published: 21 April 2021

Abstract: Magnetized quark nuggets (MQNs) are a recently proposed dark-matter candidate consistent with the Standard Model and with Tatsumi's theory of quark-nugget cores in magnetars. Previous publications have covered their formation in the early universe, aggregation into a broad mass distribution before they can decay by the weak force, interaction with normal matter through their magnetopause, and a first observation consistent MQNs: a nearly tangential impact limiting their surface-magnetic-field parameter B_0 from Tatsumi's $\sim 10^{12 +/-1}$ T to 1.65×10^{12} T $+/- 21\%$. The MQN mass distribution and interaction cross section strongly depend on B_0. Their magnetopause is much larger than their geometric dimensions and can cause sufficient energy deposition to form non-meteorite craters, which are reported approximately annually. We report computer simulations of the MQN energy deposition in water-saturated peat, soft sediments, and granite, and report the results from excavating such a crater. Five points of agreement between observations and hydrodynamic simulations of an MQN impact support this second observation being consistent with MQN dark matter and suggest a method for qualifying additional MQN events. The results also redundantly constrain B_0 to $\geq 4 \times 10^{11}$ T.

Keywords: dark matter; quark nugget; magnetized quark nugget; MQN; nuclearite; magnetar; strangelet; slet; Macro

1. Introduction

We report results of computer simulations and observations from field work that indicate that at least one non-meteorite impact crater was formed by an impactor with mass density comparable to nuclear density, with mass ~5 kg and with energy deposition of ~80 MJ/m. We show that these results are consistent with ferromagnetic Magnetized Quark Nuggets (MQNs), which are a relatively new candidate for dark matter. These observations from non-meteorite craters may also be consistent with some other dark-matter candidates. Non-meteorite craters are reported in the popular press approximately once per year and they may offer an opportunity to test hypotheses for dark matter.

This is the fifth paper on MQNs. For your convenience, the introduction will summarize the basic characteristics of MQNs that were demonstrated in previous papers and place MQNs in the context of current research on nuclearites, which are non-magnetic quark nuggets.

1.1. From Dark Matter to Quark Nuggets, Nuclearites, and MQNs

Dark matter [1–6] comprises approximately 85% of the mass in the universe. After decades of searches for experimental and observational evidence supporting any of many candidates for dark matter, the nature of dark matter is still a mystery [6,7]. A quark nugget is a dark-matter candidate that is composed of up, down, and strange quarks [8–15]. Quarks are constituents of many particles in the Standard Model of Particle Physics [16]. Quark nuggets are a Standard-Model candidate for dark matter that has not been excluded by observations [13–15]. A current summary of quark-nugget research can be found in reference [17].

Many physicists know of Witten's [8] 1984 proposal that a Quantum Chromodynamics (QCD) phase transition would have permitted stable quark nuggets to have formed in the early universe and function as dark matter. They also know that the QCD phase transition is not currently supported. Consequently, some physicists concluded quark nuggets are no longer viable. However, Aoki et al. [18] showed that an analytic crossover process would have formed quark nuggets in the early Universe without a phase transition. Recent simulations conducted by T. Bhattacharya et al. [19] support the crossover process. Both of these papers and others assert that quark nuggets could be formed without the phase transition and they are still a viable candidate for dark matter.

The search for quark-nugget dark matter continues primarily as the search for nuclearites, which are quark nuggets with an interaction cross section equal to their geometric cross section. Burdin, et al. [14] included nuclearites in his review of non-collider searches for stable massive particles in 2015. The work he reviewed analyzed data to find the upper limit to the nuclearite flux while assuming that 100% of the local dark-matter density is composed of nuclearites of a single mass. They concluded that the combined results of all the searches do not exclude nuclearites from 0.1 to 10 kg, 10^6 to 10^{14} kg and 10^{19} to 10^{21} kg with the single-mass model. The Joint Experiment Missions for Extreme Universe Space Observatory (JEM-EUSO) is the next big step in looking for nuclearites, which will feature a 2.5-m telescope with a wide ($60°$) field of view operating from the International Space Station. In addition to its other missions, JEM-EUSO could close the 0.1-to-10 kg gap in the single-mass nuclearite model with just 1 to 100 days of data for 0.1-kg to 10-kg nuclearites, respectively [20]. All of these results assume that 100% of the dark matter density is composed of nuclearites of a single (or average) mass and their interaction cross section is their geometric cross section at nuclear mass density.

These reviews and plans did not consider MQNs, which were first published in 2017 [21] after twenty years of research to explain the anomalies that we now associate with non-meteorite impacts, the subject of this paper. MQNs differ from nuclearites in that: (1) MQNs are ferromagnetic, as explained in the next subsection, (2) are predicted to have a broad mass distribution between $\sim 10^{-24}$ kg and $\sim 10^{+6}$ kg [15], as illustrated in Figure 1 instead of single-mass nuclearites, and (3) interact with normal matter through their magnetopause [21] which may be millions of times larger than their geometric cross section, as quantified by Equation (1). The large mass distribution means that capabilities like JEM-EUSO would require ~ 9 years of continuous and dedicated observations to test the MQN hypothesis, as discussed in Section 1.5, below. The much larger interaction cross section per unit mass means that lower-mass and more abundant MQNs do not reach deeply buried ancient mica or space-based track recorders behind spacecraft walls that are discussed in reference [14] or the scintillators in underground observatories [22]. The enhanced cross section also means that the energy of larger MQNs impacting Earth or the Moon is deposited in many kilometers instead of passing through the body [14], as usually assumed; the extremely high energy deposition excites strongly attenuated shear modes in rock that complicate seismic detection, in contrast to the elastic modes assumed. The broad mass distribution and the large interaction cross section both arise from MQN ferromagnetism.

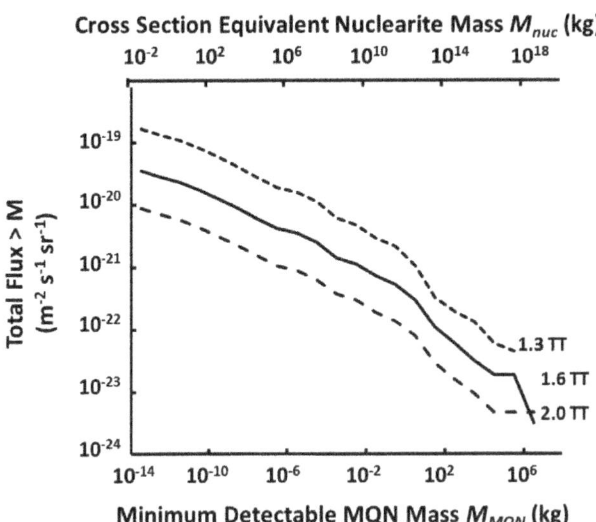

Figure 1. The integral of the MQN number density from minimum detectable MQN mass M_{MQN} to infinity times 230 km/s, the velocity of the solar system through the galactic halo, gives the predicted number flux for mass $\geq M_{MQN}$ plotted on the x-axis for range of currently allowed values of B_o. For $B_o = 1.65 \times 10^{12}$ T and atmospheric mass density ($\sim10^{-3}$ kg/m^3) at the 50 km altitude to be probed for nuclearites by JEM-EUSO, the nuclearite mass M_{nuc} with the same interaction cross section as M_{MQN} is plotted above the graph.

1.2. Theoretical Basis of Ferromagnetism in MQNs

MQN ferromagnetism is based on the existence of magnetars, which are pulsars with magnetic field ~300 times the magnetic field of neutron stars. The much larger magnetic field implies a different physical nature for magnetars, such as a quark nugget core. Xu [23] has shown that the low electron density, as permitted in stable quark nuggets, limits surface magnetic fields from ordinary electron ferromagnetism to ~2 × 10^7 T. Tatsumi [24] examined ferromagnetism from a One Gluon Exchange interaction in quark nuggets and concluded that the surface magnetic field could be ~10$^{12+/-1}$ T, which is sufficient for explaining the magnetic field inferred for magnetar cores. The result needs to be confirmed with relevant observations and/or advances in QCD calculations because the result depends on the currently unknown value of the QCD coupling strength [16] α_c at the ~90 MeV energy scale of the strange quark.

We are exploring the implications of such Magnetized Quark Nuggets (MQNs) to explain the anomaly of non-meteorite impacts that started our investigations and the anomaly of dark matter because magnetars exist [25] with a magnetic field that is ~300 times that of neutron-star pulsars [26] and since such a large magnetic field is perhaps uniquely consistent with ferromagnetic quark nuggets.

1.3. From Ferromagnetism to MQN Stability and Mass Distribution

The previously published theoretical results [15] show that MQNs would have originated at time t ~65 μs when the Universe had a temperature of ~100 MeV according to the Standard Model of Cosmology [2]. At that temperature, Λ^0 particles (consisting of one up, one down, and one strange quark) could form [27]. The simulations of their aggregation as a ferromagnetic liquid under the long-range magnetic force, similar to simulations of inelastic collisions of particles in nucleosynthesis under the short-range nuclear force, showed that MQNs magnetically aggregate [15] into a broad mass distribution of stable ferromagnetic MQNs before they could decay. In the extremely high mass density of the early Universe, aggregation happened with an initial time scale of ~1.5 ps, so aggregation dominated the ~10 ps decay by the weak interaction. Because current day accelerator experiments have a much lower mass density, aggregation does not compete with decay, so stable MQNs are not formed in those experiments.

After time t ≈ 66 µs after the big bang, the simulated mean of the MQN mass distribution is between ~10^{-6} kg and ~10^4 kg, depending on the surface magnetic field B_0. The corresponding mass distribution is sufficient for MQNs to meet the requirements [13,15] of dark matter in the subsequent processes, including those that determine the Large Scale Structure (LSS) of the Universe and the Cosmic Microwave Background (CMB).

Throughout this paper, we will use B_0 as a key parameter that defines the mass distribution of MQNs. The value of B_0 equals Tatsumi's surface magnetic field B_S if the mass density of MQNs $\rho_{QN} = 10^{18}$ kg/m^3 and density of dark matter was $\rho_{DM} = 1.6 \times 10^8$ kg/m^3 when the temperature of the universe was ~100 MeV [15,27]. If better values of ρ_{QN} and ρ_{DM} are found, then the corresponding values of B_S can be calculated by multiplying the B_0 from our results by $(1 \times 10^{-18} \rho_{QN})$ $(6.25 \times 10^{-9} \rho_{DM})$.

The previously published MQN papers, especially the observational paper [17], narrowed the allowed range of B_0 from Tatsumi's ~$10^{12+/-1}$ T to 1.65×10^{12} T $+/-$ 21% to be consistent with the maximum mass of MQNs that is allowed for a given value of B_0. The integrated MQN number flux for MQN mass $\geq M_{MQN}$ has been derived from the mass distribution assuming a constant velocity of 230 km/s, the velocity of the solar system through the galactic halo. The integrated flux is useful for evaluating event rates and it is shown in Figure 1 for the current range of B_0 that is consistent with observations.

1.4. Comparison of MQN and Nuclearite Interaction Cross Sections

Like Earth, MQNs have a dipole magnetic field. Additionally, like Earth, they also have a magnetopause, which interacts with inflowing plasma as Earth's magnetopause interacts with the solar wind. We assume the extremely high MQN velocity relative to the surrounding matter assures the plasma temperature is sufficient to fully ionize the inflowing matter. Under that assumption, Equation (1), derived from reference [21], gives the ratio of MQN cross section to its geometric cross section, and Equation (2) gives the nuclearite mass M_{nuc} with the same interaction cross section as the MQN of mass M_{MQN}.

$$\frac{\sigma_{MQN}}{\sigma_{nuc}} \approx \left(\frac{2B_0^2}{\mu_0 \rho_p v^2} \right)^{\frac{1}{3}} \text{ and} \tag{1}$$

$$M_{nuc} \approx \left(\frac{2B_0^2}{\mu_0 \rho_p v^2} \right)^{\frac{1}{2}} M_{MQN} \tag{2}$$

for ρ_p = the mass density of surrounding plasma, μ_0 = permeability of free space, B_0 = MQN surface magnetic field parameter, and v = velocity of MQN relative to the surrounding plasma. The ratio of interaction cross sections in Equation (1) and equivalent nuclearite mass in Equation (2) depend on the mass density of surrounding material. The top scale shown in Figure 1 shows the equivalent nuclearite mass for JEM-EUSO observations at the appropriate altitude.

1.5. Estimated Observation Time for JEM-EUSO to Test MQN Hypothesis

Reference [20] indicates that only 24 hours of observation will be needed for JEM-EUSO to determine whether nuclearites of single-mass $M_{nuc} = 10^{26}$ GeV/c^2 = 0.16 kg have a flux consistent with the Galactic dark-matter limit. Equation (2) gives $M_{nuc} = 1.8 \times 10^{12}$ M_{MQN} for the same cross section and for $\rho_p = 10^{-3}$ kg/m^3 appropriate for the 50 km altitude to be observed by JEM-EUSO. Conversely, the MQN mass that is equivalent to a nuclearite with M_{nuc} = 0.16 kg nuclearite mass is M_{MQN} = ~10^{-13} kg. Therefore, JEM-EUSO will be able to make the same judgement regarding MQNs with $M_{MQN} > 10^{-13}$ kg if the acceptance for MQN mass > M_{MQN} is the same as it is for M_{nuc} = 0.16 kg. However, the required time scales vary inversely with flux, which is 10^{-16} m^{-2} s^{-1} sr^{-1} for single-mass nuclearites and 3×10^{-20} m^{-2} s^{-1} sr^{-1} for distributed mass MQNs with $B_0 = 1.65 \times 10^{12}$ T. Therefore, testing the MQN hypothesis with JEM-EUSO would take ~3000 days or ~9 years of dedicated observation time. Because JEM-EUSO has many demands for its observation time, it is impractical.

1.6. Definition of Non-Meteorite Craters and Their Utility for Testing Dark-Matter Hypotheses

This paper investigates the anomaly of non-meteorite craters as a means to test the MQN hypothesis and explore other dark-matter candidates. Non-meteorite craters are defined as craters formed:

(1)   without an observable luminous streak,
(2)   without breakup in an air shower [28] and dispersal, so no crater is formed,
(3)   without meteorite material found in and/or around the impact site, and
(4)   without evidence of causation by human or other natural causes (e.g., the energetic release of methane from global warming), but
(5)   with sufficient energy deposition to form an impact crater.

Criteria 2 through 4 are straightforward, but criterion 1 requires some explanation, since energy is being deposited in the atmosphere by a hypervelocity object. MQNs interact with surrounding matter [21] through their magnetopause, which is the boundary between their compressed magnetic field and the plasma pressure from inflowing matter. Their magnetopause is much larger than their nuclear-density core, but it is still quite small. For example, a 5-kg MQN moving through air at sea level with speed of 230 km/s has a nuclear density core with radius of ~7.5×10^{-10} m and a magnetopause radius of 1.5×10^{-6} m. At a distance of 15 km, which is typical for a city, the apparent magnitude [10] of the corresponding luminosity at a distance of 15 km would be -4.4, which is approximately that of Venus on a clear night. The characteristic radius of the shock wave and characteristic cooling time of the shock temperature [29] are ~8×10^{-4} m and ~2×10^{-6} s, respectively. Even after expansion and cooling for a hundred characteristic times, the angular diameter at 15 km is only 2.2 arc seconds, which is only ~3.3% of the angular diameter of Venus at closest approach. In addition, the transit through the 8-km e-folding distance of the atmospheric density is only 0.034 s. It takes 0.25 s for humans to perceive an unexpected object in their field of view as a thing [30], so, even if it could be seen, it would not be recognized before it had gone. Therefore, non-meteorite craters are not associated with human-observable events.

This paper focuses on an observational test of the theory, not on advancing the theory of MQNs. We simulate the interaction of an MQN with a geophysical three-layer witness plate, and then test the resulting signature of an MQN impact with observations from one non-meteorite impact crater. The simulations connect the previously published theory to the observations at the crater for a test of the MQN hypothesis.

We use the terms meteors and meteorites to refer to bodies that are composed of normal matter, i.e., atoms that are held together by the electromagnetic force. The material strength that is associated with the electromagnetic force is weak. Thus, meteors and meteorites (as defined in this paper) must be quite large to survive intact and they do not make small craters. Nuclear-density quark nuggets are held together by the strong-nuclear force, so all of them survive passage through the atmosphere.

Craters that show no evidence of meteorite impact are reported in the press approximately once per year. Therefore, the event rate for non-meteorite craters is sufficient to allow the phenomenon to be studied, if access to the craters can be obtained. Three events in three years have been recently reported.

1.   ~12-m diameter crater near Managua, Nicaragua, on 6 September 2014 [31].
2.   ~1-cm diameter crater in Rhode Island, USA, on 4 July 2015 [32].
3.   ~60-cm diameter crater in Tamil Nadu, India, on 6 February 2016 [33].

None of these impacts was preceded by a luminous track in the sky, no meteorite material was found in or near any of the craters, and experts who reviewed the news reports concluded that these events were not meteorite impacts, as reported in references [31–33]. In the absence of a scientific basis for impacts that form craters without luminous tracks and without meteorite fragments, these experts attributed them to human-caused explosions by default. Our results indicate that MQNs can also cause non-meteorite craters and they should be considered in future investigations.

Each non-meteorite event provides a large-target opportunity to test the MQN dark-matter hypothesis. Because a multi-layer witness plate provides more information than a single-layer one, peat bogs on top of soft sediments and bedrock offer particularly useful opportunities. Section 3.1 through 3.2 report the results of hydrodynamic simulations and analyses of MQNs interacting with a three-layer witness plate of peat, clay-sand mixture, and granite bedrock in County Donegal, Ireland. Other peat bogs could also provide such opportunities. However, County Donegal has the advantages of: (1) a granite bedrock within excavation range of the surface, (2) a friendly and supportive population, (3) a governing authority over peat bogs that can grant permits for exploration, and (4) maximum exposure to the directed flux of dark matter. The last advantage arises because dark matter streams into Earth as the solar system moves around the galactic center and through the dark-matter halo. That direction of motion is right ascension 18 h 36 m 56.33635 s, declination +38°47′01.2802″ and it is always above the horizon for latitudes that are greater than that declination, including the latitude of County Donegal.

1.7. Significance of This Paper

MQN interaction provides at least three measurable signatures of MQN dark matter:

(1) hypervelocity (which generally refers to velocities >3000 m/s at which the material strength is much less than internal stresses) atmospheric transit without luminous streak and without breakup in an air shower [28], but with energetic (>1 kJ/m) energy deposition and with multi-meter transit through solid-density matter,

(2) electromagnetic emissions (kHz to GHz) from the rotating magnetic dipole after transit through matter [34], and

(3) magnetic levitation of rotating MQN magnetic dipole after transit through matter [17] by induced currents in adjacent conducting material or magnetic levitation of static magnetic dipole above a superconductor.

A systematic attempt to detect MQNs through the first signature was attempted [21] by looking for acoustic signals from MQN impacts in the Great Salt Lake in Utah, USA. Even though the first method monitored ~30 sq-km (i.e., ~30 times the cross section of the IceCube Neutrino Observatory at the South Pole) for impacts with MQN mass $\geq 10^{-4}$ kg, none were detected [15]. The null results implied that analysis that is based on average mass, which is the usual assumption for dark-matter candidates, may be inadequate and motivated the detailed computation of the MQN mass distribution. Those results show a much larger detector and/or much longer observation times are required.

Reference [34] describes how MQNs passing through matter spin up to MHz frequencies and emit radiofrequency energy (the basis of the second signature), and proposes using Earth's magnetosphere as a sufficiently large detector to obtain enough events. The large uncertainty in B_0 from Tatsumi's theory is a major impediment to designing and fielding such an experiment.

Reference [17] describes an extremely rare episodic observation and supporting simulations of the third method. That method is too rare to provide enough data to measure the mass distribution and it provides sufficient statistics for discovery of MQNs. However, those results do limit the key parameter B_0 to 1.3×10^{12} T $\leq B_0 \leq 2 \times 10^{12}$ T and they permit the design of the systematic experiment based on the second signature with Earth's magnetosphere as the detector area.

Such a project requires major investment. Additional episodic data would help to justify the project to obtain systematic data. The current paper provides additional episodic observational and supporting computational results of the first method, based on non-meteorite crater impacts on Earth. We also discuss extending the results to find a statistically significant number of MQN impacts.

1.8. Organization of the Paper

Section 3.4 through 3.8 report the observations and associated analyses from excavating a non-meteorite impact that occurred in County Donegal in May of 1985. We find the crater and subsurface damage to be consistent with an MQN impact that deposited ~80 MJ/m in the water-saturated peat. The corresponding MQN mass depends on the value of B_0. MQN mass distributions that are consistent with ~80 MJ/m energy deposition provide an additional exclusion of $B_0 < 4 \times 10^{11}$ T. For the most likely range of 1.3×10^{12} $T \leq B_0 \leq 2 \times 10^{12}$ T, ~80 MJ/m energy deposition corresponds to MQN mass of 5 +/− 1 kg.

Section 4 discusses the results, alternative explanations, and a potential method for more efficiently determining whether a candidate impact is consistent with a deeply penetrating MQN impact or with some surface phenomenon. In principle, such a method could locate additional MQN events.

1.9. Declaration of Controversial Topic

The editorial policy for this journal requires authors to declare whether an article is controversial. For the last four decades, searches for dark-matter candidates have focused on particles that are Beyond the Standard Model (BSM) of Particle Physics. Because MQNs are composed of Standard Model quarks and the theory for the ferromagnetic state of MQNs is an approximate solution within the Standard Model, the MQN hypothesis for dark matter does not require a BSM particle and it may be considered controversial by the BSM community. However, a potential solution to dark matter for the Standard Model of Cosmology and not requiring a hypothetical BSM particle may be less controversial to others because it is a modest extension of well-established physics.

1.10. Limitations of These Results

We report the results from just one episodic event, which we calculate in Section 4.2 to have a <2% of arising randomly from unknown effects. Nevertheless, we conclude that many more instances are required to determine whether or not MQNs exist and contribute to dark matter. The event that is presented in this paper suggests a method for adding more instances of MQN interaction by qualifying other non-meteorite impacts, which occur approximately annually.

2. Materials and Methods

This study used

1. computational 2D and 3D hydrodynamic simulations, as described in the results section and in movies of pressure, mass density, and temperature in supplementary dataset at https://doi.org/10.5061/dryad.cc2fqz641 (accessed on 21 April 2021) and associated analyses of energy deposition in the multi-layers of peat bog in the results section;
2. original field work at the location of a non-meteorite impact in May of 1985 in County Donegal, Ireland, and associated analyses in the results section;
3. additional details in Appendix A: Excavations, to assist an independent team to extend our findings; and,
4. potential sites in Appendix B: coordinates and description of deformations in peat-bog survey, for future investigations if suitable sensor technology can be developed.

3. Results

3.1. Hydrodynamic Simulation of MQN Impact in Three-Layer Witness Plate

Two- and three-dimensional simulations with the CTH hydrodynamics simulation software [35,36] were conducted to investigate MQN interactions with a three-layer witness plate of peat-bog, clay-sand mixture, and granite bedrock. Two-dimensional simulations examined the craters that formed by MQN impacts as a function of MQN mass and the B_0 parameter. Three dimensional simulations investigated the circularity of the crater as a function of angle relative to vertical and guided the excavation of an impact site in County Donegal, Ireland.

In both cases, the MQN deposits energy along its path and within the magnetopause radius [21] (e.g., ~1 mm radius for a 30 MJ/m energy deposition from a 1-kg MQN for $B_0 \approx 10^{12}$ T). The interaction produces a plasma with an initial temperature of ~2100 eV. We did not have access to any computer program that could reliably resolve the dynamics of the initial channel (requiring <<1-ns and ~100-μm resolution) and still simulate acoustic propagation in two dimensions over many meters and for many milliseconds. Other studies have shown that turbulent mixing with colder material dominates the early dynamics of the interaction and produces a channel of ~ 1 eV temperature within a radius that preserves the energy per unit length. Therefore, we approximated the post-turbulence phase of the plasma channel as a cylinder with the mass density of the peat, clay-sand, or granite. The temperature was varied from 0.5 to 1.55 eV and the radius was varied to give the specified energy per unit length. The equation of state was imported from SESAME4 [36] data. The results were essentially independent of temperature over that range and they validated the assumption that energy/length is the dominant variable.

The fluid above the peat was atmosphere at standard temperature and pressure. The simulated depth of the peat was the actual 0.7 m of the Irish peat bog with an initial density of 1.12×10^3 kg/m^3 and sound speed of 1.46×10^3 m/s. The 4.7 m-thick clay-sand layer was simulated with a 1.0 meter thick layer and with an initial density of 2.02×10^3 kg/m^3 and sound speed of 2.2×10^3 m/s. The granite layer was simulated with a 0.3-m layer with initial density of 2.6×10^3 kg/m^3 and sound speed of 5.0×10^3 m/s. The bottom of each simulation was unmovable, and material could freely exit from the other boundaries.

Two-dimensional hydrodynamic simulations were conducted with 1, 3, 9, 27, 81, and 243 MJ/m energy deposition. They show that a shock wave reflects off the mass discontinuities and it propagates radially outward in all three layers. Low-density and high-temperature material in the central channel is ejected into the atmosphere. Lower temperature material behind the shock wave moves radially and almost one-dimensionally outward. Finally, the peat distorts two-dimensionally in response to the velocity field that it has acquired and the shear planes that have developed within the peat.

Figure 2 shows representative results of the density maps for times when the material velocities are well below their peak values.

The 81 MJ/m case shown in Figure 2c is especially relevant to the crater discussed below. The shear planes and voids form relatively smooth sides of the peat crater. Most of the peat is ejected in small fragments into the atmosphere. Larger pieces of peat are shown as vertical pieces about to be ejected radially away from the crater. The channel is almost one-dimensional in the clay-sand and granite.

Figure 3 shows the summary results of simulations for 1 MJ/m to 243 MJ/m.

The central channel in the granite that is shown in Figure 2 is caused by the compressive pulse that decreases rapidly with increasing distance from the high-energy-density center. When the compressive pulse reflects at the clay-sand boundary, it becomes a tensile pulse and breaks the rock in tension. Because the tensile strength of granite is only ~1.5% of the compressive strength [38], the diameter of fractured granite is much larger than the diameter of the compressed channel [37]. The fracture diameter depends on the geometry [38] and composition [37,38] of the explosive, shock impedances of both materials, and distance [37] to material with lower shock impedance. However, those effects are secondary to the main trend, as shown by the scatter in fracture data presented in Figure 3. Over a wide range of parameters, the fracture diameter from the tensile strain is approximately a factor of 30 larger than the diameter of the channel that is caused by compressive strain.

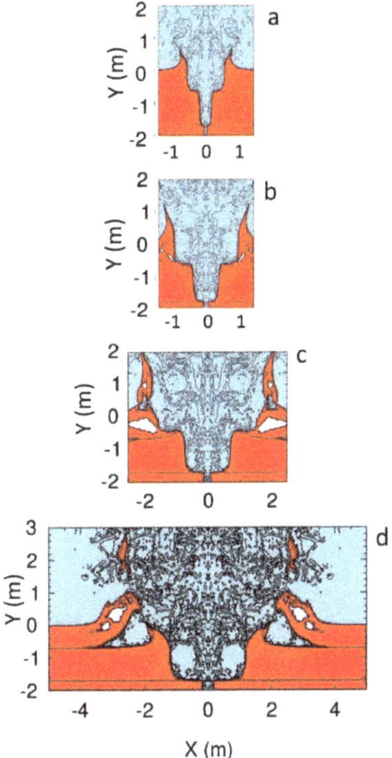

Figure 2. Representative density maps are shown for times when radial expansion of the channel in the clay-sand layer is approaching its maximum: (**a**) 9 MJ/m at t = 20 ms, (**b**) 27 MJ/m at t = 25 ms, (**c**) 81 MJ/m at t = 40 ms, and (**d**) 243 MJ/m at t = 55 ms. The red material represents three layers separated by black lines. From the bottom up, the layers are 0.3 m of granite, 1.0 m of clay-sand, and 0.7 m of peat with initial density of 1.12×10^3 kg/m^3. The blue area is atmospheric air. The white spaces are voids at shear planes. Movies of pressure, density, and temperature are available online (accessed on 14 February 2021: https://datadryad.org/stash/share/Lt7dMvxEAUWNnkfKt2xPg2l5 TuWz7Bbec67iY4Kvazg.

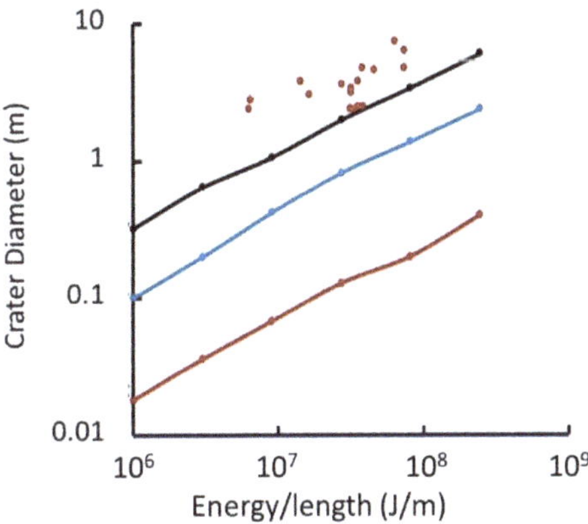

Figure 3. Solid lines show crater diameter in granite (red), clay-sand (blue), and peat (black) as a function of the energy/length from CTH simulations. The data points show the diameter of fractured granite from reference [37].

3.2. Potential for Liquefaction and Flow of the Clay-Sand Layer

Peat is ejected into the atmosphere and it leaves a crater with smooth sides that formed from the shear planes. The granite layer is fractured around the path of an MQN; however, the channel in the fully water-saturated clay-sand of the County Donegal peat bog is very likely to undergo liquefaction and close the channel within tens of seconds after the passage of an MQN. Fine-grained soils, i.e., silts and clays for which the percent of dried soil passing through a No. 200 standard sieve (i.e., 0.074-mm diameter openings) exceeds 50%, require careful testing and analysis to determine whether or not they will undergo liquefaction under an impulse or shaking, according to Boulanger and Idriss [39]. Conversely, soils with much less than 50% passing through a No. 200 sieve are much more likely to liquefy when they are saturated with water. We analyzed the clay-sand layer in the County Donegal peat bog and found that it is composed of ~10% rock of ~1 cm diameter, ~20% soil that does not pass a 1 mm screen, and 11% soil passing through a No 200 sieve. Therefore, the fine-grained portion is only ~11% of total mass or, more conservatively, 19% of the sub-mm sized content and well within the < 50% criterion for susceptibility to liquefaction.

Owen and Moretti [40] identified five conditions that contribute to liquefaction-induced soft-sediment deformation in sands under a transient increase in pore fluid pressure: (1) fine to medium-sized grains of sand, (2) high porosity, (3) high percent saturation with water, (4) low overburden pressure (<10 m of overburden), and (5) no previous liquefaction. The clay-sand layer between the peat and granite satisfies all five conditions. In addition, Owen and Moretti cite impact by extra-terrestrial objects as a likely trigger for liquefaction. Therefore, we conclude that the clay-sand layer is very likely to have undergone liquefaction and obscured the channel within tens of seconds after impact.

3.3. Simulations on Circularity of MQN Crater as a Function of Entrance Angle

In addition to identifying the signature of MQN impacts, CTH simulations examined the circularity of the crater in the peat bog as a function of entrance angle relative to vertical. The information is helpful in identifying the likely path of the MQN through the liquefied intermediate layer to the bedrock.

Simulations modeled channels at $0°$, $15°$, $30°$, $45°$, and $60°$ to the vertical and between the surface and an immovable solid at 1.0-m depth. The simulated MQN instantaneously deposited 30 MJ/m energy density, as described above. Because of the low strength of the peat, the crater continues to grow for an extended period of time. Each simulation was stopped at 10 ms in order to reasonably simulate the relative effects of the impact angle on the crater dynamics. Figure 4 shows representative profiles.

At $y = -0.5$ m, the ratio of major to minor axes is approximately $\cos^{-1}(\theta)$, as expected for a cylinder intersecting a plane at angle θ. However, Figure 4 shows that the peat on the right-hand edges is forced against low-density air, while the peat on the opposite side is forced against higher-density peat. The less-impeded peat moves more. Therefore, the asymmetry is enhanced near the rim of the crater. Using the crater shape to estimate θ gives a maximum angle for the trajectory.

3.4. Non-Meteorite Crater in May 1985 Near Glendowan, County Donegal, Ireland

A non-meteorite impact occurred in the middle of May of 1985, on Stramore Upper, near Glendowan, County Donegal, Ireland at 54°58.257′ N, 8°0.408′ W. It was reported in the *Donegal People's Press*, 31 May 1985. The article said that it occurred "when people were walking their dog"; that would be about 18:00 hours GMT.

The site is on Common Land with rights assigned to a group of nearby landowners, who kindly allowed our research. The National Parks and Wildlife Service has authority over the land and it granted us a permit to excavate the site.

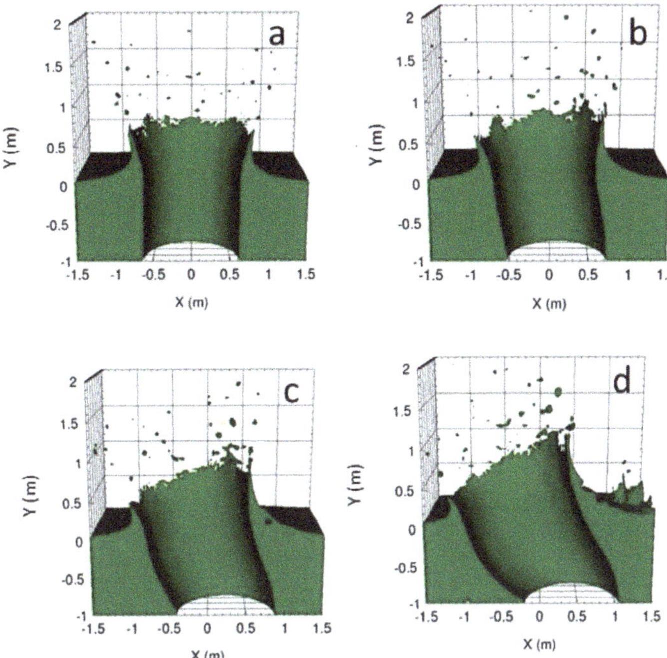

Figure 4. Mass density profiles of peat at t = 10 ms after the start of a simulated MQN interaction depositing 30 MJ/m on a trajectory inclined at (**a**) 0°, (**b**) 15°, (**c**) 30°, and (**d**) 45° from the vertical.

Two of us (D. D. and S. McG.) investigated the crater the day after the event, as part of our duties as Park Rangers. The inside sloped surface of the crater was very smooth. A few-centimeter diameter hole or "pointy" depression was present in the dirt at the center of the crater bottom. There was a distinct, ~2 cm high lip on the peat edge of the crater. Pieces of the bog were scattered up to 10 m away. A visual search of the crater and the surrounding area did not find any meteorite material. The crater filled with water within two days. The water prevented sub-surface searches and, to our knowledge, preserved the site until our team excavated it.

Figure 5 shows the relatively smooth sides, which are very unusual for craters that are produced by surface explosives in peat bogs. In addition, large pieces of peat were scattered approximately 10 m away. The smooth sides, diameter of the crater, and energetically-detached ejecta were consistent with the CTH simulation and they confirm that the simulation was accurately modeling the crater formation.

The crater has a diameter of 3.984 $\pm$ 0.065 m in 2006. The yield strength of the peat was measured and found to be 530 $\pm$ 120 kN m^{-2}. Figures 2c and 3 give an energy/meter of ~80 MJ/m for a 4.0-m diameter crater. Uncertainties in the equation of state variables imply a fidelity of +/− 20%.

The shape of the crater was measured in 2006, before it was distorted by investigations. The best fit to an ellipse gives a 1.030 $\pm$ 0.005 ratio of major to minor axes and it corresponds to $\theta \leq 15°$, as shown in Figure 4a or Figure 4b. The major axis aligned east-west. Therefore, the excavation was planned to explore the volume within 15° of vertical and optimized for the east or west of center.

Figure 5. (**a**) Photograph of the crater soon after the event in 1985 illustrate good agreement in the actual and simulated profile of crater sides. (**b**) Photograph of the crater in March 2005 showing the full diameter and circularity before excavation.

3.5. Excavations of the 1985 Non-Meteorite Crater in County Donegal, Ireland

Field work a third of the way around the world and in a protected wilderness area is challenging at best. However, it is the least expensive way to test the MQN dark-matter hypothesis. The additional information presented in Appendix A should assist independent groups in learning from our experiences and re-excavating the site.

The site was excavated in three stages, as shown in Figure 6.

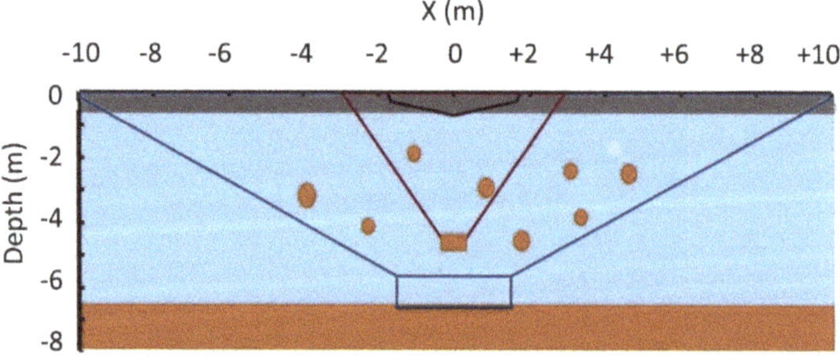

Figure 6. Cross-sectional view of the three excavations: 2017 (black line), 2018 (red line), and 2019 (blue line), and the three layers: peat (gray), clay-sand (blue), and granite (brown). Brown ellipsoids represent the two granite boulders found to be distributed within the clay-sand volume of the 2018 excavation and the ten found in the 2019 excavation. The brown rectangle shows the location of the only ensemble of fractured rock found in the excavations.

The 2017 expedition excavated the volume that is defined by the black line in Figure 6, and found that the bottom was composed of compacted clay-sand mixture at a depth 0.6 ± 0.1 m. The compacted, post-liquefaction material was too hard to continue excavating by hand.

In 2018, a six-ton excavator was used in an attempt to reach the bedrock. The volume bounded by the red line in Figure 6 was excavated, with the sides sloping an average of 0.5:1, i.e., 0.5 m horizontal for every 1.0 m vertical, or ~27° from vertical, in accordance with local experience in this soil. At 4.7 ± 0.1-m depth, a grouping of fractured rock was discovered just northeast of the center line. After an hour of observing the stability of the sides, the principal investigator was cleared by the civil engineer safety officer to enter the pit. He scooped accumulated water into a bucket and found that the rock was closely packed shards of granite with dimensions varying between 0.02 m and 0.1 m.

The excavation had to be quickly abandoned because the sides of the water-saturated clay-sand mixture showed signs of fracture and sliding at various points down the slope. No samples were removed because we did not have time to do a careful and well-documented investigation. The dimension of the rocky bottom was at least the ~0.5 m of the cleared bottom, but the horizontal dimension of the rocky area could not be determined; it could be an extensive layer of fractured rock, fractured bedrock, or a localized deposit.

The 2019 expedition used two 14-ton excavators to dig the hole that is defined by the blue line shown in Figure 6. The slope of the sides averaged 1.5:1, i.e., 1.5 m horizontal for every 1.0 m vertical, or ~55° from vertical, to assure they would not collapse. Ten boulders were found throughout the excavation. Figure 7 shows two of these.

Figure 7. Two of the ten boulders found within the excavated volume are shown. Their diameters are approximately 0.4 m and 0.6 m, and they were distributed throughout the 633 m^3 of the 2019 excavation, but similar boulders were not observed on the surface of the peat bog.

Because the material above the rocky grouping of interest had been back-filled after the 2018 excavation, the precise positions of the boulders were not relevant to the 1985 event and they were not recorded by the excavator operators.

The operators were requested to excavate to the rocky layer at -4.7 ± 0.1 meters, stop, and alert the team. They did so; however, by the time they stopped and measured the depth, they had removed the volume of fractured rock in just one bucket load, demonstrating that it was a localized deposit, and then discharged it through the relocation process to a pile where it spread out. Although they showed us where that load lay, its relational context was lost. We encourage another group to re-excavate the site and look for fractured granite in the bedrock below our excavation; extreme care is recommended to preserve the context of fractured rock.

Water was pumped from the excavation. The muddy bottom was explored by hand. The rocks shown in Figure 8 were found between 5.0-m and 6.3-m depth. The smaller ones are consistent with those that were observed at 4.7 m in the 2018 excavation. The larger rocks have slightly rounded edges and they may not have been from that grouping.

Figure 8. Granite shards from the volume below the grouping of fractured granite at 4.7 ± 0.1 m are shown. The rocks are covered with the fines from the clay-sand mixture which distorts their natural colors. Although rocks that are similar to the larger samples in Figure 8 were found on the surface of the peat, no collection of rocks similar to the single shattered boulder was found on the surface.

These granite rocks were examined with Energy Dispersive Spectroscopy [41] for evidence of large pressure gradients having altered the quartz in the granite. Streaks of darkened mineral was determined to be natural feldspar. No damage that was attributable to extreme pressures was found.

The excavation continued to a depth of 5.7 m, as illustrated by the rectangle outlined in blue in Figure 6. The west face of the crater, just west of the grouping of shards found in 2018, was washed with a pressure washer to better reveal its composition. Figure 9 shows a photo of the washed face.

Figure 9. Pressure-washed face of the excavation's west side, adjacent to the grouping of shards at the 4.7 ± 0.1 m depth found in the 2018 excavation.

Figure 9 shows no evidence of a horizontal layer of shards or bedrock, demonstrating that the ensemble of fractured granite was an isolated one, approximately the size of a shattered boulder. This group of shards was in the projected path of the impactor and it was the only such grouping found in the excavation. Because boulders that are closer to the surface, but outside the projected path, were not shattered, we conclude that the ensemble of fractured granite was not shattered by pressure waves originating from energy deposited near the surface.

The decrease of pressure with increasing distance in the movies of the CTH simulations in Supplementary Material and the diameter of the fractured granite shown in Figure 3 indicate that a direct hit by an MQN may be required, and would be sufficient, to fracture a whole boulder. If so, and if the boulders were randomly distributed within the excavated volume, the probability of even one boulder being intercepted and fractured was only ~7%. Consequently, it is not surprising that only one collection of shards was found, and that it was well within the projected path of the penetrator. Therefore, we infer the hypervelocity object shattered the granite boulder after passing through 0.7 m of peat and 3.9 m water-saturated soft sediments.

We found at ~6.3-m depth, irregular boulders and large flat slabs of granite, with the vector normal to a slab inclined at ~30° to the vertical on the south, ~60° to the vertical on the north, and ~90° to the vertical in the middle, as shown in Figure 10. We did not find a uniform slab of bedrock that would have been a perfect witness plate of a quark-nugget passage by showing a cylinder of fractured granite extending into the earth. Broken slabs in disarray might be expected because our simulations give ~160 MJ/m (the equivalent of ~160 sticks of dynamite per meter) in granite to match the crater.

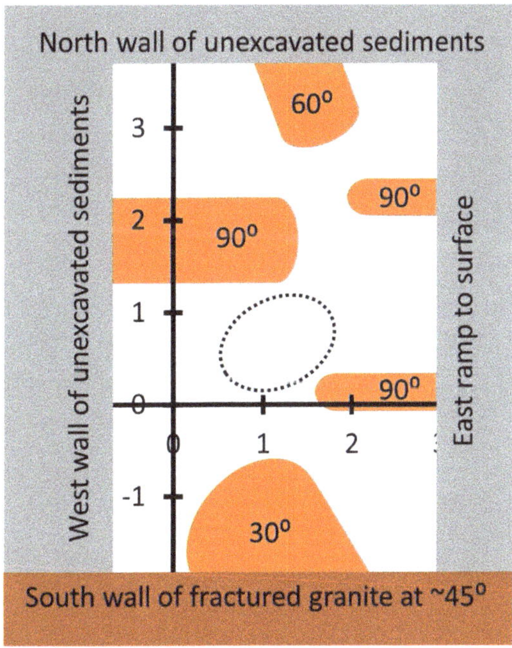

Figure 10. The layout of granite rocks (brown) and sediment walls (gray) at depth of ~6.3 to 6.5 m. The scale is in meters and the angles are between the vertical and the normal to the largest-area surface. The origin is directly below the center of the original impact crater on the surface with an estimated accuracy of +/−0.3 m. The dotted ellipse shows the approximate projection of the shattered rock found in the 2018 excavation at 4.7-m depth to the 6.4-m depth shown here.

We could not determine whether the mixture of rocks and slabs at different angles to the horizontal, as shown in Figure 10, were characteristic of the site before the 1985 event or were caused by that event. Additional excavation directly beneath the grouping of fractured rock at ~4.7-m depth was blocked by large boulders or displaced slabs around that volume. These obstacles were too large to move with available equipment. In addition, the excavation from 4.8 m to 6.3 m had nearly vertical walls, which introduced a safety risk and precluded more excavation within the limitations of the project.

3.6. Potential for Independent Validation of the 1985 Event

The force equation for a high-velocity body with instantaneous radius r_m, mass m, and velocity v, moving through a fluid of density ρ_p with a drag coefficient $K \approx 1$, is

$$F_e \approx K\pi r_m^2 \rho_p v^2 \tag{3}$$

MQNs have a velocity-dependent interaction radius [21] that is equal to the radius of their magnetopause

$$r_m \approx \left(\frac{2B_0^2 r_{QN}^6}{\mu_0 K \rho_p v^2} \right)^{\frac{1}{6}} \tag{4}$$

in which r_{QN} is the radius of the MQN of mass m and mass density ρ_{QN}:

$$r_{QN} = \left(\frac{3m}{4\pi \rho_{QN}} \right)^{\frac{1}{3}} \tag{5}$$

The interaction radius of an MQN varies as velocity $v^{-1/3}$ in Equation (4). Including that velocity dependence in the calculation with initial velocity v_0 gives velocity as a function of depth x yields

$$v = v_0 \left(1 - \frac{x}{x_{\max}} \right)^{\frac{3}{2}} \tag{6}$$

in which x_{max} is the stopping distance for an MQN:

$$x_{\max} = \left(\frac{3m}{2\pi r_{QN}^2} \right) \left(\frac{\mu_0 v_0^2}{2K^2 \rho_p^2 B_0^2} \right)^{\frac{1}{3}} \tag{7}$$

The ~10 kg MQN that is inferred for the 1985 crater penetrates to x_{max} = 3572 m for ρ_p = 2020 kg/m^3.

Because we only explored the 1985 event to a depth of 6.5 m, it is possible for an independent team to re-excavate the site of the 1985 event to the bedrock and look for an extended volume of fractured granite. We marked the site to facilitate such an independent examination.

3.7. Additional Limit on Bo to $\geq 4 \times 10^{11}$ T

Comparing simulation results with observations from the crater implies that the crater was formed with 80 +/− 16 MJ/m energy deposition in the peat. The MQN mass that can deposit that energy density and the corresponding number of events per year were computed as a function of B_o [15]. Table 1 summarizes the results to compare with observations.

Table 1 shows the MQN mass that is necessary to deposit 80 MJ/m in water-saturated peat as a function of B_0. We exclude B_0 <4 $\times$ 10^{11} T because the maximum MQN mass in distributions with $B_0 < 4 \times 10^{11}$ T cannot deliver that energy deposition.

Table 1. Maximum mass, mass capable of delivering 80 MJ/m and 1 kJ/m, and an estimate of the corresponding event rates for interstellar dark-matter mass density ~7 × 10^{-22} kg/m^3 [1,6] and 250 km/s impact velocity, the relative velocity of the solar system through the dark matter halo. NA = Not Applicable and means that there was no solution.

B_o (10^{12} T)	Max Mass (kg)	Mass of MQNs Depositing ~80 MJ/m (kg)	Expected Events/y on Earth Capable of ~80 MJ/m	Mass of MQNs Depositing >1 kJ/m (kg)	Expected Events/y Depositing >1 kJ/m	Ratio of Rate >1 KJ to Rate ~80 MJ/m
0.1	7×10^{-3}	NA	0	7×10^{-6}	100,000,000	NA
0.2	2×10^{-1}	NA	0	2×10^{-5}	3,000,000	NA
0.3	3	NA	0	4×10^{-5}	800,000	NA
0.4	2×10^1	16.9	2000	5×10^{-5}	600.000	350
0.5	1×10^2	13.9	3000	8×10^{-5}	200,000	50
0.6	1×10^2	11.8	3000	1×10^{-4}	50,000	20
0.7	7×10^2	10.3	900	1×10^{-4}	20,000	30
0.8	1×10^3	9.1	300	2×10^{-4}	20,000	50
0.9	7×10^3	8.2	20	2×10^{-4}	1000	50
1.0	2×10^4	7.5	20	2×10^{-4}	1000	60
1.1	4×10^4	6.9	20	2×10^{-4}	900	50
1.2	1×10^5	6.4	40	3×10^{-4}	200	60
1.3	1×10^5	5.9	20	3×10^{-4}	100	50
1.4	6×10^5	5.6	30	4×10^{-4}	100	40
1.5	7×10^5	5.2	20	4×10^{-4}	90	40
1.6	1×10^6	4.9	0.9	4×10^{-4}	20	30
1.7	3×10^6	4.7	0.4	5×10^{-4}	10	30
1.9	6×10^6	4.2	0.3	6×10^{-4}	10	30
2.0	9×10^6	4.0	0.2	6×10^{-4}	7	30
2.1	1×10^7	3.9	0.5	7×10^{-4}	2	30
2.3	5×10^7	3.6	0.06	8×10^{-4}	2	30
2.4	2×10^8	3.4	0.06	8×10^{-4}	0.4	10
2.6	2×10^8	3.2	0.003	9×10^{-4}	0.2	10
2.8	6×10^8	3.0	0.003	1×10^{-3}	0.2	10
3.0	1×10^9	2.8	0.005	1×10^{-3}	0.03	10
3.1	2×10^9	2.7	0.0003	1×10^{-3}	0.02	10
10.0	8×10^{14}	1.0	3×10^{-8}	7×10^{-3}	2×10^{-7}	10

The last column shown in Table 1 gives the ratio of event rate with enough energy deposition (~1 kJ/m) to leave some geophysical evidence to the event rate that is sufficient for producing the crater in 1985 (~80 MJ/m). The ratio varies from 10 to 350 for the allowed range of B_o, and it indicates that there could be a sufficient number of events to study, if they can be identified and if access to the sites can be obtained.

3.8. Event Rate of Non-Meteorite Cratering Events and Duplicative Constraint on Bo

In addition to comparisons by MQN mass, an observed event rate can be compared to theoretical predictions of the event rate in Table 1. Three non-meteorite events in three years were cited in the Introduction. The estimated energy/meter deposited from Figure 3 above for the 2016 event [33] in Tamil, India that killed a man was ~80 MJ/m, which is comparable to the energy deposition in the County Donegal event. The 2015 event [32] in Rhode Island, USA, is consistent with ~1 kJ/m energy deposition. The 2014 event [31] in Managua, Nicaragua, is consistent with ~30 GJ/m deposited in soft sediment. The approximately annual event rate can be associated with MQN impacts delivering $\geq$1 kJ/m. Table 1 summarizes the event rate for that mass range as a function of B_o. An annual event rate appears to exclude $B_o > 2.3 \times 10^{12}$ T.

However, we need to interpret these results cautiously. We do not know what fraction of all events are observed and reported. If that fraction is small, then some values of $B_o \leq 2.3 \times 10^{12}$ T would also be excluded. On the other hand, MQNs can certainly survive transit through a portion of the solar chromosphere and photosphere and be decelerated (by the MQN magnetopause interaction with solar plasma) to less than the escape velocity from the solar system. A very small fraction of these trapped MQNs can receive sufficient angular momentum, by subsequent interaction with a planet, so that they are not absorbed into the sun. In principle, these captured MQNs can accumulate and enhance the dark-matter density inside the solar system, as compared to that of interstellar space. Our preliminary estimates provide an enhancement factor of ~300. Until adequate simulations of this aerocapture process are completed, we refrain from excluding B_o values that are compatible with an enhancement of 300. Therefore, the upper excluded value that is based on event rate remains $B_o > 2 \times 10^{12}$ T and it is less restrictive than the constraint that is based on MQN mass in Section 3.7.

4. Discussion

4.1. Consistency with MQN Impact

Five points of agreement between theory, as interpreted by the simulations, and data from the three witness-plate layers combine to provide the second of many needed observations that are consistent with MQN dark matter.

1. The 4-m diameter crater is consistent with CTH simulations of ~80 MJ/m energy deposition. That energy/length is consistent with a $10 +/- 7$ kg MQN with 4×10^{11} T $\leq B_o \leq 3 \times 10^{12}$ T. It is not consistent with a meteorite, because no meteorite material was found and because the crater diameter is much too small to be within the range of meteorite craters. Meteorites must be either very aerodynamically shaped, which is very unlikely, or be at least ~20-m diameter to be large enough to survive the transit through the atmosphere and create an impact crater. Meteorite craters are typically an order of magnitude larger in diameter than the meteorites that make them, so meteorites are usually found in craters with diameter ~200 m or larger. The smallest diameter crater associated with a meteorite in the last century impacted in 2007 at Caranacas, Peru. It was 13.5-m in diameter. The crater was at an altitude of 3500 m. Its small diameter may be attributed to its not having to survive the densest part of the atmosphere. Non-meteorite craters are reported in the press approximately annually and are less than 12-m diameter, as noted in the Introduction. The lack of overlapping size and event rate suggests that craters, like the one studied in this paper, must be caused by a phenomenon other than a meteorite.

2. The CTH simulations show that the crater sides are formed by shear-planes and are smooth, as independently reported by the two Rangers investigating the day after the event. We found that smooth sides are in stark contrast to the irregular sides of craters that are produced by large explosives on the surface of the peat bog, so smooth crater sides are a distinguishing point of comparison.

3. The CTH simulations show that chunks of ejecta have sufficient velocity to be thrown clear of the site. Rangers reported the ejecta landed $\geq$10 m from the crater. The photograph presented in Figure 5 shows no ejecta near the crater, which confirms their report.

4. The "pointy depression" at the center of the crater bottom is consistent with the computed channel through the soft sediments and subsequent flow of material. Because the water-saturated soft sediments below the peat met all of the requirements for

liquefaction [39] by the impact, the soft sediments must have liquefied and flowed back into the void to refill the central channel that is shown in Figure 3. Refilling would have occurred from the bottom, where the pressure from the overburden is the greatest, to the top. When the overburden pressure becomes too small to overcome viscosity, a "pointy" depression should remain, as observed on the day following the impact.

5. A volume of shattered granite was only found at 4.7-m depth and within the projected impact trajectory at $15°$ from vertical. All 10 boulders that were found outside the trajectory were intact. The uniqueness of the shattered granite and its location indicates the hypervelocity body that caused the crater in the peat layer also shattered the granite boulder at 4.7-m depth. Passage through the 0.7-m peat layer and 4.0-m soft-sediment layer with sufficient residual velocity to shatter the granite requires material strength that is much greater than the material strength of normal matter. The electromagnetic force holds normal matter together. The strong nuclear force is the only alternative. It holds quark nuggets together. Strong-force material strength, the corresponding nuclear mass density, and energy deposition in the MJ/m range in solid density matter are uniquely consistent with quark nuggets. Therefore, hypervelocity penetration through many meters or kilometers of solid or liquid density normal matter and energy deposition in the MJ/m range are a unique signature of an MQN. Therefore, the deeply buried and shattered granite is consistent with an MQN impact.

4.2. Probability of Fractured Granite Attributable to the Impactor That Made the 1985 Crater

The fifth point of comparison in Section 4.1 assumes that the fractured-granite deposit was caused by the impactor that produced the crater. We only found one fractured-granite deposit in the 633 m^3 excavation. It was at 4.7-m depth and within the calculated trajectory of the crater-forming impactor. The state of subsurface rock before the 1985 impact is uncertain, as with all non-meteorite impacts. Consequently, we cannot be certain that the highly localized and uniquely fractured granite was not fractured before the impact. Its association with the impact is only based on its location and uniqueness.

The null hypothesis is that the fractured granite at 4.7-m depth and the crater on the surface were not caused by the same event. If the probability of the null hypothesis is <0.05, then the results are usually considered to be worthy of further investigation as possible evidence for a new phenomenon. We estimate the probability P_{null} that the null hypothesis is true. P_{null} has two factors:

1. P_1 = probability of the single shattered boulder being randomly located within the effective range R_{eff} of the impactor trajectory for fracturing granite. P_1 = volume_ratio = $\pi R_{eff}^2 L / (633$ m^3 volume of excavation), where L~5 m depth of excavation.

2. P_2 = probability of 10 intact boulders being outside the effective range R_{eff}. Because the probability that one boulder is outside R_{eff} is 1 − the probability it is inside R_{eff} and since all 10 are assumed to be independently located, $P_2 = (1 - \text{volume\_ratio})^{10}$.

   Therefore, $P_{null} = (0.025\ R_{eff}^2) \times (1 - 0.025\ R_{eff}^2)^{10}$.

Figure 10 shows that the impactor trajectory is within one meter of granite slabs that are still intact, so $R_{eff} \leq 1$ m if the energy deposition in the clay is effective in fracturing granite. Granite fractures from the tensile stress after compression waves that originate inside the granite reflect off of the interface with lower-impedance media, as summarized in Section 3.1 from references [37,38]. If fracturing into shards requires energy being directly deposited inside the granite, then $R_{eff} \leq 0.45$ m, the boulders' mean diameter. The two estimates of R_{eff} give probability P_{null} between 0.005 and 0.02. Because we did not measure the exact location of each boulder as it was excavated, then the P_2 term is less certain but it is not sensitive to this number. Setting the less certain P_2 term to 1 still gives P_{null} between 0.005 and 0.025. The probability that the fractured boulder is associated with the impact is $1.0 - P_{null}$ is >98%. The high probability of association supports the consistency of the 1985 event with an MQN impact.

4.3. Less Than 20-m Diameter Craters Are Incompatible with Normal-Matter Impactors

The 80 MJ/m that was deposited in the 1985 event requires the impactor to have been a hypervelocity body, in which the material strength is much less than the internal stresses. Hydrodynamic simulations [28] of the disintegration of large (1 m to 1 km in size) meteoroids in Earth's atmosphere show aerodynamic force, which is proportional to atmospheric density times the square of the velocity, causes it to decelerate, and produces a strong shock wave in front of it. The interaction compresses, heats, and ionizes atmospheric gas. Plasma temperatures can reach 25,000–30,000 K. The associated thermal radiation is absorbed by the surface material of the impactor and causes rapid ablation and vaporization. Rayleigh–Taylor instabilities strongly deform the body, which first breaks up in the center and then completely breaks into many small fragments that quickly slow to subsonic terminal velocity incapable of making a crater. The results are validated by comparison of the predicted light signatures with satellite-based observations and they are consistent with meteorites distributed over a wide area without a crater, as observed in Antarctica.

If the meteor is sufficiently large and sufficiently aerodynamic, then it reaches the ground intact with a significant fraction of its mass and it is still moving at hypervelocity speeds. It then produces an impact crater that is accompanied by meteorite material at the impact site. The dynamics that are associated with passage through Earth's atmosphere assure zero to a very small fraction of meteors with <20-m diameter survive and maintain sufficient speed to cause an impact crater [28]. Because the diameter of an impact crater is typically ~5% the diameter of the impactor, impact craters that are less than ~100 m in diameter are inconsistent with normal-matter impactors. Therefore, the 3.5-m diameter crater from the 1985 event is inconsistent with the normal-matter impactor.

4.4. Normal-Matter Impactors Delivering the Inferred Energy to the Peat Are Incompatible with Shattering Granite 4.7 m below the Surface

The impactor in the 1985 event delivered ~80 MJ/m to the 0.7 m of peat, penetrated 4.0 m of water-saturated soft sediments, and still had enough momentum to shatter the granite boulder with an observed diameter of $\geq$0.6 m. Any MQN that deposits ~80 MJ/m in the peat will also deposit, proportional to its mass density, ~160 MJ/m in the granite. This is well in excess of the ~1 MJ/m required in granite to shatter the boulder, as shown in Figure 3.

Transit through solid or liquid density media would require surviving dynamic forces more than 1000 times those that were experienced in the atmosphere and discussed in Section 4.3, so such transits are prohibited for normal matter. In addition, conservation of linear momentum assures an approximately spherical body (not a long rod penetrator) is decelerated with an e-folding distance of approximately its diameter times the ratio of impactor density to media density. Meteorites typically make a crater approximately 20 times the meteor diameter. Even if the impactor is not vaporized upon impact, a normal-matter impactor would lose most of its velocity within a few tenths of the crater diameter. Consequently, we can rule out normal-matter impactors as the cause of shattered granite boulders that are 4.7 m below the 3.5-m diameter crater in Ireland.

However, the strong nuclear force determines the material strength of an MQN. They are indestructible in interactions at even 250 km/s. The corresponding mass density is nuclear density $>7 \times 10^{17}$ kg/m^3 and this assures that their momentum will let them penetrate many meters or even kilometers into Earth, as discussed in Section 3.6.

4.5. Alternative Explanations

Quark nuggets, neutronium, and black holes have mass densities that are greater than the required value. However, neutronium is not stable outside of neutron stars, and black holes are small enough to provide the local density of dark matter and provide at least one impact per year reported in the press, i.e., ~10 kg mass, would have evaporated in about 150 y, which is much shorter than the time over which the effects of dark matter have been stable. Therefore, crater formation by quark-nugget impact is the only explanation that we have found that fits the data and is consistent with established physics.

These results are also consistent with any other hypervelocity nuclear-density impactor of mass ~5 kg and interaction physics that are capable of depositing ~80 MJ/m. Axion Quark Nuggets (AQNs) [11] have been proposed as an extension to the Standard Model. Their predicted characteristics also satisfy these requirements. Therefore, the results that are reported in this paper also support the AQN candidate for dark matter and may be used to test other dark-matter hypotheses.

4.6. Limitations to Evidence and Need for Systematic Study

The 1985 impact is only the second reported event that is consistent with the MQN hypothesis. Many more are needed to conclude that MQNs exist. The three non-meteorite events cited in the Introduction could be investigated as additional quark-nugget events. Additional candidates are listed in the Supplementary Results: Additional candidate sites for MQN impacts in County Donegal. However, the dates of these potential events are unknown, and there may be competing processes for producing crater-like holes in otherwise flat peat bogs. Gaining physical and administrative access to investigate these additional sites may be difficult. Our investigation of the 1985 impact in Ireland required fifteen years, even with a supportive local community and national authority.

Additional and independent excavation of the 1985 event in County Donegal is lower risk and it could independently confirm or invalidate our result by determining if the bedrock shows the expected cylindrical hole of fractured granite with radius of fracture decreasing with increasing depth. In addition, the expedition could determine whether the tilted granite slabs and granite rocks at 6.3-m depth are a universal feature of the bedrock in the area or were caused by the 1985 event. The latter case would provide additional evidence of large and local energy deposition at depth. The information in Appendix A should be helpful to such an expedition.

A systematic study is necessary, even with additional evidence from non-meteorite craters. Obtaining the results presented in this paper required fifteen years, including the time to obtain permission to excavate from supportive land owners and Irish national authorities. Although such events apparently occur annually on Earth, obtaining permission and excavating each one is impractical. If remote acoustic sensing [42] of the subsurface interface between granite bedrock and soft sediments could be further developed to provide a profile on the interface with ~10-cm resolution through ~10 m of soft sediments, then the pattern that is shown in Figure 10 might be identified as uniquely associated with the impact event. If so, new events could be explored if access to the site can be secured. In addition, the rest of the peat bog in County Donegal could be non-destructively mapped to identify additional sites that occurred over the last 3500 years. Appendix B presents a list of candidate sites for MQN impacts in County Donegal. With this method, a statistically significant set of data might be obtainable.

Whether or not such a technology can be developed, the event that is reported in this paper motivates developing and deploying a constellation of three satellites at 51,000 km altitude to look for RF signatures of MQNs after they transit the magnetosphere [34]. Such a space-based system would provide a real-time search for MQNs based on their predicted Doppler-shifted-radiofrequency signature and it is the best approach for the necessary and systematic study of the MQN hypothesis for dark matter.

5. Conclusions

We report computer simulations of the MQN energy deposition in water-saturated peat, soft sediments, and granite, and report the results from excavating such a crater. The >98% probability that the fractured boulder is associated with the impact (Section 4.2) and the five points of agreement between the simulation results and the observations (Section 4.1) support the inference that the 1985 event is consistent with an MQN impact. This is the second event found to be consistent with MQNs. The first is described in reference [17]. However, many additional non-meteorite impacts with a similar effect on deeply buried rock and/or additional tests that stress different aspects of the MQN hypothesis are needed in order to conclude whether or not MQNs exist and contribute to dark matter. The results also redundantly constrain B_0 to $\geq 4 \times 10^{11}$ T, which is consistent with the previously published most likely values of $B_0 = 1.65 \times 10^{12}$ T $+/- 21\%$.

Although these results are consistent with MQNs, they are also consistent with any other hypervelocity nuclear-density impactor of mass ~5 kg and interaction physics that are capable of depositing ~ 80 MJ/m, such as Axion Quark Nuggets. The results may be also be consistent with some other phenomenon unknown to us. If such candidates are found, they may also be viable candidates for dark matter.

Non-meteorite craters are reported in the popular press approximately once per year. That frequency of reported events suggests a much larger event rate that may offer an opportunity to test hypotheses for dark matter.

Supplementary Materials: Movies of pressure, mass density, and temperature from the CTH simulations available at VanDevender, J. Pace; Schmitt, Robert; Simulations of magnetized quark nugget dark matter in three-layer witness plate, Dryad, Dataset, 2020, https://doi.org/10.5061/dryad.cc2fqz641, accessed on 21 April 2021.

Author Contributions: Conceptualization, A.P.V.; Data curation, J.P.V. and N.M.; Formal analysis, J.P.V. and R.G.S.; Funding acquisition, J.P.V.; Investigation, J.P.V., R.G.S., N.M., D.G.D., S.M., A.P.V., P.W., D.D., H.G. and J.M.; Methodology, J.P.V., R.G.S., N.M., A.P.V. and P.W.; Project administration, J.P.V.; Resources, J.P.V.; Software, R.G.S.; Supervision, J.P.V.; Validation, R.G.S. and A.P.V.; Visualization, R.G.S.; Writing–review & editing, R.G.S., N.M., D.G.D., S.M., A.P.V., P.W., D.D. and H.G. All authors have read and agreed to the published version of the manuscript.

Funding: This research was primarily funded by VanDevender Enterprises and volunteers. In addition, New Mexico Small Business Assistance to Sandia National Laboratories funded the CTH hydrodynamic simulations.

Institutional Review Board Statement: Not applicable.

Informed Consent Statement: Not applicable.

Data Availability Statement: All final analyzed data generated during this study are included in this published article and associated supplementary information.

Acknowledgments: We gratefully acknowledge S. V. Greene for first suggesting we consider quark nuggets and Mark Boslough for critical review and suggestions. Ranger David Dugan provided greatly appreciated guidance and assistance in obtaining the necessary permit from the National Parks and Wildlife Service. Charlie Callahan of Glendowan provided essential guidance and assistance from the point of view of local residents with rights to use this land. Cathal Moy and his associates were invaluable; without their knowledge of local soil conditions and exceptional skill with operating their excavators, the project could not have been accomplished. Volunteers Rebecca Stair and Nathan Girard assisted in the 2018 and 2019 excavations respectively. Finally, Jesse A. Rosen edited this paper and improved the structure, logic, and flow. This work was supported by VanDevender Enterprises, LLC. The CTH simulations were supported by the New Mexico Small Business Assistance (NMSBA) Program through Sandia National Laboratories, a multi-program laboratory operated by National Technology and Engineering Solutions of Sandia (NTESS), a wholly owned subsidiary of Honeywell International, with support from Northrup Grumman, Universities Research Association and others, for the U.S. Department of Energy's National Nuclear Security Administration. By policy, work performed by Sandia National Laboratories for the private sector does not constitute endorsement of any commercial product.

Conflicts of Interest: The authors declare no conflict of interest.

Appendix A. Excavations

To assist another team to extend the excavation into the bedrock and independently test and extend our findings, the three excavations are described in this section. Please check the Acknowledgements for the names of essential team members from County Donegal.

The 2017 expedition cleared out debris and plant growth by hand. The bottom, at depth -0.6 ± 0.1 m, was compacted clay-sand mixture.

The 2018 expedition employed a single Hitachi EX-60, 6-ton excavator shown in Figure A1 with the 4-m diameter crater drained by the channel on the right edge. The site was excavated with the sides sloping at 27° to the vertical, in accord with local experience in this soil. However, the excavation had to be quickly abandoned because the sides of the water-saturated clay-sand mixture showed signs of fracture and sliding at various points down the 27° slope.

Figure A1. View of the 4-m diameter crater from 1985, drained with a channel shown to the down sloping terrain on the right.

The 2019 expedition employed two Doosan 140LC, 14-ton excavators. One was on a ramp inside the excavation and moving material to the surface. The second excavator relocated each scoop of material to a safe distance from the hole to avoid increasing pressure on the soil adjacent to the hole and provide a flat surface for the second excavator to traverse. The slope of the sides was approximately 55° from vertical as shown in Figure A2. That slope held.

An additional 1.5 m of material, plus a water-collecting hole for the submersible pump, was excavated to look for bedrock. From 4.8-m to 6.3-m depth, we found irregular boulders, smaller rocks, and large flat slabs of granite with their normal vector inclined at 30° to the vertical on the south side, and 60° to the vertical on the north, as shown in Figure A3.

We did not find a uniform slab of bedrock and could not determine if the mixture of rocks and slabs at different angles to the horizontal were characteristic of the site before the 1985 event or were caused by that event. Additional excavation directly beneath the grouping of fractured rock at ~4.5-m to ~4.8-m depth was blocked by two large boulders or displaced slabs to either side of that volume. These obstacles were too large to move with available equipment. In addition, the excavation from 4.8 m to 6.3 m had nearly vertical walls, which introduced a safety risk that precluded more excavation within the limitations of the project.

If another group re-excavates the site to examine the bedrock and search for the signature of MQN passage, i.e., a cylinder of fractured granite extending well into the earth, extreme care is recommended below the 6.3-m depth to preserve the context of fractured rock.

Figure A2. Excavation in 2019 to depth of 5.7 m and showing access ramp, submersible water pump and hose, and exit ladder. Examining the bottom and recovering rock fragments had to be done by feel at this stage.

Figure A3. Photo of the collection of granite rocks found at approximately 6.3-m depth to the southeast of the center of the crater. The normal to the slab on the right is inclined at 30° to the vertical.

In accord with our permit, the site was first filled with the rock and clay-sand mixture and topped with the peat layer. A wooden pole was driven into the peat to mark the center of the original 1985 impact crater. Three orange plastic stakes are located on elevated mounds at (1) 21.9 m to the south, (2) 26.76 m to the west, and (3) at 40.12 m at 61.5° north of west. Surveyor's lines from each stake connect the stake to the center post in hopes that another expedition could easily find and re-excavate the site.

Figure A4. Wooden stake and three survey lines mark the center of the original impact crater for subsequent expeditions to extend the investigation into the granite bedrock below 6.5-m depth.

Appendix B. Additional Candidate Sites for MQN Impacts in County Donegal

In 2006, we conducted an aerial survey of about 600 km^2 of peat bog to look for other evidence of deformations from massive objects. We found at least two additional holes and inspected them on the ground: (1) approximately 4-m by 5-m diameter and 2 m deep at 54°55.362′ N and 8°15.260′ W and (2) approximately 4.8-m by 5.1-m diameter and 2.1 m deep at 54°55.434′ N and 8°15.002′ W. Since no one reported witnessing their being formed, we could not confirm that they were associated with impacts.

Since the aerial search in 2006, the resolution in the Google maps covering the western portion of the peat bog has been improved to the point that the maps are useful for a survey. Water flowing below the peat can create multiple holes aligned along the flow in peat bogs. Other mechanisms may also produce holes. Therefore, a survey of isolated round holes, like the 1985 event but without eye witnesses, will only give an upper limit to the event rate. A survey that was informed by the examination of the two holes found in 2006 was conducted in 2014. The survey consisted of 200 randomly selected areas in a square defined by the GPS coordinates of the opposing corners (54.918855, −8.222008) and (54.977614, −8.421822). The chosen area had adequate resolution and did not include any human structures. It was a peat bog with reeds growing on top of the older peat. The total area surveyed in the 200 samples was 3 km2. The survey identified 33 circular depressions like the two we qualified in the ground-based survey. The 33 positions are shown in Table A1.

Table A1. Coordinates and description of deformations in peat-bog survey.

GPS Coordinates	Diameter	Description
54.920869, −8.398206	6 m ± 2 m	very circular, more faded than other shapes
54.921385, −8.296381	6 m ± 2 m	circular, lighter strip running through circle
54.926524, −8.310256	6 m ± 2 m	circular, faded, extension from bottom right of circle
54.927851, −8.34372	4 m ± 2 m	very circular, more faded than other shapes
54.929398, −8.261827	3 m ± 2 m	circular, nodule on top right and left of circle
54.931636, −8.270241	2 m ± 1 m	circular, little nodule on top right of circle
54.931716, −8.225976	4 m ± 2 m	circular, dip on bottom right of circle
54.93359, −8.355017	4 m ± 2 m	circular, tiny dip at top left corner, surrounded by white
54.935795, −8.269017	3 m ± 2 m	circular, sort of flat on top and bottom
54.93674, −8.265257	2 m ± 1 m	very circular, surrounded by white
54.937784, −8.267118	3 m ± 2 m	circular, nodule on top left of circle
54.938974, −8.274889	4 m ± 2 m	circular, dip on top left of circle
54.939262, −8.267507	4 m ± 2 m	circular, diagonal oval shape
54.940387, −8.322002	3 m ± 2 m	very circular, more faded than other shapes
54.942222, −8.319636	4 m ± 2 m	circular, white in center of circle
54.943955, −8.276018	2 m ± 1 m	very circular, surrounded by white
54.944405, −8.278464	2 m ± 1 m	very circular, surrounded by white
54.946506, −8.292173	2 m ± 1 m	very circular, small
54.947903, −8.23144	6 m ± 2 m	circular, two small rounded extensions at bottom
54.949308, −8.399971	4 m ± 2 m	circular, nodule on bottom left of circle
54.949494, −8.298809	5 m ± 2 m	circular, upright oval looking, faded
54.951723, −8.30839	5 m ± 2 m	circular, diagonal oval shape
54.95572, −8.265512	3 m ± 2 m	circular, slightly greater width than height
54.956526, −8.24444	3 m ± 2 m	circular, small dip on bottom of circle
54.962613, −8.273741	1 m ± 0.5 m	circular, tiny nodule on right side of circle
54.964019, −8.305927	4 m ± 2 m	very circular, more faded than other shapes
54.964546, −8.269087	2 m ± 1 m	circular, slightly greater height than width
54.964943, −8.235828	2 m ± 1 m	very circular, more faded than other shapes
54.965639, −8.237676	5 m ± 2 m	circular, nodule on bottom right of circle
54.969051, −8.261765	3 m ± 2 m	circular, bit cut off bottom right of circle
54.969271, −8.277284	3 m ± 2 m	circular, nodule on top right of circle
54.970346, −8.377996	1 m ± 0.5 m	circular, diagonal oval shape
54.971057, −8.320993	3 m ± 2 m	circular, extension from bottom of circle

Poisson statistics gives a 95% confidence for an upper limit of 11 ± 3.7 events per km^2. Their diameters ranged from 2 ± 1 m to 9 ± 2 m. The crater from the 1985 event has changed little in 33 years and should last at least 100 years under the same environmental stresses. The extrema of 100 and 200 years for the time period give an estimated event rate of 0.1 to 0.05 km^{-2} yr^{-1}. Since the area of the earth is ~5×10^8 km^2, the corresponding global event rate is between 30×10^6 and 60×10^6 events per year. Such a large number of potential events illustrates the likelihood of other phenomena forming holes in peat bogs and the importance of eyewitnesses to impacts.

References

1. Aghanim, N.; Akrami, Y.; Ashdown, M.; Aumont, J.; Baccigalupi, C.; Ballardini, M.; Barreiro, R.B.; Bartolo, N.; Basak, S.; Battye, R.; et al. (Planck Collaboration), Planck 2018 Results. VI. Cosmological Parameters. Available online: https://arxiv.org/pdf/1807.06209.pdf (accessed on 14 February 2021).
2. Peebles, P.J.E. *Cosmology's Century, an Inside Story of Our Modern Understanding of the Universe*; Princeton University Press: Princeton, NJ, USA, 2020; pp. 216–297.
3. Iocco, F.; Pato, M.; Bertone, G. Evidence for dark matter in the inner Milky Way. *Nat. Phys.* **2015**, *11*, 245–248. [CrossRef]
4. Steinmetz, M.; Navarro, J.F. The hierarchical origin of galaxy morphologies. *New Astron.* **2002**, *7*, 155–160. [CrossRef]
5. Douglas, C.; Bradac, M.; Gonzalez, A.H.; Markevitch, M.; Randall, S.W.; Jones, C.; Zaritsky, D. A direct empirical proof of the existence of dark matter. *Astrophys. J.* **2006**, *648*, L109–L113. [CrossRef]
6. Salucci, P. The distribution of dark matter in galaxies. *Astron. Astrophys. Rev.* **2019**, *27*, 2. [CrossRef]
7. Patrignani, C. Review of particle properties (Particle Data Group). *Chin. Phys.* **2016**, *C40*, 100001. [CrossRef]
8. Witten, E. Cosmic separation of phases. *Phys. Rev. D* **1984**, *30*, 272–285. [CrossRef]
9. Farhi, E.; Jaffe, R.L. Strange matter. *Phys. Rev. D* **1984**, *30*, 2379–2391. [CrossRef]
10. De Rújula, A.; Glashow, S.L. Nuclearites—a novel form of cosmic radiation. *Nature* **1984**, *312*, 734–737. [CrossRef]
11. Zhitnitsky, A. "Non-baryonic" dark matter as baryonic color superconductor. *JCAP* **2003**, *0310*, 010. [CrossRef]
12. Xia, C.J.; Peng, G.X.; Zhao, E.G.; Zhou, S.G. From strangelets to strange stars: A unified description. *Sci. Bull.* **2016**, *61*, 172. [CrossRef]
13. Jacobs, D.M.; Starkman, G.D.; Lynn, B.W. Macro dark matter. *MNRAS* **2015**, *450*, 3418–3430. [CrossRef]
14. Burdin, S.; Fairbairn, M.; Mermod, P.; Milstead, D.; Pinfold, J.; Sloan, T.; Taylor, W. Non-collider searches for stable massive particles. *Phys. Rep.* **2015**, *582*, 1–52. [CrossRef]
15. VanDevender, J.P.; Shoemaker, I.; Sloan, T.; VanDevender, A.P.; Ulmen, B.A. Mass distribution of magnetized quark-nugget dark matter and comparison with requirements and direct measurements. *Sci. Rep.* **2020**, *10*, 17903. [CrossRef] [PubMed]
16. Oerter, R. *The Theory of Almost Everything: The Standard Model, the Unsung Triumph of Modern Physics*; Penguin Group: New York, NY, USA, 2006; pp. 203–218.
17. VanDevender, J.P.; VanDevender, A.P.; Wilson, P.; Hammel, B.F.; McGinley, N. Limits on Magnetized Quark-Nugget Dark Matter from Episodic Natural Events. *Universe* **2021**, *7*, 35. [CrossRef]
18. Aoki, Y.; Endrődi, G.; Fodor, Z.; Katz, S.D.; Szabó, K.K. The order of the quantum chromodynamics transition predicted by the standard model of particle physics. *Nature* **2006**, *443*, 675–678. [CrossRef]
19. Bhattacharya, T.; Buchoff, M.I.; Christ, N.H.; Ding, H.-T.; Gupta, R.; Jung, C.; Karsch, F.; Lin, Z.; Mawhinney, R.D.; McGlynn, G.; et al. QCD phase transition with chiral quarks and physical quark masses. *Phys. Rev. Lett.* **2014**, *113*, 082001. [CrossRef]
20. The JEM-EUSO Collaboration; Adams, J.H., Jr.; Ahmad, S.; Albert, J.-N.; Allard, D.; Anchordoqui, L.; Andreev, V.; Anzalone, A.; Arai, Y.; Asano, K.; et al. JEM-EUSO: Meteor and nuclearite observations. *Exp. Astron.* **2015**, *40*, 253–279. [CrossRef]
21. VanDevender, J.P.; VanDevender, A.P.; Sloan, T.; Swaim, C.; Wilson, P.; Schmitt, R.G.; Zakirov, R.; Blum, J.; Cross, J.L.; McGinley, N. Detection of magnetized quark nuggets, a candidate for dark matter. *Sci. Rep.* **2017**, *7*, 8758. [CrossRef]
22. Guo, W.-L.; Xia, C.-J.; Lin, T.; Wang, Z.-M. Exploring detection of nuclearites in a large liquid scintillator neutrino detector. *Phys. Rev. D* **2017**, *95*, 015010. [CrossRef]
23. Xu, R.X. Three flavors of a triangle. *Sci. China Phys. Mech. Astron.* **2020**, *63*, 119531. [CrossRef]
24. Tatsumi, T. Ferromagnetism of quark liquid. *Phys. Lett. B* **2000**, *489*, 280–286. [CrossRef]
25. Olausen, S.A.; Kaspi, V.M. The McGill magnetar catalog. *Astrophys. J. Sup.* **2014**, *212*, 6. [CrossRef]
26. Taylor, J.H.; Manchester, R.N.; Lyne, A.G. Catalog of 558 pulsars. *Astrophys. J. Sup.* **1993**, *88*, 529–568. [CrossRef]
27. Scherrer, R.J.; Turner, M.S. On the relic, cosmic abundance of stable, weakly interacting massive particles. *Phys. Rev. D* **1986**, *33*, 1585–1589, Erratum in **1986**, *34*, 3263. [CrossRef]
28. Svetsov, V.V.; Nemtchinov, I.V. Disintegration of large meteoroids in Earth's atmosphere: Theoretical models. *Icarus* **1995**, *116*, 131–153. [CrossRef]
29. Steiner, H.; Gretler, W. The propagation of spherical and cylindrical shock waves in real gases. *Phys. Fluids* **1994**, *6*, 2154–2164. [CrossRef]
30. Koch, C. *A Quest for Consciousness*; Roberts: Greenwood Village, CO, USA, 2004; p. 260.
31. Cooke, W. Did a meteorite cause a crater in Nicaragua? 2014. Available online: http://blogs.nasa.gov/Watch_the_Skies/2014/09/08/did-a-meteorite-cause-a-crater-in-nicaragua/; http://www.cnn.com/2014/09/08/tech/innovation/nicaragua-meteorite/; (accessed on 29 March 2021).
32. Shapiro, E.; Cathcart, C.; Donato, C. Bomb squad, ATF investigating mysterious explosion at Rhode Island beach. 2015. Available online: http://abcnews.go.com/US/explosion-report-prompts-evacuation-rhode-island-beach/story?id=32384143 (accessed on 29 March 2021).
33. Hauser, C. That wasn't a meteorite that killed a man in India, NASA says. 2016. Available online: http://www.nytimes.com/2016/02/10/world/asia/that-wasnt-a-meteorite-that-killed-a-man-in-india-nasa-says.html?_r=0 (accessed on 29 March 2021).
34. VanDevender, J.P.; Buchenauer, C.J.; Cai, C.; VanDevender, A.P.; Ulmen, B.A. Radio frequency emissions from dark-matter-candidate magnetized quark nuggets interacting with matter. *Sci. Rep.* **2020**, *10*, 13756. [CrossRef]

35. McGlaun, J.M.; Thompson, S.L.; Elrick, M.G. CTH: A three-dimensional shock wave physics code. *Int. J. Impact Eng.* **1990**, *10*, 1–4, 351–360. [CrossRef]
36. Crocket, S. SESAME Database: Equation-of-State tabular data for the thermodynamic properties of materials. 1999. Available online: https://www.lanl.gov/org/ddste/aldsc/theoretical/physics-chemistry-materials/sesame-database.php (accessed on 13 February 2021).
37. Duvall, W.I.; Atchison, T.C. Rock breakage by explosives. In *Report of Investigations 5356*; United States Department of the Interior: Washington, DC, USA, 1957; pp. 1–52.
38. Starfield, A.M.; Pugliese, J.M. compression waves generated in rock by cylindrical explosive charges: A comparison between a computer model and field measurements. *Int. J. Rock. Mech. Min. Sci.* **1968**, *5*, 65–77. [CrossRef]
39. Boulanger, R.W.; Idriss, I.M. Liquefaction susceptibility criteria for silts and clays. *J. Geotech. Geoenviron. Eng.* **2006**, *132*, 1413–1426. [CrossRef]
40. Owen, G.; Moretti, M. Identifying triggers for liquefaction-induced soft-sediment deformation in sands. *Sediment. Geol.* **2011**, *235*, 141–147. [CrossRef]
41. Ngo, P.D. Energy Dispersive Spectroscopy. In *Failure Analysis of Integrated Circuits*; Wagner, L.C., Ed.; The Springer International Series in Engineering and Computer Science; Springer: Boston, MA., USA, 1999; Volume 494, pp. 205–215. [CrossRef]
42. Gao, D. Latest developments in seismic texture analysis for subsurface structure, facies, and reservoir characterization: A review. *Geophysics* **2011**, *76*, W1–W13. [CrossRef]

Article

Multi-Modal Clustering Events Observed by Horizon-10T and Axion Quark Nuggets

Ariel Zhitnitsky

Physics and Astronomy Department, University of British Columbia, Vancouver, BC V6T 1Z1, Canada; arz@physics.ubc.ca

Abstract: The Horizon-10T collaboration have reported observation of Multi-Modal Events (MME) containing multiple peaks suggesting their clustering origin. These events are proven to be hard to explain in terms of conventional cosmic rays (CR). We propose that these MMEs might be result of the dark matter annihilation events within the so-called axion quark nugget (AQN) dark matter model, which was originally invented for completely different purpose to explain the observed similarity between the dark and the visible components in the Universe, i.e., $\Omega_{\mathrm{DM}} \sim \Omega_{\mathrm{visible}}$ without any fitting parameters. We support this proposal by demonstrating that the observations, including the frequency of appearance, intensity, the spatial distribution, the time duration, the clustering features, and many other properties nicely match the emission characteristics of the AQN annihilation events in atmosphere. We list a number of features of the AQN events which are very distinct from conventional CR air showers. The observation (non-observation) of these features may substantiate (refute) our proposal.

Keywords: dark matter; axion; cosmic rays

check for
updates

Citation: Zhitnitsky, A. Multi-Modal Clustering Events Observed by Horizon-10T and Axion Quark Nuggets. *Universe* **2021**, *7*, 384. https://doi.org/10.3390/universe7100384

Academic Editor: Kazuharu Bamba

Received: 18 August 2021
Accepted: 12 October 2021
Published: 15 October 2021

Publisher's Note: MDPI stays neutral with regard to jurisdictional claims in published maps and institutional affiliations.

1. Introduction

In this work we discuss two naively unrelated stories. The first one deals with the recent puzzling observations [1–5] by the Horizon-10T (H10T) collaboration of the Multi-Modal Events (MME).

The second one is the study of a specific dark matter (DM) model, the so-called axion quark nugget (AQN) model [6], see a brief overview of this model below. We overview the corresponding events below in details. We also highlight some difficulties in interpretation of these events in terms of the standard CR air showers. The unusual features of MME include:

1. *"clustering puzzle":* Two or more peaks separated by $\sim 10^2$ ns are present in several detection points, while entire event may last $\sim 10^3$ ns. It can be viewed as many fronts separated by $\sim (10^2$–$10^3)$ ns, instead of a single front;
2. *"particle density puzzle":* The number density of particles recorded at different detection points apparently weakly dependent on distance from Extensive Air Showers (EAS) axis;
3. *"pulse width puzzle":* The width of each individual pulse is around (20–35) ns and apparently does not depend on distance from EAS axis;
4. *"intensity puzzle":* The observed intensity of the events (measured in units of a number of particles per unit area) is of order $\rho \sim (100$–$300)$ m^{-2}

When measured at distances (300–800) m from EAS axis. Such intensity would correspond to the CR energy of the primary particle on the level $E_p \gtrsim 10^{19}$ eV which would have dramatically different event rate in comparison with observed rate recorded by H10T.

Before we proceed with our explanation of the proposal to view the MMEs as the AQN events one should mention that similar unusual features of the CR air showers have been noticed long ago for the first time by Jelley and Whitehouse [7] in 1953. Later, EAS exhibiting the unusual time structures were studied by several independent experiments, see e.g., [8]. We refer to ref. [2] for overview and references of the older observations and studies of such unusual events.

The considerable recent advances in rising the resolution (on the level of a few ns) has allowed the H10T collaboration dramatically improve the collection and analysis of the MMEs. In particular, during $\sim$3500 h of operation the H10T collaboration collected more than 10^3 MMEs.

In what follows we overview the observation [1–5] by emphasizing the puzzling nature of these events if interpreted as conventional EAS events. At the same time the same observations can be explained in very natural way if interpreted in terms of the AQN annihilation events as we shall argue in this work. One should mention here that a similar conclusion has been also reached for a different type of unusual CR-like events. First, it has been argued in [9,10] that the Telescope Array (TA) "mysterious bursts" [11,12] can be naturally interpreted as the AQN events. Secondly, it has been also argued in [13] that the Antarctic Impulse Transient Antenna (ANITA) observation [14–16] of two anomalous events with noninverted polarity might be also a consequence of the AQN annihilation events. Important comment here that in all those cases the basic parameters of the AQN model remain the same as they have been fixed long ago from dramatically different observations in a very different context.

Our presentation is organized as follows. In next Section 2 we explain why the observation [1–5] listed as **1–4** *puzzles* are very mysterious events if interpreted as conventional EAS events. In Section 3 we give a brief overview of the AQN framework with emphasize on the key elements relevant for the present studies. In Section 4 we formulate our proposal on identification of the unusual Multi-Modal Events with the upward moving AQN events. In Section 5 we estimate the event rate. Our main section is Section 6 where we estimate a variety of relevant time scales and the intensity of the events. In the same section we also confront our proposal with observations and argue that puzzles **1–4** can be naturally understood within the AQN framework. Section 7 is our conclusion where we suggest several tests which may substantiate or refute our proposal on identification of the Multi-Modal Cluster Events with the upward moving AQNs.

2. Conventional CR Picture Confronts the MME Observations

This section is devoted to the first part of our story where we describe the MME observations [1–5] and argue why the observed events are inconsistent with conventional CR interpretation, while the next Section 3 is devoted to the second part of our story, the AQN dark matter model.

We start by reviewing the conventional picture of the EAS. It is normally assumed that EAS can be thought as a disk -pancake with well defined EAS axis. It is also assumed that EAS represents an uniform, without any breaks structure. The standard picture also assumes that the particle density drops smoothly with distance when moving away from the core, while the thickness of the EAS pancake increases with the distance from the core. Now we want to see why this conventional picture is in dramatic contradiction with observations of the MME events.

1.   Indeed, a typical MME is shown on Figure 1 where a complicated temporal features are explicitly seen. Several peaks separated by $\sim10^2$ ns in a single detector represent the *"clustering puzzle"*, listed above. In conventional EAS picture one should see a single pulse in each given detector with the amplitude which depends on the distance from the EAS axis. It is not what actually observed by H10T.

2.   The manifestation of the *"particle density puzzle"* is as follows. On Figure 2 we show the particle density distribution $\rho(R)$ in simulated EAS disk versus distance from axis for different energies shown by solid lines with different colours, depending on energy of the CR. In particular, for energy of the primary particle on the level of 10^{17} eV one should expect a strong suppression $\sim10^3$ when distance R to the EAS axis changes from $R \simeq 100$ m to $R \simeq 700$ m. It is not what actually observed by H10T:

the density of the particles $\rho(R)$ is not very sensitive to the distance to the EAS axis, and remains essentially flat for the entire region of observations. Furthermore, the magnitude of the density $\rho(R)$ is much higher than it is normally expected for CR energies $\sim(10^{17}\text{--}10^{18})$ eV.

3. The manifestation of the *"pulse width puzzle"* is as follows. On Figure 3 we show the simulated EAS disk width versus distance from axis for different energies shown by solid lines in different colours. As explained above the thickness of the EAS pancake increases with the distance from the core. Therefore, the pulse width must also increase correspondingly as shown by sold lines on Figure 3 for different energies. It is not what actually observed by H10T: all MME events show similar duration of the pulse width on the level of (20–35) ns irrespective to the distance to the AES axis. This observation is in dramatic conflict with conventional picture as outlined above.

4. The manifestation of the *"intensity puzzle"* is as follows. The Figure 2 suggests that the charged particle density $\rho(R)$ varies between (30–300) particles per m^2 at the distances (500–700) m from the EAS axis. This is at least factor 10^2 above the expected $\rho(R)$ for the primary particle with energy in the interval $(10^{17}\text{--}10^{18})$ eV. Only the particles with energies well above 10^{19} eV could generate such enormous particle density as shown on Figure 2. However the frequency of appearance for such highly energetic particles is only once every few years. This represents the *"intensity puzzle"* when the intensity of the events estimated from $\rho(R)$ for MME is several orders of magnitude higher than the energy estimated by the event rate. This shows a dramatic inconsistency between the *measured* intensity and *observed* event rate carried out by one and the same detector.

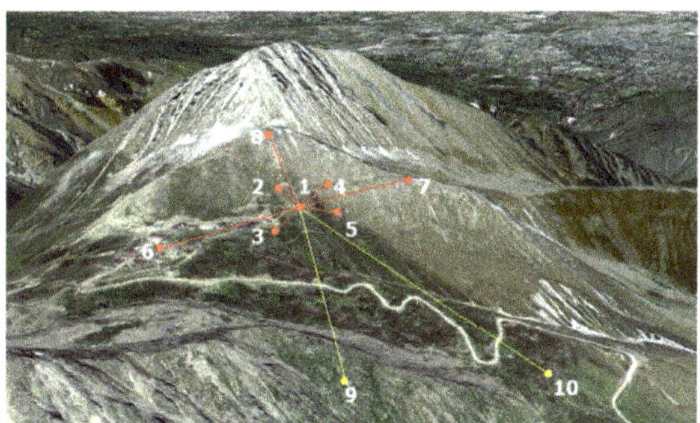

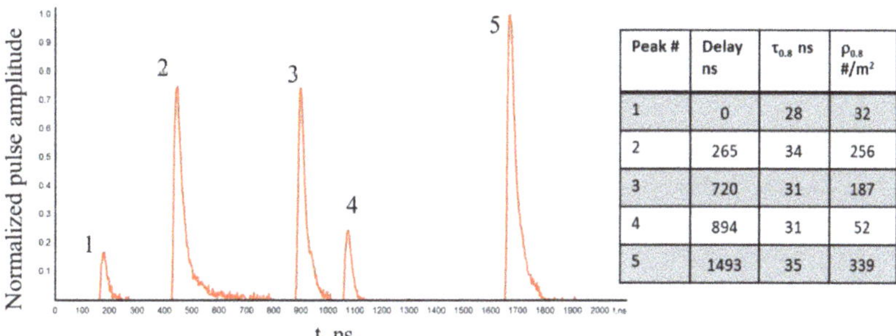

Peak #	Delay ns	$\tau_{0.8}$ ns	$\rho_{0.8}$ #/m^2
1	0	28	32
2	265	34	256
3	720	31	187
4	894	31	52
5	1493	35	339

Figure 1. (**top**): Aerial view and geometry of H10T instrument with location of 10 detectors, adapted from Beznosko et al., 2019 [4]; (**bottom**): A typical MME event recorded on 6 April 2018 by H10T instrument at point #9, adapted from Beznosko et al., 2019 [4]. All pulses are recorded at a single detection point. Delay times, the width of each peak $\tau_{0.8}$ in ns, and the particle density $\rho_{0.8}$ per m^{-2} within $\tau_{0.8}$ are also shown in the table.

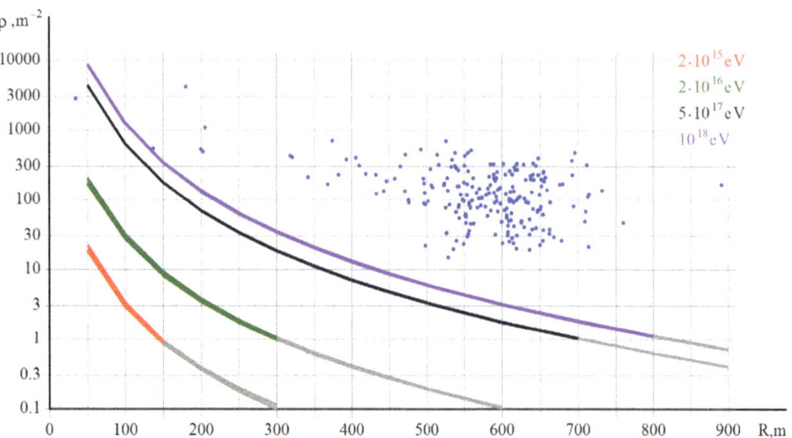

Figure 2. Solid lines: the particle density distribution $\rho(R)$ in simulated EAS disk versus distance from axis for different energies shown by different colours, depending on energy of the CR. Blue dots: cumulative particle density for each bimodal pulse vs. distance to EAS axis, adapted from Beznosko et al., 2019 [4].

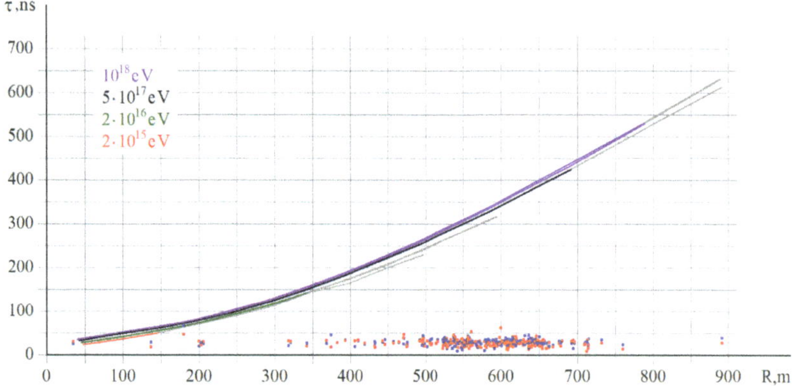

Figure 3. Solid lines: simulated EAS disk width versus distance from axis for different energies shown by different colours. Pulse width $\tau_{0.8}$ in ns for the first pulse (in blue) and second pulse (in red) for the bimodal pulses versus distance to the EAS axis, adapted from Beznosko et al., 2019 [4].

We conclude this section with the following comment. From the puzzles listed above it must be obvious that the events detected by H10T are not the conventional CR air showers as they demonstrate enormous inconsistency with standard CR interpretation. What it could be? Before we put forward our proposal we would like to briefly overview in next Section 3 the second part of our story- the AQN dark matter model.

3. The AQN Dark Matter Model

We start with few historical remarks and motivation of the AQN model in Section 3.1, while in Section 3.2 we overview recent observations of CR-like events (such as puzzling bursts observed by the TA experiment and ANITA's two anomalous events with non-inverted polarity) of some mysterious events which could be explained by the AQN events hitting the Earth. Finally, in Section 3.3 we briefly overview some specific features of the AQNs traversing the Earth (such as internal temperature, level of ionization, etc.). These characteristics will be important for the present study interpreting the MMEs as the AQN events.

3.1. The Basics

The AQN dark matter model [6] was invented long ago with a single motivation to explain in natural way the observed similarity between the dark matter and the visible densities in the Universe, i.e., $\Omega_{DM} \sim \Omega_{visible}$ without any fitting parameters. We refer to recent brief review [17] on the AQN model. Here we want to mention few key elements which are important for this work.

The AQN construction in many respects is similar to the Witten's quark nuggets, see [18–20], and review [21]. This type of DM is "cosmologically dark" not because of the weakness of the AQN interactions, but due to their small cross-section-to-mass ratio, which scales down many observable consequences of an otherwise strongly-interacting DM candidate.

There are two additional elements in the AQN model compared to the original models [18–21]:

(a). First new element is the presence of the axion domain walls which are copiously produced during the QCD transition. The axion field $\theta(x)$ plays a dual role in this framework: it serves as an additional stabilization factor for the nuggets, which helps to alleviate a number of problems with the original nugget construction [18–21]. Furthermore, the same axion field $\theta(x)$ generates the strong and coherent CP violation in the entire visible Universe. This is because the $\theta(x)$ axion field before the QCD epoch could be thought as classical CP violating field correlated on the scale of the entire Universe.

One should comment here that the axion field had been introduced into the theory to resolve the so-called the strong CP problem which is related to the fundamental initial parameter $\theta_0 \neq 0$. The axion remains the most compelling resolution of the strong CP problem and it also represents a DM candidate, see original papers on the axion [22–28], and recent reviews [29–37]. The most recent constraints on the axion mass for entire broad classical window $m_a \in (10^{-6}\text{–}10^{-3})$ eV is set by the CAST collaboration [38], while the world best constraints in specific narrow bands around $m_a \sim 10^{-5}$ eV are imposed by different cavity type detectors, see review articles mentioned above for references and details.

One should also comment here that in recent years many researchers consider the so-called axion-like models when the axion coupling constant f_a and its mass m_a are completely decoupled, in contrast with the original models [22–28] when these parameters are linked: $m_a f_a \sim \Lambda_{QCD}^2$. In axion-like models the strong CP problem is not resolved. However, the axion itself remains a feasible DM candidate. It may produce a number of interesting observable effects. In particular, ultra light axion DM may significantly affect structure formation at small scales [39].

We should emphasize that in this work we assume a classical window for the axion mass $m_a \in (10^{-6}\text{–}10^{-3})$ eV. Furthermore, the axions which are produced from the AQNs have dramatically different spectral features, and therefore, the corresponding axion searches should have very different strategies, see recent brief review [17] covering this topic. In the present work we do not study the axions and their possible detections. Instead, this work is devoted to very different observables, other than the axions, which emerge as a result of the AQN annihilation events as discussed in Section 3.2.

(b). Another feature of the AQN model which plays absolutely crucial role for the present work is that nuggets can be made of *matter* as well as *antimatter* during the QCD transition. Precisely the coherence of the CP violating field on large scale mentioned above provides a preferential production of one species of nuggets made of *antimatter* over another species made of *matter*. The preference is determined by the initial sign of the θ field when the formation of the AQN nuggets starts. The direct consequence of this feature along with coherent CP violation in entire Universe is that the DM density, Ω_{DM}, and the visible density, $\Omega_{visible}$, will automatically assume the same order of magnitude densities $\Omega_{DM} \sim \Omega_{visible}$ without any fine tuning as they both proportional to one and the same dimensional parameter Λ_{QCD}. We refer to the brief review [17] devoted to the specific questions related to the nugget's formation, generation of the baryon asymmetry, and survival pattern of the nuggets during the evolution in early Universe with its unfriendly environment.

One should emphasize that AQNs are absolutely stable configurations on cosmological scales. Furthermore, the antimatter which is hidden in form of the very dense nuggets is unavailable for annihilation unless the AQNs hit the stars or the planets.

However, when the AQNs hit the stars or the planets it may lead to observable phenomena. The strongest direct detection limit[1] is set by the IceCube Observatory's, see brief review [17] for the details:

$$\langle B \rangle > 3 \cdot 10^{24} \quad [\text{direct (non)detection constraint}], \tag{1}$$

where we formulated the constraints in term of the AQN's baryon charge B rather than in terms of its mass $M \approx m_p B$. Similar limits are also obtainable from the ANITA and from geothermal constraints which are also consistent with (1) as estimated in [41].

While ground based direct searches offer the most unambiguous channel for the detection of quark nuggets the flux of nuggets is inversely proportional to the nugget's mass and consequently even the largest available conventional dark matter detectors are incapable to exclude the entire potential mass range of the nuggets. Instead, the large area detectors which are normally designed for analyzing the high energy CR are much better suited for our studies of the AQNs as we discuss in next Section 3.2.

3.2. When the AQNs Hit the Earth. . .

For our present work, however, the most relevant studies are related to the effects which may occur when the AQNs made of antimatter hit the Earth and continue to propagate in deep underground in very dense environment. In this case the most of the energy deposition will occur in the Earth's interior. The corresponding signals are very hard to detect as the photons, electrons and positrons will be quickly absorbed by surrounding dense material deep underground, while the emissions of the very weakly interacting neutrinos and axions are hard to recover. In this short subsection we want to mention several observed phenomena which could be related to the AQN annihilation events when the nuggets propagate in the Earth's atmosphere.

We start with the Telescope Array (TA) experiment which has recorded [11,12] several short time bursts of air shower like events. These events are very unusual and cannot be interpreted in terms of conventional CR single showers. In particular, if one tries to fit the observed bursts (cluster events) with conventional code, the energy for CR events should be in 10^{13} eV energy range to match the frequency of appearance, while the observed bursts correspond to $(10^{18}–10^{19})$ eV energy range as estimated by signal amplitude and distribution. Therefore, the estimated energy from individual events within the bursts is five to six orders of magnitude higher than the energy estimated by event rate [11,12]. It has been argued in [9,10] that these bursts represent the direct manifestation of the AQN annihilation events.

This feature, in many aspects, is very similar to the *intensity puzzle* listed in previous Section 2, where analogous dramatic inconsistency (between the observed intensity and measured event rate) is recorded by H10T instrument.

Our next example is the observed seasonal variations of the X ray background in the near-Earth environment. To be more precise, the XMM-Newton at 11σ confidence level [42] had recorded the seasonal variations in the 2–6 keV energy range at distances $r \gtrsim 8R_\oplus$ where the measurements are effectively had been performed. The authors [42] argue that conventional astrophysical sources have been ruled out. Furthermore, the XMM-Newton's operations exclude pointing at the Sun and at the Earth directly, diminishing a possible direct X ray background from the Earth and the Sun. It has been argued in [43] that this seasonal variation may be naturally explained by the AQNs exiting the Earth. The AQNs continue the emission of the X rays at $r \gtrsim 8R_\oplus$, long after the nuggets exit the Earth's atmosphere. It has been also shown that the spectrum and the intensity computed in [43] almost identically match the observed spectrum [42].

The final example we want to mention here is the observation by ANITA [14–16] of two anomalous events with noninverted polarity. These two events had been identified as the Earth emergent upward going CR-like events with exit angles of -27^0 and -35^0 relative to the horizon. These anomalous events are in dramatic tension with the standard model because neutrinos are exceedingly unlikely to traverse through Earth at a distance of $\gtrsim 5 \times 10^3$ km with such ultrahigh energy, even accounting for the ν_τ regeneration [14]. The analysis [44] reviewed the high-energy neutrino events from the IceCube Neutrino Observatory and inferred that the ν_τ interpretation is excluded by at least 5σ confidence level. We advocated in [13] that these two anomalous events with noninverted polarity observed by ANITA could be interpreted as the upward going AQNs.

The applications [43] to the X ray emission in the near-Earth environment and the ANITA anomalous events [13] are especially relevant for the present work because both these phenomena are originated from the AQN upward going (Earth emergent) events when the AQNs traversed through the Earth interior and exit the Earth surface. As we shall argue in this work the MME events could be also the consequence of the same upward moving AQNs.

However, there is an additional technical challenging problem to be addressed for the application to MME analysis. The point is that the ANITA anomalous events are generated by AQNs which just crossed the surface and entered the Earth's atmosphere (upward-going events). Exactly at this instant large number of electrons instantaneously released into atmosphere and generated the radio pulse measured by ANITA. In contrast, the X rays are observed by XMM-Newton at very large distances $r \gtrsim 8R_\oplus$ from Earth. In this case the impact of the atmospheric material at $r \lesssim 60$ km can be ignored, and the cooling of the AQN is dominated by nugget's propagation in an empty space. The application to MME, which is the topic of the present work, deals with an intermediate stage between these two cases: it is not the first instant when the AQN enters the atmosphere emerging from the interior, and it is not the asymptotically far away when the AQN propagates in empty space. In other words, we need to know the emission pattern and the energy deposition rate for the AQNs when they propagate in the atmosphere, and the impact of the atmospheric surrounding material cannot be ignored. We formulate our proposal in Section 4 where we argue that emission of the electrons by the AQN in form of the well-separated "bunches" may explain the unusual features of the MME observations.

However, before we present our arguments supporting this interpretation of the MME unusual events (as highlighted in Section 2) we should overview the basic characteristics of the AQNs traversing the Earth, which represents the topic of the next Section 3.3.

3.3. Internal Structure of the AQN

The goal here is to explain the basic features of the AQNs when they enter the dense regions of the surrounding material and annihilation processes start. The related computations originally have been carried out in [45] in application to the galactic environment with a typical density of surrounding visible baryons of order $n_{\text{galaxy}} \sim 300$ cm^{-3} in the galactic center, in dramatic contrast with dense region in the Earth's interior when $n_{\text{rock}} \sim 10^{24}$ cm^{-3} and atmosphere with $n_{\text{air}} \sim 10^{21}$ cm^{-3}. We review these computations with few additional elements which must be implemented in case of propagation in the Earth's atmosphere and interior when the density of the environment is much greater than in the galactic environment.

The total surface emissivity from electrosphere has been computed in [45] and it is given by

$$F_{\text{tot}} \approx \frac{16}{3} \frac{T^4 \alpha^{5/2}}{\pi} \sqrt[4]{\frac{T}{m}}, \tag{2}$$

where $\alpha \approx 1/137$ is the fine structure constant, $m = 511$ keV is the mass of electron, and T is the internal temperature of the AQN. One should emphasize that the emission from the electrosphere is not thermal, and the spectrum is dramatically different from black body radiation, see [45], see also Appendix A with more details.

A typical internal temperature of the AQNs for very dilute galactic environment can be estimated from the condition that the radiative output of Equation (2) must balance the flux of energy onto the nugget

$$F_{\text{tot}}(4\pi R^2) \approx \kappa \cdot (\pi R^2) \cdot (2\,\text{GeV}) \cdot n \cdot v_{\text{AQN}}, \tag{3}$$

where n represents the number density of the environment. The left hand side accounts for the total energy radiation from the AQN's surface per unit time as given by (2) while the right hand side accounts for the rate of annihilation events when each successful annihilation event of a single baryon charge produces $\sim 2m_p c^2 \approx 2\,\text{GeV}$ energy. In Equation (3) we assume that the nugget is characterized by the geometrical cross section πR^2 when it propagates in environment with local density n with velocity $v_{\text{AQN}} \sim 10^{-3} c$.

The factor κ is introduced to account for the fact that not all matter striking the AQN will annihilate and not all of the energy released by an annihilation will be thermalized in the AQNs by changing the internal temperature T. In particular, some portion of the energy will be released in form of the axions, neutrinos[2]. The high probability of reflection at the sharp quark matter surface lowers the value of κ. The propagation of an ionized (negatively charged) nugget in a highly ionized plasma (such as solar corona) will increase the effective cross section. As a consequence, the value of κ could be very large as reviewed in [17] in application to the solar corona heating problem.

The internal AQN temperature had been estimated previously for a number of cases. It may assume dramatically different values, mostly due to the huge difference in number density n entering (3). In particular, for the galactic environment $T_{\text{galaxy}} \approx 1\,\text{eV}$, while in deep Earth's interior it could be as high as $T_{\text{rock}} \approx (100\text{–}200)\,\text{keV}$. Precisely this value of T had been used as initial temperature of the nuggets in the proposal [43] explaining the seasonal variations of the X rays observed the XMM-Newton at 11σ confidence level [42] at distances $r \sim (6\text{–}10)\,R_\oplus$ from the Earth surface.

The crucial element for the study [43] was the emission rate by AQNs at very high temperatures $T \approx (100\text{–}200)\,\text{keV}$. In this case the emission is not determined by simple Bremsstrahlung radiation given by (2) which is only valid for relatively low temperatures. In the high temperature regime a number of many-body effects in the electrosphere, that were previously ignored, become important. In particular, it includes: the generation of the plasma frequency in electrosphere, which suppresses the low frequency emission. Also, the the AQNs get ionized which dramatically decreases the density of positrons in electrosphere, and consequently suppresses the emission, among many others effects. All these complications have been carefully considered in [43] in applications to the seasonal variations of the X rays as observed the XMM-Newton, and we refer to that paper for the details. Here we quote the result of these studies. The rate of emission can be effectively represented as follows:

$$\frac{dE_{\text{emiss}}}{dt} \sim (4\pi R^2)\eta(T,R)F_{\text{tot}}(T) \tag{4}$$

where factor $\eta(T,R)$ is a result of strong ionization of the electrosphere, which leads to the corresponding suppression of the emission. The details of this suppression are explained in Appendix A and given by (A12). This suppression is a direct consequence of high internal temperature T when a large number of weakly bound positrons are expanded over much larger distances order of R rather than distributed over much shorter distances of order m^{-1} around the nugget's core. This basically determines the suppression factor $\eta \sim (mR)^{-1} \sim 10^{-6}$.

In both previously considered applications (to ANITA [13] and to XMM-Newton [43] experiments) when the upward moving AQNs with high $T \approx (100\text{–}200)\,\text{keV}$ play the key role in the explanations of the observations, we ignored an additional[3] annihilation events with atmospheric material as we already mentioned at the end of the previous Section 3.2. Now we want to include the corresponding physics in our analysis.

First of all we want to estimate the number of direct head-on collisions of the atmospheric molecules with AQN per unit time. It can be estimated as follows:

$$\frac{dN}{dt} \simeq (\pi R^2) N_m v_{\text{AQN}} \simeq 0.8 \cdot 10^{18} \left(\frac{n_{\text{air}}}{10^{21}\ \text{cm}^{-3}} \right) \text{s}^{-1}, \tag{5}$$

where $N_m \simeq 2.7 \cdot 10^{19}$ cm^{-3} is the molecular density in atmosphere when each molecule contains approximately 30 baryons such that the typical density of surrounding baryons in air is $n_{\text{air}} \simeq 30 \cdot N_m \simeq 10^{21}$ cm^{-3}. The dominant portion of these collisions are the elastic scattering processes rather than successful annihilation events suppressed by parameters κ as discussed above. The energy being deposited to the AQN per unit time as a result of annihilating processes can be estimated as follows

$$\begin{aligned}
\frac{dE_{\text{deposit}}}{dt} &\approx (2\ \text{GeV}) \cdot \kappa \cdot \frac{dN}{dt} \\
&\approx 1.6 \cdot 10^{18} \cdot \kappa \cdot \left(\frac{n_{\text{air}}}{10^{21}\ \text{cm}^{-3}} \right) \frac{\text{GeV}}{\text{s}},
\end{aligned} \tag{6}$$

One should emphasize that this energy is being deposited into the AQN which is already at full capacity with very high temperature $T \simeq (100\text{–}200)$ keV accumulated during a long journey in much denser Earth's interior.

What is the mechanism to release this extra energy? In other words: In what form the energy (6) could be emitted into the surrounding atmosphere as it cannot be easily transferred to the AQN's quark core (as it is already saturated, see footnote 3)?

There are two qualitatively different regimes which can be normally realized in such circumstances. If a typical time scale $\tau_{deposit}$ to deposit a specific amount of energy (let us say, 1 GeV) determined by (6) is longer than the time scale τ_{cool} to release the same amount of energy, i.e., $\tau_{deposit} \gtrsim \tau_{\text{cool}}$ than a continuous cooling process takes place and the temperature slowly decreases. This is a thermodynamically equilibrium behaviour which can be analyzed using conventional technical tools[4].

The second option is realized when the time scale to deposit the energy $\tau_{deposit}$ is much shorter than the time scale of cooling processes, i.e., $\tau_{deposit} \ll \tau_{\text{cool}}$, in which case the burst-like (explosion like, eruption-like blasts) non- equilibrium processes must occur. These eruption-like events will release the extra accumulated energy in form of very short pulses, which cannot be described as conventional cooling equilibrium processes. These short pulses must alternate with much longer periods of accumulation energy which inevitably end with subsequent eruption-like events. In other words, in this case, the non-equilibrium fast equilibration process can be thought as the eruption-like event[5].

Now we give a numerical estimation for the cooling rate determined by (4) to find out what regime is realized in given circumstances. The order of magnitude estimate can be expressed as follows:

$$\frac{dE_{\text{emiss}}}{dt} \sim 10^{16} \left(\frac{T}{100\ \text{keV}} \right)^{\frac{17}{4}} \frac{\text{GeV}}{\text{s}}. \tag{7}$$

Comparing this estimate for cooling rate with estimate for the deposition rate (6) one concludes that the rate of energy deposition is much shorter than the rate at which the system is capable to release the same amount of energy in form of the photon's emission. As a result, we arrive to the conclusion that second option when $\tau_{deposit} \ll \tau_{\text{cool}}$ is realized. This conclusion inevitably implies the eruption like events must occur, and these very short explosions must be alternated with much longer periods of time when the energy is being accumulated by the AQNs. These longer periods of the energy accumulations also must end with consequent eruption-like blasts. This alternating pattern continues as long as following condition holds:

$$\tau_{deposit} \ll \tau_{\text{cool}} \quad \rightarrow \quad (\text{short eruptions occur}). \tag{8}$$

With these preliminary comments with estimates from the previous studies on the AQN features we are now in position to formulate the proposal interpreting the MME events observed by X10T collaboration in terms of the upward moving AQNs, which is the topic of the next Section 4.

4. MME as the AQN Event

In this Section we formulate the basic idea of our proposal on identification of the unusual Multi-Modal Events with the upward moving AQN events, while supporting estimates on the event rate, the intensity, and the variety of time scales will be presented in following Sections 5 and 6.

The AQNs which propagate in the Earth's interior are very hot. Their temperature could be as hot as $T \approx (100\text{--}200)$ keV at the moment of exit on the Earth's surface as we discussed in [43] in the application to studies of the seasonal variations of the X rays observed by XMM-Newton at very large distances $r \sim (6\text{--}10)\, R_\oplus$ from the Earth. At such large distances the AQN's cooling can be described by conventional thermodynamically equilibrium processes with well defined spectrum, see footnote 4. It should be contrasted with our application to ANITA anomalous events [13] when the same upward moving AQNs just crossed the surface and entered the Earth's atmosphere. In this case the dominant cooling mechanism is realized in form of short burst-like events according to (8). The outcome of these short burst-like events is emission of numerous bunches (clumps) of relativistic electrons. These eruption like events are analogous to lightning flashes from footnote 5.

In our studies [13] we focused on an estimation of the number of emitted electrons N and their energy $\langle E \rangle \sim 10$ MeV which, according to the proposal [13], essentially determine the intensity and the spectral features of the ANITA anomalous events. These electrons will be always accompanied by the emission of much more numerous number of photons with typical energy of order $\omega \sim T \approx (100\text{--}200)$ keV, which in fact represent the dominant fast non- equilibrium mechanism of cooling of the system in these circumstances (8).

Precisely these multiple explosion-like emissions of photons and electrons will be identified with Multi-Modal Events with their unusual features reviewed in Section 2. The mean free path of the photons with $\omega \sim T$ is relatively short, measured in meters. Therefore, they cannot be directly observed. In contrast, the mean free path of the energetic electrons at sea level with $\langle E \rangle \sim 10$ MeV is around several kilometres in atmosphere (and even longer at the elevation of 3346 m where H10T is located). The propagation of these ultra-relativistic electrons takes place in the background of the geomagnetic field $\mathcal{B} \sim 0.5$ G which determines the instantaneous curvature $\rho \sim 3$ km of the electron's trajectories with such energies, see Section 5 with relevant estimates. Due to the large component of magnetic field parallel to the Earth's surface the electrons may move in downward direction even if they are initially emitted in upward direction. Our proposal is that precisely these downward moving energetic electrons could mimic the CR and could be responsible for the Multi-Modal pulses as detected by H10T instrument.

Any precise computation during a short period of time of emission is very hard problem of non-equilibrium dynamics, which is beyond the scope of the present work. Fortunately, the observable intensity, the event rate, the time delays between the pulses are determined in the AQN framework by the basic characteristics of the model and are not very sensitive to the details of this non-equilibrium mechanism of production and its corresponding time scales. This is because all these observables depend on several parameters such as typical energy $\langle E \rangle \sim 10$ MeV of the emitted electrons, typical number of emitted electrons N, the deposition rate (6) and the cooling rate (7) which had been previously estimated for very different purposes in different circumstances. We shall use the same parameters in the present work.

In particular, the number of electrons $N \sim (10^8\text{--}10^9)$ emitted by AQNs is very hard to compute from the first principles. However it can be fixed by the observed intensity and spectral features of the anomalous ANITA radio pulses assuming, of course, that these observed radio pulses with inverted polarities are due to the upward moving AQNs [13]. The typical energy $\langle E \rangle \sim 10$ MeV of these electrons is also consistent with duration of the observed pulses [13]. The next Section 5 is devoted to the estimation of the event rate of MME, while Section 6 is devoted to explanations of the *puzzles* which demonstrate a dramatic deviation from conventional CR picture as formulated in Section 2. We shall argue that the observed MME features are consistent with our AQN-based interpretation.

5. Event Rate of MME

This section is devoted to estimate of the event rate of MMEs within the AQN framework. We anticipate that this evaluation is expected to be very qualitative estimation due to large uncertainties in parameters and rare occurrence of the observed MMEs. In particular, it is known that key parameter, the DM density locally may dramatically deviate from the well established average global value $\rho_{DM} \approx 0.3\,\text{GeV cm}^{-3}$. Nevertheless we would like to present such estimate to demonstrate that our interpretation of MMEs as an outcome of upward-going AQNs is at least a self-consistent proposal. In what follows we use the same formulae we used for the estimates of the frequency of appearance of the ANITA Anomalous Events [13] such that the numerous uncertainties related to the local DM density or/and the AQN size distribution would not dramatically affect our estimates below if they are normalized to the ANITA event rate.

We start with the same formula from [13] for the expected number $\mathcal{N}$ of the MMEs assuming that they are induced by the AQNs:

$$\mathcal{N} \approx \mathcal{A}_{\text{eff}} \mathcal{T} \Delta\Omega \frac{d\Phi}{dAd\Omega}, \tag{9}$$

where $\Delta\Omega \approx 2\pi$ for the isotropic AQN flux, and the expression for the local rate of upward-going AQNs per unit area is given by:

$$\frac{d\Phi}{dAd\Omega} \approx \frac{\Phi}{4\pi R_\oplus^2} \approx 4 \cdot 10^{-2} \left(\frac{10^{25}}{\langle B \rangle}\right) \frac{\text{events}}{\text{yr} \cdot \text{km}^2}, \tag{10}$$

$$\Phi \approx \frac{2 \cdot 10^7}{\text{yr}} \left(\frac{\rho_{DM}}{0.3\,\text{GeV cm}^{-3}}\right) \left(\frac{v_{AQN}}{220\,\text{km s}^{-1}}\right) \left(\frac{10^{25}}{\langle B \rangle}\right),$$

where ρ_{DM} is the local density of DM and $R_\oplus = 6371$ km is the radius of the Earth and Φ is the total hit rate of AQNs on Earth [48]. In Formula (9) the $\mathcal{T}$ is the time of operation while $\mathcal{A}_{\text{eff}}$ is the effective area to be estimated below.

The estimation of the effective area $\mathcal{A}_{\text{eff}}$ is a complicated task as it is not simply determined by the area of the detectors similar to standard CR analysis. Instead, it is determined by the area along the AQN's path where it can emit the bunches of electrons in form of short pulses which can mimic the CR as outlined in previous Section 4. The area $\mathcal{A}_{\text{eff}} \sim L_{AQN} \cdot D_{AQN}$ can be thought as a strip of length L_{AQN} and width D_{AQN}.

To estimate these parameters let us consider an AQN moving in upward direction with angle θ_{AQN} with respect to the Earth's surface such that upward velocity component is $(v_{AQN} \cdot \sin\theta_{AQN})$. This vertical velocity component determines the maximal height h of the atmosphere where the atmospheric density is still sufficiently high such that condition (8) holds. We estimate $h \approx (30\text{–}50)$ km where the atmospheric density falls by two order of magnitude such that conventional X-ray emission (which had been used in computations [43] to explain the seasonal variations observed by XMM Newton) becomes the dominant cooling process at higher altitudes. This condition determines the maximal length L_{AQN} and time scale τ_{AQN} when the AQNs can emit the bunches of electrons in form of pulses which we identify with MMEs and which can mimic the CR air showers as outlined in previous Section 4. Numerically these parameters are estimated as follows:

$$\tau_{AQN} \sim \frac{h}{(v_{AQN} \cdot \sin\theta_{AQN})} \sim 0.2\text{s}, \tag{11}$$

$$L_{AQN} \sim h \cot\theta_{AQN} \sim (50\text{–}90)\text{km},$$

where we used $\theta_{\mathrm{AQN}} \simeq 30°$ for an order of magnitude numerical estimates.

To estimate the width of the strip D_{AQN} we recall that the geomagnetic field $\mathcal{B}$ in location of the H10T instrument can be characterized by two numbers: it has strong component in up to down direction $\mathcal{B}_{\mathrm{down}} \simeq 0.49$ G, and strong component from south to north direction $\mathcal{B}_{\mathrm{north}} \simeq 0.24$ G, see e.g., [49].

Assuming that the energetic electrons with $\langle E \rangle \sim 10$ MeV will be emitted along the $\mathbf{v}_{\mathrm{AQN}}$-direction one can estimate instantaneous radius of curvature ρ, see e.g., Jackson [50]:

$$\rho \approx \frac{\gamma mc}{e\mathcal{B}\sin\theta_B} \approx 2.8\,\mathrm{km}\left(\frac{\gamma}{20}\right), \quad \gamma \equiv \frac{\langle E \rangle}{m} \tag{12}$$

where $\mathcal{B} \approx \mathcal{B}_{\mathrm{north}} \approx 0.24$ G is the local magnetic field strength, which changes the direction of the electrons from upward to downward moving such that they can mimic the CR air showers. The angle θ_B is the angle between the particle electron velocity $\mathbf{v}$ and magnetic direction. We choose $\theta_B \approx 30°$ in (12) for the numerical estimates.

Now we estimate the relevant scale λ which determines the survival pattern of the electron's bunches. It is mostly determined by the Coulomb elastic scattering with cross section σ_{Coul}

$$\sigma_{\mathrm{Coul}} \approx \frac{\alpha^2}{E^2\theta^4} \approx 3 \cdot 10^{-27}\left(\frac{1/2}{\theta}\right)^4\left(\frac{20}{\gamma}\right)^2 \mathrm{cm}^2. \tag{13}$$

The electrons with $\theta \gtrsim 1/2$ may strongly deviate from their main paths and cease to stay with majority of particles forming the bunch (which eventually becomes the pulse being interpreted as MME). The corresponding length scale $\lambda(h)$ at altitude h (accounting for the air-density $n_{\mathrm{air}}(h)$ variation) is estimated as follows

$$\lambda(h) \sim \frac{1}{\sigma_{\mathrm{Coul}}n_{\mathrm{air}}(h)} \sim 3 \cdot \exp\left(\frac{h}{8\,\mathrm{km}}\right)\mathrm{km}. \tag{14}$$

The physical meaning of λ is the length distance particles propagate at which the majority of the particles in the bunch remain within the bunch before the dispersing to much larger distances when they cease to be a part of the pulse (burst). Important comment here is that $\lambda \gtrsim \rho$ such that the majority of electrons within the bunch do not strongly re-scatter, and therefore survive the change in orientation due to the geomagnetic field $\mathcal{B}$ (from upward to downward direction).

The width of the strip D_{AQN} now can be roughly estimated as follows:

$$D_{\mathrm{AQN}} \sim \lambda \sim 3 \cdot \exp\left(\frac{h}{8\,\mathrm{km}}\right) \cdot \left(\frac{\gamma}{20}\right)\mathrm{km}. \tag{15}$$

Combining all the estimates above one arrives to the following order of magnitude estimation for the expected number of MME events:

$$\mathcal{N} \approx 0.5 \cdot 10^2 \mathrm{events}\left(\frac{\mathcal{A}_{\mathrm{eff}}}{500\,\mathrm{km}^2}\right)\left(\frac{\mathcal{T}}{0.4\,\mathrm{yr}}\right). \tag{16}$$

This estimate carries enormous internal uncertainty due to a number of reasons. First, as we already mentioned the parameters entering (10) for the basic normalization, the flux Φ, are in fact not precisely known. Indeed, the fundamental parameter such as ρ_{DM} may dramatically deviate locally (by factor 2 or even 3) from the well established average global value $\rho_{\mathrm{DM}} \approx 0.3\,\mathrm{GeV}\,\mathrm{cm}^{-3}$. Furthermore, the size distribution factor entering (10) in form $\langle B \rangle^{-1}$ had been fixed from dramatically different physics (including solar corona heating puzzle) and can easily deviate by similar factor 2 or 3. Furthermore, the effective area $\mathcal{A}_{\mathrm{eff}}$ could also deviate by factor 2 or 3 in comparison with expression (16) because the important parameter h which essentially determines the effective area $\mathcal{A}_{\mathrm{eff}}$ cannot be precisely evaluated as it depends on condition (8) which itself is determined by complex internal dynamics of the nuggets. As a result we consider the Formula (16) as an order of magnitude estimate at the very best. However, our main arguments of the identification of the MMEs with the AQN events are not based on the estimate (16). We present this estimate here exclusively for illustrative purposes. Our main arguments are based on specific qualitative features (formulated as *puzzles*) which have been listed in Section 1, and which cannot be explained in terms of the conventional CR air showers as highlighted in Section 2.

This order of magnitude estimate should be compared with observed frequency of appearance of the MMEs. According to [2] the observations from start until August 2016 during $\mathcal{T} \approx 3500$ h ≈ 0.4 yr of operation the H10T instrument had detected $\mathcal{N}_{\mathrm{obs}} \simeq 10^3$ MMEs. According to [4] the observations from 15 February 2018 to 12 May 2018 (which corresponds to $\mathcal{T} \approx 0.25$ yr) the H10T instrument had detected $\mathcal{N}_{\mathrm{obs}} = 217$ MMEs.

The AQN based estimations (16), if literally taken, suggest that number of events $\mathcal{N}$ is almost one order of magnitude lower than $\mathcal{N}_{\mathrm{obs}}$ events observed by H10T. Nevertheless, we consider this order of magnitude estimation (16) being consistent with our proposal due to many uncertainties mentioned above.

It is interesting to note that the numerical suppression factor $\sim$0.1 (between computed $\mathcal{N}$ and observed $\mathcal{N}_{\mathrm{obs}}$ values) which appears in our order of magnitude estimates (16) is very similar to suppression factor $\sim$0.1 which occurred in analogous estimates for the mysterious TA bursts [9] and the ANITA anomalous events [13]. This similarity hints on a common origin for all these phenomena. Therefore, if one normalize $\mathcal{N}$ to the observed ANITA anomalous events (assuming that all three phenomena originated from the same AQN based physics) one arrives to correct order of magnitude estimate as the suppression factor $\sim$0.1 is approximately the same for all three phenomena estimated for the mysterious TA bursts in [9], the ANITA anomalous events in [13], and Multi-Modal Events estimated in (16). The main uncertainty related to the basic normalization of the flux Φ cancels out if such relative ratios are considered.

As we mentioned above, our main arguments leading to the identification of the MMEs with the AQN events are not based on an order of magnitude estimate of the frequency of appearance (16). Rather, our main arguments are based on specific qualitative features (formulated as *puzzles*) and which cannot be explained in terms of the conventional CR air showers. We consider a qualitative agreement between the observations and our theoretical estimates (to be discussed in next Section 6) as the strong arguments supporting our identification.

6. AQN Proposal Confronts the MME Observations

We start this section by estimating the particle number density $\rho(R)$ which appears in formulation of the *puzzles* in Section 2 describing the observations [1–5]. The particle number density in the AQN framework is determined by the number of particles which could be observed at the detector site. Assuming that the total number of electrons being emitted in form of a pulse (as a result of erupted burst) is similar to the number of particles which was used in our analysis [13] of the ANITA anomalous events $N \simeq (10^8$–$10^9)$ we arrive to the following estimate for $\rho(R)$:

$$\rho(R) \sim \frac{N}{(\lambda \Delta \theta)^2} \sim \frac{(10^8 - 10^9)}{(2.5 \text{ km})^2} \sim \frac{(20 - 200)}{\text{m}^2} \tag{17}$$

where the area which is hit by the bunch of particles from a single burst is estimated as $(\lambda \Delta \theta)^2$ with $\lambda(h)$ given by (14) at the detector site with $h \simeq 4$ km. It is assumed that the particles are propagating within the cone with angle $\Delta \theta \lesssim 1/2$.

Our next task is the estimation of the time delay between the pulses within the same cluster. The corresponding time delay can be estimated from the condition that the energy deposited into the AQN during time τ_{delay} must be released in form of the eruption when $N \simeq (10^8$–$10^9)$ electrons being emitted in form of a single pulse. As we mentioned previously, the dominant portion of the energy in such eruption is released in form of the photons with energy $\omega \sim T$ which always accompany the electron's emission as a result of these bursts. The relative ratio r between these two components is estimated in Appendix B, see Equation (A14). Numerically, the estimate can be presented as follows:

$$r \equiv \frac{\langle E_{e^+e^-} \rangle}{\langle E_\gamma \rangle} \sim \frac{\alpha}{2\pi} \exp\left(-\frac{2m}{T}\right) \sim 0.7 \cdot 10^{-5}, \tag{18}$$

where for numerical estimates we use $T \simeq 200$ keV. We must emphasize that (18) is really an order of magnitude estimate as mechanism of emission is determined by a very complex physics as highlighted in Appendix B. Furthermore, the corresponding eruption event is not a thermodynamically equilibrium process as we already mentioned previously. The dominant contribution in form of the photon's emission during a single pulse can be estimated in terms of these parameters as follows

$$\langle E_\gamma \rangle \sim N \cdot (2m) \cdot r^{-1} \sim 1.4 \cdot (10^{10} - 10^{11}) \text{ GeV}, \tag{19}$$

where we use the numerical value for $N \in (10^8 - 10^9)$ extracted from analysis of the ANITA anomalous events [13] and parameter r is given by (18). In this estimate we assumed that the energy of the pair $\langle E_{e^+e^-} \rangle \sim 2mN$ computed at the moment of the pair creation. The released energy during a single pulse (19) determines the time delay τ_{delay} between the pulses within the same cluster:

$$\tau_{\text{delay}}^{-1} \sim \frac{dE_{\text{deposit}}/dt}{\langle E_\gamma \rangle} \sim \kappa \frac{10^{18} \cdot \text{GeVs}^{-1}}{(10^{10} - 10^{11}) \text{ GeV}}, \tag{20}$$

where we used numerical value for (dE_{deposit}/dt) from (6) with density of atmosphere at $h \simeq 4$ km where the detector H10T is located. Our final, the order of magnitude estimate for the time delay τ_{delay} between the pulses can be represented as follows

$$\tau_{\text{delay}} \sim \left(\frac{0.1}{\kappa}\right) \cdot \left(10^2 - 10^3\right) \text{ns}, \tag{21}$$

where we use $\kappa \simeq 0.1$ in (21) for numerical estimates which was previously extracted from studies of the emission in the dilute galactic environment, see footnote 2.

We would like to make few comments on uncertainties related to the estimate (21). The main uncertainties are related to the parameters r and N entering (19). The parameter r as defined by (18) is very hard to estimate as discussed in Appendix B where we mentioned that the related question for much simpler problem on emission from the bare quark stars remains to be a matter of debates. Our system is much more complex than the bare quark star because of the small size of the nuggets when the thermal equilibrium cannot be established, in contrast with quark stars. This is precisely the reason why we use parameter N from analysis of the ANITA anomalous events [13], rather than from theoretical estimates. Another source of uncertainty is parameter κ entering (21). This parameter was defined in Equation (3) with explanation of many uncertainties related to its numerical value[6]. Therefore, the time delay τ_{delay} as given by (21) should be considered as an order of magnitude estimate at the very best. It obviously cannot be computed from the first principles similar to any other parameters when one deals with any complex systems, see also footnote 6.

However, the important point here is that the time scale τ_{delay} may dramatically vary from event to event as it strongly depends on many parameters entering the problem as mentioned above. In particular, it obviously strongly depends on intensity (number of emitted particles N) of a previous pulse. Essentially the parameter (21) should be interpreted as a preparation time the system requires for the next eruption.

The strong variation of the τ_{delay} should be contrasted with another parameter which is the time duration of a single pulse τ_{pulse}. In the AQN framework this parameter must not vary much from one event to another as it is entirely determined by internal dynamics of the AQN during the eruption irrespective to the intensity of the previous events, and irrespective to the prehistory of the AQN propagation as long as condition (8) is met[7]. In different words, our proposal suggests that

$$\tau_{\text{pulse}} \approx \text{constant}, \tag{22}$$

where the constant cannot be computed from first principles as it is entirely determined by the non-equilibrium dynamics of the AQN at the instant of eruption[8].

Now we are prepared to confront the basic consequences of our proposal (identifying the MMEs with the AQN annihilation events) with observations.

1 . *"clustering puzzle"*: The multiple number of events is a very generic feature of the system as explained at the very end of Section 3.3 as long as condition (8) is met. The time delays between the pulses τ_{delay} dramatically fluctuate between different events. The window for these variations is huge according to (21). In fact, it could be even well outside of this window. The time delays τ_{delay} may considerably vary even for different events within the same cluster at the same detection point, which is consistent with observed cluster shown on Figure 1. Furthermore, the AQN itself remains almost at the same location as the displacement Δl_{AQN} during entire cluster of events is very tiny

$$\Delta l_{\text{AQN}} \sim v_{\text{AQN}} \cdot \tau_{\text{delay}} \sim 20 \text{ cm}, \tag{23}$$

which implies that all individual bunches making the cluster are likely to be emitted by the AQN along the same direction, and can be recorded and classified as MME by H10T. Each event can be viewed as an approximately uniform front as mere notion of the "EAS axis" does not exist in this framework, see also next item. However, each individual event may appear to arrive from slightly different direction due to the inherent spread of the emitted electrons at the moment of eruption.

2. *"particle density puzzle"*: Particle density distribution $\rho(R)$ in the AQN framework is estimated by (17) and it shows strong fluctuation from one event to another event. These variations are mostly related to the intensity of the individual bursts being expressed by the number of electrons in the bunch N. However, the distinct feature of the distribution in the AQN framework is that it does not depend on R, as the notion of the "EAS axis" does not exist in this framework as we already mentioned. All these generic features of the AQN framework are perfectly consistent with H10T observations as presented on Figure 2. However, these observed features are in dramatic contradiction with conventional EAS prediction shown by solid lines on Figure 2 with different colours, depending on energy of the CR.

Furthermore, the magnitude of the density $\rho(R)$ in the AQN framework (which is mostly determined by parameter N) is much higher than one normally expects for CR energies $\sim(10^{17}$–$10^{18})$ eV. The corresponding parameter N representing the number of electrons in the bunch was not fitted for the present studies to match the observations. Instead, it was extracted from different experiment in dramatically different circumstances (in proposal [13] to explain the ANITA anomalous events as the AQN events).

3. *"pulse width puzzle"*: In the AQN framework the width of the pulse (22) cannot vary much from one event to another. It is a fundamental feature of the framework because the duration of the pulse is entirely determined by internal dynamics of the AQN during the blast as long as condition (8) is met, see footnote 7 with a comment. This feature is in perfect agreement with observations [4]

$$\tau_{\text{pulse}} \approx (20\text{–}35) \text{ ns} \quad \Leftarrow \quad [\text{observations}] \tag{24}$$

for all recorded MMEs. At the same time, this feature is in dramatic conflict with conventional picture when the duration of the pulse must depend on the distance to the EAS axis as shown by sold lines on Figure 3. This basic prediction of the conventional CR analysis is due to increase of the thickness of the EAS pancake with the distance from the EAS axis.

As we already mentioned the mere notions such as the "EAS axis" and the "thickness of the EAS pancake" do not exist in the AQN framework.

4. *"intensity puzzle"*: Particle density distribution $\rho(R)$ in the AQN framework is estimated by (17). The corresponding event to event fluctuations do not depend on the distance to the EAS axis as we already mentioned. Such intensity of the events as given by (17) in the AQN framework is consistent with observations shown on Figure 2. However, the observations are in dramatic conflict with conventional CR analysis when such huge intensity could be generated by a primary particle with energy well above 10^{19} eV with dramatically lower event rate on the level of once every few years. The frequency of appearance in the AQN framework is estimated in Section 5, and it is consistent with observed event rate.

We conclude this section with the following comment. All our formulae presented in this section are the order of magnitude estimations at the very best, as they include many inherent uncertainties which are inevitable features of any composite system (such as AQN) propagating in a complex environment (such as Earth atmosphere) with very large Mach number M, see footnote 6.

Nevertheless, the emergent picture suggests that all the *puzzles* formulated in Sections 1 and 2 can be naturally understood within the AQN framework as explained above in items **1–4**. Needless to say that the crucial phenomenological parameters used in the estimates had been fixed long ago for dramatically different observations in different circumstances for different environments as overviewed in Section 3.

7. Conclusions and Future Development

Our basic results can be summarized as follows. We have argued that all the *puzzles* formulated in Sections 1 and 2 can be naturally understood within the AQN framework as explained in items **1–4** in previous Section, and we do not need to repeat these arguments again in this Conclusion. Instead, we want to discuss the drastic differences between the events induced by conventional CR showers and the AQNs. These dramatic distinct features can be tested by the future experiments, such that our proposal can be discriminated from any other proposals and suggestions. We list below the following typical features of the AQN events and contrast them with any other possible mechanisms which could be responsible for MMEs.

1.   The events which are generated by the bunches of electrons as a result of eruption of the propagating AQN in the Earth atmosphere suggests an enormous number of possibilities to generate different clusters when each event within a given cluster may have very different intensity from a previous and consequent events with very different time delays between the events. In other words, the AQN proposal suggests that there should be large variety of shapes and delays between the events with very different patterns due to the complexity of the AQN system. It should be contrasted, for example, with hypothesis of "delayed particles" (which was originally suggested to explain the MMEs) in which case all clusters must be the same as they should be determined by a specific pattern of decaying fundamental particle of unknown nature.

2.   A "rule of thumb" suggests that a typical number of charged particles (mostly electrons and positrons) in CR air shower is E_{CR}/GeV, which implies that $N \approx (10^8\text{–}10^9)$ for the energy of the primary particle $E_{CR} \approx (10^{17}\text{–}10^{18})$ eV. This estimate suggests that any detector which is designed to study the EAS with energies $E_{CR} \approx (10^{17}\text{–}10^{18})$ eV are, in principle, capable to study MMEs if the resolution of the detectors is in $\sim$10-ns level, similar to H10T, see also item 4 below as an alternative option to properly select and discriminate the MMEs.

3.   In particular, we expect that the extension of the H10T detector would produce more multiple pulses (at each given detector) instead of simple bimodal pulses. We also expect that more detectors in the area will be recording MMEs because the area covered by each individual pulse is relatively large (few kilometres) according to (17), which is well above the present size of H10T instrument.

4. A large number of charged particles $N \approx (10^8\text{--}10^9)$ in the background of the geomagnetic field $\mathcal{B} \sim 0.5$ gauss will produce the radio pulse in both cases: the CR-induced radio pulse [51,52] as well as AQN-induced radio pulse [13]. However, these pulses can be easily discriminated from each other as argued in [13].

The main reason for the dramatic differences between these two radio pulses is that the AQN event could be viewed as an (approximately) uniform front of size $\lambda \Delta \theta \sim$ km with a constant width, while EAS is characterized by central axis. In different words, the number of particles per unit area $\rho(R)$ in the AQN case does not depend on the distance from the central axis, in huge contrast with conventional CR air showers when $\rho(R)$ strongly depends the distance from the central axis. The width of the "pancake" in CR air shower also strongly depends on R. As a result, the effective number of coherent particles contributing to the radio pulse is highly sensitive to the width of the "pancake" when it becomes close to the wavelength of the radio pulse. These distinct features lead to very different spectral properties of the radio pulses in these two cases, which can be viewed as an independent characteristic of MMEs. In fact, this unique feature can be used in future studies for purpose of discrimination and proper selection of the Multi-Modal Clustering Events.

If future studies and tests (including the detecting of the synchronized radio pulses with MMEs as suggested above) indeed substantiate our proposal it would be a strong argument supporting the AQN nature of the MMEs.

We conclude this work with the following comment. We estimated the event rates for three dramatically different puzzling CR-like events: for mysterious TA bursts [9], for the ANITA anomalous events [13], and finally for the Multi-Modal Events in this work as estimated in (16). All three puzzling phenomena are proportional to one and the same AQN flux (10). The self-consistency between all three estimates hints on a common nature for these puzzling CR-like events. We interpret this self-consistency in the event rates as an additional indirect argument supporting the AQN nature for all three mysterious phenomena, while our direct arguments are presented in Section 6 and listed as item **1–4**. We finish on this optimistic note.

Funding: The APC was funded by the Natural Sciences and Engineering Research Council of Canada.

Acknowledgments: I am very thankful to Vladimir Shiltsev of FNAL who brought to my attention the original references [1–5] and asked me many questions on possible nature of MMEs, which essentially motivated and initiated this work.

Conflicts of Interest: The author declares no conflict of interest.

Appendix A. On Suppression of Emissivity Due to the Ionization

The main goal of this section is to produce simple estimates of the suppression parameter $\eta(T, R)$ which enters the cooling rate (4) in case of high temperature and large ionization. It was introduced in [43] as a suppression parameter accounting for the ionization of the AQN after it travelled through the Earth. It was also shown that $\eta(T, R) \sim 10^{-6}$ produces very reasonable results consistent with intensity of the radiation observed by XMM-*Newton* [42].

The original computations of the Bremsstrahlung radiation were performed in [45] in case of low temperature and low ionization as given by (2). The key element of the computations was the electrosphere density

$$n(z) = \frac{T}{2\pi\alpha} \frac{1}{(z + \bar{z})^2}, \tag{A1}$$

with

$$\bar{z}^{-1} = \sqrt{2\pi\alpha} \cdot m \cdot \left(\frac{T}{m}\right)^{1/4}, \quad n(z = 0) \simeq (mT)^{3/2},$$

where $n(z)$ is the positron number density of the electrosphere, $z = 0$ corresponds to the surface of the nuggets, and $n(z = 0)$ reproduces an approximate formula for the plasma density in the Boltzmann regime at the temperature T as computed in [45]. The 1d approximation formulated in terms of distance z from the surface was more than sufficient sufficient to study the low temperature behaviour.

However, after an AQN crosses the Earth, it acquires a very high internal temperature of the order $T \simeq (100\text{–}200)$ keV, and approximation (A1) is not sufficient. This is because temperature's rise will cause the electrosphere to expand well beyond the thin layer $\sim \bar{z}$ surrounding the nugget surface. Some positrons will leave the system but the majority will stay in close vicinity of the moving, negatively charged nugget core. Consequently, the nugget will acquire a negative charge of approximately $-|e|Q$ with the number of positrons Q estimated as:

$$Q \simeq 4\pi R^2 \int_0^\infty n(z)dz \sim \frac{4\pi R^2}{\sqrt{2\pi\alpha}} \cdot (mT) \cdot \left(\frac{T}{m}\right)^{1/4}. \tag{A2}$$

The distance ρ at which the positrons remain attached to the nugget is given by the capture radius $R_{\text{cap}}(T)$, determined by the Coulomb attraction:

$$\frac{\alpha Q(\rho)}{\rho} > \frac{mv^2}{2} \approx T \quad \text{for} \quad \rho \lesssim R_{\text{cap}}(T). \tag{A3}$$

In order to estimate η entering (4) we start our analysis with an estimation of the positron density $n(\rho, T)$ when the AQNs enter the Earth atmosphere moving in upward direction. Formula (A3) shows that the capture radius $R_{\text{cap}}(T)$ could be much larger than R, i.e., we have $R \lesssim \rho \lesssim R_{\text{cap}}$. The expression (A1), which is valid when $z = |\rho - R| \ll R$, in close vicinity to the core, breaks down for $\rho \gtrsim R$. In that case the curvature of the nugget surface cannot be neglected and one should use a truly 3-dimensional formula for $n(\rho, T)$ instead of the 1-dimensional approximation (A1). We will assume that the density $n(\rho, T)$ behaves as a power law for $\rho \gtrsim R$ with exponent p:

$$n(\rho, T) \simeq n_0(T)\left(\frac{R}{\rho}\right)^p, \quad \rho \gtrsim R, \tag{A4}$$

where $n_0(T) \equiv n(\rho = R, T)$ is a normalization factor and p a free parameter. Equation (A4) is consistent with our previous numerical studies [53] of the electrosphere with $p \simeq 6$. It is also consistent with the conventional Thomas-Fermi model[9] at $T = 0$ [54]. While keeping p as a free parameter, we will show below that our main claim is not very sensitive to the precise value of p. The normalization $n_0(T)$ can be estimated from the condition that the finite portion of the positrons satisfying Equation (A3) is such that the total number of positrons surrounding the nugget is approximately equal to the ionization charge Q determined by (A2). Therefore, we arrive to the following expression for the normalization factor $n_0(T)$:

$$4\pi \int_R^\infty \rho^2 d\rho \, n(\rho, T) \simeq Q(T), \tag{A5}$$

Note that the integrant $\rho^2 n(\rho, T)$ is mostly dominated by the inner shells $\rho \sim R$ such that the intgration can be extended to infinity with very high accuracy, instead of cutting off at R_{cap}. The resulting estimate for $n_0(T)$ assumes the form:

$$n_0(T) \simeq \frac{(mT)}{\sqrt{2\pi\alpha}} \cdot \frac{(p-3)}{R} \cdot \left(\frac{T}{m}\right)^{1/4}, \tag{A6}$$

which, as expected, is smaller than $n(z = 0) \simeq (mT)^{3/2}$ because the positrons now are distributed over a distance of order R from the nugget core surface, rather than over a distance of order $\bar{z}$. This suppression is convenient to represent in terms of the dimensionless ratio:

$$\frac{n_0(T)}{(mT)^{3/2}} \simeq \frac{(p-3)}{\sqrt{2\pi\alpha}} \cdot \frac{1}{(mR)} \cdot \left(\frac{m}{T}\right)^{1/4} \ll 1. \tag{A7}$$

The next step is the calculation of the Bremsstrahlung emissivity by the nuggets due to the strong suppression of the plasma density (A7) as a result of the electrosphere's expansion. The spectral surface emissivity is denoted as $dF/d\omega = dE/dt\,dA\,d\omega$, representing the energy emitted by a single nugget per unit time, per unit area of nugget surface and per unit frequency. For low temperature, when approximation (A1) for the flat geometry is justified, the corresponding expression assumes the form [45]:

$$\frac{dF}{d\omega}(\omega, T) = \frac{1}{2} \int_0^\infty dz\, n^2(z)\, \mathcal{K}(\omega, T) \tag{A8}$$

where $n(z)$ is the local density of positrons at distance z from quark nugget surface and function $\mathcal{K}(\omega)$ does not depend on z and describes the spectral dependence of the system:

$$\mathcal{K}(\omega, T) = \frac{4\alpha}{15}\left(\frac{\alpha}{m}\right)^2 2\sqrt{\frac{2T}{m\pi}}\left(1 + \frac{\omega}{T}\right)e^{-\omega/T}h\left(\frac{\omega}{T}\right). \tag{A9}$$

The dimensionless function $h\left(\frac{\omega}{T}\right)$ in (A9) is a slowly varying logarithmic function of ω/T which was explicitly computed in [45]. In order to calculate the emissivity with the positron density (A4), when the nugget core is *ionized*, we have to replace the integral $\int dz$ in (A8) by the spherical integration over ρ:

$$4\pi R^2 \int_0^\infty dz\, n^2(z) \;\;\Rightarrow\;\; 4\pi n_0^2(T) \int_R^\infty \rho^2 d\rho \left(\frac{R}{\rho}\right)^{2p} \tag{A10}$$

where the normalization $n_0(T)$ is determined by (A6). Using the positrons distributed according to Equation (A4) the nugget surface emissivity can be calculated as follows:

$$\frac{dF}{d\omega}(\omega, T) = \frac{\mathcal{K}(\omega, T)}{2R}\frac{(mT)^2}{2\pi\alpha}\left[\frac{(p-3)^2}{2p-3}\right]\left(\frac{T}{m}\right)^{\frac{1}{2}}. \tag{A11}$$

The key result here is the strong suppression factor R^{-1} which was not present in the no-ionization case (A8) when the positrons are localized in a thin layer. It is instructive to compare the emissivity given by Equation (A11) accounting for ionization of the nugget core with the original Formula (A8). The dimensionless ratio is given by

$$\eta \equiv \frac{\frac{dF}{d\omega}^{(\text{ion})}}{\frac{dF}{d\omega}^{(\text{no ion})}} = \frac{\left[\frac{(p-3)^2}{2p-3}\right]}{3(mR)}\frac{1}{\sqrt{2\pi\alpha}}\left(\frac{m}{T}\right)^{\frac{1}{4}} \sim 10^{-6}. \tag{A12}$$

The total emissivity integrated over all frequencies in case of low temperature without ionization has been computed in [45] and it is given by (2). In case of strong ionization, Equation (A12) implies:

$$F_{\text{tot}}^{(\text{ion})}(T) = \eta(T, R) \cdot F_{\text{tot}}^{(\text{no ion})}(T). \tag{A13}$$

The critical element leading to the suppression factor (A12) is the very small quantity $(mR)^{-1} \sim 10^{-6}$. It is a direct result of the expanding electrosphere and high level of ionization at very high temperature T.

The expression (A13) with numerical suppression (A12) is precisely the Formula (4) which we used in main body of the text to proceed with our key arguments.

Appendix B. On e^+e^- Emission at High Temperature in High Density QCD Phases

The main goal of this Appendix is to overview the photon emission and the e^+e^- production at high temperature $T \gtrsim 10^2$ keV from high density QCD phases. The corresponding studies [55–61] have been carried out in the past in context of the quark stars. In context of the present work all the key ingredients relevant for e^+e^- production are also present in the system. Indeed, the AQN is characterized by very high temperature $T \gtrsim 10^2$ keV, the quark core is assumed to be in CS dense phase, and a strong internal electric field is also present in the system. However, we cannot literally use the results from the previous studies obtained in context of the quark stars because the size of the AQN is much smaller than the relevant mean free paths for all elementary processes as discussed in details in [13]. As a result the thermal equilibration cannot be achieved in the AQN system, and entire physics is determined by non-equilibrium dynamics in high temperature regime. It should be contrasted with large size quark stars where the thermal equilibrium is maintained.

Nevertheless, it is very instructive to review the relevant results from the previous studies [55–61] on quark stars due to the following reasons. First, it explicitly shows the role of the main ingredients of the system, such as temperature and the large electric field. Secondly, it demonstrates the complexity of the problem when even a much simpler case of the bare quark star remains to be a matter of debates.

The idea on possibility of the e^+e^- emission at high temperature from quark stars was originally suggested in [55,56]. The temperature range considered in [55,56] includes the typical temperatures $T \gtrsim 10^2$ keV which is expected to occur in our case when the AQN exits the Earth's surface as mentioned in Section 3.3. In refs. [57,58] the authors argue that bremsstrahlung radiation from the electrosphere could be much more important than e^+e^- emission. It has been also argued that a number of effects such as the boundary effects, inhomogeneity of the electric field, and the Landau-Pomeranchuk-Migdal (LPM) suppression may dramatically modify the emission rate. In [59] it has been argued that the Pauli blocking will strongly suppress the bremsstrahlung emission. Finally, in refs. [60,61] it has been argued that the so-called mean field bremsstrahlung could be the dominant mechanism.

It is not the goal of the present work to critically analyze all these suggested mechanisms of the emission. Rather, our goal is to demonstrate that even a relatively simple system of the bare quark star remains to be a matter of debates. Our case of the emission from AQN at high temperature is even more complicated as it is determined by non-equilibrium dynamics.

In the present work we do not even attempt to solve this very ambitious problem of estimating the absolute intensity rate of the e^+e^- and γ emissions. Rather, the absolute number $N \simeq (10^8$–$10^9)$ of the e^+e^- pairs in a single pulse was extracted from ANITA anomalous events as explained in Section 6. In particular, this number N enters the estimate (17) for the density of the observed particles by H10T experiment [1–4]. Our goal here is much less ambitious as we try to estimate the relative ratio r between the e^+e^- pair production and the γ radiation assuming that the emission of both components is mostly originated from the region in electrosphere where the Boltzmann regime is justified, and the Pauli blocking suppression is not very dramatic and can be ignored.

In this case, the relative ratio r between these two components contributing to the emission can be estimated as follows. It is assumed that the dominant contribution to the γ emission comes from the Bremsstrahlung radiation ($e^+e^+ \to e^+e^+ + \gamma$) resembling the low temperature case considered in [45]. The e^+e^- pair is produced by a similar mechanism through the virtual photon ($e^+e^+ \to e^+e^+ + \gamma^*$) which consequently decays to the pair: $\gamma^* \to e^+e^-$. If this is the dominant mechanism of radiation the pair production is expected to be suppressed by a factor $\alpha/2\pi$ as a result of conversion of the virtual photon $\gamma^* \to e^+e^-$

to pair[10]. It must be also suppressed by a factor $\exp(-2m/T)$ describing the suppression in the density distribution because the virtual photons must have sufficient energy $\gtrsim 2m$ to produce e^+e^- pair. This oversimplified estimate leads to the following expression

$$r \equiv \frac{\langle E_{e^+e^-}\rangle}{\langle E_\gamma\rangle} \sim \frac{\alpha}{2\pi}\exp\left(-\frac{2m}{T}\right) \sim 0.7\cdot 10^{-5}, \tag{A14}$$

where for numerical estimates we use $T \simeq 200$ keV, and ignored a possible complicated function which could depend on parameter $T/m \sim 0.4$, which is order of unity for our case. We must emphasize that (A14) is really an order of magnitude estimate as mechanism of emission is not a thermodynamically equilibrium process as we already previously mentioned. Formula (A14) is used in the main body of the text in (18) for our estimate (21) for τ_{delay} which is measured by H10T experiment [1–5] as it describes a typical time delay between subsequent pulses representing the Multi-Modal Clustering Events.

Notes

1.   Non-detection of etching tracks in ancient mica gives another indirect constraint on the flux of dark matter nuggets with mass $M < 55$ g [40]. This constraint is based on assumption that all nuggets have the same mass, which is not the case as we discuss below. The nuggets with small masses represent a tiny portion of all nuggets in this model, such that this constraint is easily satisfied with any reasonable nugget's size distribution.

2.   In a neutral dilute galactic environment considered previously [45] the value of κ was estimated as $\kappa \approx 0.1$.

3.   We use the term "additional" to emphasize that the AQN had accumulated a huge amount of energy during the transpassing the Earth interior as the capacity of the quark core nugget is very large [43]. Precisely this accumulated energy will be emitted in form of the X rays when the AQN propagates in empty space at distances $r \sim (6-10)R_\oplus$ from the Earth surface.

4.   For example, the AQNs moving in empty space and slowly cooling by emitting the X rays as computed in [43] belongs to this class.

5.   An analogy for such eruption-like event is the lightning flash under thunderclouds. The clouds accumulate the electric charge in form of the ionized molecule very efficiently. The corresponding time scales plays the role of $\tau_{deposit}$ in our system. The neutralization of these ions is less efficient process, which is analogous to our τ_{cool}. If some conditions are met (the so-called runaway breakdown conditions are satisfied, see [46,47] for review) the discharge occurs in form of the eruption which is the lightning strike in our analogy. The system is getting neutralized in form of the non- equilibrium lightning event (eruption) on the time scales which are much shorter than any time scales of the problem. This analogy is in fact, quite deep, and will be used in Section 6 when we discuss different time scales of MMEs as observed by H10T detector.

6.   One should keep in mind that the AQN propagates in atmosphere with very large velocity, much larger than the speed of sound c_s such that the Mach number $M \equiv v_{\text{AQN}}/c_s \gg 1$. This obviously implies the presence of the turbulence with accompanying shock waves. The corresponding effects which are hard to compute from first principles obviously modify the value of κ and, as a consequence, the corresponding estimate for the time delay τ_{delay}.

7.   This approximately constant parameter τ_{pulse} can be understood using the analogy with lightnings mentioned in footnote 5. Indeed, the time scale between lightning flashes may dramatically vary (measured in minutes) during the same thunderstorm while the lightning strikes themselves are much shorter and characterized approximately by the same duration (measured in ms).

8.   The analogy with lightnings mentioned in footnotes 5 and 7 can be useful here: the instability in form of the runaway breakdown mechanism in lightnings which is responsible for the flashes is similar to non-equilibrium dynamics of the AQN at the instant of eruption. Furthermore, in both cases there must be some trigger which initiates the eruption. In case of lightning events the trigger is thought to be related to the cosmic rays, though this element remains a part of controversy, see [46,47] for review. In case of the AQN eruption events such triggers could be several consequent successful annihilation events with atmospheric material.

9.   In [54], the dimensionless function $\chi(x)$ behaves as $\chi \sim x^{-3}$ at large x. The potential $\phi = \chi(x)/x$ behaves as $\phi \sim x^{-4}$. The density of electrons in Thomas-Fermi model scales as $n \sim \phi^{3/2} \sim x^{-6}$ at large x.

10.   The cross section for pair production as a result of collisions of two particles is known exactly [62]. It contains, of course factor $\alpha/2\pi$ entering (A14) along with many other numerical factors reflecting a complex kinematics of the process.

References

1. Beznosko, D.; Beisembaev, R.U.; Baigarin, K.A.; Batyrkhanov, A.; Beisembaeva, E.A.; Beremkulov, T.; Dalkarov, O.D.; Iakovlev, A.; Ryabov, V.A.; Suleimenov, N.S.; et al. Horizon-T experiment and detection of Extensive air showers with unusual structure. *J. Phys. Conf. Ser.* **2020**, *1342*, 012007. [CrossRef]
2. Beznosko, D.; Beisembaev, R.; Baigarin, K.; Beisembaeva, E.; Dalkarov, O.; Ryabov, V.; Sadykov, T.; Shaulov, S.; Stepanov, A.; Vildanova, M.; et al. Extensive Air Showers with unusual structure. *Eur. Phys. J. Web Conf.* **2017**, *145*, 14001. [CrossRef]
3. Beisembaev, R.; Beznosko, D.; Baigarin, K.; Batyrkhanov, A.; Beisembaeva, E.; Dalkarov, O.; Iakovlev, A.; Ryabov, V.; Sadykov, T.; Shaulov, S.; et al. Extensive Air Showers with Unusual Spatial and Temporal Structure. *Eur. Phys. J. Web Conf.* **2019**, *208*, 06002. [CrossRef]
4. Beznosko, D.; Beisembaev, R.; Beisembaeva, E.; Dalkarov, O.D.; Mossunov, V.; Ryabov, V.; Shaulov, S.; Vildanova, M.; Zhukov, V.; Baigarin, K.; et al. Spatial and Temporal Characteristics of EAS with Delayed Particles. In Proceedings of the ICRC 2019, Madison, WI, USA, 24 July–1 August 2019; p. 195. [CrossRef]
5. Beisembaev, R.U.; Beisembaeva, E.A.; Dalkarov, O.D.; Mosunov, V.D.; Ryabov, V.A.; Shaulov, S.B.; Vildanova, M.I.; Zhukov, V.V.; Baigarin, K.A.; Beznosko, D.; et al. Unusual Time Structure of Extensive Air Showers at Energies Exceeding 10^{17} eV. *Phys. Atom. Nucl.* **2019**, *82*, 330–333. [CrossRef]
6. Zhitnitsky, A.R. 'Nonbaryonic' dark matter as baryonic colour superconductor. *JCAP* **2003**, *10*, 010. [CrossRef]
7. Jelley, J.V.; Whitehouse, W.J. The Time Distribution of Delayed Particles in Extensive Air Showers using a Liquid Scintillation Counter of Large Area. *Proc. Phys. Soc. Sect. A* **1953**, *66*, 454–466. [CrossRef]
8. Linsley, J.; Scarsi, L. Arrival Times of Air Shower Particles at Large Distances from the Axis. *Phys. Rev.* **1962**, *128*, 2384–2392. [CrossRef]
9. Zhitnitsky, A. The Mysterious Bursts observed by Telescope Array and Axion Quark Nuggets. *J. Phys. G Nucl. Part. Phys.* **2020**, *48*, 065201. [CrossRef]
10. Liang, X.; Zhitnitsky, A. Telescope Array Bursts, Radio Pulses and Axion Quark Nuggets. *arXiv* **2021**, arXiv:2101.01722.
11. Abbasi, R.U.; Abe, M.; Abu-Zayyad, T.; Allen, M.; Anderson, R.; Azuma, R.; Barcikowski, E.; Belz, J.W.; Bergman, D.R.; Blake, S.A.; et al. The bursts of high energy events observed by the telescope array surface detector. *Phys. Lett. A* **2017**, *381*, 2565–2572. [CrossRef]
12. Okuda, T. Telescope Array observatory for the high energy radiation induced by lightning. *J. Phys. Conf. Ser.* **2019**, *1181*, 012067. [CrossRef]
13. Liang, X.; Zhitnitsky, A. The ANITA Anomalous Events and Axion Quark Nuggets. *arXiv* **2021**, arXiv:2105.01668.
14. Gorham, P.W.; Nam, J.; Romero-Wolf, A.; Hoover, S.; Allison, P.; Banerjee, O.; Beatty, J.J.; Belov, K.; Besson, D.Z.; Binns, W.R.; et al. Characteristics of Four Upward-pointing Cosmic-ray-like Events Observed with ANITA. *Phys. Rev. Lett.* **2016**, *117*, 071101. [CrossRef] [PubMed]
15. Gorham, P.W.; Rotter, B.; Allison, P.; Banerjee, O.; Batten, L.; Beatty, J.J.; Bechtol, K.; Belov, K.; Besson, D.Z.; Binns, W.R.; et al. Observation of an Unusual Upward-going Cosmic-ray-like Event in the Third Flight of ANITA. *Phys. Rev. Lett.* **2018**, *121*, 161102. [CrossRef] [PubMed]
16. Gorham, P.W.; Ludwig, A.; Deaconu, C.; Cao, P.; Allison, P.; Banerjee, O.; Batten, L.; Bhattacharya, D.; Beatty, J.J.; Belov, K.; et al Unusual Near-Horizon Cosmic-Ray-like Events Observed by ANITA-IV. *Phys. Rev. Lett.* **2021**, *126*, 071103. [CrossRef]
17. Zhitnitsky, A. Axion quark nuggets. Dark matter and matter–antimatter asymmetry: Theory, observations and future experiments. *Mod. Phys. Lett. A* **2021**, *36*, 2130017. [CrossRef]
18. Witten, E. Cosmic separation of phases. *Phys. Rev. D* **1984**, *30*, 272–285. [CrossRef]
19. Farhi, E.; Jaffe, R.L. Strange matter. *Phys. Rev. D* **1984**, *30*, 2379–2390. [CrossRef]
20. De Rujula, A.; Glashow, S.L. Nuclearites—A novel form of cosmic radiation. *Nature* **1984**, *312*, 734–737. [CrossRef]
21. Madsen, J. Physics and Astrophysics of Strange Quark Matter. In *Hadrons in Dense Matter and Hadrosynthesis*; Cleymans, J., Geyer, H.B., Scholtz, F.G., Eds.; Lecture Notes in Physics; Springer: Berlin, Germany, 1999; Volume 516, p. 162. [CrossRef]
22. Peccei, R.D.; Quinn, H.R. Constraints imposed by CP conservation in the presence of pseudoparticles. *Phys. Rev. D* **1977**, *16*, 1791–1797. [CrossRef]
23. Weinberg, S. A new light boson? *Phys. Rev. Lett.* **1978**, *40*, 223–226. [CrossRef]
24. Wilczek, F. Problem of strong P and T invariance in the presence of instantons. *Phys. Rev. Lett.* **1978**, *40*, 279–282. [CrossRef]
25. Kim, J.E. Weak-interaction singlet and strong CP invariance. *Phys. Rev. Lett.* **1979**, *43*, 103–107. [CrossRef]
26. Shifman, M.A.; Vainshtein, A.I.; Zakharov, V.I. Can confinement ensure natural CP invariance of strong interactions? *Nucl. Phys. B* **1980**, *166*, 493–506. [CrossRef]
27. Dine, M.; Fischler, W.; Srednicki, M. A simple solution to the strong CP problem with a harmless axion. *Phys. Lett. B* **1981**, *104*, 199–202. [CrossRef]
28. Zhitnitsky, A.R. On Possible Suppression of the Axion Hadron Interactions. *Sov. J. Nucl. Phys.* **1980**, *31*, 260. (In Russian)
29. Van Bibber, K.; Rosenberg, L.J. Ultrasensitive Searches for the Axion. *Phys. Today* **2006**, *59*, 30. [CrossRef]
30. Asztalos, S.J.; Rosenberg, L.J.; van Bibber, K.; Sikivie, P.; Zioutas, K. Searches for Astrophysical and Cosmological Axions. *Annu. Rev. Nucl. Part. Sci.* **2006**, *56*, 293–326. [CrossRef]
31. Sikivie, P. Axion Cosmology. In *Axions*; Kuster, M., Raffelt, G., Beltrán, B., Eds.; Lecture Notes in Physics; Springer: Berlin, Germany, 2008; Volume 741, p. 19.

32. Raffelt, G.G. Astrophysical Axion Bounds. In *Axions*; Kuster, M., Raffelt, G., Beltrán, B., Eds.; Lecture Notes in Physics; Springer: Berlin, Germany, 2008; Volume 741, p. 51.
33. Sikivie, P. Dark Matter Axions. *Int. J. Mod. Phys. A* **2010**, *25*, 554–563. [CrossRef]
34. Rosenberg, L.J. Dark-matter QCD-axion searches. *Proc. Natl. Acad. Sci. USA* **2015**, *112*, 12278–12281. [CrossRef]
35. Marsh, D.J.E. Axion cosmology. *Phys. Rep.* **2016**, *643*, 1–79. [CrossRef]
36. Graham, P.W.; Irastorza, I.G.; Lamoreaux, S.K.; Lindner, A.; van Bibber, K.A. Experimental Searches for the Axion and Axion-Like Particles. *Annu. Rev. Nucl. Part. Sci.* **2015**, *65*, 485–514. [CrossRef]
37. Irastorza, I.G.; Redondo, J. New experimental approaches in the search for axion-like particles. *Prog. Part. Nucl. Phys.* **2018**, *102*, 89–159. [CrossRef]
38. Anastassopoulos, V.; Aune, S.; Barth, K.; Belov, A.; Cantatore, G.; Carmona, J.M.; Castel, J.F.; Cetin, S.A.; Christensen, F.; Collar, J.I.; et al. New CAST Limit on the Axion-Photon Interaction. *Nat. Phys.* **2017**, *13*, 584–590. [CrossRef]
39. Zhang, J.; Tsai, Y.L.S.; Kuo, J.L.; Cheung, K.; Chu, M.C. Ultralight Axion Dark Matter and Its Impact on Dark Halo Structure in N-body Simulations. *Astrophys. J.* **2018**, *853*, 51. [CrossRef]
40. Jacobs, D.M.; Starkman, G.D.; Lynn, B.W. Macro Dark Matter. *Mon. Not. R. Astron. Soc.* **2015**, *450*, 3418–3430. [CrossRef]
41. Gorham, P.W. Antiquark nuggets as dark matter: New constraints and detection prospects. *Phys. Rev. D* **2012**, *86*, 123005. [CrossRef]
42. Fraser, G.W.; Read, A.M.; Sembay, S.; Carter, J.A.; Schyns, E. Potential solar axion signatures in X-ray observations with the XMM?Newton observatory. *Mon. Not. R. Astron. Soc.* **2014**, *445*, 2146–2168. [CrossRef]
43. Ge, S.; Rachmat, H.; Siddiqui, M.S.R.; Van Waerbeke, L.; Zhitnitsky, A. X-ray annual modulation observed by XMM-Newton and Axion Quark Nugget Dark Matter. *arXiv* **2020**, arXiv:2004.00632.
44. Fox, D.B.; Sigurdsson, S.; Shandera, S.; Mészáros, P.; Murase, K.; Mostafá, M.; Coutu, S. The ANITA Anomalous Events as Signatures of a Beyond Standard Model Particle, and Supporting Observations from IceCube. *arXiv* **2018**, arXiv:1809.09615.
45. Forbes, M.M.; Zhitnitsky, A.R. WMAP haze: Directly observing dark matter? *Phys. Rev. D* **2008**, *78*, 083505. [CrossRef]
46. Gurevich, A.V.; Zybin, K.P. Runaway breakdown and electric discharges in thunderstorms. *Physics-Uspekhi* **2001**, *44*, 1119–1140. [CrossRef]
47. Dwyer, J.R.; Uman, M.A. The physics of lightning. *Phys. Rep.* **2014**, *534*, 147–241. [CrossRef]
48. Lawson, K.; Liang, X.; Mead, A.; Siddiqui, M.S.R.; Van Waerbeke, L.; Zhitnitsky, A. Gravitationally trapped axions on the Earth. *Phys. Rev. D* **2019**, *100*, 043531. [CrossRef]
49. NOAA, National Centers for Environmental Information. Available online: https://www.ngdc.noaa.gov/geomag/calculators/magcalc.shtml#igrfgrid (accessed on 15 October 2021).
50. Jackson, J.D. *Classical Electrodynamics*, 3rd ed.; Wiley: New York, NY, USA, 1999.
51. Huege, T.; Falcke, H. Radio emission from cosmic ray air showers: Coherent geosynchrotron radiation. *Astron. Astrophys.* **2003**, *412*, 19–34. [CrossRef]
52. Huege, T.; Falcke, H. Radio emission from cosmic ray air showers: Simulation results and parametrization. *Astropart. Phys.* **2005**, *24*, 116–136. [CrossRef]
53. Forbes, M.M.; Lawson, K.; Zhitnitsky, A.R. Electrosphere of macroscopic "quark nuclei": A source for diffuse MeV emissions from dark matter. *Phys. Rev. D* **2010**, *82*, 083510. [CrossRef]
54. Landau, L.D.; Lifshitz, E.M. *Course of Theoretical Physics*; Butterworth-Heinemann: Oxford, UK, 1975.
55. Usov, V.V. Bare quark matter surfaces of strange stars and e^+e^- emission. *Phys. Rev. Lett.* **1998**, *80*, 230–233. [CrossRef]
56. Usov, V.V. Thermal emission from bare quark matter surfaces of hot strange stars. *Astrophys. J. Lett.* **2001**, *550*, L179. [CrossRef]
57. Jaikumar, P.; Gale, C.; Page, D.; Prakash, M. Bremsstrahlung photons from the bare surface of a strange quark star. *Phys. Rev. D* **2004**, *70*, 023004. [CrossRef]
58. Harko, T.; Cheng, K.S. Photon Emissivity of the Electrosphere of Bare Strange Stars. *Astrophys. J.* **2005**, *622*, 1033–1043. [CrossRef]
59. Caron, J.F.; Zhitnitsky, A.R. Bremsstrahlung emission from quark stars. *Phys. Rev. D* **2009**, *80*, 123006. [CrossRef]
60. Zakharov, B.G. Photon emission from bare quark stars. *J. Exp. Theor. Phys.* **2011**, *112*, 63–76. [CrossRef]
61. Zakharov, B.G. Effect of electric field of the electrosphere on photon emission from quark stars. *Phys. Lett. B* **2010**, *690*, 250–254. [CrossRef]
62. Berestetskii, V.B.; Pitaevskii, L.P.; Lifshitz, E.M. *Course of Theoretical Physics*; Butterworth-Heinemann: Oxford, UK, 1982.

Article

Sensitivity to Axion-like Particles with a Three-Beam Stimulated Resonant Photon Collider around the eV Mass Range

Kensuke Homma, Fumiya Ishibashi, Yuri Kirita and Takumi Hasada

Graduate School of Advanced Science and Engineering, Hiroshima University, Kagamiyama, Higashi-Hiroshima 739-8526, Japan
* Correspondence: khomma@hiroshima-u.ac.jp

Abstract: We propose a three-beam stimulated resonant photon collider with focused laser fields in order to directly produce an axion-like particle (ALP) with the two beams and to stimulate its decay by the remaining one. The expected sensitivity around the eV mass range has been evaluated. The result shows that the sensitivity can reach the ALP-photon coupling down to $\mathcal{O}(10^{-14})\,\mathrm{GeV}^{-1}$ with 1 J class short-pulsed lasers.

Keywords: dark matter; axion; axion-like particle; ALP; inflaton; laser; stimulated resonant photon collider; four-wave mixing

Citation: Homma, K.; Ishibashi, F.; Kirita, Y.; Hasada, T. Sensitivity to Axion-like Particles with a Three-Beam Stimulated Resonant Photon Collider around the eV Mass Range. *Universe* **2023**, *9*, 20. https://doi.org/10.3390/universe9010020

Academic Editor: Kazuharu Bamba

Received: 29 November 2022
Revised: 22 December 2022
Accepted: 25 December 2022
Published: 29 December 2022

1. Introduction

CP violation is rather naturally expected from the topological nature of the QCD vacuum, θ-vacuum, which is required, at least, to solve the $U(1)_A$ anomaly. Nevertheless, the θ-value evaluated from the measurement of the neutron dipole moment indicates the CP conserving nature in the QCD sector. This so-called strong CP problem is one of the most important problems yet unresolved in the standard model of particle physics. Peccei and Quinn advocated the introduction of a new global $U(1)_{PQ}$ symmetry [1] in order to dynamically cancel out the finite θ-value expected in the QCD sector with a counter θ-value around which a massive axion appears as a result of the symmetry breaking. If the PQ-symmetry breaking scale is much higher than that of the electroweak scale, the coupling of axion to ordinary matter may be feeble. This invisible axion can thus be a reasonable candidate for dark matter as a byproduct.

In addition to axion, axion-like particle (ALP) not necessarily requiring the linear relation between mass and coupling such as in the QCD axion scenario [2–5], is also important in the context of inflation as well as dark matter in the universe. Among many possible ALPs, the *miracle* model [6] which unifies inflaton and dark matter within a single ALP attracts laser-based experimental searches, because the preferred ranges of the ALP mass (m_a) and its coupling to photons (g/M) are $0.01 < m_a < 1$ and $g/M \sim 10^{-11}\,\mathrm{GeV}^{-1}$, respectively, based on the viable parameter space consistent with the CMB observation.

So far, we have advocated a method to directly produce axion-like particles and simultaneously stimulate their decays by combining two-color laser fields in collinearly focused geometry [7]. This quasi-parallel photon–photon collision system has been dedicated to sub-eV axion mass window and the searches have been actually performed [8–12]. Given the axion mass window above eV and a typical laser photon energy of $\sim$1 eV, stimulated photon–photon collisions with different collision geometry has a potential to be sensitive to a higher mass window. In addition to the well-known axion helio- and halo-scopes, the proposed method can cooperatively provide unique test grounds totally independent of any of implicit theoretical assumptions on the axion flux in the Sun as well as in the universe. Therefore, if any of the helio- or halo-scopes detects a hint on an ALP, this method can unveil the nature of the ALP via the direct production and its stimulated decay in laboratory-based experiments by tuning the sensitive mass range to that specific mass window. In this sense, it is indispensable for us to prepare the independent method for expanding its sensitive mass window as wide as possible.

In this paper we propose a three-beam laser collider and discuss its expected sensitivity to an unexplored domain for the *miracle* model as well as the benchmark models of the QCD axion based on a realistic set of beam parameters available at world-wide high-intensity laser systems.

2. Formulation Dedicated for a Stimulated Three Beam Collider

We focus on the following effective Lagrangian describing the interaction of an ALP as a pseudoscalar field ϕ_a with two photons

$$-\mathcal{L} = \frac{1}{4}\frac{g}{M}F_{\mu\nu}\tilde{F}^{\mu\nu}\phi_a.\tag{1}$$

As illustrated in Figure 1, in the averaged or approximated sense, we consider a coplanar scattering with four-momenta $p_i(i = 1\text{–}4)$

$$< p_c(p_1) > + < p_c(p_2) > \rightarrow p_3 + < p_i(p_4) >,\tag{2}$$

where two focused laser beams, $< p_c >$, create an ALP with a symmetric incident angle θ_c and the produced ALP simultaneously decays into two photons due to an inducing laser beam in the background, $< p_i >$, incident with a different angle θ_i. As a result of the stimulated decay, emission of a signal photon p_3, is induced. $<>$ symbols reflect the fact that all three beams contain energy and momentum (angle) spreads at around the focal point. The energy uncertainty is caused by Fourier limited short pulsed lasers such as femtosecond lasers with the optical frequency, while the momentum uncertainty and fluctuations on angle of incidence are unavoidable due to focused fields. Thus, p_1, p_2 and p_4 must be stochastically selected from individual beams, while p_3 is generated as a result of energy–momentum conservation via $p_1 + p_2 \rightarrow p_3 + p_4$.

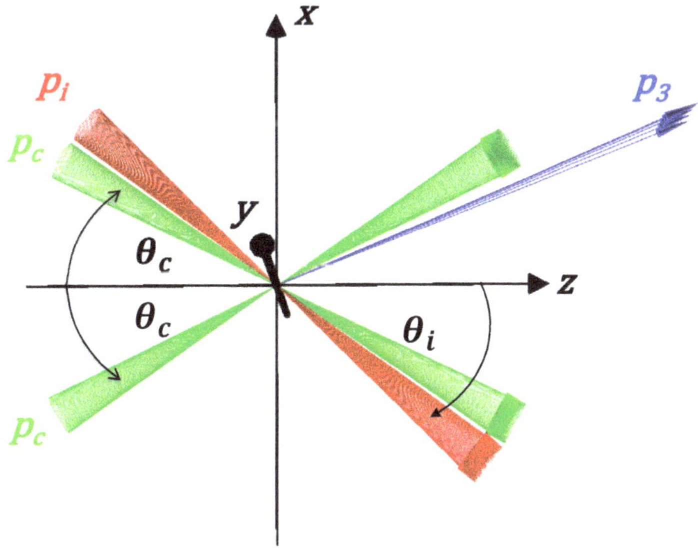

Figure 1. Schematic view of a three-beam stimulated photon collider.

We then assume a search for ALPs by scanning θ_c, equal incident angles of the two creation beam axes, to look for an enhancement of the interaction rate when the resonant condition

$$m_a = E_{cms} = 2\omega_c \sin\theta_c \tag{3}$$

is satisfied, where m_a is the ALP mass, ω_c is the central value of single photon energies in the incident creation laser beams and E_{cms} is center-of-mass system (cms) collision energy between two incident photons. Because individual incident photons fluctuate around the average beam energy ω_c and also around the average incident angle θ_c, this resonance condition has to be evaluated via weighted integral (averaging) over proper fluctuation distributions as we discuss below. In the following subsections, we thus review necessary formulae to numerically evaluate the interaction rate by taking generic collision geometry with asymmetric incident energies and asymmetric incident angles into account. This asymmetric treatment is essentially required, because unless we implement the degrees of the spreads at fixed θ_c and ω_c depending on experimental parameters, we cannot determine reasonable discretized steps for the scanning over the ALP mass range of interest.

In our previous work [13], we introduced a theoretical interface allowing the asymmetric treatment in the case where a single focused beam is used for creation of an ALP resonance state and the other focused beam sharing the same optical axis as the creation beam is co-moving for inducing the decay. However, if the sensitive mass range must be increased, we have to introduce two separated incident beams for the creation part. Thus, a modified geometrical treatment for the three separated beams must be reconsidered. In [13] we provided formulae only for the case of the scalar field exchange. In order to discuss ALPs, we have further extended the formulae to the pseudoscalar exchange case with the proper treatment of polarizations affecting the vertex factors [14]. In the following subsections we will provide necessary formulae developed in [13,14] with necessary modifications for the purpose of this paper.

2.1. Expression for Signal Yield in Stimulated Resonant Scattering

Figure 2 explains the relation between theoretical coordinates with the primed symbol and laboratory coordinates to which laser beams are physically mapped. The z'-axis is theoretically obtainable so that stochastically selected two incident photons satisfying the resonance condition have zero pair transverse momentum (p_T) with respect to z'. The Lorentz invariant scattering amplitude is calculated on the primed coordinates where rotation symmetries of the initial and final state reaction planes around z' are maintained. Definitions of four-momentum vectors p'_i and four-polarization vectors $e'(\lambda_{p'_i})$ with polarization states $\lambda_{p'_i}$ for the initial-state ($i = 1, 2$) and final-state ($i = 3, 4$) plane waves are given. The conversion between the two coordinates is possible via a simple rotation $\mathcal{R}$ as explained below. In the following, unless confusion is expected, the prime symbol associated with the momentum vectors is omitted.

We start by reviewing a spontaneous yield of the signal p_3, $\mathcal{Y}$, in the scattering process $p_1 + p_2 \to p_3 + p_4$ only with two incident photon beams having densities ρ_1 and ρ_2. The concept of *cross section* is useful for fixed p_1 and p_2 beams. In a situation where p_1 and p_2 largely fluctuate within beams, however, its convenience is lost. Thus we apply the following factorization of *volume-wise interaction rate* $\overline{\Sigma}$ [15,16] instead of *cross section* with units of length L and time s in [ ]

$$\mathcal{Y} = N_1 N_2 \left(\int dt d\mathbf{r} \rho_1(\mathbf{r}, t) \rho_2(\mathbf{r}, t) \right) \times \tag{4}$$

$$\left(\int dQ W(Q) \frac{c}{2\omega_1 2\omega_2} |\mathcal{M}_s(Q')|^2 dL'_{ips} \right)$$

$$\equiv N_1 N_2 \mathcal{D} \left[s/L^3 \right] \overline{\Sigma} \left[L^3/s \right] \tag{5}$$

where the probability density of cms-energy, $W(Q)$, is multiplied for averaging over the possible range. $W(Q)$ is a function of the combinations of photon energies (ω_α), polar (Θ_α) and azimuthal (Φ_α) angles in laboratory coordinates, denoted as

$$Q \equiv \{\omega_\alpha, \Theta_\alpha, \Phi_\alpha\} \quad \text{and} \quad dQ \equiv \Pi_\alpha d\omega_\alpha d\Theta_\alpha d\Phi_\alpha \tag{6}$$

for the incident beams $\alpha = 1, 2$. The integral with the weight of $W(Q)$ implements the resonance enhancement by including the off-shell part as well as the pole in the s-channel amplitude including the Breit–Wigner resonance function [13,14].

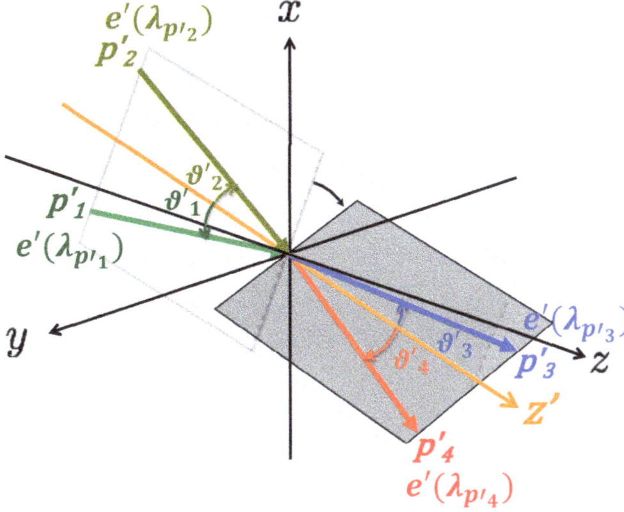

Figure 2. Relation between theoretical coordinates with the primed symbol and laboratory coordinates to which laser beams are physically mapped. The z'-axis is theoretically obtainable so that stochastically selected two incident photons satisfying the resonance condition have zero pair transverse momentum (p_T) with respect to z'. The Lorentz invariant scattering amplitude is calculated on the primed coordinates where rotation symmetries of the initial and final state reaction planes around z' are maintained. Definitions of four-momentum vectors p'_i and four-polarization vectors $e'(\lambda_{p'_i})$ with polarization states $\lambda_{p'_i}$ for the initial state ($i = 1, 2$) and final state ($i = 3, 4$) plane waves are given. This figure is extracted from [14].

As illustrated in Figure 2, $Q' \equiv \{\omega_\alpha, \vartheta_\alpha, \phi_\alpha\}$ are kinematical parameters in a rotated coordinates Q' constructed from a pair of two incident wave vectors so that the transverse momentum of the pair with respect to a z'-axis becomes zero. The primed coordinates are convenient because the axial symmetry around the z'-axis allows simpler calculations for the following solid angle integral. The conversions from Q to Q' are thus expressed as rotation matrices on polar and azimuthal angles: $\vartheta_\alpha \equiv \mathcal{R}_{\vartheta_\alpha}(Q)$ and $\phi_\alpha \equiv \mathcal{R}_{\phi_\alpha}(Q)$.

By adding an inducing beam with the central four-momentum p_4 having normalized density ρ_4 with the average number of photons N_4, we extend the *spontaneous* yield to the *induced* yield, $\mathcal{Y}_{c+i}$, with the following extended set of kinematical parameters,

$$Q_I \equiv \{Q, \omega_4, \Theta_4, \Phi_4\} \quad \text{and} \quad dQ_I \equiv dQ d\omega_4 d\Theta_4 d\Phi_4. \tag{7}$$

as follows

$$\mathcal{Y}_{c+i} = N_1 N_2 N_4 \left(\int dt d\mathbf{r} \rho_1(\mathbf{r},t) \rho_2(\mathbf{r},t) \rho_4(\mathbf{r},t) V_4 \right) \times \tag{8}$$

$$\left(\int dQ_I W(Q_I) \frac{c}{2\omega_1 2\omega_2} |\mathcal{M}_s(Q')|^2 dL'^I_{ips} \right)$$

$$\equiv N_1 N_2 N_4 \mathcal{D}_{three} \left[s/L^3 \right] \overline{\Sigma}_I \left[L^3/s \right],$$

where the factor $\rho_4(\mathbf{r},t) V_4$ is a probability corresponding to a degree of spacetime overlap of the p_1 and p_2 beams with the inducing beam p_4 for a given volume of the p_4 beam, V_4. dL'^I_{ips} describes an inducible phase space in which the solid angles of p_3 balance solid angles of p_4 via energy–momentum conservation within the distribution of the given inducing beam after conversion from p_4 in the primed coordinate system to the corresponding laboratory coordinate where laser beams are physically mapped. With Gaussian distributions G, $W(Q_I)$ is explicitly defined as

$$W(Q_I) \equiv \Pi_\beta G_E(\omega_\beta) G_p(\Theta_\beta, \Phi_\beta) \tag{9}$$

over $\beta = 1,2,4$, where G_E reflecting an energy spread via Fourier transform limited duration of a short pulse and G_p in the momentum space, equivalent to the polar angle distribution, are introduced based on the properties of a focused coherent electromagnetic field with an axial symmetric nature for an azimuthal angle Φ around the optical axis of a focused beam, as we discuss below.

2.1.1. Evaluation of Spacetime Overlapping Factor $\mathcal{D}_{three}$ with Three Beams

The factor $\mathcal{D}_{three}$ in Equation (8) expresses a spatiotemporal overlapping factor of the focused creation beams (subscript $c1$ and $c2$) with the focused inducing beam (subscript i) in laboratory coordinates. The following photon number densities $\rho_{k=c1,c2,i}$ deduced from the electromagnetic field amplitudes based on the Gaussian beam parameterization [17] corresponding to the black pulse in Figure 3 are integrated over spacetime $(t,\mathbf{r})$:

$$\rho_k(t,\mathbf{r}) = \left(\frac{2}{\pi}\right)^{\frac{3}{2}} \frac{1}{w_k^2(ct) c\tau_k} \times \tag{10}$$

$$\exp\left(-2\frac{x^2+y^2}{w_k^2(ct)}\right) \exp\left(-2\left(\frac{z-ct}{c\tau_k}\right)^2\right),$$

where w_k are the beam radii as a function of time t whose origin is set at the moment when all the pulses reach the focal point, and τ_k are the time durations of the pulsed laser beams with the speed of light c and the volume for the inducing beam V_i is defined as

$$V_i = (\pi/2)^{3/2} w_{i0}^2 c\tau_i, \tag{11}$$

where w_{i0} is the beam waist (minimum radius) of the inducing beam. As a conservative evaluation, the integrated range for the overlapping factor is limited in the Rayleigh length

$$z_{iR} = \frac{\pi w_{i0}^2}{\lambda_i} \tag{12}$$

with the wavelength of the inducing beam λ_i only around the focal point where the induced scattering probability is maximized.

Figure 3 illustrates spacetime pulse functions propagating along individual optical axes of the three beams that are defined by rotating coordinates in Equation (10) around y-axis. ρ_{c1}, ρ_{c2} and ρ_i are defined with the rotation angles: θ_c, $-\theta_c$ and $-\theta_i$, respectively. That is, we assume symmetric incident angles between the two creation laser beams and supply the inducing laser so that photon four-momenta satisfy energy–momentum conservation with respect to a fixed central value for signal photon four-momenta.

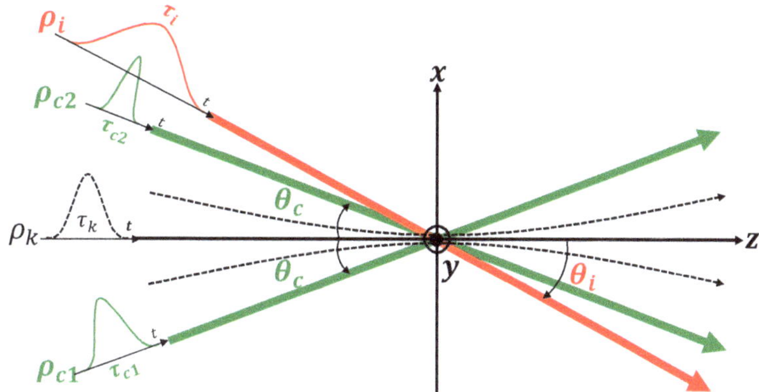

Figure 3. Collision geometry between three short pulsed laser beams to define the spacetime overlapping factor $\mathcal{D}_{three}$.

The overlapping factor with units of $[s/L^3]$ can be analytically integrated over spatial coordinates and is eventually obtained by numerically integrating over time from $-z_{iR}/c$ to 0 as follows:

$$
D_{three}\,[s/L^3] = \left(\frac{1}{\pi}\right)^{\frac{3}{2}} 2^3 w_{i0}^2 \int_{-\frac{z_{iR}}{c}}^{0} dt\, \frac{1}{w_c^3 w_i c^2 \tau_{c1} \tau_{c2}} \sqrt{\frac{1}{2\left(2w_i^2 + w_c^2\right)}}
$$
$$
\sqrt{\frac{J}{HS}} \exp\left[\frac{T^2}{4S} - R\right] \exp\left[-2\frac{\tau_{c2}^2 \tau_i^2 + \tau_{c1}^2 \tau_i^2 + \tau_{c1}^2 \tau_{c2}^2}{\tau_{c1}^2 \tau_{c2}^2 \tau_i^2} t^2\right].
$$
(13)

The individual variables in Equation (13) are summarized as follows, where we use abbreviations $\mathbf{C_k} = \cos\theta_k$ and $\mathbf{S_k} = \sin\theta_k$ for $k = c1, c2, i$ and we assume $\mathbf{C_c} \equiv \mathbf{C_{c1}} = \mathbf{C_{c2}}$, $\mathbf{S_c} \equiv \mathbf{S_{c1}} = \mathbf{S_{c2}}$, $w_c \equiv w_{c1} = w_{c2}$, $d_c \equiv d_{c1} = d_{c2}$ and $f_c \equiv f_{c1} = f_{c2}$, because two creation beams are incident with a symmetric angle and focused with equal beam diameters and focal lengths.

$$
J\,[L^6 \cdot s^4] = w_c^2 w_i^2 c^2 \tau_{c1}^2 \tau_{c2}^2 \tau_i^2, \qquad H\,[L^4 \cdot s^4] = 2\left(2C\mathbf{C_c}^2 + D\mathbf{C_i}^2 + E\mathbf{S_c}^2 + F\mathbf{S_c}^2 + G\mathbf{S_i}^2\right),
$$
$$
S\,[1/L^2] = \frac{O}{J} - P, \qquad\qquad T\,[1/L] = -(N + Q),
$$
(14)
$$
R\,[1] = -\frac{4}{HJ}(B_c M - B_i G)^2.
$$

The parameters B, C, D, E, F and G are

$$
B_k\,[L] = 2ct\mathbf{S_k}, \qquad C\,[L^4 \cdot s^4] = w_i^2 c^2 \tau_{c1}^2 \tau_{c2}^2 \tau_i^2, \qquad D\,[L^4 \cdot s^4] = w_c^2 c^2 \tau_{c1}^2 \tau_{c2}^2 \tau_i^2,
$$
$$
E\,[L^4 \cdot s^4] = w_c^2 w_i^2 \tau_{c2}^2 \tau_i^2, \qquad F\,[L^4 \cdot s^4] = w_c^2 w_i^2 \tau_{c1}^2 \tau_i^2, \qquad G\,[L^4 \cdot s^4] = w_c^2 w_i^2 \tau_{c1}^2 \tau_{c2}^2.
$$
(15)

The parameters M, N, O, P and Q are

$$M\left[L^4 \cdot s^4\right] = E - F, \qquad N\left[1/L\right] = 4t\left(\frac{\tau_{c2}^2\tau_i^2\mathbf{C_c} + \tau_{c1}^2\tau_i^2\mathbf{C_c} + \tau_{c1}^2\tau_{c2}^2\mathbf{C_i}}{c\tau_{c1}^2\tau_{c2}^2\tau_i^2}\right),$$

$$O\left[L^4 \cdot s^4\right] = 2\left(2C\mathbf{S_c}^2 + D\mathbf{S_i}^2 + E\mathbf{C_c}^2 + F\mathbf{C_c}^2 + G\mathbf{C_i}^2\right),$$

$$P\left[1/L^2\right] = \frac{4}{HJ}\left\{(\mathbf{C_iS_i}D + \mathbf{C_cS_c}M)^2 - 2\mathbf{C_cS_cC_iS_i}MG + \mathbf{C_i}^2\mathbf{S_i}^2G^2 - 2\mathbf{C_i}^2\mathbf{S_i}^2DG\right\},$$

$$Q\left[1/L\right] = \frac{4}{HJ}\left\{\mathbf{C_cS_c}M(B_cM + B_iG) - \mathbf{C_iS_i}\left(B_cMG - B_iG^2 + DB_iG - DB_cM\right)\right\}.$$

(16)

The beam parameters relevant to focused geometry used above are expressed as

$$w_k = w_{k0}\sqrt{1 + \frac{c^2t^2}{z_{kR}^2}}, \qquad w_{k0} = \frac{\lambda_k}{\pi\vartheta_{k0}}, \qquad \vartheta_{k0} = \arctan\left(\frac{d_k}{2f_k}\right)$$

(17)

with $k = c1, c2, i$.

2.1.2. Evaluation of Inducible Volume-Wise Interaction Rate, $\overline{\Sigma}_I$

Performing the analytical integral for $\overline{\Sigma}_I$ in Equation (8) is not practical and we are forced to evaluate it with the numerical integral. The expression for $|\mathcal{M}_s(Q')|^2$ in the inducible volume-wise interaction rate is fully explained in [14]. In this paper, we focus on how to implement the numerical integral configured for a three-beam collider with focused beams. Figure 4 illustrates the entire flow of the calculation. The left figure depicts the initial state of two scattering photons with incidence of two creation beams (green) and an inducing beam (red), while the right figure indicates the final state photons, that is, the inducing beam photons and signal photons (blue) in the laboratory coordinates by omitting the outgoing two creation beams. The top figure is to remind of the scattering amplitude calculation in the primed coordinates. Probability distribution functions in momentum space G_p as a function of polar angles Θ_i and azimuthal angles Φ_i in the laboratory coordinates and those in energy $G_E(\omega_i)$ for individual photons $i = 1, 2, 4$ are assigned to individual focused beams by denoting the normalized Gaussian distributions as G. The actual steps for the calculations are as follows:

1. Select a finite-size segment of p_1 from given $G_E(\omega_1)G_p(\Theta_1, \Phi_1)$ distributions.
2. Find p_2 which satisfies the following resonance condition

$$m_a = E_{cms} = 2\sqrt{\omega_1\omega_2}\sin\left(\frac{\vartheta_1 + \vartheta_2}{2}\right)$$

(18)

with respect to the selected p_1 and to a finite energy segment in $G_E(\omega_2)$ for a given mass parameter m_a. The possible p_2 candidates satisfying the resonance condition form the yellow thin cone around the p_1-axis reflecting the width of the Breit–Wigner function as shown in the left figure.
3. Form a z'-axis so that the pair transverse momentum, p_T, becomes zero, which is defined as zero-p_T coordinates (primed coordinates) in contrast to the laboratory coordinates to which the three beams are physically mapped. Only a portion of the creation beam prepared for p_2 overlapping with the yellow cone can effectively contribute to the resonance production and, hence, the field weight for the pair can be eventually evaluated by properly reflecting $G_E(\omega_1)G_p(\Theta_1, \Phi_1)$ and $G_E(\omega_2)G_p(\Theta_2, \Phi_2)$.
4. Convert the polarization vectors $e_i(\lambda_i)$ as well as the momentum vectors from the laboratory coordinates to the zero-p_T coordinates through the coordinate rotation $\mathcal{R}(Q \rightarrow Q')$.
5. The axial symmetric nature of possible final-state momenta p_3' and p_4' around z' is represented by the light-blue and magenta vectors in the right figure. A spontaneous

scattering probability with the vertex factors using the primed polarization vectors in the planes containing the four photon wave vectors is calculated in the given zero-p_T coordinates as illustrated in the top figure. By using the axial symmetric nature around z', the probability can be integrated over possible final state planes containing p'_3 and p'_4.

6. In order to estimate the inducing effect for given $G_E(\omega_4)G_p(\Theta_4, \Phi_4)$ distributions fixed in the laboratory coordinates, a matching fraction of p_4 is calculated after rotating the primed vectors back to those in the laboratory coordinates from the zero-p_T coordinates via the inverse rotation $\mathcal{R}^{-1}(Q' \to Q)$. Based on the spread of $G_E(\omega_4)$, the weights along the overlapping belt between the magenta and red vectors in the right figure are taken into account as the enhancement factor for the stimulation of the decay.

7. Due to energy–momentum conservation, p_3 must balance with p_4. Thus a signal energy spread via $\omega_s \equiv \omega_3 = \omega_1 + \omega_2 - \omega_4$ and also the polar-azimuthal angle spreads by taking the $G_E(\omega_4)G_p(\Theta_4, \Phi_4)$ distributions into account are automatically determined. The volume-wise interaction rate $\overline{\Sigma}_I$ is then integrated over the inducible solid angle of p_3 reflecting all the energy and angular spreads included in the focused three beams.

8. With Equation (8) the signal yield $\mathcal{Y}_{c+i}$ can be evaluated.

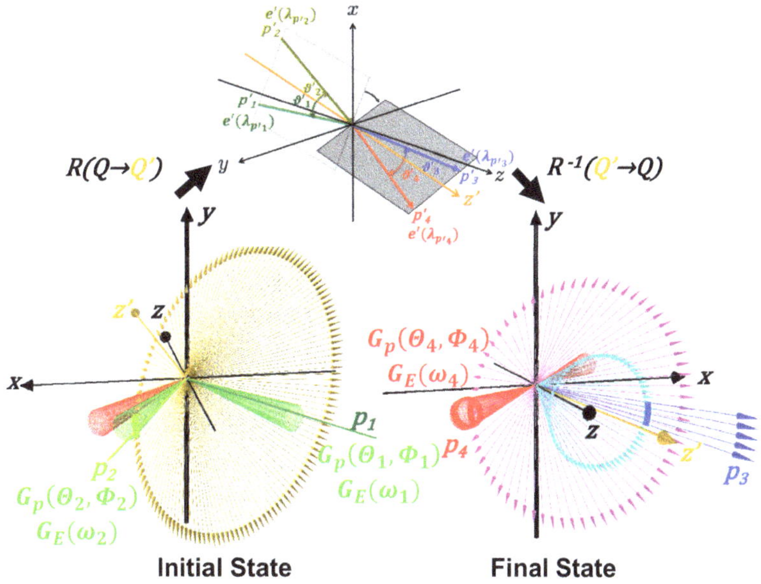

Figure 4. Flow of the numerical calculation. The left figure depicts the initial state of two scattering photons with incidence of two creation beams (green) and an inducing beam (red), while the right figure indicates the final state photons, that is, the inducing beam photons and signal photons (blue) in the laboratory coordinates by omitting the outgoing two creation beams. The top figure is to remind of the scattering amplitude calculation in the primed coordinates. Probability distribution functions in momentum space G_p as a function of polar angles Θ_i and azimuthal angles Φ_i in the laboratory coordinates and those in energy $G_E(\omega_i)$ for individual photons $i = 1, 2, 4$ are assigned to individual focused beams by denoting the normalized Gaussian distributions as G.

3. Expected Sensitivity

We evaluate search sensitivities based on the concept of a three-beam stimulated resonant photon collider (tSRPC) with variable incident angles for scanning ALP masses around the eV range as illustrated in Figure 1. By assuming high-intensity femtosecond lasers such as Titanium:Sapphire lasers with 1 J pulse energy for simplicity, we consider that two identical creation beams with the central photon energy ω_c and the time duration τ_c are symmetrically incident with the same beam incident angle θ_c and an inducing laser with the central photon energy $\omega_i \equiv u\omega_c$ with $0 < u < 1$ is incident with the corresponding angle which satisfies energy–momentum conservation by requiring a common signal photon energy $(2 - u)\omega_c$ independent of various incident angle combinations. The central wavelength is around 800 nm and then we assume the ability to produce high harmonic waves from the fundamental wavelength for creation beams and to generate an inducing beam with a non-integer number u based on the optical parametric amplification (OPA) technique in order to discriminate signal waves against the integer number high harmonic waves originating from the creation beams. Since the OPA technique cannot achieve the perfect conversion from the fundamental wavelength, we assume 0.1 J pulse energy and also the elongation of the pulse duration for the inducing beam compared to that in the creation beam. Table 1 summarizes assumed parameters for two identical creation laser beams and an including laser beams as well as the common focusing and statistical parameters.

Table 1. Experimental parameters used to numerically calculate the upper limits on the coupling–mass relations.

Parameter	Value
Central wavelength of creation laser λ_c	800 nm (ω)/400 nm (2ω)/267 nm (3ω)
Relative linewidth of creation laser, $\delta\omega_c / <\omega_c>$	10^{-2}
Duration time of creation laser, τ_c	30 fs
Creation laser energy per τ_c, E_c	1 J
Number of creation photons (ω), N_c	4.03×10^{18} photons
Number of creation photons (2ω), N_c	2.01×10^{18} photons
Number of creation photons (3ω), N_c	1.34×10^{18} photons
Beam diameter of creation laser beam, d_c	60 mm
Polarization	linear (P-polarized state)
Central wavelength of inducing laser, λ_i	1300 nm
Relative linewidth of inducing laser, $\delta\omega_i / <\omega_i>$	10^{-2}
Duration time of inducing laser beam, τ_i	100 fs
Inducing laser energy per τ_i, E_i	0.1J
Number of inducing photons, N_i	6.54×10^{17} photons
Beam diameter of inducing laser beam, d_i	30 mm
Polarization	circular (left-handed state)
Focal length of off-axis parabolic mirror, $f_c = f_i$	600 mm
Overall detection efficiency, ϵ	1%
Number of shots, N_{shots}	10^5 shots
δN_S	100

Given a set of three-beam laser parameters P in Table 1, the number of stimulated signal photons, N_{obs}, is expressed as

$$N_{obs} = \mathcal{Y}_{c+i}(m_a, g/M; P)N_{shot}\epsilon, \tag{19}$$

which is a function of ALP mass m_a and coupling g/M, where N_{shot} the number of laser shots and ϵ the overall efficiency of detecting p_3. For a set of m_a values with an assumed N_{obs}, a set of coupling g/M can be estimated by numerically solving Equation (19).

Based on parameters in Table 1, Figure 5 shows the reachable sensitivities in the coupling–mass relation for the pseudoscalar field exchange at a 95% confidence level by ^{t}SRPC. The red, blue and magenta solid/dashed curves show the expected upper limits by ^{t}SRPC when we assume $\omega_c = 800$ nm (fundamental wavelength ω), 400 nm (second harmonic 2ω) and 267 nm (third harmonic 3ω), respectively. The ALP mass scanning is assumed to be performed with the step of 0.1 eV. Thanks to energy and momentum fluctuations at around the focal point, the same order sensitivities are maintained within the assumed scanning step (the local minima of the parabolic behavior in the coupling correspond to different incident angle setups in Figure 5). For easy viewing, the solid and dashed curves are drawn alternatively. These assumed photon sources are all available within the current technology [18] in terms of the photon wavelength and energy per pulse.

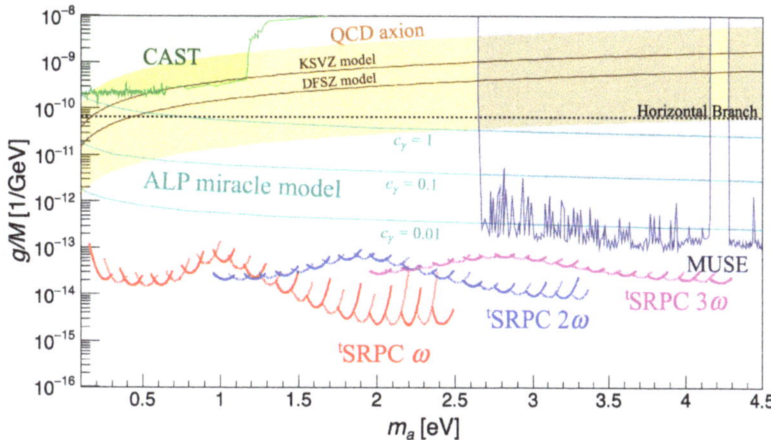

Figure 5. Expected sensitivities in the coupling–mass relation for the pseudoscalar field exchange at a 95% confidence level by a three-beam stimulated resonant photon collider (^{t}SRPC) with focused short-pulsed lasers based on the beam parameters in Table 1.

These sensitivity curves are obtained based on the following condition. In this virtual search, the null hypothesis is supposed to be fluctuations on the number of photon-like signals following a Gaussian distribution whose expectation value, μ, is zero for the given total number of collision statistics. The photon-like signals implies a situation where photons-like peaks are counted by a peak finder based on digitized waveform data from a photodevice [11], where electrical fluctuations around the baseline of a waveform cause both positive and negative numbers of photon-like signals. In order to exclude this null hypothesis, a confidence level $1 - \alpha$ is introduced as

$$1 - \alpha = \frac{1}{\sqrt{2\pi}\sigma} \int_{\mu-\delta}^{\mu+\delta} e^{-(x-\mu)^2/(2\sigma^2)} dx = \mathrm{erf}\left(\frac{\delta}{\sqrt{2}\sigma}\right), \tag{20}$$

where μ is the expected value of an estimator x following the hypothesis, and σ is one standard deviation. In this search, the estimator x corresponds to the number of signal photons N_S and we assume the detector-acceptance-uncorrected uncertainty δN_S as the one standard deviation σ around the mean value $\mu = 0$. For setting a confidence level of 95%, $2\alpha = 0.05$ with $\delta = 2.24\sigma$ is used, where a one-sided upper limit by excluding above $x + \delta$ [19] is considered. For a set of experimental parameters P in Table 1, the upper limits on the coupling–mass relation, m_a vs. g/M, are then estimated by numerically solving the following equation

$$N_{obs} = 2.24\delta N_S = \mathcal{Y}_{c+i}(m_a, g/M; P)N_{shots}\epsilon. \tag{21}$$

The horizontal dotted line shows the upper limit from the Horizontal Branch (HB) observation [20]. The purple area shows bounds by the optical MUSE-faint survey [21]. The green area is excluded by the helioscope experiment CAST [22]. The yellow band shows the QCD axion benchmark models with $0.07 < |E/N - 1.95| < 7$ where KSVZ($E/N = 0$) [4,5] and DFSZ($E/N = 8/3$) [23] are shown with the brawn lines. The cyan lines show predictions from the ALP *miracle* model [6] with its intrinsic model parameters $c_\gamma = 1.0$, $0.1, 0.01$, respectively.

4. Conclusions

We have evaluated expected sensitivities to axion-like particles coupling to photons based on the concept of a three-beam stimulated resonant photon collider with focused short-pulse lasers. Within the current high-intensity laser technology reaching the pulse energy 1 J, we found that the searching method can probe ALPs in the eV mass range down to $g/M = \mathcal{O}(10^{-14})$ GeV^{-1}. This sensitivity is sufficient to test the unexplored domain motivated by the *miracle* model as well as the benchmark QCD axion models.

Author Contributions: Software, F.I. and Y.K.; Investigation, T.H.; Writing—original draft, K.H. All the authors have read and agreed to the published version of the manuscript.

Funding: This work was funded by the Collaborative Research Program of the Institute for Chemical Research at Kyoto University (Grant Nos. 2018-83, 2019-72, 2020-85, 2021-88, and 2022-101); Grants-in-Aid for Scientific Research Nos. 17H02897, 18H04354, 19K21880, and 21H04474 from the Ministry of Education, Culture, Sports, Science and Technology (MEXT) of Japan; JST, the establishment of university fellowships for the creation of science technology innovation, Grant No. JPMJFS2129; a Grant-in-Aid for JSPS fellows No. 22J13756 from the Ministry of Education, Culture, Sports, Science and Technology (MEXT) of Japan.

Institutional Review Board Statement: Not applicable.

Informed Consent Statement: Not applicable.

Data Availability Statement: Not applicable.

Acknowledgments: K. Homma acknowledges the support of the Collaborative Research Program of the Institute for Chemical Research at Kyoto University (Grant Nos. 2018-83, 2019-72, 2020-85, 2021-88, and 2022-101) and Grants-in-Aid for Scientific Research Nos. 17H02897, 18H04354, 19K21880, and 21H04474 from the Ministry of Education, Culture, Sports, Science and Technology (MEXT) of Japan. Y. Kirita acknowledges support from the JST, the establishment of university fellowships for the creation of science technology innovation, Grant No. JPMJFS2129, and a Grant-in-Aid for JSPS fellows No. 22J13756 from the Ministry of Education, Culture, Sports, Science and Technology (MEXT) of Japan.

Conflicts of Interest: The authors declare no conflict of interest.

References

1. Peccei, R.D.; Quinn, H.R. CP conservation in the presence of pseudoparticles. *Phys. Rev. Lett.* **1977**, *38*, 1440. [CrossRef]
2. Weinberg, S. A new light boson? *Phys. Rev. Lett.* **1978**, *40*, 223. [CrossRef]
3. Wilczek, F. Problem of Strong *P* and *T* Invariance in the Presence of Instantons. *Phys. Rev. Lett.* **1978**, *40*, 279. [CrossRef]
4. Kim, J.E. Weak-Interaction Singlet and Strong CP Invariance. *Phys. Rev. Lett.* **1979**, *43*, 103. [CrossRef]
5. Shifman, M.A.; Vainshtein, A.I.; Zakharov, V.I. Can confinement ensure natural CP invariance of strong interactions? *Nucl. Phys. B* **1980**, *166*, 493–506. [CrossRef]
6. Daido, R.; Takahashi, F.; Yin, W. The ALP miracle revisited. *J. High Energy Phys.* **2018**, *2018*, 104. [CrossRef]
7. Fujii, Y.; Homma, K. An approach toward the laboratory search for the scalar field as a candidate of Dark Energy. *Prog. Theor. Phys.* **2011**, *126*, 531–553; Erratum in *Prog. Theor. Exp. Phys.* **2014**, 089203. [CrossRef]
8. Homma, K.; Hasebe, T.; Kume, K. The first search for sub-eV scalar fields via four-wave mixing at a quasi-parallel laser collider. *Prog. Theor. Exp. Phys.* **2014**, *2014*, 083C01. [CrossRef]
9. Hasebe, T.; Homma, K.; Nakamiya, Y.; Matsuura, K.; Otani, K.; Hashida, M.; Inoue, S.; Sakabe, S. Search for sub-eV scalar and pseudoscalar resonances via four-wave mixing with a laser collider. *Prog. Theo. Exp. Phys.* **2015**, *2015*, 073C01. [CrossRef]
10. Nobuhiro, A.; Hirahara, Y.; Homma, K.; Kirita, Y.; Ozaki, T.; Nakamiya, Y.; Hashida, M.; Inoue, S.; Sakabe, S. Extended search for sub-eV axion-like resonances via four-wave mixing with a quasi-parallel laser collider in a high-quality vacuum system. *Prog. Theo. Exp. Phys.* **2020**, *2020*, 073C01. [CrossRef]
11. Homma, K.; Kirita, Y.; Hashida, M.; Hirahara, Y.; Inoue, S.; Ishibashi, F.; Nakamiya, Y.; Neagu, L.; Nobuhiro, A.; Ozaki, T.; et al. Search for sub-eV axion-like resonance states via stimulated quasi-parallel laser collisions with the parameterization including fully asymmetric collisional geometry. *J. High Energy Phys.* **2021**, *12*, 108.

12. Kirita, Y.; Hasada, T.; Hashida, M.; Hirahara, Y.; Homma, K.; Inoue, S.; Ishibashi, F.; Nakamiya, Y.; Neagu, L.; Nobuhiro, A.; et al. Search for sub-eV axion-like particles in a stimulated resonant photon-photon collider with two laser beams based on a novel method to discriminate pressure-independent components. *J. High Energy Phys.* **2022**, *2022*, 176.

13. Homma, K.; Kirita, Y. Stimulated radar collider for probing gravitationally weak coupling pseudo Nambu-Goldstone bosons. *J. High Energy Phys.* **2020**, *9*, 95. [CrossRef]

14. Homma, K.; Kirita, Y.; Ishibashi, F. Perspective of Direct Search for Dark Components in the Universe with Multi-Wavelengths Stimulated Resonant Photon-Photon Colliders. *Universe* **2021**, *7*, 479. [CrossRef]

15. Bjorken, J.D.; Drell, S.D . *Relativistic Quantum Mechanicsh*; McGraw-Hill, Inc.: New York, NY, USA, 1964.

16. Greiner, W. *Quantum Electrodynamics*, 2nd ed.; Greiner, W., Reinhardt, J., Eds.; Springer: Berlin/Heidelberg, Germany, 1994.

17. Yariv, A. *Optical Electronics in Modern Communications*; Oxford University Press: Oxford, UK, 1997.

18. Available online: https://eli-laser.eu (accessed on 1 December 2022).

19. Beringer, J.; Arguin, J.F.; Barnett, R.M.; Copic, K.; Dahl, O.; Groom, D.E. Review of particle physics. *Phys. Rev.-Part Fields Gravit. Cosmol.* **2012**, *86*, 010001. [CrossRef]

20. Ayala, A.; Domínguez, I.; Giannotti, M.; Mirizzi, A.; Straniero, O. Revisiting the bound on axion-photon coupling from Globular Clusters. *Phys. Rev. Lett.* **2014**, *113*, 191302. [CrossRef] [PubMed]

21. Regis, M.; Taoso, M.; Vaz, D.; Brinchmann, J.; Zoutendijk, S.L.; Bouché, N.F.; Steinmetz, M. Searching for light in the darkness: Bounds on ALP dark matter with the optical MUSE-faint survey. *Phys. Lett. B* **2021**, *814*, 136075. [CrossRef]

22. Anastassopoulos, V. et al. [CAST Collaboration]. New CAST Limit on the Axion-Photon Interaction. *Nat. Phys.* **2017**, *13*, 584–590.

23. Dine, M.; Fischler, W.; Srednicki, M. A simple solution to the strong CP problem with a harmless axion. *Phys. Lett.* **1981**, *104*, 199–202. [CrossRef]

Article

Pilot Search for Axion-Like Particles by a Three-Beam Stimulated Resonant Photon Collider with Short Pulse Lasers

Fumiya Ishibashi [1], Takumi Hasada [1], Kensuke Homma [1,*], Yuri Kirita [1], Tsuneto Kanai [2], ShinIchiro Masuno [2], Shigeki Tokita [2,3] and Masaki Hashida [2,4]

[1] Graduate School of Advanced Science and Engineering, Hiroshima University, Kagamiyama, Higashi-Hiroshima 739-8526, Japan
[2] Institute for Chemical Research, Kyoto University Uji, Kyoto 611-0011, Japan
[3] Graduate School of Science, Kyoto University, Sakyouku, Kyoto 606-8502, Japan
[4] Research Institute of Science and Technology, Tokai University, 4-1-1 Kitakaname, Hiratsuka 259-1292, Japan
* Correspondence: khomma@hiroshima-u.ac.jp

Abstract: Toward the systematic search for axion-like particles in the eV mass range, we proposed the concept of a stimulated resonant photon collider by focusing three short pulse lasers into a vacuum. In order to realize such a collider, we have performed a proof-of-principle experiment with a set of large incident angles between three beams to overcome the expected difficulty to ensure the space–time overlap between short pulse lasers and also established a method to evaluate the bias on the polarization states, which is useful for a future variable–incident–angle collision system. In this paper, we present a result from the pilot search with the developed system and the method. The search result was consistent with null. We thus have set the upper limit on the minimum ALP-photon coupling down to 1.5×10^{-4} GeV^{-1} at the ALP mass of 1.53 eV with a confidence level of 95%.

Keywords: dark matter; axion; axion-like particle; ALP; inflaton; laser; stimulated resonant photon collider; four-wave mixing

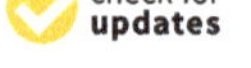

Citation: Ishibashi, F.; Hasada, T.; Homma, K.; Kirita, Y.; Kanai, T.; Masuno, S.; Tokita, S.; Hashida, M. Pilot Search for Axion-Like Particles by a Three-Beam Stimulated Resonant Photon Collider with Short Pulse Lasers. *Universe* **2023**, *9*, 123. https://doi.org/10.3390/ universe9030123

Academic Editor: Kazuharu Bamba

Received: 31 January 2023
Revised: 18 February 2023
Accepted: 22 February 2023
Published: 28 February 2023

1. Introduction

Present space observations consistently estimate that 95% of the energy density balance of the Universe is occupied by dark matter and dark energy. Among the dark components, axion [1–4] is one of the most rational candidates for cold dark matter (CDM) [5–7], which is supposed to be created via spontaneous breaking of the Peccei–Quinn symmetry [8] in order to solve the strong CP problem [9]. Furthermore, axion-like particles (ALPs), which set free the relation between mass and coupling unlike the QCD axion, are also widely discussed. Some of them are scalar-type of fields such as dilaton [10] and chameleon [11] in the context of dark energy.

In this paper, we focus on the following interaction Lagrangian between a pseudoscalar-type ALP, ϕ_a, and two photons

$$-\mathcal{L} = \frac{1}{4}\frac{g}{M}F_{\mu\nu}\tilde{F}^{\mu\nu}\phi_a \tag{1}$$

where $F^{\mu\nu} = \partial^\mu A^\nu - \partial^\nu A^\mu$ is the field strength tensor and its dual $\tilde{F}^{\mu\nu} \equiv \frac{1}{2}\epsilon^{\mu\nu\alpha\beta}F_{\alpha\beta}$ with the Levi–Civita symbol ϵ^{ijkl}, and g is a dimensionless constant while M is an energy at which a global continuous symmetry is broken.

Among many types of ALPs, a model, *miracle* [12], unifying inflation and dark matter into a single pseudoscalar-type ALP predicts the ALP mass and its coupling to photons in a range overlapping with those of the benchmark QCD axion models [3,4,13,14] in the eV mass range. Moreover, very recently, a scenario of thermal production of cold "hot dark matter" [15] and a new kind of axion model from the Grand Unified Theory (GUT) based on SU(5) $\times$ U(1)$_{PQ}$ [16] predict ALPs in the eV mass range as well. Therefore, specifically, the ALP mass range in $\mathcal{O}(0.1-1)$ eV at the coupling $g/M \sim 10^{-11}$ GeV^{-1} is the intended range of this study. A typical photon energy of laser fields, $\mathcal{O}(1)$ eV, is thus suitable for a photon collider targeting this mass range.

We have proposed a method to directly produce an ALP resonance state and simultaneously stimulate its decay by combining two-color (creation and inducing) laser fields and focusing them together with a single lens element in vacuum, which is defined as stimulated resonant photon collisions (SRPC) in a quasi-parallel collision system (QPS) [17]. In order to satisfy a resonance condition for the direct production of an ALP, the range of the center-of-mass system energy, E_{cms}, between two photons selected from a focused creation laser beam must include the ALP mass, m_a. Thus, the condition is simply expressed as

$$E_{cms} = 2\omega_c \sin\theta_c = m_a \tag{2}$$

with a common creation laser photon energy ω_c and an angle $2\theta_c$ between the two photons. Since a typical photon energy in lasers is around eV, SRPC in QPS has been employed as a way to access sub-eV ALPs with a long focal length [18–23].

In order to access a higher mass range above eV, unless an unrealistically short focal length is assumed, we cannot access a higher mass in QPS if we keep the same photon energy in the beams. On the other hand, increasing photon energy by more than one order magnitude via optical nonlinear effects is a trade-off with reduction of the beam intensity, that is, the sensitivity to weak coupling domains is reduced. We thus have extended the formulation for SRPC with a single focused beam after combining two lasers in QPS [24] to SRPC with three separated focused beams (^{t}SRPC) [25] as illustrated in Figure 1. We can introduce a symmetric incident angle of θ_c for the two beam axes of creation lasers (green); however, two incident photons from the focused two beams indeed have different incident angles ϑ_1 and ϑ_2 from θ_c with different energies ω_1 and ω_2 from ω_c, respectively, in general. The incident angle fluctuations around the beam axes are caused by momentum fluctuations at around the focal point, while energy uncertainties are caused by nearly Fourier transform limited short-pulsed lasers. These fluctuations are, in principle, unavoidable due to the uncertainty principle in momentum-energy space. Accordingly the exact resonance condition is modified as

$$E_{cms} = 2\sqrt{\omega_1\omega_2}\sin\left(\frac{\vartheta_1+\vartheta_2}{2}\right) = m_a. \tag{3}$$

The inducing beam with the central energy ω_i (red) is simultaneously focused into the overlapping focal points between the two creation beams, and part of the beam represented as ω_4 enhances the interaction rate of the stimulated scattering resulting in emission of signal photons with the energy ω_3 (blue), which satisfies energy–momentum conservation. In order to reflect realistic energy and momentum distributions in the three beams, numerical calculations are eventually required to evaluate the stimulated interaction rate [25]. Thanks to the broadening of E_{cms} due to these uncertainties, however, the sensitivity to a target ALP mass will also have a wide resolution around the mass, which allows a quick mass scan if we vary θ_c with a consistent step with the mass resolution.

On the other hand, synchronization of tightly confined pulses in space–time is required for ^{t}SRPC, which increases the experimental difficulty. In a photodetector with electric amplification, the time resolution is $\mathcal{O}(10)$ ps at most. For the duration of creation laser pulses about 40 fs, such a conventional detection technique is not applicable for ensuring synchronization of creation laser beams. Therefore, we consider utilization of nonlinear optical effects in a thin BBO crystal. Second harmonic generation (SHG) via the 2nd order nonlinear optical effect in BBO can be used for the synchronization between two creation beams. As for the three-beam synchronization, the third order nonlinear optical effect, four-wave mixing (FWM), in the same crystal can be used. For the purpose of synchronization, the atomic processes are quite important, while the atomic FWM becomes the dominant background source with respect to FWM in vacuum, that is, generation of ω_3 photons via ALP-exchange in ^{t}SRPC. This is because both atomic and ALP-exchange processes require energy–momentum conservation between four photons, and the signal photon energy ω_3 becomes kinematically almost identical.

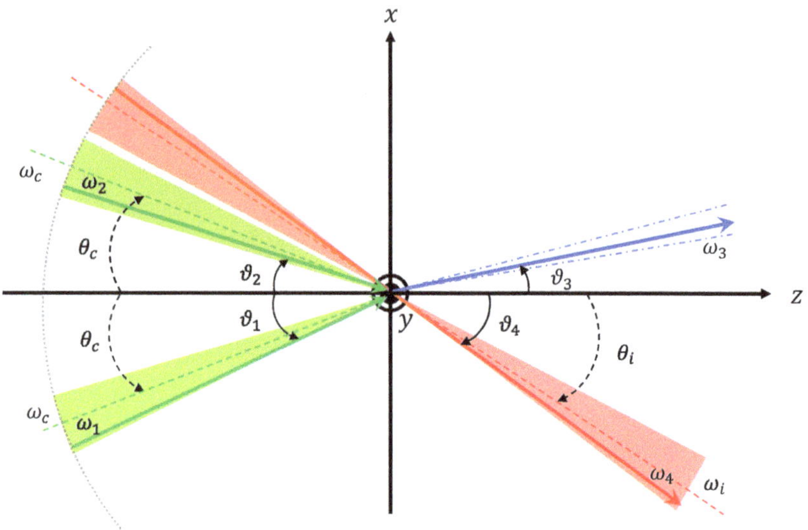

Figure 1. Concept of a three-beam stimulated resonant photon collider (^{t}SRPC) with focused coherent beams.

In this paper, we will present a result of the pilot ALP search based ^{t}SRPC in the air as a proof-of-principle experiment that demonstrates that the aforementioned method practically works to guarantee the space–time synchronization between the three beams by setting a large collision angle of the creation lasers at $\theta_c = 30$ deg to learn the real technical complications toward the continuous mass scanning by systematically varying θ_c in the near future search.

In the following sections, we describe the experimental setup and the synchronization methods in the pilot ALP search, the method for analyzing the acquired data, how to set the exclusion limits, and, finally, conclude the search results and discuss future plans toward the continuous ALP mass scanning.

2. Experimental Setup

Figures 2 and 3 show a schematic drawing of the searching setup and the photographs of the setup with the three focused laser spots at a thin cross-wire target, respectively. We used a Ti:Sapphire laser (T^6-system) with $\sim$40 fs duration and a Nd:YAG laser with 9 ns duration for the creation and inducing fields, respectively. Both of them are available in the Institute for Chemical Research in Kyoto University. The central wavelengths of these lasers were 808 nm and 1064 nm, respectively. Creation laser pulses were injected into a beam splitter (BS) and bifurcated to prepare for two creation fields with the guaranteed synchronization. In this case, one of the creation lasers transmits BS, so the duration of the pulse is slightly elongated. Therefore, in principle, there is a finite duration difference in the two pulses (τ_{c1}, τ_{c2}). The central wavelength of signal photons is expected to be 651 nm via FWM: $\omega_{c1} + \omega_{c2} - \omega_i$ with creation photon energies ω_{c1} and ω_{c2}, respectively, and inducing photon energy ω_i. In addition to energy conservation, momentum conservation requires the following angle relation: $\theta_i = 39.1°$ and the most probable $\vartheta_3 = 22.7°$ for $\theta_c = 30.0°$

resulting in the resonant mass $m_a = 1.53$ eV with respect to the given central photon energies. Spacetime synchronization at the interaction point (IP) is required between two creation pulses branched at BS. Thus, a delay line (DL) equipped with a retroreflector (RR) was constructed on an motorized-stage at one of the creation laser paths (upper green line in Figure 2). By adjusting the position of RR along DL, the timing for the two pulse incidence at IP can be synchronized. In contrast, the inducing laser pulses were electrically triggered by a clock source synchronized with an upstream oscillator dedicated for the creation laser, and the injection timing was controlled by a Q-switch based on arrival times to two fast photodiodes (PD1, PD2) for one of two creation pulses and for inducing pulses by looking at an oscilloscope.

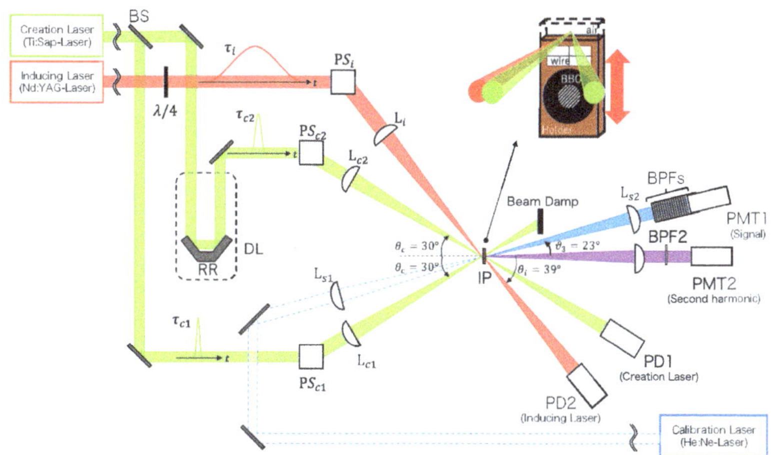

Figure 2. Schematic drawing of the search setup.

Figure 3. Photographs of the search setup (**left**) and the three focused laser spots (**right**) at a common thin cross-wire target.

Individual beams were focused into IP via periscopes (PS_{c1}, PS_{c2}, PS_i) at $30°$ for the creation lasers and $39°$ for the inducing laser as shown in Figure 2. These incident angles and signal outgoing angle were determined so that the central signal wavelength from FWM becomes 651 nm via energy–momentum conservation.

Typically, a mirror is designed to maximize reflectivity at an angle of incidence (AOI) of 45° and thus a reflection angle of 45° which can maintain linearly polarized states with respect to linearly polarized incident beams. A periscope (PS) consists of a pair of mirrors aligned vertically with AOI of 45° while it can emit a beam in any directions by changing the optical axis (beam height). Thus, in the near future, we will be able to scan collision angles between the two creation beams by the introduction of PS. However, if we use PS to rotate emission directions at arbitrarily large angles, polarization states of beams will become elliptic in general. Furthermore, one of the creation laser paths contains RR, and it can also be a source of changing elliptical polarization states. Therefore, it is necessary to introduce complex Jones vectors for representing the polarization vectors with two independent angle parameters. The two angles representing the polarization state of the two creation lasers were determined by measuring Stokes parameters as explained in Appendix A. On the other hand, the inducing laser was set to circular polarization (left-handed) using a $\lambda/4$ plate. This is because the theoretical interface is prepared for generally polarized states for the creation lasers and circularly polarized states for the inducing laser [24] in order to avoid complication on the numerical calculations to estimate inducible momentum ranges in the final state [21,24].

The two creation lasers and the inducing laser were focused at IP with lenses L_{c1}, L_{c2} and L_i, respectively, with a common focal length of $f = 300$ mm as shown in Figure 2. IP was equipped with a special holder vertically consisting of a cross-wire of 10 μm thickness, a BBO crystal which is a nonlinear crystal of 50 μm thickness, and a no target state (air) as shown in the insets of Figures 2 and 3. By attaching this special holder to the z-axis stage, cross-wire (spatial overlap), BBO (time synchronization), and no target state (search experiment) can be switched independently of the other optical elements. The camera systems (C_{c1}, C_{c2}, C_i) and photodetectors were located downstream from IP. Since individual camera systems are installed on motorized-stages, they can be moved to appropriate positions for checking the spatial overlap of the three beam spots, the time synchronization between the two or three laser pulses and performing searches, depending on the purposes. The spatial overlap was ensured by aligning the center of individual laser spots to the crossed point of the two thin wires as shown in the three pictures in Figure 3. The beam waist for the inducing laser was enlarged compared to those of the creation lasers so that the creation laser spots could be stably included in the volume of the inducing field.

After ensuring the spatial overlap between the three beams at IP, time synchronization was first performed with the two creation lasers. The duration of the creation laser pulses was $\sim$40 fs. It is impossible to ensure synchronization using a conventional photodetector due to the limited time resolution of at most $\sim$10 ps. Therefore, space–time synchronization was confirmed by observing second harmonic generation from the BBO crystal, which is known as a fast nonlinear optical effect with $\mathcal{O}$(fs) resolution when two high-intensity pulses spatiotemporally overlap. DL was actually adjusted by measuring the number of second harmonic photons as a function of RR position. In addition to the two creation pulse overlap, when the inducing laser spatiotemporally overlaps with the creation pulses, FWM in BBO may also be produced. A second harmonic from the two creation pulses and FWM from the three pulse overlap emerge at different angles. We note that FWM must conserve energy–momentum, while the second harmonic conserves energy but not necessarily momentum because translation symmetry is broken in the BBO crystal.

Second harmonic was detected using a photomultiplier tube (PMT2) by selecting a second harmonic of the creation laser wavelength, 404 nm, by a band-pass filter (BPF2). Fifteen band-pass filters (BPFs) were placed in front of the PMT for FWM detection (PMT1) in order to mainly remove residual beam photons. The BPFs were installed in multiple layers of three types of BPFs so that they eliminate wavelengths of the creation laser and the second harmonic of the creation laser, the inducing laser and its second harmonic. In this way, PMT1 can detect photons only in the proper energy band consistent with FWM. Second harmonic and FWM photons from BBO ensured the space–time synchronization of the three lasers. Since the duration of the inducing laser was 9 ns, the time resolutions of a typical photo-device, ∼40 ps, were sufficient to adjust arrival time difference between second harmonic and FWM photons, both of which were measured with photomultipliers with the same time resolution of 0.75 ns. After the space–time synchronization between the three beams was ensured, the vertical position of the holder was set at the no-target position, and we conducted the search experiment.

3. Space–Time Synchronization

As shown in the pictures of the three beam spots in Figure 3 (right), the centers of the three beams' focal spots were adjusted at the crossed point of the crossed wires of 10 μm thickness. This guarantees the spatial overlap between the three beams.

Figure 4 shows a picture of oscilloscope waveforms when space–time synchronization between three beams was satisfied, where photodiode signals PD1 (creation laser), PD2 (inducing laser), and signals from photomultipliers PMT2 (second harmonic generation from BBO), PMT1(four-wave mixing from BBO) in Figure 2 are simultaneously displayed. When the BBO crystal was inserted to the position of IP, we clearly confirmed the time synchronization between the three beams.

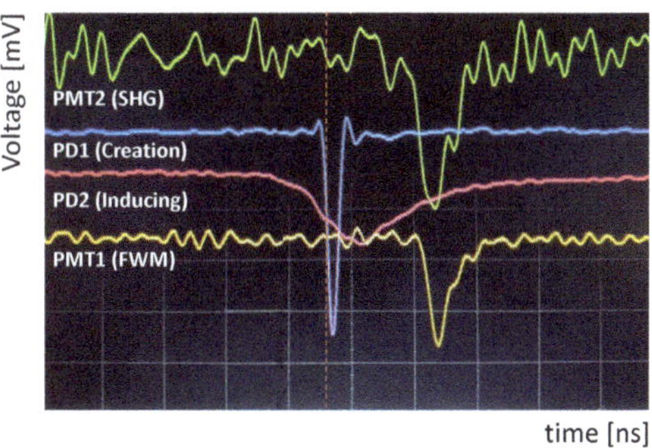

Figure 4. Photograph of oscilloscope waveforms from the four photodetectors in Figure 2. Four-wave mixing (FWM) photons were clearly observed when a thin BBO crystal was positioned at IP.

For a fine timing tuning between the two creation short pulses, we took a look at the number of FWM photons detected by PMT1 as a function of stage position in the delay line (DL) in Figure 2. Figure 5 shows the clear peak structure at the best synchronization point.

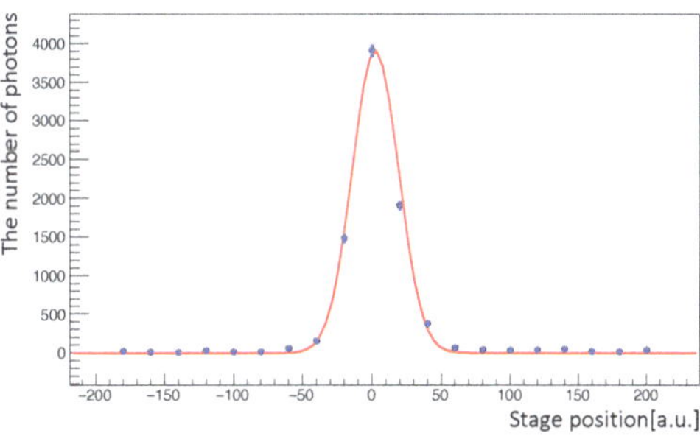

Figure 5. Observed number of FWM photons as a function of stage position in the delay line for a fine timing tuning between two creation laser pulses when a thin BBO crystal was positioned at IP.

4. Data Analysis

PMT1 detects photons from various background sources in addition to signal photons via FWM. The number of photons detected by PMT1 contains photons or photon-like events in the following four categories: the number of signal photons, n_{sig}, originating from the combination between the creation and inducing laser pulses, the number of background photons, n_c, originating from only the creation laser pulses, the number of background photons, n_i, originating from only the inducing laser pulses, and the number of noise photons, n_p, when no beam exists, that is, pedestal. In order to extract the number of signal photons, the number of photons in the above three background categories must be subtracted. Therefore, in the search experiment, both the creation and inducing laser pulses were injected at different irregular intervals of 5 Hz as illustrated in Figure 6 in order to successively form the four patterns. The number of measured photons in each pattern is expressed as Equation (4). The number of photons detected in P-pattern, N_P, is the pedestal component from environmental noises including thermal noise from PMT1. The number of photons detected in C- and I-pattern, N_C, and N_I, respectively, include the number of photons originating from individual laser focus such as plasma creation on top of the pedestals. The number of photons detected in S-pattern, N_S, includes the number of signal photons on top of all the other background sources:

$$
\begin{aligned}
N_S &= n_{sig} + n_c + n_i + n_p \\
N_C &= n_c + n_p \\
N_I &= n_i + n_p \\
N_P &= n_p
\end{aligned}
\tag{4}
$$

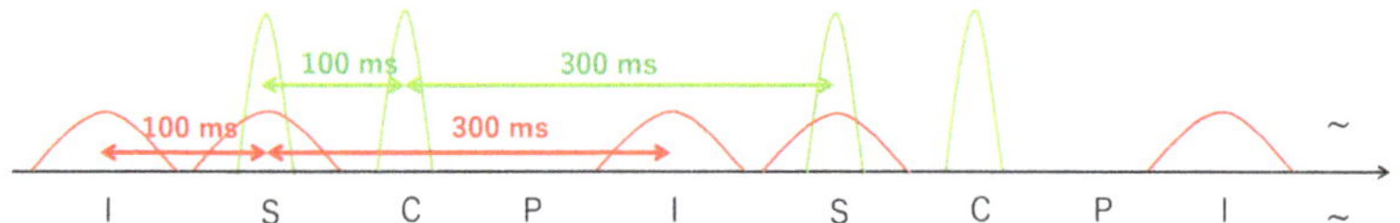

Figure 6. Four patterns of the beam combination between the two laser pulses where the green and red pulses are respectively creation and inducing laser pulses. The classifications are: S for two laser pulses, C for only the creation laser pulses, I for only the inducing laser pulses, and P for pedestals without laser pulses.

These four patterns were substituted into Equation (5), in order to extract the observed number of FWM photons

$$
n_{obs} = N_S - (N_C - N_P) - (N_I - N_P) - N_P.
\tag{5}
$$

In the search, the two photodiodes (PD1, PD2) were placed downstream of the interaction point (IP) in Figure 2. Four patterns, S, I, C, and P, were defined based on analog waveforms obtained from PD1 and PD2 assigned for the creation and inducing lasers, respectively. The number of photons was reconstructed from the voltage–time relation of analog signals from PMT1 with a waveform digitizer and applying a peak-finding algorithm to simultaneously determine the number of photons and their arrival times from falling edges of amplitudes of waveforms. The details of these instruments and the peak analysis method are described in [22].

In advance of the search, the expected arrival time of FWM photons in vacuum was determined by the arrival timing of FWM photons in BBO, which ensures space–time synchronization between focused three laser pulses. Figure 7 shows the arrival time distribution of FWM photons from BBO, where 1000 shots in S-pattern without background subtraction from the other patterns are shown. In the following analysis, n_{obs} always implies the number of observed FWM photons by integrating photon-like charges in PMT1 within the arrival time window of 2.5 ns, which is indicated by the two vertical lines in Figure 7.

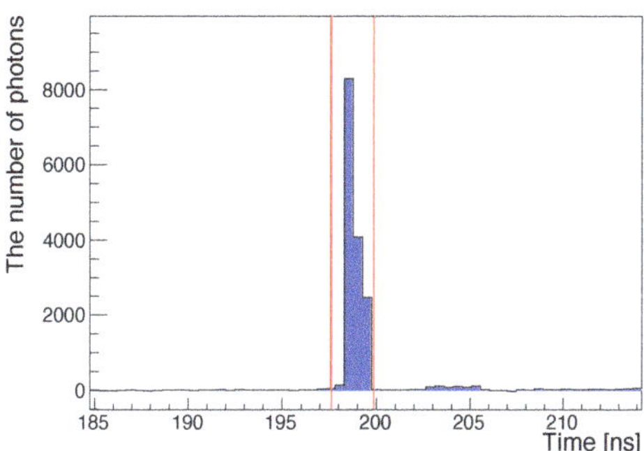

Figure 7. Arrival time distribution of FWM photons via the atomic process when a BBO crystal was placed at IP and space–time synchronization was ensured by PMT1. The red lines thus provide the expected time window for FWM photons via ALP-exchange to arrive.

5. Search Results

Figure 8 shows arrival time distributions of photons in individual patterns. The histograms in S, C, I, and P patterns are shown in the upper left (blue), upper right (green), lower left (pink), and lower right (gray), respectively. The expected arrival time windows were indicated by the two vertical lines. The total number of laser shots was 48,000 in the four patterns and thus the valid statistics in the S-pattern was 12,000 shots. Figure 9 shows arrival time distributions after subtraction with Equation (5). The interval between the two vertical lines represents the expected arrival time window of FWM photons. Thus, the number of FWM photons was evaluated by summing charges in PMT1 within this window and dividing the sum by a single-photon equivalent charge. As a result, the observed number of FWM photons, n_{obs}, was null within the error size as follows:

$$n_{obs} = -17.4 \pm 28.4(\text{stat.}) \pm 9.8(\text{syst.I}) \pm 5.4(\text{syst.II}) + 22.4 - 15.2(\text{syst.III}). \qquad (6)$$

The first systematic error (syst.I) was estimated by calculating the root-mean-square of the number of photon-like noise excluding the expected arrival time window of FWM photons. This corresponds to the baseline uncertainty of the PMT1 connected to the waveform digitizer in the real noise environment. The second systematic error (syst.II) was obtained by changing the default internal threshold -1.3 mV in the peak finder from -1.2 to -1.4 mV with the assumption of the uniform distribution. The details of the peak finding method are explained in [20,22]. The third systematic error (syst.III) was evaluated by changing the expected arrival time window size for FWM photons from 1.5 ns to 3.5 ns with respect to the most likely arrival time window of 2.5 ns.

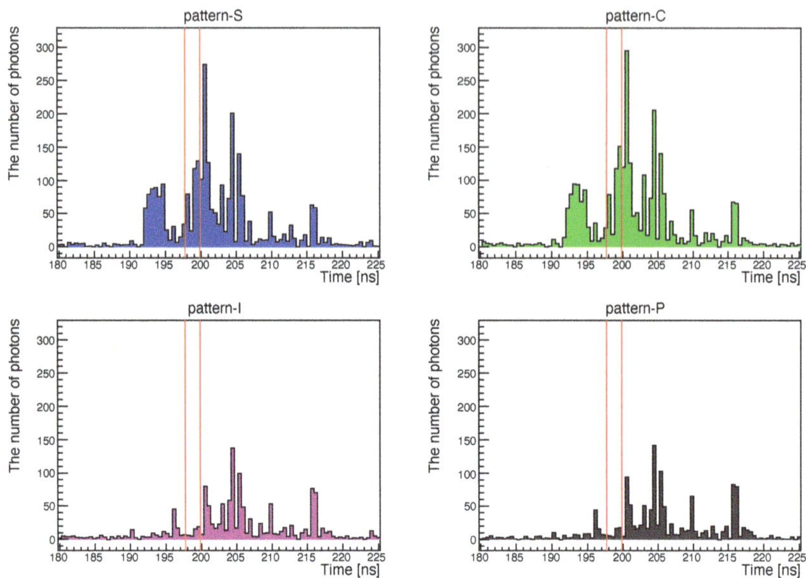

Figure 8. Arrival time distributions of photons with no target state (air) at IP. The histograms in the upper left, upper right, lower left, and lower right correspond to S, C, I, and P patterns of beam combinations, respectively. The interval between the two red lines in the S-pattern indicates the expected time windows for FWM photons via ALP-exchange to arrive.

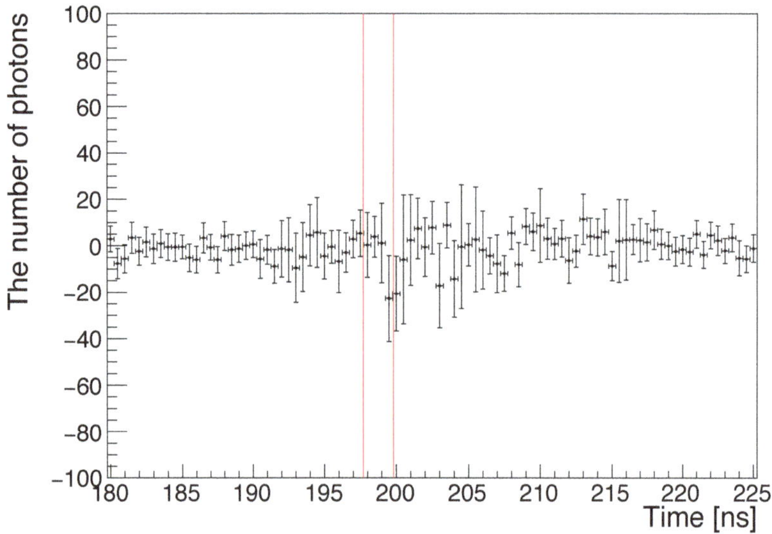

Figure 9. Arrival time distribution of reconstructed photons after subtraction between the four patterns based on Equation (5) over the entire time range in Figure 8. The interval between the two red lines indicates the expected time windows for FWM photons via ALP-exchange to arrive.

6. Exclusion Region in the Coupling-Mass Relation for ALP-Exchange

Since we have obtained the null result in Section 5, we set an exclusion region in the coupling-mass relation for the ALP exchange based on the formulation for the signal photon yield given in [25] and the measured total error size as follows. The signal photons yield in stimulated resonant scattering per pulse collision, $\mathcal{Y}_{c+i}$, is expressed as [25]

$$\mathcal{Y}_{c+i} \equiv N_1 N_2 N_4 \mathcal{D}_{three} \left[s/L^3 \right] \bar{\Sigma}_I \left[L^3/s \right], \tag{7}$$

where $N_1 (= N_{c1})$, $N_2 (= N_{c2})$, and $N_4 (= N_i)$ are the average numbers of photons containing individual lasers, respectively, $\mathcal{D}_{three}$ is a factor representing the space–time overlap of focused three beams at the interaction point [25], and $\bar{\Sigma}_I$ is the volume-wise interaction rate [21,24]. Individual units are given in [ ] with length L and time s.

Based on the set of experimental parameters P summarized in Table 1, the observed number of FWM photons via an ALP exchange with the mass m_a and the coupling g/M to two photons is expressed as

$$n_{obs} = \mathcal{Y}_{c+i}(m_a, g/M; P) N_{shot} \epsilon, \tag{8}$$

where N_{shot} is the number of shots in the S-pattern, and ϵ is the overall detection efficiency. A coupling constant g/M can be evaluated by solving Equation (8) for an ALP mass m_a and a given observed number of photons n_{obs}.

Table 1. Experimental parameters used to obtain the upper limit.

Parameter	Value
Central wavelength of creation laser, λ_{c1}	808 nm
Relative linewidth of creation laser, $\delta\omega_{c1}/<\omega_{c1}>$	1.7×10^{-2}
Duration time of creation laser, τ_{c1}	(38.8 ± 1.4) fs (FWHM)
Measured creation laser energy per τ_{c1}, E_{c1}	(1.21 ± 0.13) μJ
Creation energy fraction within 3 σ_{xy} focal spot, f_{c1}	0.82
Effective creation energy per τ_{c1} within 3 σ_{xy} focal spot	$E_{c1}f_{c1} = 1.0$ μJ
Effective number of creation photons, N_{c1}	4.0×10^{12} photons
Beam diameter of creation laser beam, d_{c1}	(5.0 ± 0.5) mm
Polarization (see Appendix A)	$\epsilon_{c1} = 0.41$ rad, $\theta_{c1} = 0.30$ rad
Central wavelength of creation laser, λ_{c2}	808 nm
Relative linewidth of creation laser, $\delta\omega_{c2}/<\omega_{c2}>$	1.7×10^{-2}
Duration time of creation laser, τ_{c2}	(39.2 ± 1.7) fs (FWHM)
Measured creation laser energy per τ_{c2}, E_{c2}	(1.52 ± 0.14) μJ
Creation energy fraction within 3 σ_{xy} focal spot, f_{c2}	0.85
Effective creation energy per τ_{c2} within 3 σ_{xy} focal spot	$E_{c2}f_{c2} = 1.3$ μJ
Effective number of creation photons, N_{c2}	5.2×10^{12} photons
Beam diameter of creation laser beam, d_{c2}	(5.0 ± 0.5) mm
Polarization (see Appendix A)	$\epsilon_{c2} = 0.91$ rad, $\theta_{c2} = -0.31$ rad
Central wavelength of inducing laser, λ_i	1064 nm
Relative linewidth of inducing laser, $\delta\omega_i/<\omega_i>$	1.0×10^{-4}
Duration time of inducing laser beam, τ_{ibeam}	9 ns (two standard deviation)
Measured inducing laser energy per τ_{ibeam}, E_i	(1.58 ± 0.05) μJ
Linewidth-based duration time of inducing laser, $\tau_i/2$	$\hbar/(2\delta\omega_i) = 2.8$ ps
Inducing energy fraction within 3 σ_{xy} focal spot, f_i	0.88
Effective inducing energy per τ_i within 3 σ_{xy} focal spot	$E_i(\tau_i/\tau_{ibeam})f_i = 0.87$ nJ
Effective number of inducing photons, N_i	4.7×10^9 photons
Beam diameter of inducing laser beam, d_i	(3.0 ± 0.5) mm
Polarization	circular (left-handed state)
Common focal length of lens, f	300.0 mm
Single-photon detection efficiency, ϵ_{det}	1.4%
Efficiency of optical path from IP to PMT, ϵ_{opt}	53%
Total number of shots in trigger pattern S, N_{shot}	12,000 shots
δn_{obs}	37.9

When counting photon-like peaks by the peak finding algorithm in waveforms, fluctuations of the baseline may produce both positive and negative amplitudes resulting in negative numbers of photon-like peaks as well as positive ones. Thus, even if the mean value is zero, we assume a Gaussian distribution to be the most natural null hypothesis. The confidence level for this null hypothesis is defined as

$$1 - \alpha = \frac{1}{\sqrt{2\pi}\sigma} \int_{\mu-\delta}^{\mu+\delta} e^{-(x-\mu)^2/(2\sigma^2)} dx = \mathrm{erf}\left(\frac{\delta}{\sqrt{2}\sigma}\right), \tag{9}$$

where μ is the expected value of x according to the hypothesis and σ is the standard deviation. In the search, the expected value x corresponds to the number of FWM photons n_{obs}, and σ is one standard deviation δn_{obs}. Based on (6) which indicates $\mu = 0$ because of the null result, we determined the acceptance-uncorrected uncertainty δn_{obs} around $n_{obs} = 0$ as the root mean square of all the error components as follows:

$$\delta n_{obs} = \sqrt{28.4^2 + 9.8^2 + 5.4^2 + 22.4^2} \simeq 37.9 \tag{10}$$

where the larger error on the positive side (+22.4) was used for syst.III in (6). In order to set a 95% confidence level, $2\alpha = 0.05$ with $\delta = 2.24$ was used to obtain the one-sided upper limit by excluding $x + \delta$ [26]. The upper limit in the relation m_a vs. g/M was estimated by numerically solving Equation (11) with δn_{obs} in (10) for the set of experimental parameters P in Table 1

$$2.24 \delta n_{obs} = \mathcal{Y}_{c+i}(m_a, g/M; P) N_{shot} \epsilon, \tag{11}$$

where $N_{shot} = 12{,}000$, and the overall efficiency $\epsilon \equiv \epsilon_{opt}\epsilon_{det}$ with the optical path acceptance from IP to PMT1, ϵ_{opt}, and the single photon detection efficiency of PMT1, ϵ_{det}, were substituted. ϵ_{opt} was obtained by using the continuous He:Ne laser mimicking the path of signal photons as indicated in Figure 2 by taking the ratio between the laser intensity at IP and that measured at the PMT1 position with a common CCD camera. ϵ_{det} was measured in advance using another pulse laser combined with a beam splitter system so that an equal number of photons were prepared between the two paths. By taking the ratio between the number of incident photons in one path and the number of counted photons by PMT1 in the other path, ϵ_{det} was determined.

Figure 10 shows the upper limit in the coupling-mass relation from this search, the three-beam stimulated resonant photon collider (^{t}SRPC00) enclosed by the red solid curve. The limit was set at a 95% confidence level by assuming only pseudoscalar-type ALP exchanges. The most sensitive ALP mass in this search is expected to be $m_a = 1.53$ eV because the creation lasers have a fixed collision angle of 30°. In reality, however, the sensitivity is not limited to $m_a = 1.53$ eV because of energy and momentum uncertainties of focused short pulse lasers. These uncertainties are exactly taken into account in the numerical calculation based on Equation (7) [25]. The magenta area indicates the excluded range based on SRPC in quasi-parallel collision geometry (SAPPHIRES01) [23]. The purple areas are excluded regions by the Light Shining through a Wall (LSW) experiments (ALPS [27] and OSQAR [28]). The gray area shows the excluded region by the vacuum magnetic birefringence (VMB) experiment (PVLAS [29]). The light-cyan horizontal solid line indicates the upper limit from the search for eV (pseudo)scalar penetrating particles in the SPS neutrino beam (NOMAD) [30]. The horizontal dotted line indicates the upper limit from the Horizontal Branch observation [31]. The blue areas indicate exclusion regions from the optical MUSE-faint survey [32]. The green area is the excluded region by the helioscope experiment CAST [33–36]. We also put predictions from the benchmark QCD axion models. The yellow band and the upper solid brown line are the predictions from the KSVZ model [3,4] with $0.07 < |E/N - 1.95| < 7$ and $E/N = 0$, respectively, while the bottom dashed brown line is the prediction from the DFSZ model [13,14] with $E/N = 8/3$. The cyan lines are the predictions from the ALP *miracle* model [12] with the intrinsic parameters $c_\gamma = 1, 0.1, 0.01$.

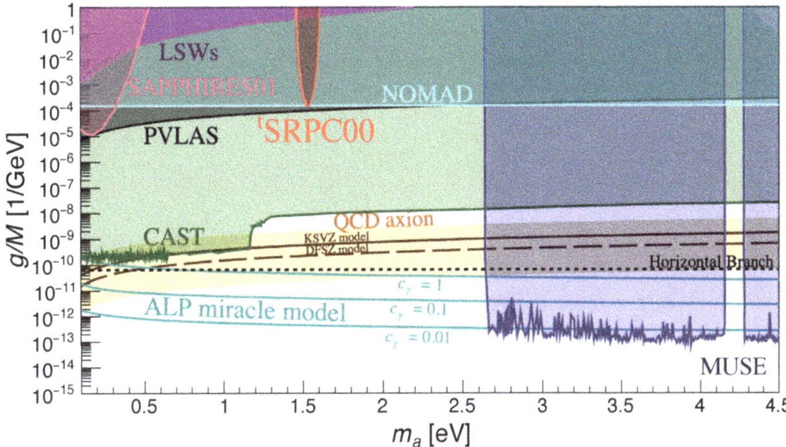

Figure 10. Upper limit in the parameter space of the coupling-mass relation (region enclosed by the red solid curve) evaluated at a 95% confidence level for the pseudoscalar field exchange achieved by the three-beam stimulated resonant photon collider (ᵗSRPC00). The magenta area indicates the excluded range based on SRPC in quasi-parallel collision geometry (SAPPHIRES01) [23]. The purple areas are excluded regions by the Light Shining through a Wall (LSW) experiments (ALPS [27] and OSQAR [28]). The gray area shows the excluded region by the vacuum magnetic birefringence (VMB) experiment (PVLAS [29]). The light-cyan horizontal solid line indicates the upper limit from the search for eV (pseudo)scalar penetrating particles in the SPS neutrino beam (NOMAD) [30]. The horizontal dotted line indicates the upper limit from the Horizontal Branch observation [31]. The blue areas indicate the excluded regions from the optical MUSE-faint survey [32]. The green area is the excluded region by the helioscope experiment CAST [33–36]. The yellow band and the upper solid brown line are the predictions of QCD axion by the KSVZ model [3,4] with $0.07 < |E/N - 1.95| < 7$ and $E/N = 0$, respectively. The bottom dashed brown line is the prediction from the DFSZ model [13,14] with $E/N = 8/3$. The cyan lines are the predictions from the ALP *miracle* model [12] with the intrinsic parameter values $c_\gamma = 1, 0.1, 0.01$, respectively.

7. Conclusions and Future Prospects

We presented a result of the pilot ALP search by a three-beam stimulated resonant photon collider (ᵗSRPC) with focused short pulse lasers in the air as a proof-of-principle experiment. We demonstrated that the space–time synchronization between a pair of short creation laser pulses with a large incident angle of 30 deg, and a relatively long-duration inducing laser pulse can be ensured by atomic four-wave mixing with a thin BBO crystal positioned at the interaction point. The search result was consistent with null, and we could successfully obtain an exclusion region in the minimum coupling $g/M = 1.5 \times 10^{-4}$ GeV^{-1} at $m = 1.53$ eV based on the formulation dedicated for ᵗSRPC [25]. We found the solutions to technical complications to handle three focused short-pulsed beams and the impact on the physics, in particular, on the polarization states of creation beams by the introduction of periscopes, which is an important optical element to realize variable incident angles at a ᵗSRPC.

The pilot search was indeed performed at a narrow mass range indicated by the angle points as a function of ALP mass as shown in Figure 11. Our prospect is to cover the broad mass range in the eV scale [25]. Toward the continuous mass scanning over the eV range with much higher laser intensity in the near future, the technical solutions developed in this pilot search will enable a realistic designing for a more compact ᵗSRPC operational in a vacuum chamber. Although the pilot search with the low laser intensity looks dominated by the large systematic uncertainty, the uncertainty is actually dominated by electric noise in the experimental environment, which is independent of the increase of laser intensity, while the signal yield is increased by the cube of laser intensity. Therefore, the method we demonstrated opens up a new window toward very feeble coupling of ALPs to photons by increasing laser intensity in the near future.

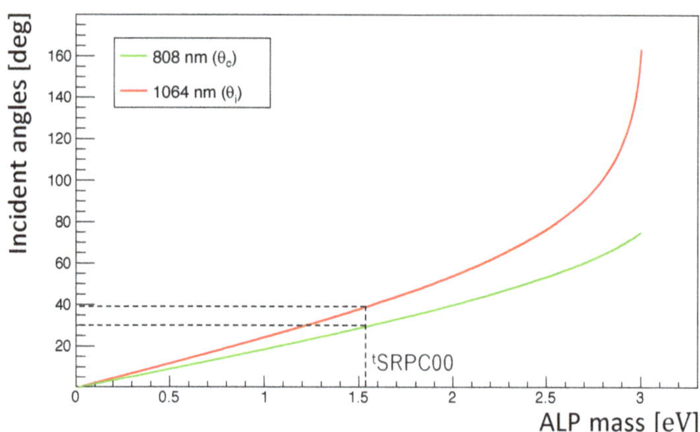

Figure 11. Expected incident angles of creation and inducing lasers, θ_c and θ_i, respectively, as a function of ALP mass when two wavelengths of creation (808 nm) and inducing lasers (1064 nm) are assumed, resulting in the fixed wavelength of FWM signals, 651 nm, in vacuum.

Author Contributions: Conceptualization, K.H.; methodology, K.H.; software, Y.K. and F.I.; validation, Y.K., T.K. and S.M.; formal analysis, F.I. and Y.K.; investigation, T.K.; resources, S.T. and M.H.; data curation, F.I., T.H. and Y.K.; writing—original draft preparation, F.I. and K.H.; writing—review and editing, Y.K., T.H. and K.H.; visualization, F.I. and K.H.; supervision, K.H.; project administration, K.H.; funding acquisition, K.H. All authors have read and agreed to the published version of the manuscript.

Funding: This research was funded by Grants-in-Aid for Scientific Research Nos. 21H04474 from the Ministry of Education, Culture, Sports, Science, and Technology (MEXT) of Japan.

Institutional Review Board Statement: Not applicable.

Informed Consent Statement: Not applicable.

Data Availability Statement: Not applicable.

Acknowledgments: The T^6 system was financially supported by the MEXT Quantum Leap Flagship Program (JPMXS0118070187) and the program for advanced research equipment platforms (JPMXS0450300521). Kensuke Homma acknowledges the support of the Collaborative Research Program of the Institute for Chemical Research at Kyoto University (Grant Nos. 2018-83, 2019-72, 2020-85, 2021-88, and 2022-101) and Grants-in-Aid for Scientific Research Nos. 17H02897, 18H04354, 19K21880, and 21H04474 from the Ministry of Education, Culture, Sports, Science, and Technology (MEXT) of Japan. Yuri Kirita acknowledges support from the JST, the establishment of university fellowships for the creation of science and technology innovation, Grant No. JPMJFS2129, and a Grant-in-Aid for JSPS fellows No. 22J13756 from the Ministry of Education, Culture, Sports, Science, and Technology (MEXT) of Japan.

Conflicts of Interest: The authors declare no conflict of interest.

Appendix A. Two Angle Parameters for Jones Vectors Representing General Polarization States

As mentioned in Section 2, incident angles of individual lasers were set at 30° for the two creation lasers with respect to the horizontal dashed line including IP as shown in

Figure A1 through periscopes PS_{c1} and PS_{c2}, respectively. We adopted PS to introduce a large collision angle because it could reflect a laser beam to any angles changing the beam height thanks to a vertical pair of mirrors with an incident angle of $45°$ and a reflection angle of $45°$ inside PS. However, if the output direction is rotated by PS to a large angle, a linear polarization state of an incident laser beam becomes elliptically polarized. Therefore, it is necessary to measure Stokes parameters to obtain ellipticity angle ϵ and tilt angle θ of a complex Jones vector defined as follows that represents elliptically polarized states in general

$$\begin{pmatrix} e_1 \\ e_2 \end{pmatrix} = \begin{pmatrix} \cos\theta & -\sin\theta \\ \sin\theta & \cos\theta \end{pmatrix} \begin{pmatrix} \cos\epsilon \\ -i\sin\epsilon \end{pmatrix} = \begin{pmatrix} \cos\theta\cos\epsilon + i\sin\theta\sin\epsilon \\ \sin\theta\cos\epsilon - i\cos\theta\sin\epsilon \end{pmatrix}. \tag{A1}$$

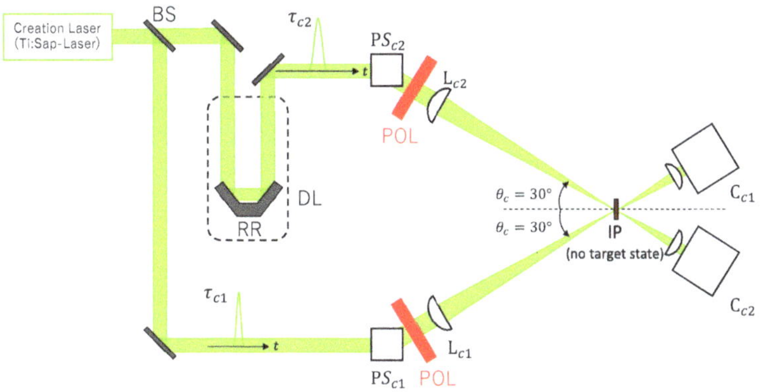

Figure A1. Schematic view of the setup to evaluate two angle parameters to define Jones vectors for the two creation lasers based on Stokes parameters. A polarizer (POL) was placed between a periscope (PS) and a lens (L) in each beamline. Stokes parameters were obtained by measuring transmitted laser intensity through POL at four different rotation angles by using individual cameras C_{c1} and C_{c2} assigned for the two creation lasers.

Complex Jones vectors were actually implemented to polarization vectors in four-vector form $e \equiv (0, e_1, e_2, 0)$ to define vertex factors for the ALP-photon coupling in the numerical calculation to obtain volume-wise interaction rates $\bar{\Sigma}_I$ in Equation (7) (see Ref. [24] in more detail).

To obtain these two angles in complex Jones vectors for the two creation lasers, a polarizer (POL) was placed between a periscope (PS) and a lens (L) for each of the two creation laser lines as shown in Figure A1. Rotation angles of POL around the optical axis were set to select a linear polarization direction of $0°$(horizontal), $90°$(vertical), $45°$, and $135°$. The Stokes parameters, which can be converted into the two angle parameters for complex Jones vectors, were obtained by measuring laser intensities monitored by cameras (C_{c1}, C_{c2}) after laser lights pass through a rotated POL set at the four rotation angles above. A set of Stokes parameters can be related to two angle parameters of a complex Jones vector as follows:

$$\begin{pmatrix} S_0 \\ S_1 \\ S_2 \\ S_3 \end{pmatrix} = \begin{pmatrix} |E_H|^2 + |E_V|^2 \\ |E_H|^2 - |E_V|^2 \\ |E_{45°}|^2 - |E_{135°}|^2 \\ |E_L|^2 - |E_R|^2 \end{pmatrix} = S_0 \begin{pmatrix} 1 \\ \cos 2\epsilon \cos 2\theta \\ \cos 2\epsilon \sin 2\theta \\ \sin 2\epsilon \end{pmatrix} \tag{A2}$$

where $E_H, E_V, E_{45°}, E_{135°}, and E_L, E_R$ are the amplitudes for linear polarization cases with the polarization direction of horizontal, vertical, $45°$, $135°$, and for left- and right-handed circular polarization cases, respectively. In the search experiment, we did not measure the right-handed and left-handed laser amplitudes because we only have to obtain the two angle parameters: ϵ_k and θ_k for $k = c1, c2$ from S_0, S_1 and S_2.

The two angle parameters for the Jones vector in the creation laser path without retroreflector (RR) (lower side of the green optical path in Figure A1) were $\epsilon_{c1} = 0.41$ rad, $\theta_{c1} = 0.30$ rad, while those in the path with RR were $\epsilon_{c2} = 0.91$ rad, $\theta_{c2} = -0.31$ rad as summarized in the following relation:

$$\begin{pmatrix} S_{c10} \\ S_{c11} \\ S_{c12} \\ S_{c13} \end{pmatrix} = A_{c1}^2 \begin{pmatrix} 1 \\ \cos 2\epsilon_{c1} \cos 2\theta_{c1} \\ \cos 2\epsilon_{c1} \sin 2\theta_{c1} \\ \sin 2\epsilon_{c1} \end{pmatrix} = A_{c1}^2 \begin{pmatrix} 1 \\ \cos (0.82) \cos (0.60) \\ \cos (0.82) \sin (0.60) \\ \sin (0.82) \end{pmatrix} \tag{A3}$$

$$\begin{pmatrix} S_{c20} \\ S_{c21} \\ S_{c22} \\ S_{c23} \end{pmatrix} = A_{c2}^2 \begin{pmatrix} 1 \\ \cos 2\epsilon_{c2} \cos 2\theta_{c2} \\ \cos 2\epsilon_{c2} \sin 2\theta_{c2} \\ \sin 2\epsilon_{c2} \end{pmatrix} = A_{c2}^2 \begin{pmatrix} 1 \\ \cos (1.82) \cos (-0.62) \\ \cos (1.82) \sin (-0.62) \\ \sin (1.82) \end{pmatrix} \tag{A4}$$

where $A_{c1,c2}^2$ correspond to intensities measured by $C_{c1,c2}$, respectively. The ellipticity angle for the creation laser containing RR was closer to $\pi/4$ than that of the other creation laser because the incident angle of $45°$ and the reflection angle of $45°$ were not guaranteed within RR, while the tilt angles were opposite to each other as expected. Therefore, the two creation lasers indeed had very different angle parameters, and these factors were taken into account for the numerical calculation to set the exclusion region.

References

1. Weinberg, S. A New Light Boson? *Phys. Rev. Lett.* **1978**, *40*, 223. [CrossRef]
2. Wilczek, F. Problem of Strong P and T Invariance in the Presence of Instantons. *Phys. Rev. Lett.* **1978**, *40*, 271. [CrossRef]
3. Kim, J.E. Weak-Interaction Singlet and Strong CP Invariance. *Phys. Rev. Lett.* **1979**, *43*, 103. [CrossRef]
4. Shifman, M.A.; Vainshtein, A.I.; Zakharov, V.I. Can confinement ensure natural CP invariance of strong interactions? *Nucl. Phys. B* **1980**, *166*, 493. [CrossRef]
5. Preskill, J.; Wise, M.B.; Wilczek, F. Cosmology of the invisible axion. *Phys. Lett. B* **1983**, *120*, 127. [CrossRef]
6. Abbott, L.F.; Sikivie, P. A cosmological bound on the invisible axion. *Phys. Lett.* **1983**, *B120*, 133. [CrossRef]
7. Dine, M.; Fischler, W. The not-so-harmless axion. *Phys. Lett.* **1983**, *B120*, 137. [CrossRef]
8. Peccei, R.D.; Quinn, H.R. CP Conservation in the Presence of Pseudoparticles. *Phys. Rev. Lett.* **1977**, *38*, 1440. [CrossRef]
9. Peccei, R.D. The strong CP problem and Axions. *Lect. Notes Phys.* **2008**, *741*, 3–17.
10. Fujii, Y.; Maeda, K. *The Scalar-Tensor Theory of Gravitation*; Cambridge University Press: Cambridge, UK, 2003.
11. Katsuragawa, T.; Matsuzaki, S.; Homma, K. Hunting dark energy with pressure-dependent photon-photon scattering. *Phys. Rev. D* **2022**, *106*, 044011. [CrossRef]
12. Daido, R.; Takahashi, F.; Yin, W. The ALP miracle revisited. *J. High Energy Phys.* **2018**, *2*, 104. [CrossRef]
13. Dine, M.; Fischler, W.; Srednicki, M. A simple solution to the strong CP problem with a harmless axion. *Phys. Lett.* **1981**, *104*, 199–202. [CrossRef]
14. Zhitnitsky, A. On Possible Suppression of the Axion Hadron Interactions. *Sov. J. Nucl. Phys.* **1980**, *31*, 260.
15. Yin, W. Thermal production of cold "hot dark matter" around eV. *arXiv* **2023**, arXiv:2301.08735.
16. Takahashi, F.; Yin, W. Hadrophobic Axion from GUT. *arXiv* **2023**, arXiv:2301.10757.
17. Fujii, Y.; Homma, K. An Approach toward the Laboratory Search for the Scalar Field as a Candidate of Dark Energy. *Prog. Theor. Phys.* **2011**, *126*, 531; Erratum in *Prog. Theor. Exp. Phys.* **2014**, *2014*, 089203. [CrossRef]
18. Homma, K.; Hasebe, T.; Kume, K. The first search for sub-eV scalar fields via four-wave mixing at a quasi-parallel laser collider. *Prog. Theor. Exp. Phys.* **2014**, *2014*, 083C01. [CrossRef]
19. Hasebe, T.; Homma, K.; Nakamiya, Y.; Matsuura, K.; Otani, K.; Hashida, M.; Inoue, S.; Sakabe, S. Search for sub-eV scalar and pseudoscalar resonances via four-wave mixing with a laser collider. *Prog. Theo. Exp. Phys.* **2015**, *2015*, 073C01. [CrossRef]
20. Nobuhiro, A.; Hirahara, Y.; Homma, K.; Kirita, Y.; Ozaki, T.; Nakamiya, Y.; Hashida, M.; Inoue, S.; Sakabe, S. Extended search for sub-eV axion-like resonances via four-wave mixing with a quasi-parallel laser collider in a high-quality vacuum system . *Prog. Theo. Exp. Phys.* **2020**, *2020*, 073C01. [CrossRef]
21. Homma, K.; Kirita, Y. Stimulated radar collider for probing gravitationally weak coupling pseudo Nambu-Goldstone bosons. *J. High Energy Phys.* **2020**, *9*, 095. [CrossRef]
22. Homma, K. et al. [SAPPHIRES]. Search for sub-eV axion-like resonance states via stimulated quasi-parallel laser collisions with the parameterization including fully asymmetric collisional geometry. *J. High Energy Phys.* **2021**, *12*, 108.
23. Kirita, Y. et al. [SAPPHIRES]. Search for sub-eV axion-like resonance states via stimulated quasi-parallel laser collisions with the parameterization including fully asymmetric collisional geometry. *J. High Energy Phys.* **2022**, *10*, 176.

24. Homma, K.; Kirita, Y.; Ishibashi, F. Perspective of Direct Search for Dark Components in the Universe with Multi-Wavelengths Stimulated Resonant Photon-Photon Colliders. *Universe* **2021**, *7*, 479. [CrossRef]
25. Homma, K.; Ishibashi, F.; Kirita, Y.; Hasada, T. Sensitivity to axion-like particles with a three-beam stimulated resonant photon collider around the eV mass range. *Universe* **2023**, *9*, 20. [CrossRef]
26. Beringer, J. et al. [Particle Data Group]. Review of Particle Physics. *Phys. Rev. D* **2012**, *86*, 010001. [CrossRef]
27. Ehret, K. et al. [ALPS]. New ALPS results on hidden-sector lightweights. *Phys. Lett. B* **2010**, *689*, 149. [CrossRef]
28. Ballou, R. et al. [OSQAR]. New exclusion limits on scalar and pseudoscalar axionlike particles from light shining through a wall. *Phys. Rev. D* **2015**, *92*, 092002. [CrossRef]
29. Ejlli, A.; Della Valle, F.; Gastaldi, U.; Messineo, G.; Pengo, R.; Ruoso, G.; Zavattini, G. The PVLAS experiment: A 25 year effort to measure vacuum magnetic birefringence. *Phys. Rep.* **2020**, *871*, 201374. [CrossRef]
30. Astier, P. et al. [NOMAD]. Search for eV (pseudo)scalar pene- trating particles in the SPS neutrino beam. *Phys. Lett. B* **2000**, *479*, 371–380. [CrossRef]
31. Ayala, A.; Domínguez, I.; Giannotti, M.; Mirizzi, A.; Straniero, O. Revisiting the Bound on Axion-Photon Coupling from Globular Clusters. *Phys. Rev. Lett.* **2014**, *113*, 191302. [CrossRef]
32. Regis, M.; Taoso, M.; Vaz, D.; Brinchmann, J.; Zoutendijk, S.L.; Bouché, N.F.; Steinmetz, M. Searching for light in the darkness: Bounds on ALP dark matter with the optical MUSE-faint survey. *Phys. Lett. B* **2021**, *814*, 136075. [CrossRef]
33. Arik, E. et al. [CAST]. Probing eV-scale axions with CAST. *J. Cosmol. Astropart. Phys.* **2009**, *2*, 8;
34. Arik, M. et al. [CAST]. Search for Sub-eV Mass Solar Axions by the CERN Axion Solar Telescope with ^{3}He Buffer Gas. *Phys. Rev. Lett.* **2011**, *107*, 261302. [CrossRef] [PubMed]
35. Arik, M. et al. [CAST]. Search for Solar Axions by the CERN Axion Solar Telescope with ^{3}He Buffer Gas: Closing the Hot Dark Matter Gap. *Phys. Rev. Lett.* **2014**, *112*, 091302.
36. Anastassopoulos, V. et al. [CAST]. New CAST limit on the axion–photon interaction. *Nat. Phys.* **2017**, *13*, 584. [CrossRef]

Article

Design and Construction of a Variable-Angle Three-Beam Stimulated Resonant Photon Collider toward eV-Scale ALP Search

Takumi Hasada [1], Kensuke Homma [1,2,*] and Yuri Kirita [1]

[1] Graduate School of Advanced Science and Engineering, Hiroshima University, Kagamiyama, Higashi-Hiroshima 739-8526, Hiroshima, Japan

[2] International Center for Quantum-Field Measurement Systems for Studies of the Universe and Particles (QUP), KEK, Tsukuba 305-0801, Ibaraki, Japan

* Correspondence: khomma@hiroshima-u.ac.jp

Abstract: We aim to search for axion-like particles in the eV mass range using a variable-angle stimulated resonance photon collider (SRPC) with three intense laser beams. By changing angle of incidence of the three beams, the center-of-mass-system collision energy can be varied and the eV mass range can be continuously searched for. In this paper, we present the design and construction of such a variable-angle three-beam SRPC (^{t}SRPC), the verification of the variable-angle mechanism using a calibration laser, and realistic sensitivity projections for searches in the near future.

Keywords: dark matter; axion; axion-like particle; ALP; inflaton; laser; stimulated resonant photon collider; four-wave mixing

Citation: Hasada, T.; Homma, K.; Kirita, Y. Design and Construction of a Variable-Angle Three-Beam Stimulated Resonant Photon Collider toward eV-Scale ALP Search. *Universe* **2023**, *9*, 355. https://doi.org/10.3390/universe9080355

Academic Editor: Kazuharu Bamba

Received: 16 June 2023
Revised: 17 July 2023
Accepted: 27 July 2023
Published: 29 July 2023

1. Introduction

Some of the unsolved problems of the Standard Model may be answered by new particles in the low-mass region, which have not yet been fully explored. Nambu–Goldstone bosons (NGBs), which are supposed to be ideally massless, may appear whenever global continuous symmetries spontaneously break [1–3]. Axions [4,5] are a representative candidate for such new fields in the low-mass range because an axion is a pseudo Nambu–Goldstone boson (pNGB) arising from spontaneous breaking of the Peccei–Quinn symmetry [6] introduced to solve the strong CP problem [7] in the context of Quantum ChromoDynamics (QCD). Moreover, if an axion weakly couples with matter fields, it can be a natural candidate for dark matter in the Universe [8–10]. More generalized low-mass particles are called Axion-Like-Particles (ALPs), some of which can also be reasonable dark matter candidates.

Many experiments have been conducted to detect ALPs focusing on their coupling to photons given by the following interaction Lagrangian $\mathcal{L} = -\frac{1}{4}\frac{g}{M}F_{\mu\nu}\tilde{F}^{\mu\nu}a$, where a is an ALP field, $F_{\mu\nu}$ is the electric field strength, $\tilde{F}^{\mu\nu}$ is its dual, and g/M is the coupling constant with dimensionless parameter g and M denoting an energy scale at which a symmetry breaking takes place. From the experimental point of view, a mass range between 0.001 and 10 eV has not been intensively explored, especially by laboratory-based experiments. In addition to the QCD axion scenario, a model *miracle* predicts the existence of an ALP in the mass range 0.01~1 eV as a possible explanation for both inflation and dark matter [11,12]. It is thus very intriguing to conduct search experiments using laser fields in the near-infrared region in order to have sensitivities to the eV range.

We have been performing the ALP search based on the concept of Stimulated Resonant Photon Collider (SRPC) [13–15]. In this method, a single pulsed creation laser is focused and two arbitrary photons included in the field collide with each other, resulting in the production of an ALP and another pulsed inducing laser simultaneously stimulates its decay. This method does not require any assumptions except the interaction Lagrangian, and is thus independent of any cosmological and astrophysical models. In our previous study of a quasi-parallel collision system, we focused two lasers into the same optical axis and collided them at a shallow incident angle to search for ALPs in the sub-eV mass range [15–19]. Recently, we have proposed and demonstrated a pilot search for heavier ALPs based on a three-beam stimulated resonant collider (^{t}SRPC) [20,21]. In the near future, we plan to search for ALPs by continuously changing collision angles in the eV mass range.

In this paper, we present the design and construction of a three-beam SRPC that can continuously scan the eV mass range by changing the incident angles of the three colliding lasers. We then show the verification of the mechanism of collision angle changes for individual mass range using a He:Neon laser for the calibration purpose. In the following sections, we first discuss a choice of the basic design to introduce the variable collision angles. Secondly, we introduce the concrete design for the variable-angle stimulated resonant photon collider by taking several aspects of calibration steps into account. Thirdly, we provide the verification of the selected mechanism using the calibration laser. We finally discuss the achievable sensitivity projections based on a realistic experimental parameter set and conclude the paper.

2. Kinematics in Three-Beam Stimulated Resonant Photon Collider, ^{t}SRPC

Figure 1 shows the conceptual drawing of the ^{t}SRPC. We consider a case that two creation laser pulses (green) are focused at the same incident angle θ_c with the same energy ω_c and similarly an inducing laser pulse (red) is focused with the energy ω_i which increases the interaction rate of stimulated scattering emitting signal photons of energy ω_s (blue) that satisfies energy-momentum conservation. Energy-momentum conservation between three-beam photons and a signal photon requires the following kinematical relations:

$$\omega_c + \omega_c = \omega_s + \omega_i$$
$$2\omega_c \cos\theta_c = \omega_s \cos\theta_s + \omega_i \cos\theta_i \tag{1}$$
$$\omega_s \sin\theta_s = \omega_i \sin\theta_i.$$

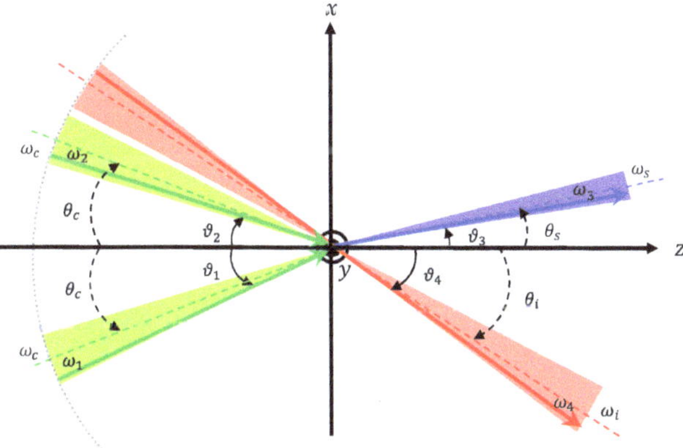

Figure 1. Concept of a three-beam stimulated resonant photon collider (^{t}SRPC) with focused coherent beams [21]. The two focused creation laser beams (green) at the incident angle θ_c produces an ALP resonance state and the focused inducing laser beam (red) stimulates its decay. The creation photons have different energies ω_1 and ω_2 from the central value of ω_c and different incident angles ϑ_1 and ϑ_2 from θ_c, respectively. Similarly, the inducing laser (red) with a central wavelength of ω_i has part of the beam with ω_4, increasing the emission probability of the signal photon of ω_3 (blue) via energy-momentum conservation.

Experimentally we first fix beam energies ω_c and ω_i among available laser wavelengths. We then target a ALP mass of m_a which coincides with the center of mass system energy E_{cms} defined as

$$m_a = E_{cms} = 2\omega_c \sin \theta_c. \tag{2}$$

From Equations (1) and (2) the angle of incidence for the inducing beam can be determined as follows

$$\theta_i = \arccos\left\{ \left(1 - \frac{m_a^2}{4\omega_c\omega_i}\right)\left(1 - \frac{m_a^2}{4\omega_c^2}\right)^{-\frac{1}{2}} \right\}. \tag{3}$$

We emphasize, however, that individual photons within focused beams indeed have different energies, ω_1 and ω_2 from ω_c and ω_4 from ω_i. These energy uncertainties are caused by the Fourier transform-limited short-pulse lasers. In addition, in the focused fields, the incident angles ϑ_1, ϑ_2, and ϑ_4 are also different from θ_c and θ_i, respectively. Fluctuations in the angle of incidence around the beam axis are caused by momentum fluctuations near the focal point. Fortunately, these uncertainties give the center-of-mass collision energy E_{cms} a finite width via the following relation

$$E_{cms} = 2\sqrt{\omega_1\omega_2} \sin\left(\frac{\vartheta_1 + \vartheta_2}{2}\right). \tag{4}$$

Thus, ALP mass scanning is possible even though central values, θ_c, are varied in a discrete step if the E_{cms} uncertainty defined by laser pulse duration and the focal parameters is consistent with the discrete angle step in θ_c in a search. For more information, see [20].

3. Basic Design to Realize Variable Collision Angles

In order to realize continuously changeable collision angles between three focused beams, the following two main ideas were considered. Figure 2 (left) shows a natural way to change collision angles by changing the incident positions of lasers on a parabolic mirror surface, while Figure 2 (right) shows a focusing system on multi-layered rotating stages where angles of incidence in the individual layers are changeable by independently rotating the individual stages.

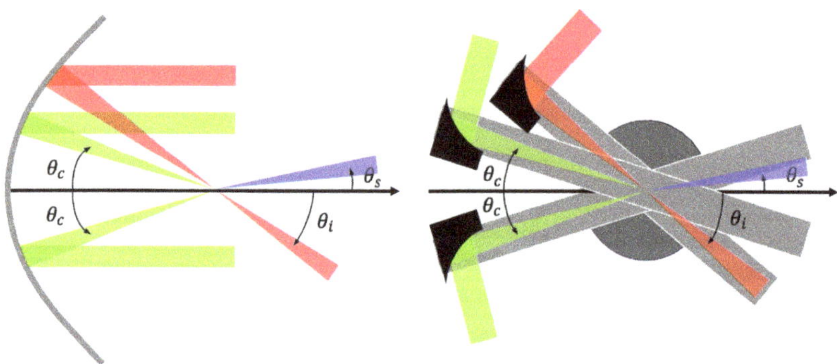

Figure 2. Two proposals for variable angle mechanisms. The green beams are creation lasers, the red beam is an inducing laser, and the blue beam indicates signal photons. **Left:** parabolic mirror type where the collision angle is changeable by changing the incident position of lasers. **Right:** rotating stage type where the collision angle is changeable by assembling a beam focusing system on multiple rotary stages.

Table 1 shows a comparison between two ideas based on the advantages and disadvantages. While the parabolic mirror type has the advantages of fewer optical components and readiness of the angle adjustment, it is necessary to prepare a custom-made large-area mirror to reach heavier mass region with large incident angles θ_c and θ_i. The search must be performed in a vacuum chamber in order to reduce four-wave mixing (FWM) originating from ambient atoms. Therefore, a large parabolic mirror is not suitable for implementing it in the vacuum chamber. We also note that the focal length must slightly change for individual incident angles.

Table 1. Comparison of variable-angle mechanisms.

Item	Parabolic Mirror Type	Rotary Stage Type
Adjustment	easy	difficult
Size	large	compact (vacuum chamber compatible)
Angle range	narrow	wide
Focal length	angle-dependently variable	fixed
Flexibility	low (custom-ordered mirror)	high (catalog items)

On the other hand, the rotary stage type allows scanning over a wide range of incident angles in a compact size. In addition, it can be constructed using a combination of commercial products, hence having the advantage of flexibility for the design so that we can replace focusing mirrors as we need. As we demonstrated in the previous pilot search [21], we indeed found that it was necessary to prepare a space for a target holder at the focal point to ensure the spatiotemporal synchronization of three pulses, a camera system to record the beam profile, and a shielding wall to suppress background from beam remnants. This requires lots of flexibility including changing the focal length, which has an impact on the sensitivity. However, there is a disadvantage of the rotary stage type in narrow incident angles because focusing optical elements spatially overlap between two incident creation beams.

Therefore, in this study, we adopted the hybrid concept combining good futures of the parabolic mirror type and the rotary stage type depending on incident angles of creation lasers. For large incident angles, the design is based on the rotary stage type with two moving stages for individual creation beams while the inducing beam is fixed at an optical table (LA collider). For narrow incident angles, on the other hand, one of the two creation beams and the inducing beam share a common parabolic focusing mirror and the mirror is fixed at the optical table while the remaining creation beam rotates together with a moving stage (NA collider). In the next section we discuss the relation between LA and NA setups in detail.

The variable-angle mechanism using the rotary stages associates additional complications. The rotary stages consist of individual aluminum plates placed on individual rotating stages. On the individual aluinum plate, a periscope (PS) and a parabolic mirror (PM) to focus a beam are assembled. The incident angles are changeable by rotating the stage. Changing incident angles must accompany changes of incident points of laser beams. However, typically the laser incident point is not readily movable because high-intensity laser systems are not compact. In order to compensate for this immobility, we introduce periscope (PS) components as shown in the picture of Figure 3. PS consists of a pair of mirrors aligned vertically with the angle of incidence (AOI) of 45°. PS can bend a beam in any directions by rotating the direction of the mirror in the upside of the PS. The height of an optical axis is changeable by adjusting the relative distance between the two mirrors in the incident and outgoing sides. The change of incident angles is compensated by the parallel movement of the mirror (M) along the x-axis rail stage in advance of injection to the PS. However, since PS reflects a beam to PM with a possibly large angle, it is necessary to evaluate the effect that the incident linearly polarized state becomes an elliptically polarized one. As we demonstrated in the pilot search, the change of the polarization state can be evaluated based on the measurement of Stokes parameters [21]. In this collision system, incident angles are changeable without moving the collision point by setting the PM's focal point at the common center of the rotating stages (RSs). The focal spots can be checked using a camera which also rotates around the center. By stacking layers consisting of an aluminum plate and a rotating stage, the collision angle of the three beams can be independently varied. We note, however, that it is not necessary to change the angles of all three colliding beams. As we discuss later, one of the three beams can be fixed and we have only to adjust the AOIs of the other two laser beams relative to the fixed one.

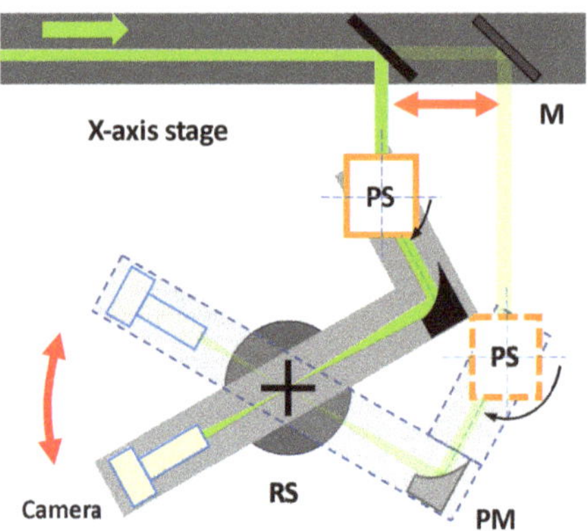

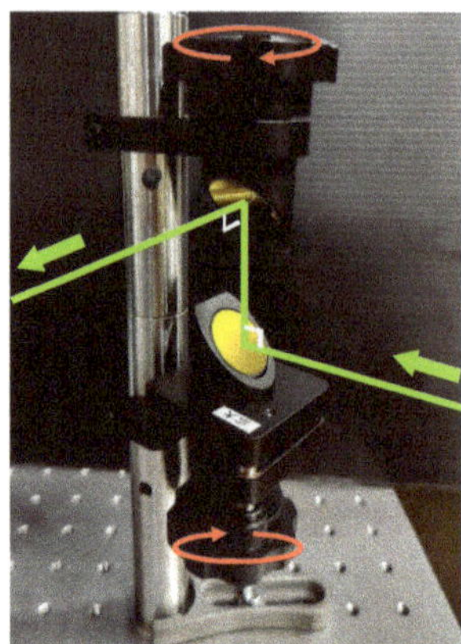

Figure 3. Variable angle mechanism using a rotary stage. The incident angle is varied by rotating a stage assembling a beam-focusing system with a periscope (PS) and a parabolic mirror (PM). By using a periscope (PS), the angle is changeable only by rotating the periscope (PS) and the mirror (M) on the x-axis rail stage in front of PS. By setting the parabolic mirror's focal point at the center of the rotary stage (RS), the collision point remains fixed even when the incident angle is varied. The focal spots can be checked using a monitoring camera.

4. Concrete Designs for the Large- and Narrow-Angle Setups

Figure 4 shows side views of the designed collision systems. First, the three beam interaction point (IP) is defined with the center of two thin crossed wires. The wire target must be immovable with respect to the multi-layer stage movements as discussed below. The common layers between the LA and NA collision cases are the bottom signal sampling layer and the top camera layer to monitor beam profiles at IP. The signal layer is necessary to move the detection point of generated signal photons because the signal direction must change depending on collision angles. We introduce two layers for the two creation laser beams in the LA collision case, while only one layer for one of the two creation laser beams is necessary in the NA collision case because the other creation and the inducing beams share a common parabolic mirror which is fixed to the optical table. The breakdown of the individual layers from the bottom are thus as follows: signal light (black), creation light 1 (cyan), creation light 2 (purple), and camera system (magenta) for LA and signal light (black), creation light 2 (purple), and camera system (magenta) for NA.

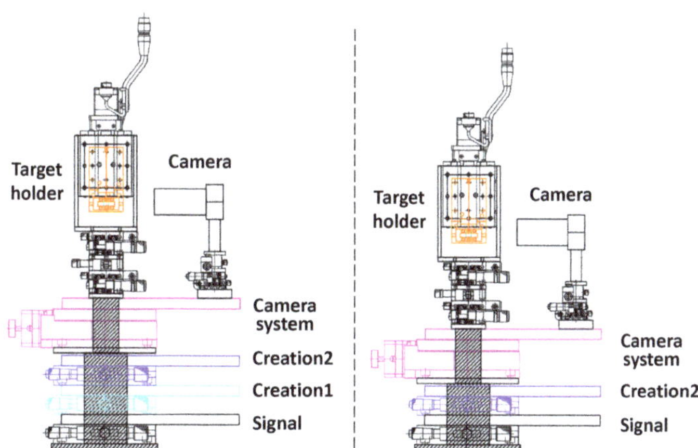

Figure 4. Side views of designed variable-angle three-beam stimulated resonant photon colliders for large-angle (**left**) and narrow-angle (**right**) setups. Detailed explanations are found in the main text.

At IP, a tower-like spacer is placed to introduce an immovable target which must be independent of movements of the four or three rotating stages. The target consists of three components with different positions aligned along a common vertical line: thin crossed wires to calibrate space overlap between the three beams, a nonlinear optical crystal, BBO, to calibrate synchronization of three laser pulses, and an empty hole to perform the search in the vacuum. The target holder is attached to an automated stage that can move vertically to select the three components depending on the purposes. The dynamic range of incident angles is determined by the geometric limitation of the optical elements. At shallow collision angles, parabolic mirrors start interfering with each other, while at wide collision angles, aluminum layers start interfering with each other.

Figure 5a shows the rotary-stage-type geometries in the LA setup covering a large-angle range from 24.8 deg (left) to 47.9 deg (right). The search eventually must be conducted in a vacuum chamber to suppress the atomic background processes. Therefore, we aim at a compact design that can be housed in a vacuum chamber. The incoming laser from the left represents the creation beam 1, creation beam 2, and inducing beam (c_1, c_2, i), respectively. The beam-like objects after the focal point represent the second harmonic generation (SHG) and the four-wave mixing photons corresponding signal photons (s). In order to guide the generated signal photons to a sensor (photomultiplier, PMT), a calibration laser mimicking signal photon trajectories and the wavelength is necessary. The red inducing beam can be fixed at the bottom optical table in the vacuum chamber, while the incident angles of the other beams can be aligned relative to this beam. The incident angles are changeable by rotating creation stages, which requires re-adjustment of PS and M on the one-axis stage as illustrated in Figure 3, respectively. Figure 5b shows the rotary-stage-type geometries combined with the parabolic-type geometry in the NA setup covering a narrow-angle range from 9.3 deg (left) to 24.8 deg (right). The inducing beam and one of the two creation beams share the common parabolic mirror fixed at the bottom optical table in the vacuum chamber. The narrow angle incidence is realizable by changing the position of incidence of the inducing beam on the surface of the common parabolic mirror by fixing the c_1 creation beam.

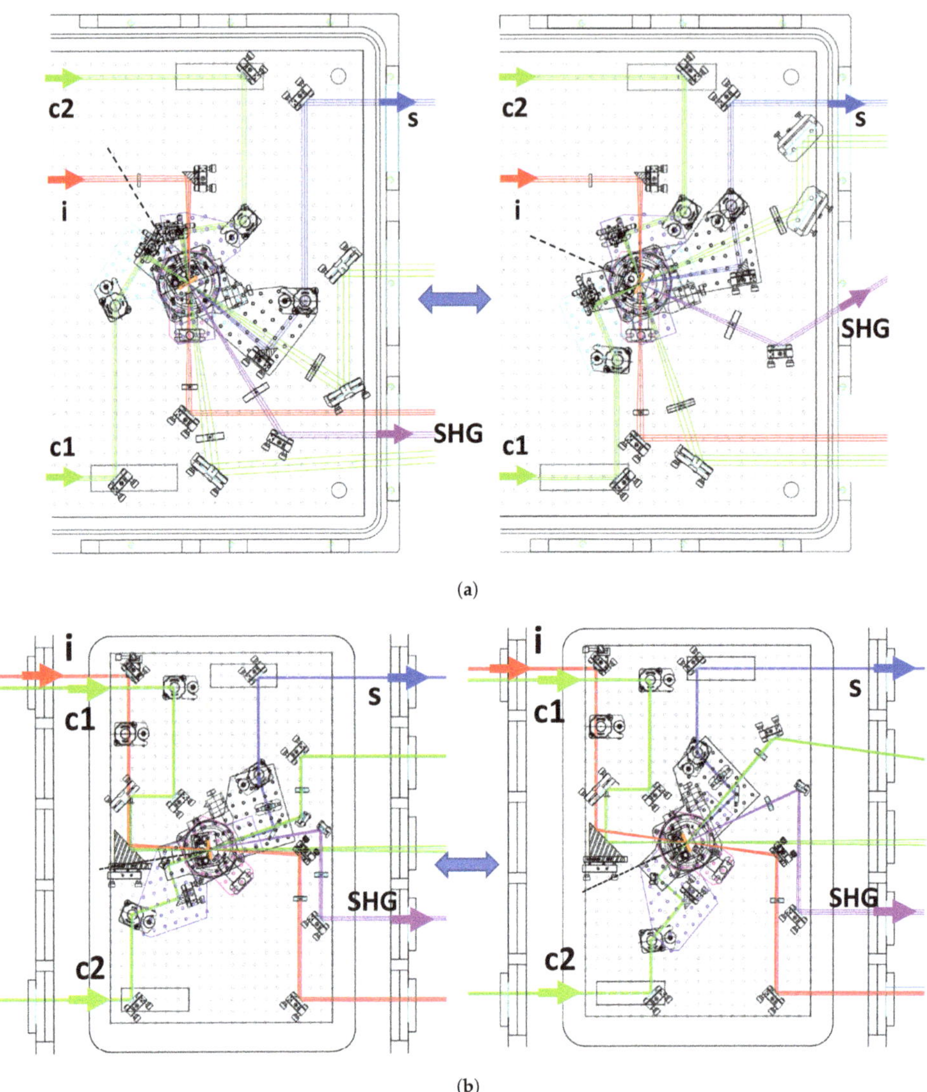

Figure 5. Collisional geometries viewed from the top. (**a**) Large-angle setup from $\theta_c = 24.8$ deg (**left**) to $\theta_c = 47.9$ deg (**right**) and (**b**) narrow-angle setup from $\theta_c = 9.3$ deg (**left**) to $\theta_c = 24.8$ deg (**right**). The details can be found in the main text.

Figure 6 shows a top view of the top common camera layer together with a picture assembling all the components. By preparing an independent layer only for the camera with a motorized rotation stage, beam profiles of all the three lasers at IP can be monitored and recorded. By reading the scale on the motorized rotation stage, the camera position can be adjusted to the incident angles $\pm\theta_c$ relative to the bisecting line (dashed line) which is set by $-\theta_i$ with respect to the inducing beam direction. We note that the camera position must be calibrated so that the camera surface is placed perpendicularly to the radial direction from the central IP position with an equal distance for any rotated positions. By changing the camera position along the radial direction aligned to IP by the local stage on which the camera is installed, one can check whether a beam profile center stays at the same pixel point in the camera or not. If there is a drift of the profile center, that is, a deviation from the perpendicular direction, one can locally fine-tune the camera positions. With this fine-tuning method, AOIs can be adjusted with sufficient accuracy of $0.1°$.

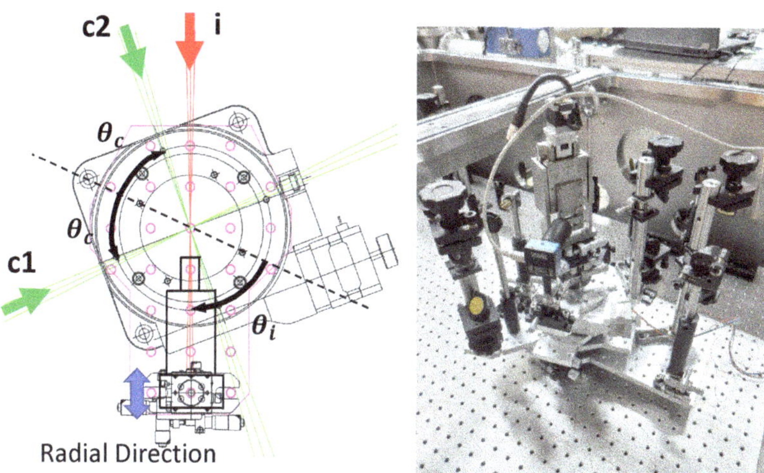

Figure 6. **Left**: top view of the camera system on the top layer of the rotary stages to monitor focal spots of all the three lasers c1, c2, and i. **Right**: picture assembling all the components for the large-angle setup. The details can be found in the main text.

5. Verification of the Rotary Stage System

A three-beam SRPC covering for the LA collision case was actually constructed after testing alignments of PS and PM on individual stages. Since the LA setup contains more rotary layers than the case of the NA setup, the verification of the LA setup guarantees the success of the NA case. The spatial overlapping was then verified using a He:Neon laser. Figures 7–9 show the three collision geometries with incident angles of the creation beam at $\theta_c = 24.8$, 35.5, and 47.9 degrees, and the respective focal images of the three beams taken by a single camera are shown. The optical paths of two creation beams, an inducing beam (c1, c2, i), and signal photons (s) are drawn for the reference. The focal images in the middle column show the spot profiles when two crossed wires with a common 10 µm diameter were placed in front of the three beams (c1, c2, i) with a smaller beam diameter of 0.8 mm for the two creation beams and 2 mm for the inducing beam to have broader focal images on purpose, while the focal images in the right column show the spot profiles at the same camera position after moving the target holder to the position for the search mode (empty hole) by changing beam diameters to a common 5 mm which will be used for the future search. We note the exact focal lengths of the common creation beams and the inducing beam were 101.6 mm 203.2 mm, respectively.

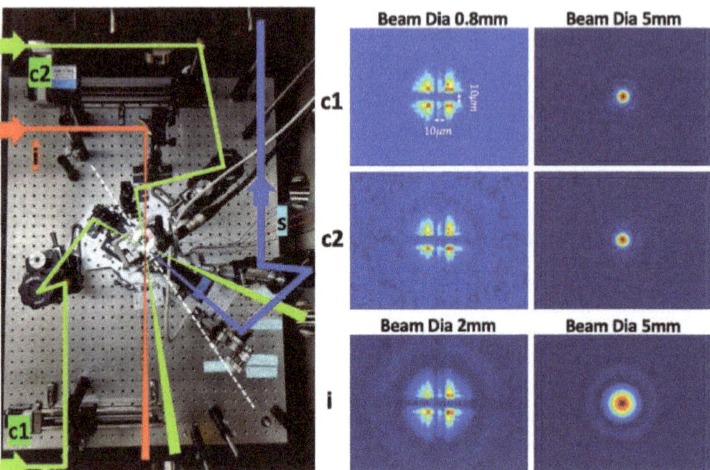

Figure 7. Picture of large-angle setup (**left**) and focal images of three individual beams (**right**) when lasers with a common beam diameter of 5 mm are focused into IP at $\theta_c = 24.8$ degree. In the picture, the optical paths of the three beams, consisting of the creation beam (c1), the creation beam (c2), and the inducing beam (i), as well as the signal photon line (s) are drawn. The middle column shows the images of three individual lasers when they hit the crossed point between two thin target wires of a 10 μm diameter.

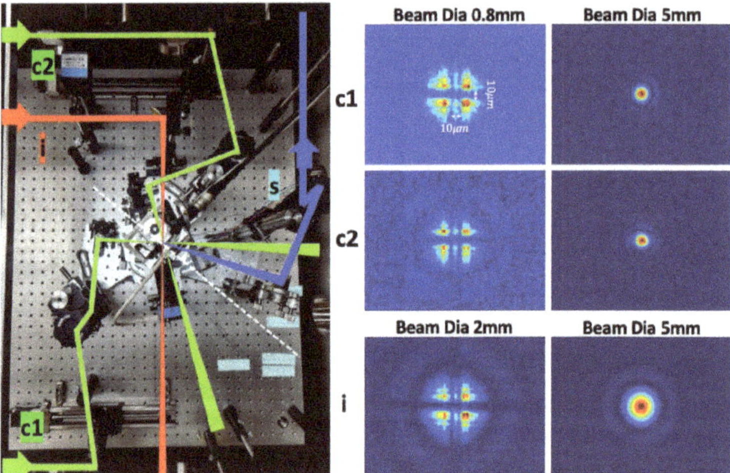

Figure 8. Picture of large-angle setup (**left**) and focal images of three individual beams (**right**) when lasers with a common beam diameter of 5 mm are focused into IP at $\theta_c = 35.5$ degree. The other details are the same as in Figure 7.

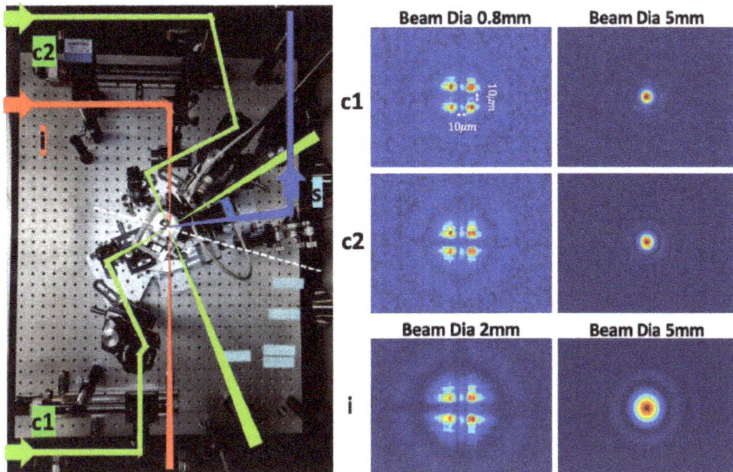

Figure 9. Picture of large-angle setup (**left**) and focal images of three individual beams (**right**) when lasers with a common beam diameter of 5 mm are focused into IP at $\theta_c = 47.9$ degree. The other details are the same as in Figure 7.

6. Realistic Sensitivity Projections

We provide sensitivity projections based on the LA and NA collision setups in the following. Figure 10 shows incident angles of the inducing beam θ_i as a function of ALP mass m_a which is determined by the incident angles of creation beams θ_c and their wavelengths. We plan to use three combinations of laser wavelengths for fundamental, second harmonic and third harmonic cases. Namely, beginning with 800 nm (Ti:Sapphire) for two creation beams and 1064 nm (Nd:YAG) for an inducing beam denoted as $800 \times 2 + 1064$ nm (red), we extend the search to those with $400 \times 2 + 532$ nm (blue) and $267 \times 2 + 355$ nm (magenta). Depending on the combinations between the two angle setups and laser wavelengths, accessible mass ranges are different. The figure shows our projections to cover from 0.5 to 6.9 eV with colored arrows corresponding to different wavelengths combinations where accessible mass ranges are specified.

Accordingly, Figure 11 shows the sensitivity projects based on parameters summarized in Table 2. Based on the set of realistic experimental parameters P in Table 2, the observed number of signal photons via an ALP exchange with the mass m_a and the coupling g/M to two photons is expressed as

$$n_{obs} = \mathcal{Y}_{c+i}(m_a, g/M; P)N_{shots}\epsilon, \tag{5}$$

where ϵ is the overall detection efficiency and N_{shots} is the number of shots. A coupling constant g/M can be numerically calculated by solving Equation (5) for an ALP mass m_a and a given observed number of photons n_{obs}. We assumed the n_{obs} as the noise originating photon-like signals δN_{noise}. NAω, NA2ω, LA2ω and LA3ω are sensitivities corresponding to the individual arrows classified in Figure 10. The details of the numerical calculations and the derivation of the upper limits on the coupling can be found in [20,21], respectively. The red shaded area shows the excluded range based on SRPC in quasi-parallel collision geometry (SAPPHIRES01) [19]. The red filled area indicates the excluded range with the fixed angle pilot search, ^{t}SRPC00 [21]. The gray area shows the excluded region by the vacuum magnetic birefringence experiment (PVLAS [22]). The purple areas are excluded regions by the Light-Shining-through-a-Wall (LSW) experiments (ALPS [23] and OSQAR [24]). The light-cyan horizontal solid line indicates the upper limit from the search for eV (pseudo)scalar penetrating particles in the SPS neutrino beam (NOMAD) [25].

The horizontal dotted line is the upper limit from the Horizontal Branch observation [26]. The blue areas are exclusion regions from the optical MUSE-faint survey [27–32]. The green area indicates the excluded region by the helioscope experiment CAST [31]. The yellow band and the upper solid brown line are the predictions from the benchmark QCD axion models: the KSVZ model [33] with $0.07 < |E/N - 1.95| < 7$ and $E/N = 0$, respectively, while the bottom dashed brown line is the prediction from the DFSZ model [34] with $E/N = 8/3$. The cyan lines are the predictions from the ALP *miracle* model [12] with the model parameters $c_\gamma = 1, 0.1, 0.01$.

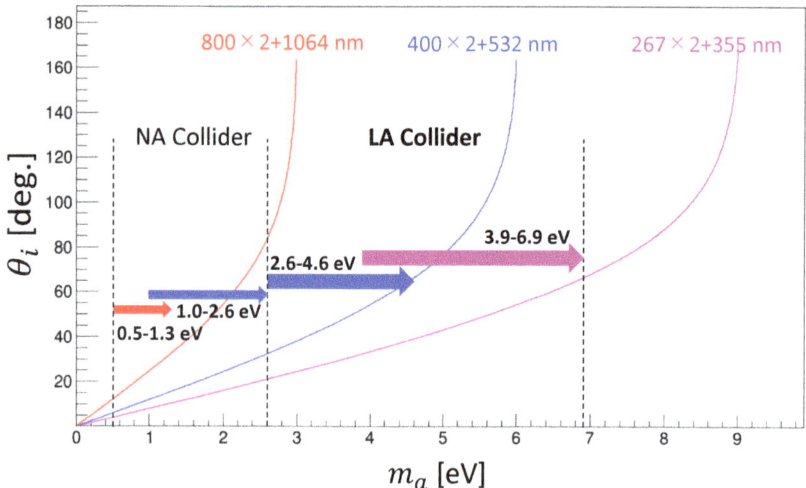

Figure 10. Incident angles of the inducing beam θ_i as a function of ALP masses m_a which are determined by incident angles of creation beams θ_c. Three combinations of laser wavelengths for fundamental, second harmonic, and third harmonic cases. Namely, beginning with 800 nm (Ti:Sapphire) for two creation beams and 1064 nm (Nd:YAG) for an inducing beam expressed as $800 \times 2 + 1064$ nm (red), we extend the search to those with $400 \times 2 + 532$ nm (blue) and $267 \times 2 + 355$ nm (magenta). Depending on the combinations between the two angle setups and laser wavelengths, accessible mass ranges are different. This figure shows projections to cover from 0.5 to 6.9 eV.

Table 2. Experimental parameters used to numerically calculate the upper limits on the coupling–mass relations. (∗) We note that the focal length of the inducing beam in the case of the narrow-angle setup must slightly vary in principle because of the nature of the parabolic mirror. However, since the incident position with respect to the focusing mirror does not vary a lot, for simplicity, we assume a common focal length to evaluate the sensitivity.

Parameters	Values
Two equal creation laser pulses	
Central wavelength of creation laser λ_c	800 nm (ω)/400 nm (2ω)/267 nm (3ω)
Relative linewidth of creation laser, $\delta\omega_c / <\omega_c>$	10^{-2}
Duration time of creation laser, τ_c	40 fs
Creation laser energy per τ_c, E_c	1 mJ
Beam diameter of creation laser beam, d_c	0.005 m
Focal length of narrow-angle setup	$f_c = 0.18$ m
Focal length of large-angle setup	$f_c = 0.10$ m
Polarization	left-handed circular polarization

Table 2. *Cont.*

Parameters	Values
One inducing laser pulse	
Central wavelength of inducing laser λ_i	1064 nm (ω)/532 nm (2ω)/355 nm (3ω)
Relative linewidth of inducing laser, $\delta\omega_i / <\omega_i>$	10^{-4}
Duration time of inducing laser beam, τ_i	9 ns
Inducing laser energy per τ_i, E_i	100 mJ
Beam diameter of inducing laser beam, d_i	0.005 m
Focal length of narrow-angle setup $(*)$	$f_i = 0.19$ m
Focal length of large-angle setup	$f_i = 0.20$ m
Polarization	right-handed circular polarization
Overall detection efficiency, ϵ	5%
Number of shots per collision angle, N_{shots}	10^4 shots
δN_{noise}	50

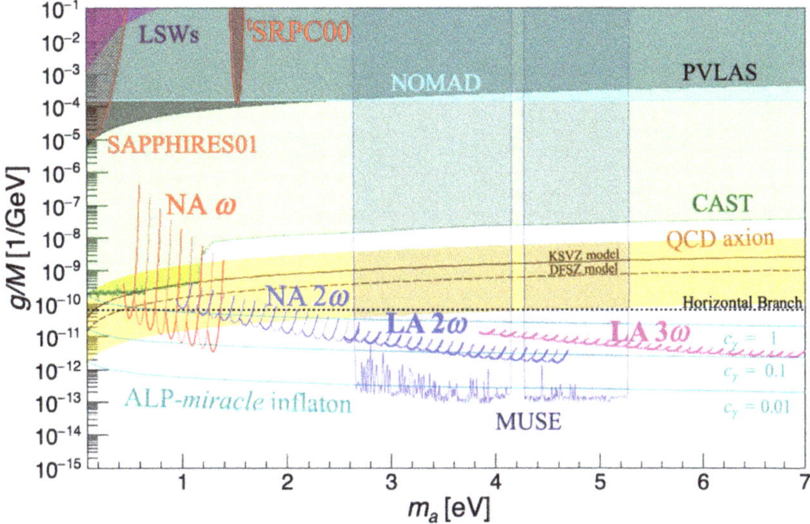

Figure 11. Sensitivity projections based on realistic parameters in Table 2. NAω, NA2ω, LA2ω, and LA3ω are sensitivities corresponding to the four arrows in Figure 10 specifying individual mass ranges. The other details can be found in the main text.

7. Conclusions and Future Plans

We have designed two types of variable-angle stimulated resonant photon colliders with three laser beams (^{t}SRPC) covering narrow and large angles, respectively. The large-angle setup sensitive to a relatively higher mass range was actually constructed, and the mechanism was verified using a He:Neon laser for the calibration. We confirmed that the incident angle can be varied by using a rotating stage and a periscope, and we ensured the spatial overlapping of three beam focal spots at multiple collision angles by developing a monitoring system that allows a single camera to check the focal spot images. As in the previous pilot search at the fixed incident angle [21], time synchronization is expected to be ensured by switching the focal point target to a nonlinear crystal, BBO, and using a delay line with a retro-reflector when a high-intensity laser is used.

Given the realistic designs for both narrow- and large-angle setups, we have provided sensitivity projections in the near future searches for ALPs based on ^{t}SRPC with possible combinations of three laser wavelengths. The sensitivity projects show that the proposed collider can reach coupling domains relevant to the QCD axion models and the *Miracle* scenario over the mass range of 0.5–6.9 eV within the present reach of laser technologies.

Author Contributions: Conceptualization, K.H.; methodology, T.H.; software, Y.K.; validation, Y.K.; formal analysis, T.H.; data curation, T.H.; writing—original draft preparation, T.H. and K.H.; writing—review and editing, K.H.; visualization, T.H. and K.H.; supervision, K.H.; project administration, K.H.; funding acquisition, K.H. All authors have read and agreed to the published version of the manuscript.

Funding: This research was funded by Grants-in-Aid for Scientific Research Nos. 21H04474 from the Ministry of Education, Culture, Sports, Science, and Technology (MEXT) of Japan.

Data Availability Statement: Not applicable.

Acknowledgments: K. Homma acknowledges the support of the Collaborative Research Program of the Institute for Chemical Research at Kyoto University (Grant Nos. 2023–101) and Grants-in-Aid for Scientific Research Nos. 21H04474 from the Ministry of Education, Culture, Sports, Science and Technology (MEXT) of Japan. Y. Kirita acknowledges support from a Grant-in-Aid for JSPS fellows No. 22J13756 from the Ministry of Education, Culture, Sports, Science and Technology (MEXT) of Japan.

Conflicts of Interest: The authors declare no conflict of interest.

References

1. Nambu, Y.; Jona-Lasinio, G. Dynamical model of elementary particles based on an analogy with superconductivity. *Phys. Rev.* **1961**, *122*, 345. [CrossRef]
2. Goldstone, J. Field theories with ≪ Superconductor≫ solutions. *Il Nuovo Cim.* **1961**, *19*, 154–164. [CrossRef]
3. Goldstone, J.; Salam, A.; Weinberg, S. Broken symmetries. *Phys. Rev.* **1962**, *127*, 965. [CrossRef]
4. Weinberg, S. A new light boson? *Phys. Rev. Lett.* **1978**, *40*, 223. [CrossRef]
5. Wilczek, F. Problem of Strong P and T Invariance in the Presence of Instantons. *Phys. Rev. Lett.* **1978**, *40*, 271. [CrossRef]
6. Peccei, R.D.; Quinn, H.R. CP Conservation in the Presence of Pseudoparticles. *Phys. Rev. Lett.* **1977**, *38*, 1440. [CrossRef]
7. Peccei, R.D. The Strong CP problem and axions. *Lect. Notes Phys.* **2008**, *741*, 3–17. [CrossRef]
8. Preskill, J.; Wise, M.B.; Wilczek, F. Cosmology of the invisible axion. *Phys. Lett. B* **1983**, *120*, 127. [CrossRef]
9. Abbott, L.F.; Sikivie, P. A cosmological bound on the invisible axion. *Phys. Lett. B* **1983**, *120*, 133–136. [CrossRef]
10. Dine, M.; Fischler, W. The not-so-harmless axion. *Phys. Lett. B* **1983**, *120*, 137–141. [CrossRef]
11. Daido, R.; Takahashi, F.; Yin, W. The ALP miracle: Unified inflaton and dark matter. *J. Cosmol. Astropart. Phys.* **2017**, *2017* , 044. [CrossRef]
12. Daido, R.; Takahashi, F.; Yin, W. The ALP miracle revisited. *J. High Energy Phys.* **2018**, *2018*, 104. [CrossRef]
13. Fujii, Y.; Homma, K. An approach toward the laboratory search for the scalar field as a candidate of Dark Energy. *Prog. Theor. Phys.* **2014**, *126*, 531–553; Erratum in *Prog. Theor. Exp. Phys.* **2014**, *2014*, 089203. [CrossRef]
14. Homma, K.; Kirita, Y. Stimulated radar collider for probing gravitationally weak coupling pseudo Nambu-Goldstone bosons. *J. High Energy Phys.* **2020**, *2020*, 95. [CrossRef]
15. Homma, K.; Hasebe, T.; Kume, K. The first search for sub-eV scalar fields via four-wave mixing at a quasi-parallel laser collider. *Prog. Theor. Exp. Phys.* **2014**, *2014*, 083C01. [CrossRef]
16. Hasebe, T.; Homma, K.; Nakamiya, Y.; Matsuura, K.; Otani, K.; Hashida, M.; Inoue, S.; Sakabe, S. Search for sub-eV scalar and pseudoscalar resonances via four-wave mixing with a laser collider. *Prog. Theor. Exp. Phys.* **2015**, *2015*, 073C01. [CrossRef]
17. Nobuhiro, A.; Hirahara, Y.; Homma, K.; Kirita, Y.; Ozaki, T.; Nakamiya, Y.; Hashida, M.; Inoue, S.; Sakabe, S. Extended search for sub-eV axion-like resonances via four-wave mixing with a quasi-parallel laser collider in a high-quality vacuum system. *Prog. Theor. Exp. Phys.* **2020**, *2020*, 073C01. [CrossRef]
18. The SAPPHIRES Collaboration; Homma, K.; Kirita, Y.; Hashida, M.; Hirahara, Y.; Inoue, S.; Ishibashi, F.; Nakamiya, Y.; Neagu, L.; Nobuhiro, A.; et al. Search for sub-eV axion-like resonance states via stimulated quasi-parallel laser collisions with the parameterization including fully asymmetric collisional geometry. *J. High Energy Phys.* **2021**, *2021*, 108. [CrossRef]
19. The SAPPHIRES Collaboration; Kirita, Y.; Hasada, T.; Hashida, M.; Hirahara, Y.; Homma, K.; Inoue, S.; Ishibashi, F.; Nakamiya, Y.; Neagu, L.; et al. Search for sub-eV axion-like particles in a stimulated resonant photon-photon collider with two laser beams based on a novel method to discriminate pressure-independent components. *J. High Energy Phys.* **2022**, *2022*, 176. [CrossRef]
20. Homma, K.; Ishibashi, F.; Kirita, Y.; Hasada, T. Sensitivity to axion-like particles with a three-beam stimulated resonant photon collider around the eV mass range. *Universe* **2023**, *9*, 20. [CrossRef]
21. Ishibashi, F.; Hasada, T.; Homma, K.; Kirita, Y.; Kanai, T.; Masuno, S.; Tokita, S.; Hashida, M. Pilot search for axion-like particles by a three-beam stimulated resonant photon collider with short pulse lasers. *Universe* **2023**, *9*, 123. [CrossRef]
22. Ejlli, A.; Della Valle, F.; Gastaldi, U.; Messineo, G.; Pengo, R.; Ruoso, G.; Zavattini, G. The PVLAS experiment: A 25 year effort to measure vacuum magnetic birefringence. *Phys. Rep.* **2020**, *871*, 1–74. [CrossRef]

23. Ehret, K.; Frede, M.; Ghazaryan, S.; Hildebrandt, M.; Knabbe, E.A.; Kracht, D.; Lindner, A.; List, J.; Meier, T.; Meyer, N.; et al. New ALPS results on hidden-sector lightweights. *Phys. Lett. B* **2010**, *689*, 149–155. [CrossRef]

24. Ballou, R.; Deferne, G.; Finger, M., Jr.; Finger, M.; Flekova, L.; Hosek, J.; Kunc, S.; Macuchova, K.; Meissner, K.; Pugnat, P.; et al. New exclusion limits on scalar and pseudoscalar axionlike particles from light shining through a wall. *Phys. Rev. D* **2015**, *92*, 092002. [CrossRef]

25. Astier, P.; Autiero, D.; Baldisseri, A.; Baldo-Ceolin, M.; Ballocchi, G.; Banner, M.; Bassompierre, G.; Benslama, K.; Besson, N.; Bird, I.; et al. Search for eV (pseudo) scalar penetrating particles in the SPS neutrino beam. *Phys. Lett. B* **2000**, *479*, 371–380. [CrossRef]

26. Ayala, A.; Domínguez, I.; Giannotti, M.; Mirizzi, A.; Straniero, O. Revisiting the bound on axion-photon coupling from Globular Clusters. *Phys. Rev. Lett.* **2014**, *113*, 191302. [CrossRef]

27. Regis, M.; Taoso, M.; Vaz, D.; Brinchmann, J.; Zoutendijk, S.L.; Bouché, N.F.; Steinmetz, M. Searching for light in the darkness: Bounds on ALP dark matter with the optical MUSE-faint survey. *Phys. Lett.* **2021**, *814*, 136075. [CrossRef]

28. CAST Collaboration. Probing eV-scale axions with CAST. *J. Cosmol. Astropart. Phys.* **2009**, *02*, 008.

29. CAST Collaboration. Search for Sub-eV Mass Solar Axions by the CERN Axion Solar Telescope with 3He Buffer Gas. *Phys. Rev. Lett.* **2011**, *107*, 261302. [CrossRef]

30. CAST Collaboration. Search for Solar Axions by the CERN Axion Solar Telescope with 3 He Buffer Gas: Closing the Hot Dark Matter Gap. *Phys. Rev. Lett.* **2014**, *112*, 091302. [CrossRef]

31. CAST Collaboration. New CAST limit on the axion-photon interaction. *Nat. Phys.* **2017**, *13*, 584. [CrossRef]

32. Kim, J.E. Weak-Interaction Singlet and Strong CP Invariance. *Phys. Rev. Lett.* **1979**, *43*, 103. [CrossRef]

33. Shifman, M.A.; Vainshtein, A.I.; Zakharov, V.I. Can confinement ensure natural CP invariance of strong interactions? *Nucl. Phys. B* **1980**, *166*, 493–506. [CrossRef]

34. Zhitnitskii, A.P. Possible suppression of axion-hadron interactions. *Sov. J. Nucl. Phys.* **1980**, *31*, 260.

Article

Statefinder and O_m Diagnostics for New Generalized Chaplygin Gas Model

Abdulla Al Mamon [1], **Vipin Chandra Dubey** [2] **and Kazuharu Bamba** [3,*]

[1] Department of Physics, Vivekananda Satavarshiki Mahavidyalaya (Affiliated to the Vidyasagar University), Manikpara 721513, India; abdulla.physics@gmail.com

[2] Department of Mathematics, Jaypee Institute of Information Technology, Noida 201309, India; vipinchandra.dubey@mail.jiit.ac.in

[3] Division of Human Support System, Faculty of Symbiotic Systems Science, Fukushima University, Fukushima 960-1296, Japan

* Correspondence: bamba@sss.fukushima-u.ac.jp

Abstract: We explore a unified model of dark matter and dark energy. This new model is a generalization of the generalized Chaplygin gas model and is known as a new generalized Chaplygin gas (NGCG) model. We study the evolutions of the Hubble parameter and the distance modulus for the model under consideration and the standard ΛCDM model and compare that with the observational datasets. Furthermore, we demonstrate two geometric diagnostics analyses including the statefinder (r, s) and $O_m(z)$ to the discriminant NGCG model from the standard ΛCDM model. The trajectories of evolution for (r, s) and $O_m(z)$ diagnostic planes are shown to understand the geometrical behavior of the NGCG model by using different observational data points.

Keywords: new GCG; statefinder diagnostic

Citation: Al Mamon, A.; Dubey, V.C.; Bamba, K. Statefinder and O_m Diagnostics for New Generalized Chaplygin Gas Model. *Universe* **2021**, *7*, 362. https://doi.org/10.3390/universe7100362

Academic Editor: Antonino Del Popolo

Received: 23 August 2021
Accepted: 23 September 2021
Published: 28 September 2021

Publisher's Note: MDPI stays neutral with regard to jurisdictional claims in published maps and institutional affiliations.

1. Introduction

Cosmic observations [1,2] indicate that the expansion of the Universe is accelerating at the present time. In this context, the most accepted idea is that a mysterious type of energy with negative pressure, dubbed as dark energy (DE), is needed to describe this acceleration mechanism (see [3–5] for reviews on DE). This mysterious DE is specified by an equation of state (EoS) parameter $\omega_{de} = \frac{p_{de}}{\rho_{de}}$, where p_{de} and ρ_{de} are the pressure and energy density of DE, respectively. The simplest and most popular model for DE is the concordance Lambda-Cold-Dark-Matter (ΛCDM) model and is consistent with most of the observational datasets. Although this model has successfully explained many phenomena while it indeed encounters some theoretical problems associated with cosmological constant ($\omega_{\Lambda} = -1$), namely, fine-tuning and cosmic coincidence problems [6,7]. Additionally, the local measurement of Hubble constant H_0 by Hubble Space Telescope [8,9] and the Lyman-α forest BAO measurement of Hubble parameter at redshift 2.34 by BOSS [10] are in tension with each other if the standard ΛCDM is assumed (for more details, the reader can see [11,12]). These issues motivate people to go deeper into theory for a better understanding of the unknown nature of the DE component. Therefore, some alternative DE models have been proposed in the literature, such as quintessence ($-1 < \omega_{de} < -\frac{1}{3}$) [13], phantom ($\omega_{de} < -1$) [14], k-essence [15,16], tachyon [17,18], holographic DE [19–25], and so forth. Besides these models, modified gravity theories were proposed to explain this acceleration [3–5]. However, the true nature of DE and DM is still unknown and also we do not have a concrete theoretical model that can provide a satisfactory solution to all the problems.

Among several DE models, the Chaplygin gas (CG) model as a unification of DE and DM is a good candidate [26,27]. The interesting feature of this model is that the CG behaves as a pressure-less dark matter (dust) at early times and behaves like a cosmological constant in the late stage. This dual role is at the heart of the surprising properties of the CG model. Another property of this model is that the CG model belongs to the category of dynamical DE with a time-varying EoS parameter alleviating the cosmic coincidence problem in ΛCDM cosmology. However, the CG model cannot explain the scenario of the structure formation in the Universe [28,29]. Later, the CG model is generalized into the generalized Chaplygin gas (GCG) model to solve this problem [30–32]. This model has been widely studied in the literature and has been confirmed by several observations [33]. Since the GCG model can be equal to the interacting ΛCDM model [33], a new generalized Chaplygin gas (NGCG) model which equals a kind of interacting XCDM model was proposed in [34] as a unification of cold DM and X-type DE. In this model, the interaction between DE and DM is characterized by a constant EoS parameter ω_X. The basic properties of this model are discussed in Section 2. Furthermore, the authors of [35,36] have also performed the statistical likelihood analysis using different datasets on the NGCG model and found some discrimination between the NGCG model and other DE models. In a recent work, Salahedin et al. [37] obtained tight constraints on the the the free parameters of NGCG model based on the statistical Markov Chain Monte Carlo (MCMC) method by using different combinations of the latest data samples. They also showed that the big tension between the high- and low-redshift observations appearing in the ΛCDM model to predict the present value of Hubble constant H_0 can be alleviated in the NGCG model. In this context, it should be mentioned here that, using various updated observational datasets, recently Yang et al. [38,39] investigated unified dark fluid models based on CG cosmologies. They reported that such models might be considered as a potential model in the list of cosmological models alleviating the H_0 tension.

Based on the Ref. [37], in this paper, we will extend the analysis on the NGCG model by performing the statefinder and O_m diagnostic analysis to differentiate the NGCG model from the standard ΛCDM model and other DE models. Furthermore, we study the evolutions of the Hubble parameter and the distance modulus for the present model and the ΛCDM model and compare that with the observational datasets. The paper is organized as follows. In the next section, we give a brief introduction of the NGCG model. Here, we also discuss some features of the present model. In Section 3, we performed the two geometric diagnostics analysis to a discriminant NGCG model from the standard ΛCDM model. Finally, we summarize our results in Section 4.

Throughout the paper, we use natural units such that $G = c = \hbar = 1$. In addition, the symbol overhead dot indicates a derivative with respect to the cosmic time t, the symbol prime indicates a derivative with respect to the scale factor (a), and a subscript zero refers to any quantity calculated at the present time.

2. New Generalized Chaplygin Gas Model

In this section, we briefly describe the NGCG model. For details of this model, one can look into Ref. [34]. In the framework of a flat Friedmann–Robertson–Walker (FRW) cosmology, the EoS of NGCG fluid is given by [34]

$$p_{NGCG} = -\frac{\tilde{A}(a)}{\rho_{NGCG}^{\alpha}} \tag{1}$$

where $\tilde{A}(a)$ is a function depends upon the scale factor (a) of the Universe and α is the constant parameter of the NGCG fluid. This fluid smoothly interpolates between a DM (dust) dominated phase $\rho \sim a^{-3}$ and a DE dominated phase $\rho \sim a^{-3(1+\omega_{de})}$, where ω_{de} is the EOS parameter. The energy density of the NGCG fluid can be expressed as [34]

$$\rho_{NGCG} = [Aa^{-3(1+\omega_{de})(1+\alpha)} + Ba^{-3(1+\alpha)}]^{\frac{1}{1+\alpha}} \tag{2}$$

where A and B are positive constants and the function $\tilde{A}(a)$ is defined as

$$\tilde{A}(a) = -\omega_{de}Aa^{-3(1+\omega_{de})(1+\alpha)} \tag{3}$$

Now, Equation (2) can be re-written as

$$\rho_{NGCG} = \rho_{NGCG0}a^3[1 - A_s + A_s a^{-3\omega_{de}(1+\alpha)}]^{\frac{1}{1+\alpha}} \tag{4}$$

where $\rho_{NGCG0} = (A + B)^{\frac{1}{1+\alpha}}$ indicates the present value of ρ_{NGCG} and, for simplicity, we have defined $A_s = \frac{A}{A+B}$. For the NGCG model, as a scenario of the unification of DE and DM, the NGCG fluid is decomposed into two components: the DE component and the DM component, i.e., $\rho_{NGCG} = \rho_{de} + \rho_{dm}$ and $p_{NGCG} = p_{de}$. Therefore, the energy density of the DE and the DM ingredients can be respectively obtained as [34]

$$\rho_{de} = \rho_{de0}a^{-3[1+\omega_{de}(1+\alpha)]} \times [1 - A_s + A_s a^{-3\omega_{de}(1+\alpha)}]^{\frac{1}{1+\alpha}-1} \tag{5}$$

$$\rho_{dm} = \rho_{dm0}a^{-3} \times [1 - A_s + A_s a^{-3\omega_{de}(1+\alpha)}]^{\frac{1}{1+\alpha}-1} \tag{6}$$

where ρ_{de0} and ρ_{dm0} represent the present values of ρ_{de} and ρ_{dm}, respectively. It is interesting to note that the NGCG will behave like GCG when we put $\omega_{de} = -1$. When $\alpha = 0$ and $\omega_{de} = -1$, the NGCG model reduces to the standard ΛCDM model as well. In addition, the standard ωCDM model corresponds to the case $\alpha = 0$. As shown in [34], the energy is transferred from DE to DM when $\alpha < 0$. On the other hand, the energy is transferred from DM to DE, if $\alpha > 0$. Therefore, α describes the interaction between DM and DE in the NGCG model.

We assume a homogeneous isotropic and spatially flat FRW Universe filled by NGCG fluid, baryonic matter, and radiation; then, the Friedmann equation can be expressed, in terms of redshift z, as

$$E^2(z) = \left(\frac{H(z)}{H_0}\right)^2 = (1 - \Omega_{r0} - \Omega_{b0})(1+z)^3 \times [1 - A_s(1 - (1+z)^{3\omega_{de}(1+\alpha)})]^{\frac{1}{1+\alpha}} \tag{7}$$

$$+\Omega_{r0}(1+z)^4 + \Omega_{b0}(1+z)^3$$

where H_0 is the present value of $H(z)$ and $z = \frac{1}{a} - 1$ in which the scale factor is scaled to be unity at the present epoch. In addition, Ω_{r0} and Ω_{b0} are the present values of dimensionless energy densities of radiation and baryonic matter, respectively.

Next, we have used the above expression of $H(z)$ to find the evolution of the deceleration parameter q, which is defined as

$$q = -\frac{\ddot{a}}{aH^2} = -1 + \frac{(1+z)}{H(z)}\frac{dH(z)}{dz} \tag{8}$$

Furthermore, for a comprehensive analysis, we compare our model with the standard ΛCDM model. The corresponding form of $E(z)$ is given by [13]

$$E(z) = \frac{H(z)}{H_0} = \sqrt{\Omega_{dm0}(1+z)^3 + 1 - \Omega_{dm0}} \tag{9}$$

where Ω_{dm0} denotes the DM density parameter at the present epoch. Assuming the base-ΛCDM cosmology, the Planck survey [40] put the constraints on the late-Universe parameters are as $\Omega_{dm0} = 0.315$ and $H_0 = 67.4$ km/s/Mpc.

Clearly, the cosmological characteristics of the present model given in Equation (7) strongly depends on values of the free parameters A_s, α, ω_{de} and Ω_{b0}. Given a cosmological model with a set of free parameters and using a set of observational data points, one can obtain the best fit values of the free parameters of the model. Given a set of data points D and a cosmological model, $M(x, \theta)$, where vector θ includes the free parameters of the model, the chi-squared (χ^2) function is defined as

$$\chi^2 = \sum_i \frac{[D_i - M(x_i|\theta)]^2}{\sigma_i^2} \tag{10}$$

where σ_i represents the error of the ith data point. In addition, the best fit values of the free parameters θ are calculated by minimizing the χ^2 function. It should be noted that the above equation for obtaining χ^2 function is valid when the observational data points are not correlated. On the other hand, if we use correlated data points, then we should use the following formula

$$\chi^2 = \sum_{i,j}[D_i - M(x_i|\theta)]X_{i,j}[D_j - M(x_j|\theta)] \tag{11}$$

where $X_{i,j}$ denotes the inverse of the covariance matrix.

Notice that we should sum all of the χ^2 functions, when we compute different χ^2 functions for different data sets. Therefore, we require the minimizing of the sum of all the χ^2 functions in order to find the best fit values of free parameters. In a recent work, Salahedin et al. [37] obtained the observational constraints on the free parameters of the present model by using different observational data samples including type Ia supernovae (SNIa) from the Union 2.1 [41] catalog and the Pantheon [42] catalog, Baryon acoustic oscillation (BAO), Big Bang nucleosynthesis (BBN) [43], and the Cosmic microwave background (CMB) from the results of WMAP observations and observational Hubble parameter data $H(z)$ obtained from cosmic chronometers (for a detailed discussion, see Ref. [37] and the references therein). By combining all data samples, Salahedin et al. [37] performed a likelihood analysis based on the statistical MCMC algorithm to calculate the minimum of χ^2 and the best fit values of the cosmological parameters. Firstly, they combined the SNIa (Pantheon) with $H(z)$, BAO, CMB, and BBN data and, secondly, they combined the SNIa (Union 2.1) with $H(z)$, BAO, CMB, and BBN data. For both cases, they obtained the best fit values of cosmological parameters leading to finding the minimum of χ^2 function. Notice that, for the ΛCDM model, the authors of [37] only used the $H(z)$ + BAO + CMB + BBN + SNIa (Union 2.1) sample and obtained $\Omega_{cdm0}(\equiv \Omega_{dm0} + \Omega_{b0}) = 0.2675$ and $H_0 = 71.3$ km/s/Mpc. The numerical results are presented in Table 1 and for more discussion on this topic, see Ref. [37].

Table 1. Results of statistical likelihood analysis (minimum of χ^2) obtained in [37] by using a different combination of observational datasets such as $(H(z)$ + BAO + CMB + BBN + SNIa (Pantheon)) and $(H(z)$ + BAO + CMB + BBN + SNIa (Union 2.1)), for the present model (for more details, one can look into Table 3 of [37])

Parameters	$H(z)$ + BAO + CMB + BBN + SNIa (Pantheon)	$H(z)$ + BAO + CMB + BBN + SNIa (Union 2.1)
Ω_{b0}	0.0460 ± 0.0017	0.0457 ± 0.0017
Ω_{dm0}	0.2508 ± 0.0081	$0.2353^{+0.0097}_{-0.0092}$
$\eta = 1 + \alpha$	0.9443 ± 0.0097	0.981 ± 0.0018
ω_{de}	-1.041 ± 0.045	-1.021 ± 0.055
H_0	70.15 ± 0.84	70.41 ± 0.92
A_s	0.7371 ± 0.0097	0.753 ± 0.010
$\chi^2_{minimum}$	1065.2	591.4

We have shown the evolution of $H(z)$ for the above-mentioned model in Figure 1 by considering the values of the model parameters, as given in Table 1 and compared it with that of the standard ΛCDM model. In this figure, we have also plotted the data points for $H(z)$ measurements (within 1σ error bars) which have been calculated from the latest compilation of 51 data points of $H(z)$ data (for more details, see Ref. [44]). We have observed from Figure 1 that the NGCG model reproduces the observed values of $H(z)$ quite effectively for each data point. Furthermore, in the inset diagram of Figure 1 (left panel), we observed that ΛCDM models are negligible around redshift $z \sim 0.7$. It has also been found that $H_{NGCG}(z) > H_{\Lambda CDM}(z)$ at low redshifts, while $H_{NGCG}(z) < H_{\Lambda CDM}(z)$ at relatively high redshifts. These scenarios are in good agreement with a recent work by Mamon and Saha [45], in which they have observed that the relative difference between the models (Lambert W single fluid model & ΛCDM model) are negligible around $z \sim 0.67$. Next, the best fit of distance modulus $\mu(z)$ for the present model (blue line) and the ΛCDM model (red line) are plotted in Figure 2. The 580 points of Supernovae Type Ia datasets (black dots) are also plotted in Figure 2 for comparision. From this figure, it has been observed that our model reproduces the observed values of $\mu(z)$ quite effectively.

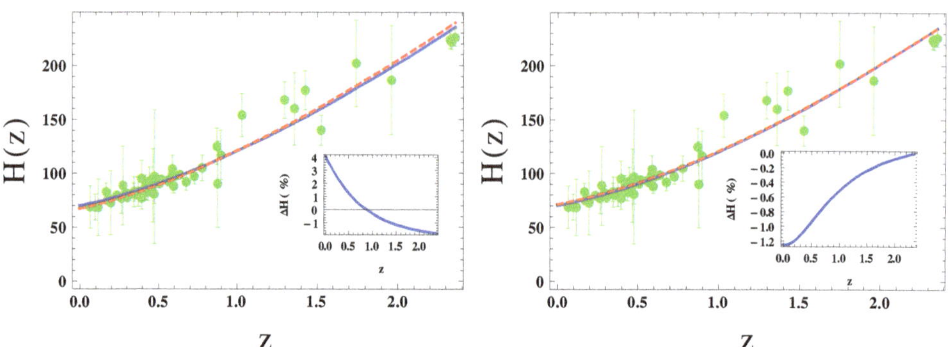

Figure 1. The evolution of the Hubble parameter (blue curve) is shown for the best-fit values of model parameters, as given in Table 1, arising from the joint analysis of $H(z)$ + BAO + CMB + BBN + SNIa (Pantheon) dataset (**left** panel) and $H(z)$ + BAO + CMB + BBN + SNIa (Union 2.1) dataset (**right** panel). Here, the red curve represents the corresponding evolution of $H(z)$ in a standard ΛCDM model with $\Omega_{cdm0} = 0.315$, $H_0 = 67.4$ km/s/Mpc [40] (**left** panel) and $\Omega_{cdm0} = 0.2675$, $H_0 = 71.3$ km/s/Mpc [37] (**right** panel). In this plot, the green dots correspond to the 51 $H(z)$ data points in the redshift range $0.07 \leq z \leq 2.36$, obtained from different surveys and the corresponding $H(z)$ values are given in [44]. In the inset diagram, the corresponding relative difference, $\Delta H(\%) = 100 \times (H_{NGCG}(z) - H_{\Lambda CDM}(z))/H_{\Lambda CDM}(z)$, is shown for the best-fit model.

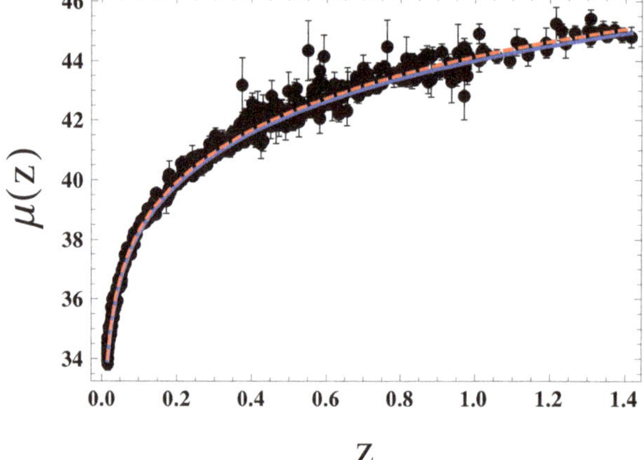

Figure 2. The evolution of $\mu(z)$ is shown for the best-fit values of model parameters, as given in Table 1, arising from the joint analysis of $H(z)$ + BAO + CMB + BBN + SNIa (Union 2.1) dataset (blue curve). The ΛCDM model ($\Omega_{cdm0} = 0.2675$ and $H_0 = 71.3$ km/s/Mpc [37]) is also shown in the red line for model comparison. Here, $\mu(z)$ represents the distance modulus, which is the difference between the apparent magnitude and the absolute magnitude of the observed supernova, is given by [3] $\mu(z) = 25 + 5\log_{10}(d_L/Mpc)$, where d_L is the luminosity distance. In this plot, the black dots correspond to the Error bar plot of 580 points of Union 2.1 compilation Supernovae Type Ia data sets [41].

3. Geometrical Diagnostics

3.1. Statefinder Diagnostics

Since various DE models have been constructed for describing or interpreting the cosmic acceleration, the problem of discriminating between the various DE candidates becomes very important. For this purpose, the authors of [46,47] have introduced a new mathematical diagnostic pair (r, s), known as a statefinder parameter. This diagnostic pair is a "geometrical" in the sense that it depends upon the scale factor directly and hence upon the metric describing space-time. The parameters r and s are defined (in terms of $H(z)$ and its derivatives) as

$$r(z) = \frac{\ddot{a}}{aH^3} = 1 - 2(1+z)\frac{H'}{H} + \left\{\frac{H''}{H} + \left(\frac{H'}{H}\right)^2\right\}(1+z)^2 \tag{12}$$

$$s(z) = \frac{(r(z) - 1)}{3(q(z) - \frac{1}{2})} \tag{13}$$

It deserves to mention here that different combinations of r and s represent different DE models [46,47]. For example,

- For ΛCDM $\rightarrow (r = 1, s = 0)$.
- For Quintessence $\rightarrow (r < 1, s > 0)$.
- For CG $\rightarrow (r > 1, s < 0)$.
- For SCDM $\rightarrow (r = 1, s = 1)$.

The evolutionary trajectories in s-r plane of holographic dark energy (HDE) model [19–25] with future event horizon as IR cut off starts from the point $s = 2/3, r = 1$ and approaches towards ΛCDM fixed point ($s = 0, r = 1$) at late time [24]. In the case of a quintessence DE model by taking constant EoS parameter [46,47] and Ricci DE (RDE) model, the curves in s-r plane are vertical [48]. The trajectory in the s-r plane in Chaplygin gas (CG) lie in the regions $s < 0, r > 1$ [49], while the phantom model with power law potential as well as the quintessence(inverse power-law) models (Q) lie in the regions $s > 0, r < 1$ [46,47] and approach the ΛCDM fixed point in both cases at a late time. The trajectory in s-r plane forms a swirl before reaching the attractor in the coupled quintessence models [50]. Both the Agegraphic DE model [51] and Polytropic gas model [52] show the ΛCDM behavior at an early time. The HDE model of DE with the model parameter $c = 1$ and the ghost DE model both show the similar behavior in (s, r) plane [53]. This behavior also matches chaplygin gas [26,27], generalized chaplygin gas [30–32,54], Yang–Mills [55], new agegraphic [51,56] and HDE [23–25] models of DE. In case of the tachyon DE model [57] and HDE model with Granda–Oliveros IR cut-off (new holographic model) [58], the curve of the s-r plane passes through the ΛCDM fixed point at the middle of the evolution of the Universe. The trajectories of the s-r plane end at the ΛCDM fixed point ($s = 0, r = 1$) at a late time, starting from matter-dominated (SCDM) $s = 1, r = 1$ through an arc segment, parabola (downward) in the case of Tsallis holographic dark energy (THDE) model [59,60]. The evolutionary curve of the s-r plane starts and ends at the ΛCDM fixed point ($s = 0, r = 1$) by making a swirl and shows the Chaplygin gas behaviour in the case of an RHDE model [61]. Recently, one of the authors has investigated the statefinder pair $r(s)$ of SMHDE model, in which it always lies in Chaplygin gas region and approaches the ΛCDM fixed point ($r = 1, s = 0$) in the late time evolution [62]. The evolutionary curve of the s-r plane starts from a cosmological constant and goes around a corner and proceeds towards another endpoint in case of the Tsallis agegraphic dark energy model [63]. In this work, we have also studied the evolution of the (s, r) pair for the NGCG model. However, one can also look into [64–66], where the authors have comprehensively discussed about the statefinder pair analysis for various DE models.

The evolution of the deceleration parameter q against the redshift parameter z, according to the values of the model parameters given in Table 1, is plotted in Figure 3 (blue curve). For comparison, the evolution of q as a function of z for a flat ΛCDM, GCG and

CG models are also shown. It is observed from Figure 3 that q gives the same prediction of the evolution of the Universe which is undergoing an accelerated expansion phase at the current epoch and experiences a transition from a decelerated expansion phase $(q > 0)$ to an accelerated expansion phase $(q < 0)$ at the transition redshift $z_t \sim 0.72$ for best-fit values of model parameters. This result is in good agreement with the current cosmological observations $(0.5 < z_t < 1)$ [67–73].

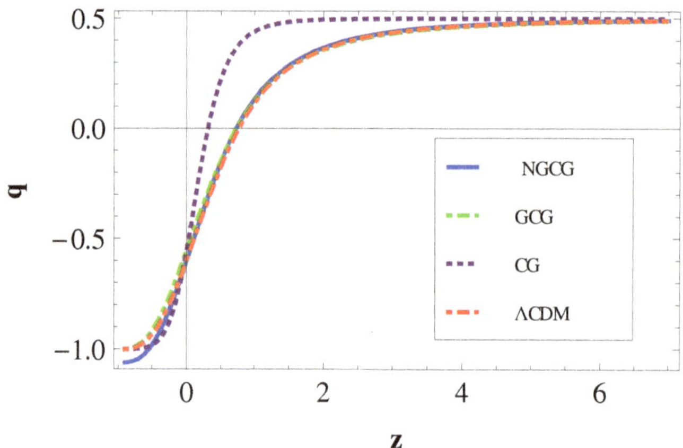

Figure 3. Plot of q as a function of z is shown by considering the values of model parameters, as given in Table 1, arising from the joint analysis of $H(z)$ + BAO + CMB + BBN + SNIa (Pantheon) dataset (blue curve). Here, the red, green, and dotted (purple) curves represent the corresponding evolution of q in a standard ΛCDM, GCG, and CG models, respectively.

We have reconstructed the evolution of the statefinder pair (s, r) according to the best fitted values of the parameters given in Table 1 for the present model. The plot of statefinder pair (s, r) is shown in the left panel of Figure 4. The evolutionary trajectories of statefinder pair of the NGCG model start its evolution along the line $r = 1$ and pass through the ΛCDM fixed point $(s = 0, r = 1)$ as time passes. After making a swirl, it lies in the Chaplygin gas region $(s < 0, r > 1)$ in the future for the best-fit values of model parameters, as given in Table 1, arising from the joint analysis of $H(z)$ + BAO + CMB + BBN + SNIa (Pantheon) dataset (blue curve). Hence, Figure 4 shows that the evolutionary trajectories of the statefinder pair of the NGCG model exhibit only the Chaplygin gas behavior and shows different behavior from other DE models. We have also shown the evolutionary trajectories of another statefinder pair (q, r) for the NGCG model in Figure 4 (right panel) for the best-fit values of model parameters, as given in Table 1, arising from the joint analysis of the $H(z)$ + BAO + CMB + BBN + SNIa (Pantheon) dataset. The fixed point $(q = 0.5, r = 1)$ corresponds to the SCDM model and the de Sitter expansion is represented by point $(q = -1, r = 1)$ in the q-r plane. The evolutionary curve of the q-r plane of NGCG model starts from the SCDM ($r = 1, q = 0.5$) in the past and reaches above the de Sitter expansion (SS) $(q = -1, r = 1)$ in the future, and it also shows the Chaplygin gas behavior throughout the evaluation. Since q changes its sign from positive to negative, it also reveals the recent phase transition of the Universe. For comparison, the evolutions of (s, r) and (q, r) pair for a NGCG, GCG, and CG models are also shown in Figure 5. Hence, these graphs (Figures 4 and 5) illustrate that, from the statefinder perspective, the NGCG model is different from various other DE models.

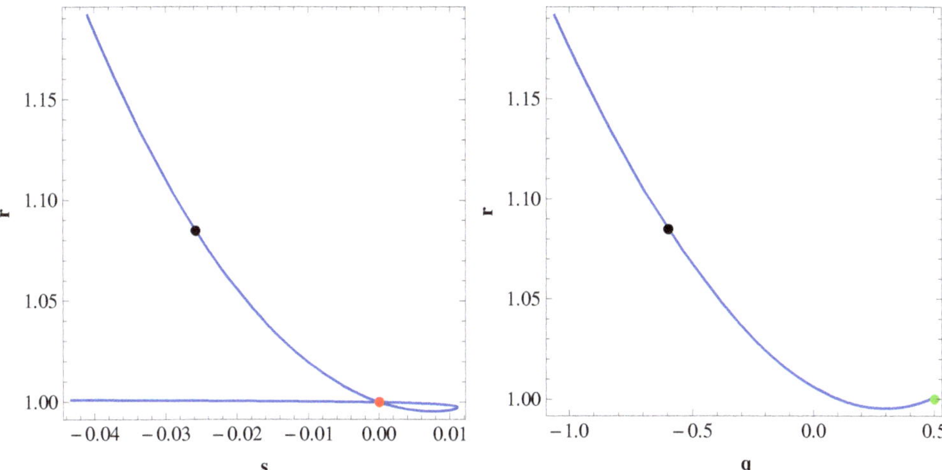

Figure 4. The time evolutions of the statefinder pair (s, r) (**left** panel) and the pair (q, r) (**right** panel) for this model are shown using the $H(z)$ + BAO + CMB + BBN + SNIa (Pantheon) dataset, as indicated in each panel. The red point $(s = 0, r = 1)$ in the left panel corresponds to the ΛCDM model, while, in the right panel, the green point $(q = 0.5, r = 1)$ represents the matter dominated Universe (SCDM). In addition, the black dots on the curves show present values (s_0, r_0) (**left** panel) and (q_0, r_0) (**right** panel) for the NGCG model.

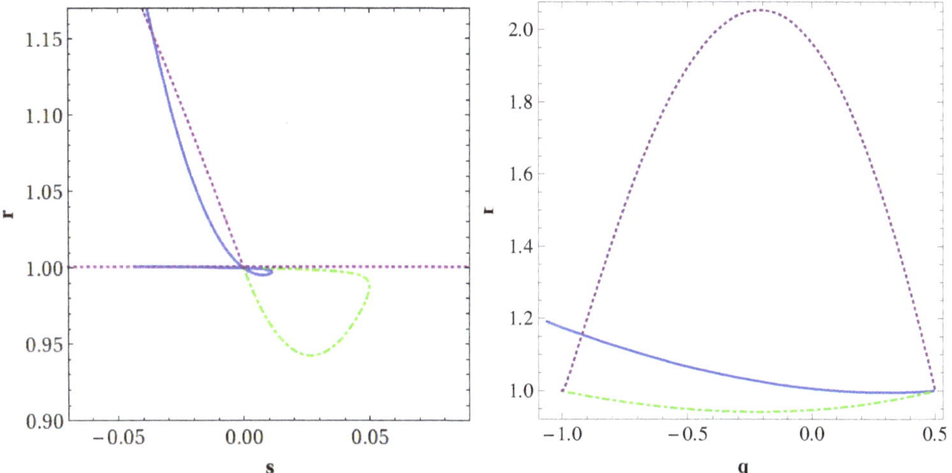

Figure 5. The time evolutions of the statefinder pair (s, r) (**left** panel) and the pair (q, r) (**right** panel) for different models are shown using the $H(z)$ + BAO + CMB + BBN + SNIa (Pantheon) dataset. Here, the blue, green, and dotted (purple) curves are for the NGCG, GCG, and CG models, respectively.

3.2. $O_m(z)$ Diagnostics

Another important and useful diagnostic tool constructed from the Hubble parameter is the O_m diagnostic parameter which provides a null test of the standard ΛCDM model. Interestingly, constant behavior of $O_m(z)$ with respect to redshift z implies that DE is a cosmological constant ($\omega_\Lambda = -1$). On the other hand, the positive slope of $O_m(z)$ signifies that DE is phantom ($\omega_{de} < -1$), whereas the negative slope implies that DE behaves like quintessence ($\omega_{de} > -1$). Following [74,75], the $O_m(z)$ parameter for a spatially flat Universe is defined as

$$O_m(z) = \frac{(\frac{H(z)}{H_0})^2 - 1}{(1+z)^3 - 1} \tag{14}$$

Note that it can differentiate a dynamical DE model from the ΛCDM model, with and without reference to matter density. For this model, $O_m(z)$ evolves as a function of z as

$$O_m(z) = \frac{(1 - \Omega_{r0} - \Omega_{b0})(1+z)^3 \times [1 - A_s(1 - (1+z)^{3\omega_{de}(1+\alpha)})]^{\frac{1}{1+\alpha}} + \Omega_{r0}(1+z)^4 + \Omega_{b0}(1+z)^3 - 1}{(1+z)^3 - 1} \tag{15}$$

It is evident that, for a spatially flat ΛCDM model $O_m(z) = \Omega_{m0}$, irrespective of the redshift, which means that, for any two distinct redshifts, say z_i and z_j, $O_m(z_i) - O_m(z_j) = 0$ is the test for ΛCDM. Currently, for any deviation from this condition, a deviation from ΛCDM is indicated. The graphical representation of $O_m(z)$ parameter of NGCG model (blue curve) is shown in Figure 6 for the values of model parameters, as given in Table 1, arising from the joint analysis of $H(z)$ + BAO + CMB + BBN + SNIa (Pantheon) dataset. It depicts that the decay of $O_m(z)$ at a lower redshift supports the flourishing DE model.

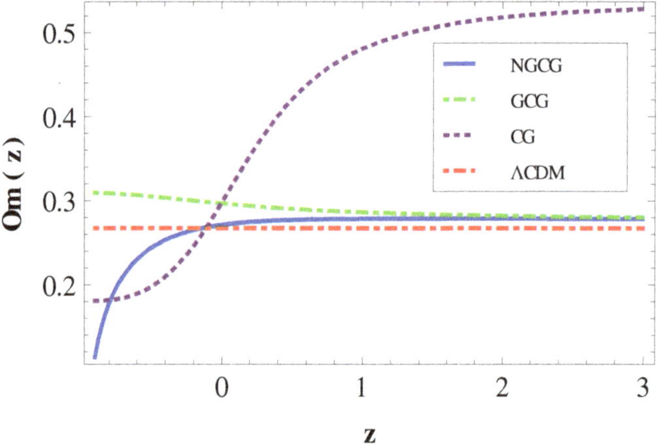

Figure 6. Evolution of $O_m(z)$ is shown for different models, as indicated in the panel.

4. Conclusions

In the present article, we have examined a new generalized Chaplygin gas (NGCG) model. The main objective of this article is to distinguish the NGCG model from other DE models through the statefinder and O_m diagnostic for the best-fit values of model parameters, as given in Table 1, arising from the joint analysis of $H(z)$ + BAO + CMB + BBN + SNIa (Pantheon) dataset. We can summarize this as:

- We have plotted the deceleration parameter q by getting its numerical solution, which exhibits a transition at $z_t \sim 0.72$, from the early decelerated phase to a late time accelerated phase. This is in good agreement with the current cosmological observations [67–73].
- The evolutionary curve in the (s, r) plane of NGCG model shows Chaplygin gas behaviour at a late time, while starting its evolution along the line $r = 1$ and passes through the ΛCDM fixed point ($s = 0, r = 1$) by making a swirl initially.
- The curve of the q-r plane of the NGCG model shows that it evolves from the matter-dominated Universe i.e., SCDM ($q = 0.5, r = 1$) initially and approaches above the de Sitter expansion (SS) ($q = -1, r = 1$) at a late time, and it always lies in the Chaplygin gas region throughout the evaluation.
- The evolutionary trajectory of $O_m(z)$ of NGCG model backs the growing DE model.

- Finally, we investigated the evolutions of the Hubble parameter and the distance modulus for the model under consideration and the standard ΛCDM model and compare that with the observational datasets (see Figures 1 and 2). For the best-fit case, it has been observed that the relative differences (ΔH) between the two models (NGCG & ΛCDM) are negligible around $z \sim 0.7$ (see inset diagram of Figure 1 (left panel)). Furthermore, we have found from Figure 2 that the present model reproduces the observed values of the distance modulus quite effectively.

We now conclude that the NGCG model provides some interesting consequences in the cosmological perspective. Furthermore, it would be interesting to investigate the effect on the growth of perturbations for the NGCG model. However, this study lies beyond the scope of the present work and is left for future works.

Author Contributions: Conceptualization, A.A.M.; methodology, A.A.M. and V.C.D.; validation, A.A.M., V.C.D. and K.B.; investigation, A.A.M. and V.C.D.; data curation, A.A.M.; writing—original draft preparation, A.A.M. and V.C.D.; writing—review and editing, K.B.; visualization, A.A.M., V.C.D. and K.B. All authors have read and agreed to the published version of the manuscript.

Funding: The work of K.B. has partially been supported by the JSPS KAKENHI Grant No. JP21K03547.

Institutional Review Board Statement: Not applicable.

Informed Consent Statement: Not applicable.

Data Availability Statement: Data sharing is not applicable to this article as no new data were created in this study. Observational data used in this paper are quoted from the cited works.

Acknowledgments: The authors thank the anonymous reviewers for some useful comments and suggestions that resulted in a significantly improved version.

Conflicts of Interest: The authors declare no conflict of interest.

References

1. Riess, A.G.; Filippenko, A.V.; Challis, P.; Clocchiatti, A.; Diercks, A.; Garnavich, P.M.; Gilliland, R.L.; Hogan, C.J.; Jha, S.; Kirshner, R.P.; et al. Observational evidence from supernovae for an accelerating universe and a cosmological constant. *Astron. J.* **1998**, *116*, 1009–1038. [CrossRef]
2. Perlmutter, S.; Aldering, G.; Goldhaber, G.; Knop, R.A.; Nugent, P.; Castro, P.G.; Deustua, S.; Fabbro, S.; Goobar, A.; Groom, D.E.; et al. Measurements of Ω and Λ from 42 high redshift supernovae. *Astrophys. J.* **1999**, *517*, 565–586. [CrossRef]
3. Copeland, E.J.; Sami, M.; Tsujikawa, S. Dynamics of dark energy. *Int. J. Mod. Phys. D* **2006**, *15*, 1753–1936. [CrossRef]
4. Amendola, L.; Tsujikawa, S. *Dark Energy: Theory and Observations*; Cambridge University Press: Cambridge, UK, 2010.
5. Bamba, K.; Capozziello, S.; Nojiri, S.; Odintsov, S.D. Dark energy cosmology: The equivalent description via different theoretical models and cosmography tests. *Astrophys. Space Sci.* **2012**, *342*, 155–228. [CrossRef]
6. Weinberg, S. The cosmological constant problem. *Rev. Mod. Phys.* **1989**, *61*, 1–23. [CrossRef]
7. Steinhardt, P.J.; Wang, L.; Zlatev, I. Cosmological tracking solutions. *Phys. Rev. D* **1999**, *59*, 123504. [CrossRef]
8. Riess, A.G.; Macri, L.M.; Hoffmann, S.L.; Scolnic, D.; Casertano, S.; Filippenko, A.V.; Tucker, B.E.; Reid, M.J.; Jones, D.O.; Silverman, J.M.; et al. A 2.4% determination of the local value of the Hubble constant. *Astrophys. J.* **2016**, *826*, 56. [CrossRef]
9. Riess, A.G.; Casertano, S.; Yuan, W.; Macri, L.M.; Scolnic, D. Large Magellanic Cloud Cepheid Standards Provide a 1% Foundation for the Determination of the Hubble Constant and Stronger Evidence for Physics beyond ΛCDM. *Astrophys. J.* **2019**, *876*, 85. [CrossRef]
10. Font-Ribera, A.; Kirkby, D.; Busca, N.; Miralda-Escudé, J.; Ross, N.P.; Slosar, A.; Rich, J.; Aubourg, É.; Bailey, S.; Bhardwaj, V.; et al. Quasar-Lyman α Forest Cross-Correlation from BOSS DR11: Baryon Acoustic Oscillations. *JCAP* **2014**, *05*, 027. [CrossRef]
11. Perivolaropoulos, L.; Skara, F. Challenges for ΛCDM: An update. *arXiv* **2021**, arXiv:2105.05208.
12. Di Valentino, E.; Mena, O.; Pan, S.; Mena, L.; Yang, W.; Melchiorri, A.; Mota, D.F.; Riess, A.G.; Silk, J. In the Realm of the Hubble tension- a Review of Solutions. *arXiv* **2021**, arXiv:2103.01183.
13. Sahni, V.; Starobinsky, A. The Case for a positive cosmological Lambda term. *Int. J. Mod. Phys. D* **2000**, *9*, 373–444. [CrossRef]
14. Caldwell, R.R.; Kamionkowski, M.; Weinberg, N.N. Phantom Energy: Dark Energy with $w < -1$ Causes a Cosmic Doomsday. *Phys. Rev. Lett.* **2003**, *91*, 071301.
15. Armendariz-Picon, C.; Mukhanov, V.; Steinhardt, P.J. A dynamical solution to the problem of a small cosmological constant and late time cosmic acceleration. *Phys. Rev. Lett.* **2000**, *85*, 4438–4441. [CrossRef]
16. Chiba, T.; Okabe, T.; Yamaguchi, M. Kinetically driven quintessence. *Phys. Rev. D* **2000**, *62*, 023511. [CrossRef]

17. Sami, M.; Chingangbam, P.; Qureshi, T. Aspects of tachyonic inflation with an exponential potential. *Phys. Rev. D* **2002**, *66*, 043530. [CrossRef]
18. Padmanabhan, T.; Choudhury, T.R. Can the clustered dark matter and the smooth dark energy arise from the same scalar field? *Phys. Rev. D* **2002**, *66*, 081301. [CrossRef]
19. Hooft, G.T. Dimensional reduction in quantum gravity. *Conf. Proc. C* **1993**, *930308*, 284–296.
20. Susskind, L. The World as a hologram. *J. Math. Phys.* **1995**, *36*, 6377–6396. [CrossRef]
21. Cohen, A.G.; Kaplan, D.B.; Nelson, A.E. Effective field theory, black holes, and the cosmological constant. *Phys. Rev. Lett.* **1999**, *82*, 4971–4974. [CrossRef]
22. Li, M. A Model of holographic dark energy. *Phys. Lett. B* **2004**, *603*, 1–5. [CrossRef]
23. Setare, M.R.; Zhang, J.; Zhang, X. Statefinder diagnosis in a non-flat universe and the holographic model of dark energy. *JCAP* **2007**, *0703*, 007. [CrossRef]
24. Zhang, X. Statefinder diagnostic for holographic dark energy model. *Int. J. Mod. Phys. D* **2005**, *14*, 1597–1606. [CrossRef]
25. Zhang, J.; Zhang, X.; Liu, H. Statefinder diagnosis for the interacting model of holographic dark energy. *Phys. Lett. B* **2008**, *659*, 26–33. [CrossRef]
26. Kamenshchik, A.; Moschella, U.; Pasquier, V. An alternative to quintessence. *Phys. Lett. B* **2001**, *511*, 265. [CrossRef]
27. Gorini, V.; Kamenshchik, A.; Moschella, U. Can the Chaplygin gas be a plausible model for dark energy? *Phys. Rev. D* **2003**, *67*, 063509. [CrossRef]
28. Bean, R.; Dore, O. Are Chaplygin gases serious contenders for the dark energy? *Phys. Rev. D* **2003**, *68*, 023515. [CrossRef]
29. Sandvik, H.B.; Tegmark, M.; Zaldarriaga, M.; Waga, I. The end of unified dark matter? *Phys. Rev. D* **2004**, *69*, 123524. [CrossRef]
30. Bento, M.C.; Bertolami, O.; Sen, A.A. Generalized Chaplygin gas, accelerated expansion and dark energy matter unification. *Phys. Rev. D* **2002**, *66*, 043507. [CrossRef]
31. Bento, M.C.; Bertolami, O.; Sen, A.A. Revival of the unified dark energy-dark matter model? *Phys. Rev. D.* **2004**, *70*, 083519. [CrossRef]
32. Fabris, J.C.; Goncalves SV, B.; de Sá Ribeiro, R. Generalized Chaplygin gas with alpha = 0 and the lambda-CDM cosmological model. *Gen. Rel. Grav.* **2004**, *36*, 211–216. [CrossRef]
33. Barreiro, T.; Bertolami, O.; Torres, P. WMAP five-year data constraints on the unified model of dark energy and dark matter. *Phys. Rev. D* **2008**, *78*, 043530. [CrossRef]
34. Zhang, X.; Wu, F.Q.; Zhang, J. New generalized Chaplygin gas as a scheme for unification of dark energy and dark matter. *JCAP* **2006**, *0601*, 003. [CrossRef]
35. Liao, K.; Pan, Y.; Zhu, Z.H. Observational constraints on new generalized Chaplygin gas model. *Res. Astron. Astrophys.* **2013**, *13*, 159–169. [CrossRef]
36. Wang, J.; Wu, Y.B.; Wang, D.; Yang, W.Q. The Extended Analysis on New Generalized Chaplygin Gas. *Chin. Phys. Lett.* **2009**, *26*, 089801.
37. Salahedin, F.; Pazhouhesh, R.; Malekjani, M. Cosmological constrains on new generalized Chaplygin gas model. *Eur. Phys. J. Plus* **2020**, *135*, 429. [CrossRef]
38. Yang, W.; Pan, S.; Paliathanasis, A.; Ghosh, S.; Wu, Y. Observational constraints of a new unified dark fluid and the H_0 tension. *arXiv* **2019**, arXiv:1904.10436.
39. Yang, W.; Pan, S.; Vagnozzi, S.; Di Valentino, E.; Mota, D.F.; Capozziello, S. Dawn of the dark: Unified dark sectors and the EDGES Cosmic Dawn 21-cm signal. *arXiv* **2019**, arXiv:1907.05344.
40. Aghanim, N.; Akrami, Y.; Ashdown, M.; Aumont, J.; Baccigalupi, C.; Ballardini, M.; Banday, A.J.; Barreiro, R.B.; Bartolo, N.; Basak, S.; et al. Planck 2018 results. VI. Cosmological parameters. *Astron. Astrophys.* **2020**, *641*, A6.
41. Suzuki, N.; Rubin, D.; Lidman, C.; Aldering, G.; Amanullah, R.; Barbary, K.; Barrientos, L.F.; Botyanszki, J.; Brodwin, M.; Connolly, N.; et al. The Hubble Space Telescope Cluster Supernova Survey: V. Improving the Dark Energy Constraints Above $z > 1$ and Building an Early-Type-Hosted Supernova Sample. *Astrophys. J.* **2012**, *746*, 85. [CrossRef]
42. Scolnic, D.M.; Jones, D.O.; Rest, A.; Pan, Y.C.; Chornock, R.; Foley, R.J.; Huber, M.E.; Kessler, R.; Narayan, G.; Riess, A.G.; et al. The Complete Light-curve Sample of Spectroscopically Confirmed SNe Ia from Pan-STARRS1 and Cosmological Constraints from the Combined Pantheon Sample. *Astrophys. J.* **2018**, *859*, 101. [CrossRef]
43. Cooke, R.J.; Pettini, M.; Steidel, C.C. One Percent Determination of the Primordial Deuterium Abundance. *Astrophys. J.* **2018**, *855*, 102. [CrossRef]
44. Magana, J.; Amante, M.H.; Garcia-Aspeitia, M.A.; Motta, V. The Cardassian expansion revisited: Constraints from updated Hubble parameter measurements and type Ia supernova data. *Mon. Not. R. Astron. Soc.* **2018**, *476*, 1036–1049. [CrossRef]
45. Al Mamon, A.; Saha, S. Testing lambert W equation of state with observational hubble parameter data. *New Astron.* **2021**, *86*, 101567. [CrossRef]
46. Sahni, V.; Saini, T.D.; Starobinsky, A.A.; Alam, U. Statefinder—A new geometrical diagnostic of dark energy. *J. Exp. Theor. Phys. Lett.* **2003**, *77*, 201–206. [CrossRef]
47. Alam, U.; Sahni, V.; Saini, T.D.; Starobinsky, A.A. Exploring the Expanding Universe and Dark Energy using the Statefinder Diagnostic. *Mon. Not. R. Astron. Soc.* **2003**, *344*, 1057. [CrossRef]
48. Feng, C.J. Statefinder Diagnosis for Ricci Dark Energy. *Phys. Lett. B* **2008**, *670*, 231–234. [CrossRef]
49. Wu, Y.B.; Li, S.; Fu, M.H.; He, J. A modified Chaplygin gas model with interaction. *Gen. Rel. Grav.* **2007**, *39*, 653–662. [CrossRef]

50. Zhang, X. Statefinder diagnostic for coupled quintessence. *Phys. Lett. B* **2005**, *611*, 1–7. [CrossRef]
51. Zhang, L.; Cui, J.; Zhang, J.; Zhang, X. Interacting model of new agegraphic dark energy: Cosmological evolution and statefinder diagnostic. *Int. J. Mod. Phys. D* **2010**, *19*, 21–35. [CrossRef]
52. Malekjani, M.; Mohammadi, A.K. Statefinder diagnostic and $w - w'$ analysis for interacting polytropic gas dark energy model. *Int. J. Theor. Phys.* **2012**, *51*, 3141–3151. [CrossRef]
53. Malekjani, M.; Mohammadi, A.K. Statefinder diagnosis and the interacting ghost model of dark energy. *Astrophys. Space Sci.* **2013**, *343*, 451–461. [CrossRef]
54. Malekjani, M.; Zarei, R.; Jafarpour, M.H. Holographic dark energy with time varying model parameter $c^2(z)$. *Astrophys. Space Sci.* **2013**, *343*, 799–806. [CrossRef]
55. Zhao, W. Statefinder diagnostic for Yang-Mills dark energy model. *Int. J. Mod. Phys. D* **2008**, *17*, 1245–1254. [CrossRef]
56. Mohammadi, A.K.; Malekjani, M. Cosmic Behavior, Statefinder Diagnostic and $w - w'$ Analysis for Interacting NADE model in Non-flat Universe. *Astrophys. Space Sci.* **2011**, *331*, 265–273. [CrossRef]
57. Shao, Y.; Gui, Y. Statefinder parameters for tachyon dark energy model. *Mod. Phys. Lett. A* **2008**, *23*, 65–71. [CrossRef]
58. Malekjani, M.; Mohammadi, A.K.; Nazari-pooya, N. Cosmological evolution and statefinder diagnostic for new holographic dark energy model in non flat universe. *Astrophys. Space Sci.* **2011**, *332*, 515–524. [CrossRef]
59. Tavayef, M.;Sheykhi, A.; Bamba, K.; Moradpour, H. Tsallis Holographic Dark Energy. *Phys. Lett. B* **2018**, *781*, 195–200. [CrossRef]
60. Sharma, U.K.; Pradhan, A. Diagnosing Tsallis holographic dark energy models with statefinder and $\omega - \omega'$. pair. *Mod. Phys. Lett. A* **2019**, *34*, 1950101. [CrossRef]
61. Sharma, U.K.; Dubey, V.C. Statefinder diagnostic for the Renyi holographic dark energy. *New Astron.* **2020**, *80*, 101419. [CrossRef]
62. Upadhyay, S.; Dubey, V.C. Diagnosing the Sharma-Mittal Holographic Dark Energy Model through the Statefinder. *Gravit. Cosmol.* **2021**, *27*, 281–291. [CrossRef]
63. Srivastava, S.; Dubey, V.C.; Sharma, U.K. Statefinder diagnosis for Tsallis agegraphic dark energy model with $\omega_D - \omega'_D$ pair. *Int. J. Mod. Phys. A* **2020**, *35*, 2050027. [CrossRef]
64. Sami, M.; Shahalam, M.; Skugoreva, M.; Toporensky, A. Cosmological dynamics of a nonminimally coupled scalar field system and its late time cosmic relevance. *Phys. Rev. D* **2012**, *86*, 103532. [CrossRef]
65. Myrzakulov, R.; Shahalam, M. Statefinder hierarchy of bimetric and galileon models for concordance cosmology. *JCAP* **2013**, *10*, 047. [CrossRef]
66. Rani, S.; Altaibayeva, A.; Shahalam, M.; Singh, J.K.; Myrzakulov, R. Constraints on cosmological parameters in power-law cosmology. *JCAP* **2015**, *03*, 031. [CrossRef]
67. Farooq, O.; Ratra, B. Hubble parameter measurement constraints on the cosmological deceleration-acceleration transition redshift. *Astrophys. J.* **2013**, *766*, L7. [CrossRef]
68. Ishida, E.E.; Reis, R.R.; Toribio, A.V.; Waga, I. When did cosmic acceleration start? How fast was the transition? *Astropart. Phys.* **2008**, *28*, 547–552. [CrossRef]
69. Magaña, J.; Cárdenas, V.H.; Motta, V. Cosmic slowing down of acceleration for several dark energy parametrizations. *J. Cosmol. Astropart. Phys.* **2014**, *10*, 017. [CrossRef]
70. Al Mamon, A.; Bamba, K.; Das, S. Constraints on reconstructed dark energy model from SN Ia and BAO/CMB observations. *Eur. Phys. J. C* **2017**, *77*, 29. [CrossRef]
71. Mamon, A. Constraints on a generalized deceleration parameter from cosmic chronometers. *Mod. Phys. Lett. A* **2018**, *33*, 1850056. [CrossRef]
72. Al Mamon, A.; Das, S. A parametric reconstruction of the deceleration parameter. *Eur. Phys. J. C* **2017**, *77*, 495. [CrossRef]
73. Al Mamon, A. Constraints on kinematic model from Pantheon SNIa, OHD and CMB shift parameter measurements. *Mod. Phys. Lett. A* **2021**, *36*, 2150049. [CrossRef]
74. Sahni, V.; Shafieloo, A.; Starobinsky, A.A. Two new diagnostics of dark energy. *Phys. Rev. D* **2008**, *78*, 103502. [CrossRef]
75. Zunckel, C.; Clarkson, C. Consistency Tests for the Cosmological Constant. *Phys. Rev. Lett.* **2008**, *101*, 181301. [CrossRef] [PubMed]

Article

Bianchi I Spacetimes in Chiral–Quintom Theory

Andronikos Paliathanasis [1,2]

1   Institute of Systems Science, Durban University of Technology, P.O. Box 1334, Durban 4000, South Africa; anpaliat@phys.uoa.gr

2   Departamento de Matemáticas, Universidad Católica del Norte, Avda. Angamos 0610, Casilla 1280, Antofagasta 1240000, Chile

Abstract: In this paper, we study anisotropic exact solutions in the homogeneous Bianchi I background geometry in a multifield theory. Specifically, we consider the Chiral–Quintom theory, which is an extension of the Chiral theory, because at least one of the scalar fields can have negative energy density. Moreover, the Quintom theory can be recovered when one of the free parameters of the theory vanishes. We find that Kasner-like and anisotropic exponential solutions exist for specific functional forms of the scalar field potential. Finally, Noether symmetry analysis is applied for the classification of the theory according to the admitted symmetries. Conservation laws are determined, while we show that the Kasner-like solution is the analytic solution for the given model.

Keywords: Bianchi spacetimes; scalar field; cosmology

1. Introduction

Cosmic inflation describes the accelerated period in the early stage of cosmological history [1–3]. Inflation has been considered a solution to long-standing problems about the structure of the universe, such as the flatness problem and horizon problems. Indeed, the inflationary mechanism surpasses the requirement for the specific initial conditions in cosmological history [4,5]. On the other hand, recent cosmological observations indicate that, at the present time, the universe is under a second accelerated phase, known as late-time acceleration attributed to the so-called dark energy [6]. In the context of general relativity, acceleration occurs when there is a matter source that has a negative equation of state parameter and provides effective "repulsive" (anti-)gravitational force.

The introduction of scalar fields in gravitational theory gives a very simple mechanism for the description of the acceleration phases of the universe. In the minimally coupled scalar field theory, the antigravitating behavior occurs when the scalar field potential dominates [7–14]. Furthermore, for the description of the late-time acceleration of the universe, phantom scalar fields with negative energy density have been proposed [15–18]. For the phantom fields, the equation of state parameter can cross the phantom divide line and take values lower than minus one. However, in order to solve the various problems, such as the appearance of ghosts, and to describe the general cosmological history, multiscalar field models have been considered.

The Quintom model [19,20] is a well studied two-scalar field cosmological model, where one of the fields is quintessence and the second field is a phantom field. The novelty of the Quintom theory is that the effective equation of state parameter can cross the phantom divide line more than once without the appearance of ghosts. Another multiscalar field model of special interest is the Chiral model [21], which has been used to describe a multifield inflation known as hyperbolic inflation [22,23]. The analytic solution for the hyperbolic inflationary model was derived recently in [24]. The equation of state parameter in the Chiral model has as a lower bound the cosmological constant limit. However, because of quantum transitions in the early universe, it can surpass that limit, and the effective equation of state parameter crosses the phantom divide line [25].

Citation: Paliathanasis, A. Bianchi I Spacetimes in Chiral–Quintom Theory. *Universe* **2022**, *8*, 503. https://doi.org/10.3390/universe8100503

Academic Editor: Kazuharu Bamba

Received: 2 August 2022
Accepted: 23 September 2022
Published: 26 September 2022

Publisher's Note: MDPI stays neutral with regard to jurisdictional claims in published maps and institutional affiliations.

Recently, a generalization of the Chiral model was proposed in [26], where the scalar fields can have negative energy density. This cosmological model has similarities with the Quintom theory. Indeed, the equation of state parameter can cross the phantom divide line more than once without the appearance of ghosts. In addition, it was found that this specific model reproduced the epoch for hyperbolic inflation [27]. Furthermore, the presence of the spatial curvature was investigated in [28,29]. In particular, it was found that this specific model solved the flatness problem.

Anisotropic and inhomogeneous exact solutions play an important role in the description of the very early universe before inflation, since they can describe the small anisotropic inhomogeneities in the cosmic observations [30,31]. The cosmic "no-hair" conjecture states that the future state of an accelerated universe is an isotropic universe [32]. In [33], anisotropic spacetimes were used to explain the CMB polarization and its implications for CMB anomalies. On the other hand, some anisotropic dark energy models were investigated in [34,35].

The first analytic result, which supported the cosmic "no-hair" conjecture was derived in [36]. Specifically, it was found that the presence of a positive cosmological constant in Bianchi anisotropic spacetimes provided expanding Bianchi spacetimes, which evolved to expanding de Sitter universes, see also the discussion in [37]. In the context of Chiral theory, anisotropic spacetimes were investigated in [38–40], while some other studies of scalar fields in anisotropic background spaces were presented in [41–43] and references therein.

In [38], exact anisotropic solutions in Chiral theory were determined, and it was found that there exist exact anisotropic solutions for Bianchi III or Kantowski–Sachs background geometry where the two scalar fields contribute to cosmological history. Moreover, anisotropic Kasner-like solutions, which belong to the Bianchi I family of spacetimes, were not supported by the Chiral theory [39]. However, when a gauge field coupled to the scalar field was introduced, anisotropic Bianchi I exact spacetimes were provided by the Chiral model [40]. In this work, we focus our analysis on the existence of anisotropic Bianchi I exact and analytic solutions for the Chiral–Quintom model proposed in [26].

Kasner spacetime [44] is one of the first anisotropic and homogeneous exact solutions derived in the literature and describes an empty Bianchi I universe. The Kasner metric depends on three parameters, which are constrained by two algebraic relations, so it is a one-parameter family of solutions. There are many applications of the Kasner metric, see for instance [45–47]. Moreover, Kasner spacetime describes the asymptotic behavior of the Mixmaster universe, Bianchi IX metric, when the effects of the spatial curvature are negligible. Kasner-like metrics [48–56], which are Kasner-like solutions with generalized Kasner-algebraic relations, are also of special interest. The structure of the paper is as follows.

In Section 2, we present the considered gravitational model, which is that of the Chiral–Quintom theory in a homogeneous and anisotropic Bianchi I background geometry. Exact solutions, which describe anisotropic geometries with power-law and exponential scale factors, are derived in Section 3. The existence of Kasner-like exact solutions are investigated. In Section 4, we perform a detailed analysis of the field equations by using the Noether symmetry approach. From this analysis, we can infer the existence of invariant functions and conservation laws for the field equations, which can be used to construct analytic solutions. Finally, in Section 5 we summarize our results.

2. Chiral–Quintom Theory

We assume the four-dimensional geometry with metric tensor $g_{\mu\nu}(x^\kappa)$ and the multiscalar field gravitational model with gravitational action integral [57]

$$S = \int dx^4 \sqrt{-g}\left(R - \frac{1}{2}g_{\mu\nu}H_{AB}\left(\Phi^C\right)\Phi^{A,\mu}\Phi^{B,\nu} - V\left(\Phi^C\right)\right), \tag{1}$$

where $R = R(x^\kappa)$ is the Ricciscalar of the background geometry $g_{\mu\nu}(x^\kappa)$.

The components of vector field $\Phi^A(x^\kappa)$ describe the scalar fields of the theory. In our analysis, we assume two scalar fields, namely $\phi(x^\kappa)$, $\psi(x^\kappa)$; that is, $\Phi^A = (\phi(x^\kappa), \psi(x^\kappa))^T$. Thus, $H_{AB}(\Phi^C(x^\kappa))$ is a two-dimensional symmetric tensor; that is, $H_{AB} = H_{BA}$ and describes the geometry in which the two scalar fields lie. The interaction of the scalar fields in the kinetic components is provided by the metric tensor H_{AB}. Finally, $V(\Phi^C(x^\kappa))$ is the potential function, which drives the dynamics and the cosmological evolution.

In the Chiral–Quintom theory, the gravitational action integral (2) is defined as follows [26]

$$S = \int dx^4 \sqrt{-g}\left(R - \frac{1}{2}g_{\mu\nu}\left(\varepsilon_1 \phi^{,\mu(x^\kappa)}\phi^{,\nu(x^\kappa)} + \varepsilon_2 e^{2\kappa\phi(x^\kappa)}\psi(x^\kappa)^{,\mu}\psi(x^\kappa)^{,\nu}\right) - V(\phi)\right), \quad (2)$$

where ε_1, ε_2 have the constraints $(\varepsilon_1)^2 = 1$ and $(\varepsilon_2)^2 = 1$. The value -1 indicates that the corresponding scalar field is phantom-like [26].

The Chiral model is recovered when ε_1 and ε_2 are positive numbers [22]. Parameter κ plays a more important role, since it is related to the curvature of the two dimensional spacetime $H_{AB}(\Phi^C(x^\kappa))$, and a nonzero value is essential in order for the hyperbolic inflation to occur [22]. Indeed, for $\kappa = 0$, the curvature of $H_{AB}(\Phi^C)$ vanishes, and the Chiral–Quintom model reduces to the Quintom theory [19]. The later model however does not reproduce the hyperbolic inflation. In this study, we consider a nonzero coupling constant parameter, κ.

For the background space, we consider that of the anisotropic and homogeneous Bianchi I spacetime

$$ds^2 = -N^2(t)dt^2 + A^2(t)dx^2 + B^2(t)dy^2 + C^2(t)dz^2, \quad (3)$$

where $A(t)$, $B(t)$, and $C(t)$ are the three scale factors, and $N(t)$ is the lapse function.

We prefer to work on the Misner variables, where the line element reads

$$ds^2 = -N^2(t)dt^2 + e^{2\alpha}\left(e^{-2\beta_+(t)}dx^2 + e^{\beta_+(t)+\sqrt{3}\beta_-(t)}dy^2 + e^{\beta_+(t)-\sqrt{3}\beta_-(t)}dz^2\right), \quad (4)$$

where now $\alpha(t)$ is the scale factor of a three-dimensional hypersurface, and $\beta_+(t)$, $\beta_-(t)$ are the two anisotropic parameters.

When $\dot{\beta}_+(t) = 0$, $\dot{\beta}_-(t) = 0$, where $\frac{d\beta_\pm}{dt} = \dot{\beta}_\pm$, the line element (4) reduces to the spatially flat Friedmann–Lemaître–Robertson–Walker spacetime. Parameter $N(t)$ is the lapse function where, without loss of generality, we select $N(t) = 1$.

For the background geometry described by the line element (4), the corresponding field equations that follow from the variational of the action integral (2) are

$$3H^2 - \frac{3}{4}\left((\dot{\beta}_+)^2 + (\dot{\beta}_-)^2\right) = \frac{1}{2}\left(\varepsilon_1\dot{\phi}^2 + \varepsilon_2 e^{2\kappa\phi}\dot{\psi}^2\right) + V(\phi), \quad (5)$$

$$2\dot{H} + 3H^2 + \frac{3}{4}\left((\dot{\beta}_+)^2 + (\dot{\beta}_-)^2\right) = -\frac{1}{2}\left(\varepsilon_1\dot{\phi}^2 + \varepsilon_2 e^{2\kappa\phi}\dot{\psi}^2\right) + V(\phi), \quad (6)$$

$$\ddot{\beta}_+ + 3H\dot{\beta}_+ = 0, \quad (7)$$

$$\ddot{\beta}_- + 3H\dot{\beta}_- = 0, \quad (8)$$

$$\varepsilon_1(\ddot{\phi} + 3H\dot{\phi}) + V_{,\phi} = e^{2\kappa\phi}\varepsilon_2\kappa\dot{\psi}^2, \quad (9)$$

$$\ddot{\psi} + 3H\dot{\psi} = -2\kappa\dot{\psi}\dot{\phi}. \quad (10)$$

where $H = \dot{\alpha}$ is the Hubble function.

3. Anisotropic Exact Solutions

In this section, we investigate the existence of exact solutions of special interest for the field equations. Firstly, let us recover the Kasner vacuum solution by assuming $\phi(t) = 0$, $\psi(t) = 0$, and $V(\phi) = 0$.

Then, for the specific functional forms

$$H(t) = \frac{H_0}{t}, \ \beta_+(t) = \beta_+^0 \ln t \text{ and } \beta_-(t) = \beta_-^0 \ln t, \tag{11}$$

from Equations (5)–(8), we obtain

$$H_0 = \frac{1}{3} \text{ and } \left(\beta_+^0\right)^2 + \left(\beta_-^0\right)^2 = \frac{4}{9}. \tag{12}$$

The later two algebraic expressions are the so-called Kasner relations expressed in the Misner variables.

3.1. Singular Solution

We consider now the case where the two scalar fields contribute to the field equations; that is, $\dot{\phi}(t)\dot{\psi}(t) \neq 0$, and the Bianchi spacetime is described by the singular solution

$$\alpha(t) = p \ln t, \ \beta_+(t) = \beta_+^0 \ln t, \text{ and } \beta_-(t) = \beta_-^0 \ln t . \tag{13}$$

This anisotropic solution corresponds to the family of Kasner-like solutions with initial cosmological singularity when $t = 0$. Recall that we have assumed the constant lapse function $N(t) = 1$.

From the field Equations (7) and (8), we derive $p = \frac{1}{3}$; thus, the remaining field equations read

$$\left(\beta_+^0\right)^2 + \left(\beta_-^0\right)^2 - \frac{4}{9} + \frac{2}{3}t^2\left(2V + \varepsilon_1\dot{\phi}^2 + \varepsilon_2 e^{2\kappa\phi}\dot{\psi}^2\right) = 0, \tag{14}$$

$$\left(\beta_+^0\right)^2 + \left(\beta_-^0\right)^2 - \frac{4}{9} - \frac{2}{3}t^2\left(2V - \varepsilon_1\dot{\phi}^2 - \varepsilon_2 e^{2\kappa\phi}\dot{\psi}^2\right) = 0, \tag{15}$$

$$\varepsilon_1\dot{\phi} + t\left(V_{,\phi} - e^{2\kappa\phi}\varepsilon_2\kappa\dot{\psi}^2 + \varepsilon_1\ddot{\phi}\right) = 0, \tag{16}$$

and

$$\ddot{\psi} + \dot{\psi}\left(\frac{1}{t} + 2\kappa\dot{\phi}\right) = 0. \tag{17}$$

Equation (17) provides $\dot{\psi} = \psi_0 \frac{e^{-2\kappa\phi}}{t}$. Hence, by replacing this in the rest of the equations, we obtain $V(\phi) = 0$, and

$$\phi(t) = \frac{1}{\kappa}\ln(\Phi(t)), \tag{18}$$

with

$$\Phi(t) = \pm\psi_0\sqrt{\frac{6\varepsilon_2}{\zeta}}\sinh\left(\Phi_1 \pm i\kappa\sqrt{\frac{\zeta}{6\varepsilon_1}}\ln t\right), \tag{19}$$

where $\zeta = 9\left(\left(\beta_+^0\right)^2 + \left(\beta_-^0\right)^2 - \frac{4}{9}\right)$, and Φ_1 is an integration constant.

Consequently, in order for the scalar field $\phi(t)$ to be a real field, $\varepsilon_1\varepsilon_2 < 0$, which means power-law solutions exist, i.e., Kasner-like solutions, only when one of the scalar fields is phantom-like, and the cosmological model is that of the Chiral–Quintom theory.

3.2. Exponential Solution

Now, we assume the nonsingular solution with

$$\alpha(t) = H_0 t, \ \beta_+(t) = \beta_+^0 t \text{ and } \beta_-(t) = \beta_-^0 t . \tag{20}$$

We substitute this into the field equations, and we find

$$\psi(t) = \psi_0 e^{-3H_0 t - 2\kappa\phi} \, , \quad V(\phi) = 3H_0^2,$$

(21)

and

$$\phi = \frac{1}{\kappa}\ln(\Phi(t)),$$

(22)

in which

$$\Phi(t) = \pm\kappa\left(i\frac{\psi_0}{3H_0}\sqrt{\frac{\varepsilon_2}{\varepsilon_1}}e^{-3H_0 t} - \Phi_1\right),$$

(23)

where Φ_1 is an integration constant.

Consequently, in order for a real solution to exist, $\frac{\varepsilon_2}{\varepsilon_1} < 0$; that is, one of the scalar fields is phantom, the other is quintessence. The cosmological model is that of Chiral–Quintom theory.

4. Noether Symmetry Analysis

The application of symmetry analysis is a powerful method for the construction of conservation laws and invariant functions necessary for the analytical study of nonlinear dynamical systems [58–60]. In the case of dynamical systems, which follow from a variational principle, Noether's theorems provide a system method for the derivation of conservation laws [58].

In cosmological studies, the Noether symmetry approach has been widely applied. For a review on the subject, we refer the reader to [61]. The analysis of the cosmological field equations with the requirement for the field equations to admit conservation laws generated by Noether's theorems has been used in two ways. Indeed, new conservation laws have been constructed for the nonlinear field equations, which led to the derivation of new analytic solutions [62–65]. Moreover, Noether symmetry analysis has been applied as a classification method for the determination of the unknown functions of the given theorem. This approach has geometric characteristics because Noether symmetries are related to the geometry where the dynamical variables are defined [66,67]. We omit the presentation of the basic properties of the Noether symmetry analysis, which can be found in [61].

For the Chiral–Quintom model of our analysis, the cosmological field equations are derived by the variation of the point-like Lagrangian

$$\mathcal{L} = \mathcal{L}\left(\alpha, \dot{\alpha}, \beta_+, \dot{\beta}_+, \beta_-, \dot{\beta}_-, \phi, \dot{\phi}, \psi, \dot{\psi}\right),$$

(24)

where

$$\mathcal{L} = e^{3\alpha}\left(6\dot{\alpha}^2 - \frac{3}{4}\left((\dot{\beta}_+)^2 + (\dot{\beta}_-)^2\right) - \frac{1}{2}\left(\varepsilon_1\dot{\phi}^2 + \varepsilon_2 e^{2\kappa\phi}\dot{\psi}^2\right) + V(\phi)\right).$$

(25)

The point-like Lagrangian (25) describes an autonomous dynamical system where, for an arbitrary potential function $V(\phi)$, it admits the Noether symmetry vector field $X_1 = \partial_t$. The corresponding conservation law is the Hamiltonian function

$$\mathcal{H} = e^{3\alpha}\left(6\dot{\alpha}^2 - \frac{3}{4}\left((\dot{\beta}_+)^2 + (\dot{\beta}_-)^2\right) - \frac{1}{2}\left(\varepsilon_1\dot{\phi}^2 + \varepsilon_2 e^{2\kappa\phi}\dot{\psi}^2\right) - V(\phi)\right),$$

(26)

where from the constraint Equation (5), $\mathcal{H} = 0$. However, for specific functional forms of $V(\phi)$ additional symmetries may exist. There exist two cases, $V_A(\phi) = V_0 e^{-\lambda\phi}$ and $V_B(\phi) = 0$, where additional Noether point symmetries for the Lagrangian (25) exist. For other forms of potential functions, symmetries may exist, including generalized symmetries, hidden symmetries, and others.

We focus on the exponential potential $V_A(\phi) = V_0 e^{-\lambda\phi}$. The admitted Noether symmetries are

$$X_2 = \partial_{\beta_+} \, , \quad X_3 = \partial_{\beta_-} \, ,$$

$$X_4 = \beta_- \partial_{\beta_+} - \beta_+ \partial_{\beta_-} \, , \; X_5 = \partial_\psi \, ,$$

and

$$X_6 = 2t\partial_t + \frac{2}{3}\partial_\alpha + \frac{4}{\lambda}\left(\partial_\phi - \kappa\psi\partial_\psi\right) .$$

The corresponding Noetherian conservation laws are

$$I_2(X_2) = e^{3\alpha}\dot{\beta}_+ \, , \; I_3(X_3) = e^{3\alpha}\dot{\beta}_- , \tag{27}$$

$$I_4(X_4) = e^{3\alpha}\left(\beta_-\dot{\beta}_+ - \beta_+\dot{\beta}_-\right), \tag{28}$$

$$I_5(X_5) = e^{3\alpha}e^{2\kappa\phi}\dot{\psi}, \tag{29}$$

$$I_6(X_6) = 2t\mathcal{H} - 4e^{3\alpha}\left(\dot{\alpha} - \frac{\varepsilon_1}{\lambda}\dot{\phi} + \frac{\varepsilon_2}{\lambda}e^{2\kappa\phi}\kappa\psi\dot{\psi}\right). \tag{30}$$

By using the constraint Equation (5), conservation law $I_6(X_6)$ reads $I_6(X_6) = -4e^{3\alpha}(\dot{\alpha} - \frac{\varepsilon_1}{\lambda}\dot{\phi} + \frac{\varepsilon_2}{\lambda}e^{2\kappa\phi}\kappa\psi\dot{\psi})$.

We can easily see that the set of conservation laws are not in involution. The conservation law $I_6(X_6)$ is written as

$$I_6(X_6) = -4e^{3\alpha}\left(\dot{\alpha} - \frac{\varepsilon_1}{\lambda}\dot{\phi} + \frac{\varepsilon_2}{\lambda}\kappa I_5\psi\right). \tag{31}$$

Thus, we can not infer the Liouville integrability property of the field equations.

A question which arises is whether we can use the invariant functions defined by the vector field X_6 in order to construct an exact solution. Indeed, the Lie invariants, which correspond to X_6, are

$$\alpha(t) = \frac{1}{3}\ln t + \alpha_0 \, , \; \phi = \frac{2}{\lambda}\ln t + \phi_0, \text{ and } \psi = \psi_0 t^{-2\frac{\kappa}{\lambda}}. \tag{32}$$

By replacing these in the field equations for the α_0, ϕ_0, and ψ_0 constants, it follows that $\psi_0 = 0$, $V_0 = 0$, and the Kasner-like relation is

$$\left(\beta_+^0\right)^2 + \left(\beta_-^0\right)^2 - \frac{4}{9\lambda^2}e^{6\alpha_0}\left(\lambda^2 - 6\varepsilon_1\right) = 0, \tag{33}$$

which is nothing other than the exact solution of a minimally coupled scalar field without any contribution of the potential function or the generalized Kasner-like solution for a five-dimensional Brane.

For the case of the zero potential function $V_B(\phi) = 0$, the admitted Noether symmetries are

$$X_2 \, , \; X_3 \, , \; X_4 \, , \; X_5 \, ,$$

$$Y_6 = \partial_\phi - \kappa\psi\partial_\psi \, , \; Y_7 = \frac{2}{3}\partial_\alpha$$

and

$$Y_8 = \psi\partial_\phi - \left(\frac{\kappa}{2}\psi - \frac{\varepsilon_1}{2\kappa}e^{-2\kappa\phi}\right)\partial_\psi.$$

The corresponding conservation laws are $I_2(X_2)$, $I_3(X_3)$, $I_4(X_4)$, $I_5(X_5)$, and

$$I_6(Y_6) = e^{3\alpha}\left(\frac{\varepsilon_1}{\lambda}\dot{\phi} - \frac{\varepsilon_2}{\lambda}e^{2\kappa\phi}\kappa\psi\dot{\psi}\right) , \tag{34}$$

$$I_7(Y_7) = 2t\mathcal{H} - 4e^{3\alpha}\dot{\alpha} \, , \tag{35}$$

$$I_8(Y_8) = e^{3\alpha}\left(\varepsilon_1\psi\dot{\phi} - \varepsilon_2 e^{2\kappa\phi}\dot{\psi}\left(\frac{\kappa}{2}\psi - \frac{\varepsilon_1}{2\kappa}e^{-2\kappa\phi}\right)\right). \tag{36}$$

In contrast to the exponential potential function, in this case, there exist at least three conservation laws, which are in involution and independent; that is, the field equations form a Liouville integrable system.

We proceed with the derivation of the analytic solution.

Analytic Solution for $V_B(\phi)$

With the use of the conservation laws $I_2(X_2)$, $I_3(X_3)$, and $I_5(X_5)$ the field equations can be written with the use of the Hamiltonian formalism as

$$\mathcal{H} = \frac{e^{-3\alpha}}{12\varepsilon_1}\left(6p_\phi^2 + \varepsilon_1\left(9\left(I_2^2 + I_3^2\right) + 6e^{-2\kappa\phi}I_5^2 - p_\alpha^2\right)\right) \equiv 0, \tag{37}$$

where

$$\dot{\alpha} = \frac{1}{6}p_\alpha e^{-3\alpha}\,,\ \dot{\phi} = -\frac{p_\phi}{\varepsilon_1}e^{-3\alpha}\,. \tag{38}$$

Consequently, the field equations are

$$\dot{p}_\alpha = 0\,,\ \dot{p}_\phi = e^{-3\alpha - 2\kappa\phi}\varepsilon_2\kappa I_5^2. \tag{39}$$

The analytic solution is derived

$$\alpha(t) = \frac{1}{3}\ln\left(\frac{p_\alpha}{2}(t - t_0)\right). \tag{40}$$

which is the Kasner-like solution found in the previous section.

We conclude that the Kasner-like solution is the analytic solution for the given cosmological model.

5. Conclusions

In this study, we determined the exact cosmological solutions for the field equations in the Chiral–Quintom theory with anisotropic and homogeneous background geometry. The Chiral–Quintom theory belongs to the multiscalar field theories, and it is an extension of the Chiral model where now at least one of the scalar fields can have a negative energy density; that is, it has a phantom behavior. The theory extends the Quintom multifield theory, where the kinetic part of the scalar fields defines a two-dimensional manifold of non-zero constant curvature.

The cosmological field equations form a nonlinear dynamical system of ordinary differential equations, the dependent variables, the scale factors $\{\alpha(t), \beta_+(t), \beta_-(t)\}$, and the two scalar fields $\{\phi(t), \psi(t)\}$. We investigated the existence of exact anisotropic solutions, which belong to the family of Kasner-like spacetimes and to the accelerated exponential geometries, by giving the explicitly functional form of the scale factors $\{\alpha(t), \beta_+(t), \beta_-(t)\}$. These two geometries, described by the latter exact solutions, were provided by the Chiral–Quintom theory for an appropriate functional form of the scalar field potential.

Moreover, because the cosmological field equations form a Hamiltonian dynamical system, and a Lagrangian function exists, we applied the Noether symmetry approach for the investigation of conservation laws. In particular, the Noether symmetry conditions were used to constrain the potential function according to the admitted Noether symmetries for the field equations. Two potential functions were derived, the exponential potential and the zero potential, where, for these two potential functional forms, extra conservation laws related to point symmetries exist. For the zero potential, we were able to infer the integrability property of the field equations, where we proved that the Kasner-like solution was the general analytic solution for this specific cosmological model. On the other hand, for the exponential potential, we were not able to prove the integrability property for the field equations and to write the analytic solution as in the case of the spatially Friedmann–Lemaître–Robertson–Walker background geometry with matter source [24].

The existence of a Kasner-family anisotropic exact solution indicates that the field equations admit actual solutions for the anisotropic initial condition. That is, anisotropy is supported by the cosmological model, and when there is not a potential function, the spacetime retains anisotropy. However, when a constant scalar field potential appears, then the resulting spacetime has exponential scale factors leading to an inflationary universe described by the isotropic de Sitter spacetime [36].

In a future work we plan to investigate the stability properties of the Chiral–Quintom theory and extend the analysis presented in [39] for the Chiral model. Such analysis is essential in order to understand the global evolution of the field equations for other potential functions, whether the Chiral–Quintom theory solves the isotropization problem, and whether anisotropic initial conditions can lead to hyperbolic inflation.

Funding: This research received no external funding.

Data Availability Statement: Not applicable.

Conflicts of Interest: The author declares no conflict of interest.

References

1. Starobinsky, A.A. A new type of isotropic cosmological models without singularity. *Phys. Lett. B* **1980**, *91*, 99. [CrossRef]
2. Guth, A. Inflationary universe. *Phys. Rev. D* **1981**, *23*, 347. [CrossRef]
3. Linde, A.D. A new inflationary universe scenario: A possible solution of the horizon, flatness, homogeneity, isotropy and primordial monopole problems. *Phys. Lett. B* **1982**, *108*, 389. [CrossRef]
4. Guth, A. Inflation and eternal inflation. *Phys. Rep.* **2000**, *333*, 555. [CrossRef]
5. Brandenberger, R. Initial Conditions for Inflation—A Short Review. *Int. J. Mod. Phys. D* **2017**, *26*, 1740002. [CrossRef]
6. Aghanim, N. et al. [Planck Collaboration]. Planck 2018 results. VI. *Cosmol. Parameters A A* **2020**, *641*, A6.
7. Ratra, B.; Peebles, P.J.E. Cosmological consequences of a rolling homogeneous scalar field. *Phys. Rev. D* **1988**, *37*, 3406. [CrossRef]
8. Liddle, A.R. Power-law inflation with exponential potentials. *Phys. Lett. B* **1989**, *220*, 502. [CrossRef]
9. Barrow, J.D.; Saich, P. Scalar-field cosmologies. *Class. Quantum Grav.* **1993**, *10*, 279. [CrossRef]
10. Basilakos, S.; Barrow, J.D. Hyperbolic inflation in the light of Planck 2015 data. *Phys. Rev. D* **2015**, *91*, 103517. [CrossRef]
11. Dimakis, N.; Karagiorgos, A.; Zampeli, A.; Paliathanasis, A.; Christodoulakis, T.; Terzis, P.A. General analytic solutions of scalar field cosmology with arbitrary potential. *Phys. Rev. D* **2016**, *93*, 123518. [CrossRef]
12. Paliathanasis, A.; Tsamparlis, M.; Basilakos, S.; Barrow, J.D. Dynamical analysis in scalar field cosmology. *Phys. Rev. D* **2015**, *91*, 123535. [CrossRef]
13. Motavali, H.; Golshani, M. Exact solutions for cosmological models with a scalar field. *Int. J. Mod. Phys. A* **2002**, *17*, 375. [CrossRef]
14. Gong, Y.; Wang, A.; Zhang, Y.-Z. Exact scaling solutions and fixed points for general scalar field. *Phys. Lett. B* **2006**, *636*, 286. [CrossRef]
15. Amendola, L. Phantom energy mediates a long-range repulsive force. *Phys. Rev. Lett.* **2004**, *93*, 181102. [CrossRef] [PubMed]
16. Catalado, M.; Arevalo, F.; Mella, P. Canonical and phantom scalar fields as an interaction of two perfect fluids. *Astrophys. Space Sci.* **2013**, *344*, 495. [CrossRef]
17. Nojiri, S.; Odintsov, S.D.; Oikonomou, V.K.; Saridakis, E.N. Singular cosmological evolution using canonical and phantom scalar fields. *J. Cosmol. Astropart. Phys.* **2015**, *2015*, 044. [CrossRef]
18. Faraoni, V. Coupled oscillators as models of phantom and scalar field cosmologies. *Phys. Rev. D* **2004**, *69*, 123520. [CrossRef]
19. Cai, Y.F.; Saridakis, E.N.; Setare, M.R.; Xia, J.-Q. Quintom cosmology: Theoretical implications and observations. *Phys. Rep.* **2010**, *493*, 1. [CrossRef]
20. Mishra, S.; Chakraborty, S. Dynamical system analysis of quintom dark energy model. *Eur. Phys. J. C* **2018**, *78*, 917. [CrossRef]
21. Chervon, S.V. Chiral non-linear sigma models and cosmological inflation. *Gravit. Cosmol.* **1995**, *1*, 91.
22. Brown, A.R. Hyperbolic Inflation. *Phys. Rev. Lett.* **2018**, *121*, 251601. [CrossRef]
23. Christodoulidis, P.; Roest, D.; Sfakianakis, E.I. Angular inflation in multi-field α-attractors. *J. Cosmol. Astropart. Phys.* **2019**, *11*, 002. [CrossRef]
24. Christodoulidis, P.; Paliathanasis, A. N-field cosmology in hyperbolic field space: Stability and general solutions. *J. Cosmol. Astropart. Phys.* **2021**, 038. [CrossRef]
25. Dimakis, N.; Paliathanasis, A. Crossing the phantom divide line as an effect of quantum transitions. *Class. Quantum Grav.* **2021**, *38*, 075016. [CrossRef]
26. Paliathanasis, A.; Leon, G. Dynamics of a two scalar field cosmological model with phantom terms. *Class. Quantum Grav.* **2021**, *38*, 075013. [CrossRef]
27. Paliathanasis, A.; Leon, G. Global dynamics of the hyperbolic Chiral-Phantom model. *Eur. Phys. J. Plus* **2022**, *137*, 165. [CrossRef]
28. Paliathanasis, A.; Leon, G. Hyperbolic inflationary model with nonzero curvature. *arXiv* **2022**, arXiv:2203.01598.
29. Tot, J.; Yildirim, B.; Coley, A.; Leon, G. The dynamics of scalar-field Quintom cosmological models. *arXiv* **2022**, arXiv:2204.06538.
30. Campanelli, L.; Cea, P.; Tedesco, L. Ellipsoidal Universe Can Solve the Cosmic Microwave Background Quadrupole Problem. *Phys. Rev. Lett.* **2006**, *97*, 131302. [CrossRef]
31. Aluri, P.K.; Cea, P.; Chingangbam, P.; Chu, M.-C.; Clowes, R.G.; Hutsemékers, D.; Kochappan, J.P.; Krasiński, A.; Lopez, A.M.; Liu, L.; et al. Is the Observable Universe Consistent with the Cosmological Principle? *arXiv* **2022**, arXiv:2207.05765.

32. Sato, K. Inflation and Cosmic No-Hair Conjecture. *J. Astrophys. Astron.* **1995**, *16*, 37.
33. Pontzen, A.; Challinor, A. Bianchi model CMB polarization and its implications for CMB anomalies. *Mon. Not. Roy. Astro. Soc.* **2007**, *380*, 1387. [CrossRef]
34. Pradhan, A.; Amirhashchi, H.; Saha, B. Bianchi Type-I Anisotropic Dark Energy Models with Constant Deceleration Parameter. *Int. J. Theor. Phys.* **2011**, *50*, 2923. [CrossRef]
35. Koivisto, T.; Mota, D.F. Anisotropic dark energy: Dynamics of the background and perturbations. *J. Cosmol. Astropart. Phys.* **2008**, *2008*, 018. [CrossRef]
36. Wald, R.M. Asymptotic behavior of homogeneous cosmological models in the presence of a positive cosmological constant. *Phys. Rev. D* **1983**, *28*, 2118. [CrossRef]
37. Barrow, J.D. Cosmic no-hair theorems and inflation. *Phys. Lett. B* **1987**, *187*, 12. [CrossRef]
38. Paliathanasis, A. New Anisotropic Exact Solution in Multifield Cosmology. *Universe* **2021**, *7*, 323. [CrossRef]
39. Giacomini, A.; Leach, P.G.L.; Leon, G.; Paliathanasis, A. Anisotropic spacetimes in chiral scalar field cosmology. *Eur. Phys. J. Plus* **2021**, *136*, 1018. [CrossRef]
40. Chen, C.-B.; Soda, J. Anisotropic Hyperbolic Inflation. *J. Cosmol. Astropart. Phys.* **2021**, *2021*, 026. [CrossRef]
41. Shanti, K.; Rao, V.U.M. Bianchi type III cosmological model in the presence of zero-mass scalar fields. *Astrophys. Space Sci.* **1990**, *173*, 157. [CrossRef]
42. Leon, G.; Gonzalez, E.; Lepe, S.; Michea, C.; Milano, A.D. Averaging Generalized Scalar Field Cosmologies I: Locally Rotationally Symmetric Bianchi III and open Friedmann-Lemaître-Robertson-Walker models. *Eur. Phys. J. C* **2021**, *81*, 414. [CrossRef]
43. Christodoulakis, T.; Grammenos, T.; Helias, C.; Kevrekidis, P.G.; Spanou, A. Decoupling of the general scalar field mode and the solution space for Bianchi type I and V cosmologies coupled to perfect fluid sources. *J. Math. Phys.* **2006**, *47*, 042505. [CrossRef]
44. Kasner, E. Geometrical theorems on Einstein's cosmological equations. *Am. J. Math.* **1921**, *43*, 217. [CrossRef]
45. Adhav, K.; Nimkar, A.; Holey, R. String Cosmology in Brans–Dicke Theory for Kasner Type Metric. *Int. J. Theor. Phys.* **2007**, *46*, 2396. [CrossRef]
46. Rasouli, S.M.M.; Farhoudi, M.; Sepangi, H.R. An anisotropic cosmological model in a modified Brans–Dicke theory. *Class. Quantum Grav.* **2011**, *28*, 155004. [CrossRef]
47. Camanho, X.O.; Dadhich, N.; Molina, A. Pure Lovelock Kasner metrics. *Class. Quantum Grav.* **2015**, *32*, 175016. [CrossRef]
48. Demaret, J.; Henneaux, M.; Spindel, P. Pure Lovelock Kasner metrics. *Phys. Lett. B* **1985**, *164*, 27. [CrossRef]
49. de Leon, J.P. 4D spacetimes embedded in 5D light-like Kasner universes. *Class. Quantum Grav.* **2006**, *26*, 185013. [CrossRef]
50. Lott, J. Kasner-like regions near crushing singularities. *Class. Quantum Grav.* **2021**, *38*, 055005. [CrossRef]
51. Paliathanasis, A.; Barrow, J.D.; Leach, P.G.L. Cosmological Solutions of f(T) Gravity. *Phys. Rev. D* **2016**, *94*, 023525. [CrossRef]
52. Binetruy, P.; Sasaki, M.; Uzawa, K. Dynamical D4-D8 and D3-D7 branes in supergravity. *Phys. Rev. D* **2009**, *80*, 026001. [CrossRef]
53. Barrow, J.D.; Hervik, S. On the evolution of universes in quadratic theories of gravity. *Phys. Rev. D* **2006**, *74*, 124017. [CrossRef]
54. Cavaglia, M.; de Ritis, G.; Gasperini, M. Relic Gravitons on Kasner-like Brane. *Phys. Lett. B* **2005**, *610*, 9. [CrossRef]
55. Rasouli, S.M.M. Kasner Solution in Brans-Dicke Theory and its Corresponding Reduced Cosmology. In *Progress in Mathematical Relativity, Gravitation and Cosmology*; Springer: Berlin/Heidelberg, Germany , 2014; Volume 60, p. 371.
56. Rasouli, S.M.M.; Moniz, P.V. Extended anisotropic models in noncompact Kaluza-Klein theory. *Class. Quantum Grav.* **2019**, *36*, 075010. [CrossRef]
57. di Marco, F.; Finelli, F.; Brandenberger, R. Adiabatic and isocurvature perturbations for multifield generalized Einstein models. *Phys. Rev. D* **2003**, *67*, 063512. [CrossRef]
58. Bluman, G.W.; Kumei, S. *Symmetries of Differential Equations*; Springer: New York, NY, USA, 1989.
59. Stephani, H. *Differential Equations: Their Solutions Using Symmetry*; Cambridge University Press: New York, NY, USA, 1989.
60. Olver, P.J. *Applications of Lie Groups to Differential Equations*; Springer: New York, NY, USA, 1993.
61. Tsamparlis, M.; Paliathanasis, A. Symmetries of Differential Equations in Cosmology. *Symmetry* **2018**, *10*, 233. [CrossRef]
62. Bhaumik, R.; Dutta, S.; Chakraborty, S. Noether symmetry analysis in chameleon field cosmology. *Int. J. Mod. Phys. A* **2022**, *37*, 2250018. [CrossRef]
63. Kucukakca, Y.; Akbarieh, A.R. Noether symmetries of Einstein-aether scalar field cosmology. *Eur. Phys. J. C* **2020**, *80*, 1019. [CrossRef]
64. Dialektopoulos, K.F.; Said, J.L.; Oikonomopoulou, Z. Classification of teleparallel Horndeski cosmology via Noether symmetries. *Eur. Phys. J. C* **2022**, *82*, 259. [CrossRef]
65. Paliathanasis, A. Analytic Solution and Noether Symmetries for the Hyperbolic Inflationary Model in the Jordan Frame. *Universe* **2022**, *8*, 325. [CrossRef]
66. Basilakos, S.; Tsamparlis, M.; Paliathanasis, A. Using the Noether symmetry approach to probe the nature of dark energy. *Phys. Rev. D* **2011**, *83*, 103512. [CrossRef]
67. Christodoulakis, T.; Dimakis, N.; Terzis, P.A. Lie point and variational symmetries in minisuperspace Einstein gravity. *J. Phys. A Math. Theor.* **2014**, *47*, 095202. [CrossRef]

Article

Nonlinear Charged Black Hole Solution in Rastall Gravity

Gamal Gergess Lamee Nashed

Centre for Theoretical Physics, The British University in Egypt, P.O. Box 43, El Sherouk City 11837, Egypt; nashed@bue.edu.eg

Abstract: We show that the spherically symmetric black hole (BH) solution of a charged (linear case) field equation of Rastall gravitational theory is not affected by the Rastall parameter and this is consistent with the results presented in the literature. However, when we apply the field equation of Rastall's theory to a special form of nonlinear electrodynamics (NED) source, we derive a novel spherically symmetric BH solution that involves the Rastall parameter. The main source of the appearance of this parameter is the trace part of the NED source, which has a non-vanishing value, unlike the linear charged field equation. We show that the new BH solution is Anti−de-Sitter Reissner−Nordström spacetime in which the Rastall parameter is absorbed into the cosmological constant. This solution coincides with Reissner−Nordström solution in the GR limit, i.e., when Rastall's parameter is vanishing. To gain more insight into this BH, we study the stability using the deviation of geodesic equations to derive the stability condition. Moreover, we explain the thermodynamic properties of this BH and show that it is stable, unlike the linear charged case that has a second-order phase transition. Finally, we prove the validity of the first law of thermodynamics.

Keywords: Rastall gravitational theory; black hole; thermodynamics and first law

Citation: Nashed, G.G.L. Nonlinear Charged Black Hole Solution in Rastall Gravity. *Universe* **2022**, *8*, 510. https://doi.org/10.3390/universe8100510

Academic Editors: Kazuharu Bamba and Júlio César Fabris

Received: 9 August 2022
Accepted: 22 September 2022
Published: 28 September 2022

Publisher's Note: MDPI stays neutral with regard to jurisdictional claims in published maps and institutional affiliations.

1. Introduction

Since the construction of Einstein's general relativity (GR), the coupling between a scalar field and the gravitational action in a geometric frame has been intensively studied. A scalar theory formulation was made in [1], and Jordan–Brans–Dicke later built a gravitational theory as an expansion of GR to investigate the variable of gravitational coupling [2–4]. Afterward, a general combination between a scalar field and its derivative, which yields second-order differential equations, is known as the Horndeski theory [5] that gained much attention. Recently, many modifications of Einstein GR have been established. Among these theories is the $f(R)$ gravitational theory, which is regarded as a natural generalization of Einstein's Hilbert action [6]. This theory could be rewritten as a GR and scalar field [7,8]. The above is a very brief summary related to the scalar fields in the frame of a gravitational context. However, there is a huge literature on this subject.

The above discussions show one way of modification of GR. However, there is another possibility that has been used to generalize the kinetic term of the scalar field that is minimally coupled to the Einstein–Hilbert action. This possibility is called the k-essence theory [9]. This theory is used as an option to the usual inflationary models that use a self-interacting scalar field [9–14]. Recently, vacuum static spherically symmetric solutions have been derived for the k-essence theories [14]. Some novel patterns have been derived that involve a study of the event horizon. Nevertheless, interpolating such solutions as black holes was difficult because it is impossible to define a distant region from the horizon. Using the no-go theorem, it has been affirmed that solutions with a regular horizon can exist but only of the type of cold black hole [15,16].

Another generalization of GR is to abound the restriction of the conservation law encoded in the zero divergence of the energy-momentum tensor. Among the theories that follow this direction is the one given by Rastall (1972), which is known as Rastall's theory [17]. In the frame of Rastall theory, the covariant divergence of the stress-energy momentum tensor is proportional to the covariant divergence of the curvature scalar, i.e., $T^{\alpha}{}_{\beta;\alpha} \propto R_{;\beta}$. Thus, any solution that has a zero or constant Ricci scalar Rastall theory will be identical to Einstein GR. Explaining the behavior of the new source of Rastall's theory is not an easy task. We can consider, phenomenologically, this new source as an appearance of quantum effects in the classical frame [18]. It is interesting to mention that the topic of non-conservation of $T^{\alpha\beta}$ is a feature that exists in diffusion models [19–23]. Furthermore, the non-conservation of the energy-momentum tensor and its link to modified gravitational theories has been analyzed in [24,25]. The variational principle in the frame of Riemannian geometry is not held due to the non-conservation of $T^{\alpha\beta}$. Nevertheless, some features such as Rastall's theory can also be discovered in the frame of Weyl geometry [26]. Moreover, external fields in the Lagrangian could give essentially the same behavior as Rastall's theory (for discussion of the external field see, for example [27]). An investigation of Rastall gravity, for an anisotropic star with a static spherical symmetry, has been discussed in [28]. The study of shadow and energy emission rates for a spherically symmetric non-commutative black hole in Rastall gravity has been carried out [29]. The quasinormal modes of black holes in Rastall gravity in the presence of non-linear electrodynamic sources have been studied [30]. Moreover, the quasinormal modes of the massless Dirac field for charged black holes in Rastall gravity have been discussed [31]. In the framework of Rastall gravity, a new black hole solution of the Ayón-Beato-García type, surrounded by a cloud of strings, is derived [32]. A solution of a static spherically symmetric black hole surrounded by a cloud of strings in the frame of Rastall gravity is derived [33]. Moreover, two classes of black hole (BH) solutions, conformally flat and non-singular BHs, are presented in [34]. A spherically symmetric gravitational collapse of a homogeneous perfect fluid in Rastall gravity has been conducted in [35]. Oliveira also presented static and spherically symmetric solutions for the Rastall modification of gravity to describe neutron stars [36].

In the frame of cosmology, Rastall's theory could degenerate into the Λ cosmological dark matter, ΛCDM, at the background and at first-order levels, which means that a viable model can be constructed in the frame of this theory. However, a few applications in the domain of astrophysics have been completed [37]. Additionally, a study of the generalized Chaplygin gas model to fit observations has been carried out in Rastall theory [38]. The quantum thermodynamics of the Schwarzschild-like black hole found in the bumblebee gravity model has been discussed in [39]. In recent years, various BH solutions, and in particular, BH solutions of the Rastall field equations, have been investigated in many scientific research papers. Among these are charged static spherically symmetric BH solutions [40,41], Gaussian BH solutions [42,43], rotating BH solutions [44,45], Abelian–Higgs strings [46], Gödel-type BH solutions [47], black branes [48], wormholes [49], BH solutions surrounded by fluid, electromagnetic field [50] or quintessence fluid [51], BH thermodynamics [52], among other theoretical efforts [53–56]. It is the aim of the present study to show the effect of the Rastall parameter in the domain of spherically symmetric spacetime using a special form of NED coupled with Einstein's GR.

This paper has the following structure: in the next section, we present a summary of Rastall's theory. In Section 2.1, we give the NED field equations of Rastall's gravity, then we apply them to a spherically symmetric spacetime with two unequal metric potentials and derive the NED differential equations. We solve this system and derive a new BH solution that involves Rastall's parameter. In Section 2.2, we extract the physical properties of the BH solution and show that the metric potentials asymptote as Anti-de-Sitter (A)dS Reissner–Nordström. Despite the applied NED field equations without cosmological constant, we obtain (A)dS Reissner–Nordström. This means that the Rastall parameter acts as a cosmological constant in this special form of NED theory. This result is consistent with the study given by Visser [57]. It is important to stress that this solution in the GR limit, i.e., when the Rastall parameter equals zero, coincides with the Reissner–Nordström solution. In Section 2.3, we derive the stability of geodesic motion using geodesic deviations. In Section 3, we study some thermodynamical quantities. In Section 3.1, we show that our BH satisfies the first law of thermodynamics. In Section 4, we discuss the output results of this study.

2. Spherically Symmetric BH Solution

Rastall's assumptions [17,58], for a spacetime with a Ricci scalar R filled by an energy-momentum $T_{\mu\nu}$, we have:

$$T^{\alpha\beta}{}_{;\alpha} = \epsilon \mathcal{R}_{;}{}^{\beta}, \tag{1}$$

where ϵ is the Rastall parameter, which is responsible for the deviation from the standard GR conservation law. Equation (1) returns to Einstein's GR when the Ricci scalar is vanishing or has a constant value.

Using the above data, we can write the Rastall field equations in the form [17,58]:

$$\mathcal{R}_{\alpha\beta} - \left[\frac{1}{2} + \lambda\right] g_{\alpha\beta}\mathcal{R} = \chi T_{\alpha\beta}, \tag{2}$$

where $\lambda = \chi\epsilon$ and χ is the Newtonian gravitational constant and the units are used so that the speed of light $c = 1$. Here, $\mathcal{R}_{\alpha\beta}$ is the Ricci tensor, $\mathcal{R}$ is the Ricci scalar, $g_{\alpha\beta}$ is the metric tensor, and $T_{\alpha\beta}$ is the energy-momentum tensor describing the material content. The constant ϵ is the Rastall parameter that is responsible for the deviation from GR and when $(\epsilon = 0)$ we obtain GR theory.

The modification in the spacetime geometry given by the L.H.S. of Equation (2) links to two modifications of different material contents of the right hand side of Equation (2):

(i) Firstly, Equation (2) is mathematically equivalent to adding new materials of the actual material sources to the right hand side of the standard GR field equations, which can be seen as an effective source accompanying the actual material sources considered in the model. For this reason, we can rewrite Equation (2) in a mathematical equivalent form as [17,58]:

$$\mathcal{R}_{\alpha\beta} - \frac{1}{2}g_{\alpha\beta}\mathcal{R} = \chi T_{1\alpha\beta}, \qquad \text{where} \qquad T_{1\alpha\beta} = T_{\alpha\beta} - \frac{\chi\epsilon}{1 + 4\chi\epsilon}g_{\alpha\beta}T. \tag{3}$$

The term $-\frac{\epsilon}{1+4\epsilon}g_{\alpha\beta}T$ is the energy-momentum tensor that represents the effective source that arises from the actual material and T is the trace of $T_{\alpha\beta}$, i.e., $T = g_{\alpha\beta}T^{\alpha\beta} = -(1 + 4\epsilon)R$. Now rewrite Equation (3) in the form: [1]

$$\mathcal{R}_{\alpha\beta} - g_{\alpha\beta}\mathcal{R}\left[\frac{1}{2} + \epsilon\right] \equiv \mathcal{R}_{\alpha\beta} + g_{\alpha\beta}T\left[\frac{1 + 2\epsilon}{2(1 + 4\epsilon)}\right] = T_{\alpha\beta}. \tag{4}$$

In this study, we will use Equation (4) but we will assume the energy-momentum tensor $T_{\alpha\beta}$ to be combined with electromagnetic field and takes the following form:

$$T_{\alpha\beta} = F_{\alpha\beta}, \qquad \text{where} \qquad E_{\alpha\beta} = F^{\mu}{}_{\alpha}F_{\mu\beta} - \frac{1}{4}g_{\alpha\beta}F, \tag{5}$$

with $F_{\mu\beta}$ being the antisymmetric Faraday tensor and $F = F^{\mu\nu}F_{\mu\nu} = d\zeta$ and $\zeta = \zeta_{\alpha}dx^{\alpha}$ is the electromagnetic gauge potential Maxwell field [59]. The tensor $F_{\mu\beta}$ satisfies the vacuum Maxwell equations:

$$F^{\alpha\beta}{}_{;\alpha} = 0, \qquad\qquad F_{\alpha\beta;\sigma} + F_{\beta\sigma;\alpha} + F_{\sigma\alpha;\beta} = 0. \tag{6}$$

(ii) Secondly, this modification implies a violation of the local conservation of the tensor $T_{1\alpha\beta}$ of an actual material source because its divergence is not necessarily vanishing.

It is important to stress that Equation (4) with the energy-momentum tensor given by Equation (6) has a contradiction since the LHS of Equation (4) has a non-vanishing covariant derivative, $\left\{\mathcal{R}^{\alpha\beta} - g^{\alpha\beta}\mathcal{R}\left[\frac{1}{2} + \frac{\epsilon}{1+4\epsilon}\right]\right\}_{;\beta} \neq 0$, while the RHS has a vanishing value, $T^{\alpha\beta}{}_{;\beta} = 0$. Thus, the only way to overcome this issue is the fact that the solution of these field equations must have a zero Ricci scalar[2], which ensures the well-known results in the literature that the Rastall parameter has no effect in the linear Maxwell field.

2.1. Nonlinear Charged Spherically Symmetric BH Solution in Rastall's Theory

In this subsection, we are going to present a special form of NED theory coupled with GR. For this aim, we are going to take into account a dual representation, i.e., imposing the auxiliary field $\mathcal{S}_{\alpha\beta}$, which is convenient to couple with GR [60,61]. Specifically, we involve the Legendre transformation:

$$\mathbb{H} = 2FL_F - L, \tag{7}$$

where $\mathbb{H}$ is an arbitrary function, $L_F \equiv \frac{\partial L}{\partial F}$ and $L(F)$ is an arbitrary function of F. If $L(F) = F$ we return to the linear case. Assuming,

$$\mathcal{S}_{\mu\nu} = L_F F_{\mu\nu}, \qquad \mathcal{S} = \frac{1}{4}\mathcal{S}_{\alpha\beta}\mathcal{S}^{\alpha\beta} = L_F^2 F, \qquad \text{with} \qquad F_{\mu\nu} = \mathbb{H}_{\mathcal{S}}\mathcal{S}_{\mu\nu}, \tag{8}$$

where $\mathbb{H}_{\mathcal{S}} = \frac{\partial \mathbb{H}}{\partial \mathcal{S}}$. The field equation of nonlinear electrodynamics yields the form [60]:

$$\partial_\nu\left(\sqrt{-g}\mathcal{S}^{\mu\nu}\right) = 0, \tag{9}$$

where the energy-momentum tensor of the NED is defined as:

$$\overset{NED}{T^\nu{}_\mu} \equiv 2(\mathbb{H}_{\mathcal{S}}\mathcal{S}_{\mu\alpha}\mathcal{S}^{\nu\alpha} - \delta^\nu_\mu[2\mathcal{S}\mathbb{H}_{\mathcal{S}} - \mathbb{H}]). \tag{10}$$

We mention that in general Equation (10) has a non-vanishing trace[3]:

$$\overset{NED}{T} = 8(\mathbb{H} - \mathbb{H}_{\mathcal{S}}\mathcal{S}) \neq 0, \tag{11}$$

and has a vanishing value in the linear theory, i.e., when $\mathbb{H} = F$ and $\mathcal{S} = F$. Finally, the electric and magnetic fields in the NED case take the form [60,61]:

$$E = \int F_{tr}dr = \int \mathbb{H}_{\mathcal{S}}\mathcal{S}_{tr}dr, \qquad\qquad B_r = \int F_{r\phi}d\phi = \int \mathbb{H}_{\mathcal{S}}\mathcal{S}_{r\phi}d\phi,$$

$$B_\theta = \int F_{\theta r}dr = \int \mathbb{H}_{\mathcal{S}}\mathcal{S}_{\theta r}dr, \qquad\qquad B_\phi = \int F_{\phi r}dr = \int \mathbb{H}_{\mathcal{S}}\mathcal{S}_{\phi r}dr, \tag{12}$$

where E and B are the components of the electric and magnetic fields, respectively. Now we are going to use the field Equation (4) with the energy-momentum tensor $T_{\alpha\beta}$, that is combined with the NED, and obtain:

$$\overset{NED}{T_{\mu\nu}} \equiv \mathbb{E}_{\mu\nu}, \qquad \text{where} \qquad \mathbb{E}_\mu{}^\nu = 2(\mathbb{H}_{\mathcal{S}}\mathcal{S}_{\mu\alpha}\mathcal{S}^{\nu\alpha} - \delta^\nu_\mu[2\mathcal{S}\mathbb{H}_{\mathcal{S}} - \mathbb{H}]). \tag{13}$$

Now, let us assume that the spherically symmetric spacetime has the form:

$$ds^2 = -\mu(r)dt^2 + \frac{dr^2}{\nu(r)} + r^2(d\theta^2 + \sin^2\theta\, d\phi^2), \tag{14}$$

where $\mu(r)$ and $\nu(r)$ are unknown functions of the radial coordinate r. For the spacetime (14), the symmetric affine connection takes the form:

$$\Gamma_{tt}{}^r = \frac{1}{2}\nu\mu', \qquad \Gamma_{tr}{}^t = \frac{\mu'}{2\mu}, \qquad \Gamma_{rr}{}^r = \frac{\nu'}{2\nu}, \qquad \Gamma_{r\theta}{}^\theta = \Gamma_{r\phi}{}^\phi = \frac{1}{r},$$

$$\Gamma_{\theta\theta}{}^r = -r\nu, \qquad \Gamma_{\theta\phi}{}^\phi = \cot\theta, \qquad \Gamma_{\phi\phi}{}^r = -r\nu\sin^2\theta, \qquad \Gamma_{\phi\phi}{}^\theta = -\sin\theta\cos\theta. \tag{15}$$

The Ricci scalar of the spacetime (14) has the form:

$$R(r) = \frac{r^2 v \mu'^2 - r^2 \mu \mu' v' - 2r^2 \mu v \mu'' - 4r\mu[v\mu' - \mu v'] + 4\mu^2(1-v)}{2r^2 \mu^2}.$$ (16)

Here, $\mu \equiv \mu(r)$, $v \equiv v(r)$, $\mu' \equiv \frac{d\mu}{dr}$, $\mu'' \equiv \frac{d^2\mu}{dr^2}$ and $v' \equiv \frac{dv}{dr}$.

Using Equation (14) in Equation (4), where the energy-momentum tensor is given by Equation (13), then:

The tt- component of Rastall field equation is:

$$\frac{1}{2r^2\mu^2[2\mu v\xi'' + \xi'(\mu v' - v\mu')]}\left\{2\mu v\xi''\left\{2\mu v\epsilon r^2\mu'' - r^2v\epsilon\mu'^2 + r\mu\mu'\epsilon[4v + rv'] + 2\mu^2[(1+2\epsilon)[rv'+v] + 1 + r^2\mathbb{H} + 2\epsilon]\right\}\right.$$

$$+\xi'\left(2r^2\mu v\mu''\epsilon[\mu v' - v\mu'] + r^2v^2\epsilon\mu'^3 - 2r\mu v\mu'^2[rv' + 2v]\epsilon + \mu^2\mu'\left[r^2\epsilon v'^2 - 2rvv' - 2v(\{1+2\epsilon\}(v-1) - r^2\mathbb{H})\right]\right.$$

$$\left.\left.+\mu^3\left\{2v'[(1+2\epsilon)\{rv'+v-1\} - r^2\mathbb{H}] + r^2v\mathbb{H}'\right\}\right)\right\} = 0,$$

The rr- component of Rastall field equation is:

$$\frac{1}{2r^2\mu^2[2\mu v\xi'' + \xi'(\mu v' - v\mu')]}\left\{2\mu v\xi''\left\{2\mu v\epsilon r^2\mu'' - r^2v\epsilon\mu'^2 + r\mu\mu'[2(1+2\epsilon)v + \epsilon rv'] + 2\mu^2[(1+2\epsilon)[v-1] + 2\epsilon rv' - r^2\mathbb{H}]\right\}\right.$$

$$+\xi'\left(2r^2\mu v\mu''\epsilon[\mu v' - v\mu'] + r^2v^2\epsilon\mu'^3 - 2r\mu v\mu'^2[rv'\epsilon + (1+2\epsilon)v] + \mu^2\mu'\left[r^2\epsilon v'^2 + 2rvv' - 2v(\{1+2\epsilon\}(v-1) - r^2\mathbb{H})\right]\right.$$

$$\left.\left.+2\mu^3\left\{2\epsilon rv'^2 + v'[(1+2\epsilon)\{v-1\} - r^2\mathbb{H}] + r^2v\mathbb{H}'\right\}\right)\right\} = 0,$$

The $\theta\theta = \phi\phi$- component of Rastall field equation is:

$$\frac{1}{4r^2\mu^2[2\mu v\xi'' + \xi'(\mu v' - v\mu')]}\left\{2\mu v\xi''\left\{2\mu v(1+2\epsilon)r^2\mu'' - r^2v(1+2\epsilon)\mu'^2 + r\mu\mu'[2(1+4\epsilon)v + (1+2\epsilon)rv'] + 2\mu^2[4\epsilon[v-1]\right.\right.$$

$$\left.+(1+4\epsilon)rv' - 2r^2\mathbb{H}]\right\} + \xi'\left(2r^2\mu v\mu''(1+2\epsilon)[\mu v' - v\mu'] + r^2v^2(1+2\epsilon)\mu'^3 - 2r\mu v\mu'^2[rv'(1+2\epsilon) + (1+4\epsilon)v]\right.$$

$$\left.\left.+\mu^2\mu'\left[r^2(1+2\epsilon)v'^2 - 4v(2\epsilon(v-1) - r^2\mathbb{H})\right] + 2\mu^3\left\{(1+4\epsilon)rv'^2 + 2v'[2\epsilon\{v-1\} - r^2\mathbb{H}] + 4r^2v\mathbb{H}'\right\}\right)\right\} = 0,$$ (17)

where $\mathbb{H}$ is an arbitrary function and ξ is the field of electric charge. Equation (17) reduces to the linear charged Einstein's field equations when $\epsilon = 0$ and $\mathbb{H} = F$ [62,63]. The exact solution of Equation (17) for the electric field takes the form[4]:

$$\mu(r) = \frac{c_2(c_3r^4 + (1+4\epsilon)[12r^2 + 12rc_4 - 3c_5])}{r^2}, \qquad v(r) = \frac{c_3r^4 + (1+4\epsilon)[12r^2 + 12rc_4 - 3c_5]}{12r^2(1+4\epsilon)},$$

$$\xi(r) = \frac{c_1}{r}, \qquad \mathbb{H} = c_3 + \frac{c_5}{r^4} \equiv c_3 + F.$$ (18)

The Rastall parameter has an effect in the NED case, as shown by Equation (18). We return to the linear charged case when $\mathbb{H} = F \equiv \frac{c_5}{r^4}$ [64]. We stress the fact that if we repeat the same above calculations taking into account the electric and magnetic fields, given by Equation (12), we can easily verify the same conclusion of the above case, i.e., Rastall's parameter has an effect and its behavior will be similar to the form given by Equation (18). If we want to derive a solution that is different from Einstein's GR, we must generalize Rastall's theory to $f(R)$-Rastall's theory [65]

2.2. The Physical Properties of the BH Solutions (18)

Now, we are going to explain the physics of the BH solution (18). For such purposes, we rewrite the components of the metric potential of the BH (18) as:

$$\mu(r) = \nu(r) = r^2 \Lambda_{eff} + 1 - \frac{2M}{r} + \frac{q^2}{r^2}, \qquad \xi = -\frac{q}{r}, \qquad \mathbb{H} = 12\Lambda_{eff.}(1 + 4\epsilon) - \frac{4q^2}{r^4}. \tag{19}$$

where we have:

$$c_1 = -q, \quad c_2 = \frac{1}{12(1 + 4\epsilon)}, \quad \Lambda_{eff.} = c_3 c_2, \quad c_4 = -2M, \quad \text{and} \quad -4q^2 = c_5, \quad c_5 = -4\sqrt{c_1}. \tag{20}$$

Equation (20) shows that we have an effective cosmological constant in the solution of the NED charged case while their field equations have no cosmological constant. This means that the Rastall parameter acts as an effective cosmological constant in the NED charged case with the fact that the Rastall parameter $\epsilon \neq -\frac{1}{4}$. From Equations (19) and (14) we obtain[5]:

$$ds^2 = -\left\{ r^2 \Lambda_{eff} + 1 - \frac{2M}{r} + \frac{q^2}{r^2} \right\} dt^2 + \frac{dr^2}{r^2 \Lambda_{eff} + 1 - \frac{2M}{r} + \frac{q^2}{r^2}} + d\Omega^2, \tag{21}$$

where $d\Omega^2 = r^2(d\theta^2 + \sin^2\theta\, d\phi^2)$ is a 2-dimensional unit sphere.

Equation (21) shows that solution (18) asymptotes as (A)dS and does not equal Reissner–Nordström spacetime due to the Rastall parameter. Equation (21) clearly investigates how the Rastall parameter acts as a cosmological constant. Equation (21) coincides with GR when $\mathbb{H} = F$, which means $c_3 = 0$, and this gives Rissner–Nordström BH solution because $\Lambda_{eff} = 0$. From Equations (19) and (16) we achieve:

$$R(r) = -12\Lambda_{\text{eff}}. \tag{22}$$

Equation (22), shows in a clear way that the Rastall parameter acts as a cosmological constant and the conservation law of both sides of Equation (2) are satisfied.

Using Equation (19) we obtain the invariants as:

$$\mathcal{R}_{\mu\nu\rho\sigma}\mathcal{R}^{\mu\nu\rho\sigma} = -24\Lambda_{\text{eff}} + \frac{48m^2}{r^6} - \frac{96mq^2}{r^7} + \frac{56q^4}{r^8}, \qquad \mathcal{R}_{\mu\nu}\mathcal{R}^{\mu\nu} = 36\Lambda_{\text{eff}} + \frac{4q^4}{r^8}, \qquad \mathcal{R} = -12\Lambda_{\text{eff}}. \tag{23}$$

Here $\left(\mathcal{R}_{\mu\nu\rho\sigma}\mathcal{R}^{\mu\nu\rho\sigma}, \mathcal{R}_{\mu\nu}\mathcal{R}^{\mu\nu}, \mathcal{R} \right)$ are the Kretschmann scalar, the Ricci tensor square, and the Ricci scalar, respectively. The Kretschmann scalar and the Ricci tensor square have a true singularity when $r = 0$. All of the above invariants are identical with the invariant of $(A)dS$-Reissner–Nordström BH solution of GR. The discussion of the invariant of (A)dS Reissner–Nordström can be applied on the invariant given by Equation (23) with the exclusion of the value $\epsilon = -\frac{1}{4}$.

Before we close this subsection we are going to calculate the trace of the NED given by Equation (11) using solution (18) as:

$$T^{NED} = c_3 \neq 0. \tag{24}$$

Equation (24) shows in a clear way that if $c_3 = 0$ we will obtain a vanishing trace and, in that case, Rastall's parameter will have no effect, which supports the above discussion.

2.3. Stability of Geodesic Motion of BH Given by Equation (19)

The equations of geodesic are given by [66]:

$$\frac{d^2 x^\gamma}{d\varepsilon^2} + \left\{ \begin{array}{c} \gamma \\ \beta\rho \end{array} \right\} \frac{dx^\beta}{d\varepsilon} \frac{dx^\rho}{d\varepsilon} = 0, \tag{25}$$

where ε is a canonical parameter. Moreover, the equations of geodesic deviation are given as [67,68]:

$$\frac{d^2\varrho^\sigma}{d\varepsilon^2} + 2\left\{\begin{matrix}\sigma\\\mu\nu\end{matrix}\right\}\frac{dx^\mu}{d\varepsilon}\frac{d\varrho^\nu}{d\varepsilon} + \left\{\begin{matrix}\sigma\\\mu\nu\end{matrix}\right\}_{,\rho}\frac{dx^\mu}{d\varepsilon}\frac{dx^\nu}{d\varepsilon}\varrho^\rho = 0, \tag{26}$$

where ϱ^α is the deviation of the four-vector.

Following the procedure in [69,70], one can get the stability condition as:

$$\frac{3\mu\nu\mu' - \sigma^2\mu\mu' - 2r\nu\mu'^2 + r\mu\nu\mu''}{\mu\nu'} > 0, \tag{27}$$

where μ and ν are given by Equation (19). Using Equation (27), one can obtain the following form of σ^2 as:

$$\sigma^2 = \frac{3\mu\nu\mu'' - 2r\nu\mu'^2 + r\mu\nu\mu''}{\mu^2\nu'^2} > 0. \tag{28}$$

Equation (28) is plotted in Figure 1 using specific values of the model. In this figure, we study $\Lambda_{eff} = 0$, Reissner−Nordström GR spacetime and $\Lambda_{eff} \neq 0$ of the BH solution (19). The two cases display the regions where the BH solution is stable/unstable by unshaded and shaded regions, respectively.

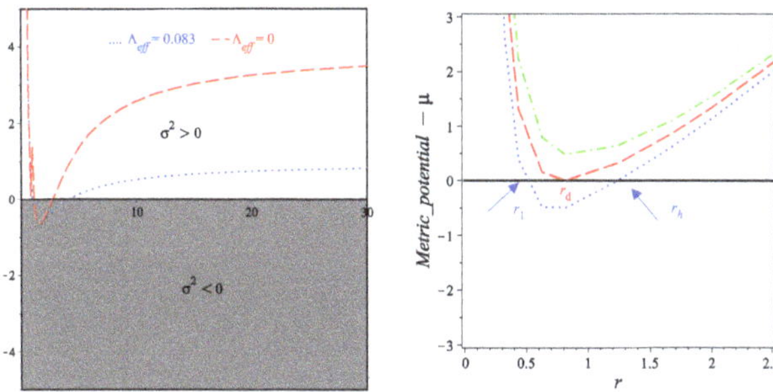

(a) Stability of the BH for the case $\Lambda_{eff} = 0$ and $\Lambda_{eff.} = 0.083$

(b) Horizons of the linear Maxwell field

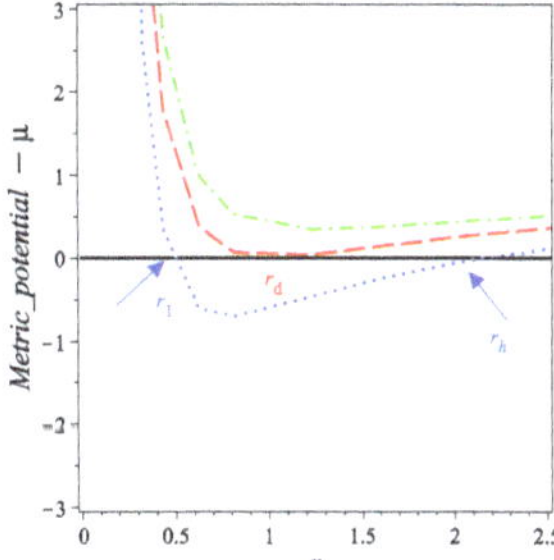

(c) Horizons of the non−linear Maxwell field

Figure 1. Plot (**a**) shows the behavior of Equation (28) viz r for BH (19). The behavior of the metric potential $\mu(r)$, which characterizes the horizons by putting $\mu(r) = 0$: (**b**) for linear Maxwell Rastall gravity theory; (**c**) for the nonlinear electrodynamics Rastall's theory. The values of m for the linear case are 1.3; 0.99; 0.8 and q = 1, while for the nonlinear case m = 1.3; 1.1 and 0.9, q = 1 and $\Lambda_{eff.} = 0.3$.

3. The Thermodynamical Properties of the of BH Given by Equation (19)

The thermodynamics of BH is considered an interesting topic in physics because it enables us to understand the physics of the solution. Two main approaches have been proposed to understand the thermodynamical quantities of the BHs: The first approach, delivered by Gibbons and Hawking [71,72] constructed to understand the thermal properties of the Schwarzschild BH through the use of Euclidean continuation. In the second approach, one has to define the gravitational surface from which we can define the Hawking temperature. Then, one can be able to study the stability of the BH [73–76]. Here, we are going to follow the second approach to investigate the thermodynamics of the (A)dS BH obtained in Equation (19) and then analyze its stability. The physical quantities characterized by the BH (19) are the mass, m, the charge, and the effective cosmological constant Λ_{eff}.

The horizons of Equation (19) are calculated by deriving the roots of $\mu(r) = 0$, which we plot in Figure 1b,c using specific values. Plots of Figure 1b,c indicate the roots of $\mu(r)$ that fix the horizons of BH (19), i.e., r_1 and r_h . We should emphasize that in the linear case, for $m > 0$, $q > 0$, and $\Lambda_{eff} = 0$, we can show that the two roots can be formed when $m > m_{min} > q$. However, when $m = m_{min}$, we fix the degenerate horizons, i.e., r_{dg}, at which $r_1 = r_h$, which is the Nariai BH whose thermodynamics is studied [77–79]. However, when $m < m_{min} < q$, there is no BH formed, which means that we have a naked singularity as shown in Figure 1b. The same discussion can be used for the NED case, where the degenerate horizon is shown in Figure 1c [78–87]. In this study, we use positive values of the effective cosmological constant because this gives two horizons. Nevertheless, it is important to mention that negative values of the effective cosmological constant create the same pattern, which is characterized by two horizons [88,89]. The stability of the BH depends on the sign of the heat capacity H_c. Now, we are going to discuss the thermal stability of the BHs through their behavior of heat capacities [10,90–92]:

$$H_c = \frac{dE_h}{dT_h} = \frac{\partial m}{\partial r_h}\left(\frac{\partial T}{\partial r_h}\right)^{-1}, \tag{29}$$

where E_h is the energy. If $H_c > 0$ or ($H_c < 0$), the BH will be thermodynamically stable or unstable, respectively. To understand this process, we suppose that at some point the BH absorbs more radiation than it emits, which yields positive heat capacity, which means that the mass is indefinitely increased. In contrast, when the BH emits more radiation than it absorbs, this yields a negative heat capacity, which means the BH mass is indefinitely decreasing until it disappears. Therefore, a BH that has a negative heat capacity is unstable thermally.

To calculate Equation (29), we need the analytical forms of $m_h \equiv m(r_h)$ and $T_h \equiv T(r_h)$. Therefore, let us calculate the mass of the BH in an event horizon r_h. Thus, we put $\mu(r_h) = 0$, given by Equation (19) and obtain:

$$m_h \Big|_{\text{Equation (19)}} = \frac{\Lambda_{eff} r_h{}^4 + r_h{}^2 + q^2}{2r_h}. \tag{30}$$

Equation (30) shows that the total mass of BH is a function of r_h, the charge and Λ_{eff}. For specific values of the charge we plot the relation of the horizon mass-radius in Figure 2a, which shows:

$$m(r_h \to 0) \to \infty, \qquad m(r_h \to \infty) \to \infty. \tag{31}$$

The temperature of BH is calculated at the outer event horizon $r = r_h$ as [93]:

$$T = \frac{\kappa}{2\pi}. \tag{32}$$

Here, κ is the surface gravity defined as $\kappa = \frac{\mu'(r_h)}{2}$. The temperatures of the BH (18) is given by:

$$T_h \underset{Equation\ (19)}{} = \frac{1}{2\pi}\left(\Lambda_{eff} r_h + \frac{1}{r_h^2}\left[m - \frac{q^2}{r_h}\right]\right), \tag{33}$$

with T_h being the temperature at r_h. For our two cases, linear and nonlinear electrodynamics, we depict the temperatures in Figure 2b for specific values. Figure 2b shows that the horizon temperature T_h has a zero value at $r_h = r_{dg}$. However, when $r_h < r_{dg}$, the horizon temperature becomes negative and forms an ultracold black hole. This result was discussed by Davies [94] who said that there are no obvious reasons from the thermodynamical viewpoint that prevent a BH temperature from becoming negative and linked this to a naked singularity. This is exactly what happened in Figure 2b when $r_h < r_{min}$ region. The case of ultracold BH is explained by the existence of a phantom energy field [95], which investigates the decrease of the mass behavior in Figure 2b. When $r_h > r_{dg}$, the temperature becomes positive. When r_h becomes larger, the temperatures of both linear and nonlinear cases change in a similar manner.

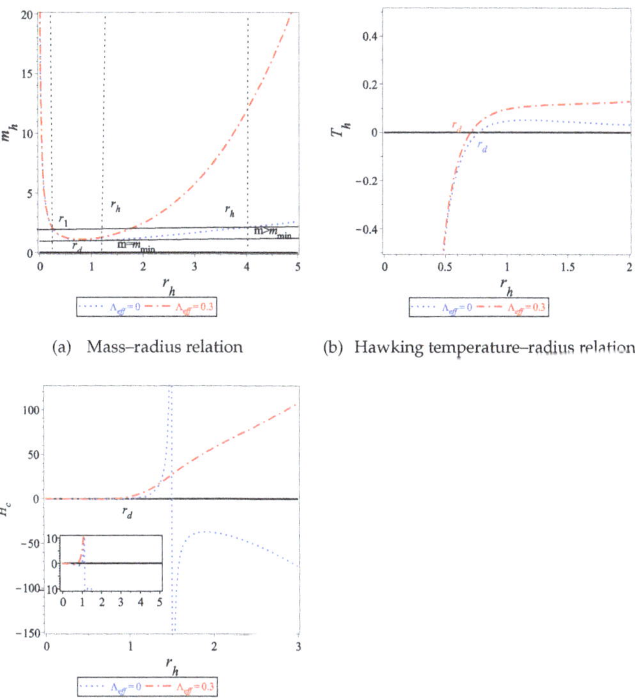

(a) Mass–radius relation

(b) Hawking temperature–radius relation

(c) Heat capacity–radius relation

Figure 2. Plots of thermodynamical quantities of BHs. (**a**) The mass-radius relation, which determines the minimal mass. (**b**) The hawking temperature, which vanishes at r_h. (**c**) The heat capacity. Moreover, the linear case investigates a second-order phase transition. All the figures are plotted for $m_h = q = 1$.

Now we are going to evaluate the heat capacity, H_c. Using Equations (29), (30), and (33) we get:

$$H_c\Big|_{Equation\ (19)} = \frac{2\pi r_h{}^2\left[3\Lambda_{eff}r_h{}^4 + r_h{}^2 - q^2\right]}{2(\Lambda_{eff}r_h{}^4 + 3q^2 - 2mr_h)}.$$

(34)

The above equation is not easy to obtain from any information; thus, we depicted it in Figure 2c with specific values of the parameters. As shown in Figure 2c, both cases of linear and nonlinear charged BH solutions, H_c vanishes at r_{dg} and also their temperatures. In the GR limit, the linear case, H_c has positive values when $r_h > r_{dg}$; however, when $r_h < r_{dg}$, it has negative values. In the NED case, the heat capacity is always positive unless $r_h < r_{dg}$.

3.1. First Law of Thermodynamics of the BH Solution (18)

Using Equation (30) we obtain:

$$M = m_h = \frac{\Lambda_{eff}r_h{}^3}{2} + \frac{r_h}{2} + \frac{q^2}{2r_h}.$$

(35)

Moreover, from the definition of entropy:

$$S = \frac{A}{4} = \pi r_h{}^2,$$

(36)

we can show that the effective cosmological constant and pressure are given as [96]:

$$P = \frac{3\Lambda_{eff}}{8\pi}.$$

(37)

Equation (35) can be rewritten in terms of pressure and entropy as:

$$M(S, q, P) = \frac{1}{6\sqrt{\pi S}}\left(3\pi q^2 + 3S + 8P S^2\right).$$

(38)

Therefore, the parameters related to S, q, and P are calculated as:

$$T = \left(\frac{\partial M}{\partial S}\right)_{P,q} = \frac{1}{4\pi r_h}\left(1 - \frac{q^2}{r_h{}^2} + 3\pi r_h{}^2\Lambda_{eff.}\right),$$

$$\zeta = \left(\frac{\partial M}{\partial q}\right)_{S,P} = \frac{q}{r_h}, \qquad V = \left(\frac{\partial M}{\partial P}\right)_{S,q} = \frac{4}{3}\pi r_h{}^3,$$

(39)

where ζ, T, and V are the electric potential, temperature, and thermodynamic volume, respectively. Using the above equations, the following Smarr relation is:

$$M = 2TS + \zeta q - 2VP,$$

(40)

from which it is easy to prove the first law of thermodynamics as:

$$dM = TdS + \zeta\, dq + VdP.$$

(41)

Equation (40) ensures the validity of the first law of the BH (19).

4. Discussion and Conclusions

In this research, we have considered spherically symmetric BH in Rastall's theory of gravity. We study the NED spherically symmetric spacetime and derive an exact solution that is affected by the Rastall parameter. This is the first time we derive a NED BH solution from the field equation of Rastall's gravitational theory. The main contribution of Rastall's parameter in this study comes from the contribution of the trace of the NED, which has a non-vanishing value in contrast to the linear Maxwell theory. We show that the effect of the Rastall parameter acts as a cosmological constant, and the BH behaves asymptotically as (A)dS Reissner–Nordström spacetime. When the Rastall parameter vanishes, we obtaiin spacetime, which asymptotes as flat Reissner–Nordström spacetime.

We have used the geodesic deviation to obtain the stability of the geodesic motion of the NED case. Furthermore, we investigated the horizons and demonstrated that the BHs presented in this study could have two horizons: the event horizon r_1 and the effective cosmological one r_h. Furthermore, we fixed the minimum value of the BH mass that occurred at the degenerate horizon. We have also studied the thermal phase transitions and showed that in the linear electrodynamics case, i.e., $\epsilon = 0$, the temperature became negative when $r_h < r_d$ and, therefore, heat capacity became negative and, thus, we have unstable BH [97–100]. The same conclusions can be applied to the NED case. However, at $r_h > r_d$, we have a positive value of the H_c, which yields a stable BH. Finally, we proved the validity of the first law of thermodynamics. It is worth noting that the result of thermodynamics presented in this study agrees with the study of thermodynamics presented in [101] when the rotation parameter a is vanishing.

In this study, we have discussed Rastall's theory using a special form of non-linear electrodynamics. This special form of non-linear electrodynamics reduces in our model to a linear form plus a cosmological constant. However, a deeper analysis is necessary, possibly regarding quantum effects in the universe. Meanwhile, the effects of Rastalls cosmology on the formation and properties of non-linear structures is a very promising research program. Furthermore, the study of $f(R)$-Rastall's theory will be extremely rich in the context of astrophysics [65]. Within the frame of $f(R)$, a BH, which is similar to Reissner–Nordström BH is presented [102] for a specific form of $f(R)$. Is it possible to derive a similar solution within Rastall's $f(R)$? This study will be carried out elsewhere.

Funding: This research received no external funding.

Conflicts of Interest: The authors declare no conflict of interest.

Sample Availability: Samples of the compounds are available from the authors.

Notes

1.   In this study we assume the relativistic units, i.e., $\chi = \frac{8\pi G}{c^4} = 1$.
2.   In the frame of Rastall theory, Reissner–Nordström is a solution since its Ricci scalar has a vanishing value.
3.   The non-vanishing of the trace is an important property in the frame of Rastall's theory so that the effect of the Rastall parameter may appear unlike Maxwell field theory.
4.   Solution (18) has been checked using Maple software 19.
5.   This result is consistent with what we have done in [57] where the author has shown that the Rastall theory is equivalent to Einstein's general relativity or equivalent to Einstein's field equation plus an arbitrary cosmological constant

References

1. Nordström, G. Relativitätsprinzip und gravitation. *Phys. Z.* **1912**, *13*, 1126.
2. Schucking, E.L. *On Einstein's Path*; Springer: Berlin/Heidelberg, Germany, 1999; pp. 1–14.
3. Brans, C.; Dicke, R.H. Mach's Principle and a Relativistic Theory of Gravitation. *Phys. Rev.* **1961**, *124*, 925. [CrossRef]
4. Dirac, P.A. The Cosmological Constants. *Nature* **1937**, *139*, 323. [CrossRef]
5. Deffayet, C.; Steer, D.A. A formal introduction to Horndeski and Galileon theories and their generalizations. *Class. Quant. Grav.* **2013**, *30*, 214006. [CrossRef]
6. Felice, A.D.; Tsujikawa, S. f(R) Theories. *Living Rev. Rel.* **2010**, *13*, 3. [CrossRef]
7. Elizalde, E.; Nashed, G.G.L.; Nojiri, S.; Odintsov, S.D. Spherically symmetric black holes with electric and magnetic charge in extended gravity: Physical properties, causal structure, and stability analysis in Einstein's and Jordan's frames. *Eur. Phys. J.* **2020**, *C80*, 109. [CrossRef]
8. Nashed, G.G.L.; Hanafy, W.E.; Odintsov, S.D.; Oikonomou, V.K. Thermodynamical correspondence of f(R) gravity in Jordan and Einstein frames. *arXiv* **2019**, arXiv:1912.03897v1.

9. Armendariz-Picon, C.; Mukhanov, V.F.; Steinhardt, P.J. Essentials of k-essence. *Phys. Rev. D* **2001**, *63*, 103510. [CrossRef]
10. Nashed, G.G.L. Rotating charged black hole spacetimes in quadratic f(R) gravitational theories. *Int. J. Mod. Phys. D* **2018**, *27*, 1850074. [CrossRef]
11. Nashed, G.G.L.; Saridakis, E.N. Rotating AdS black holes in Maxwell-f(T) gravity. *Class. Quant. Grav.* **2019**, *36*, 135005. [CrossRef]
12. Nashed, G.G.L.; Capozziello, S. Stable and self-consistent compact star models in teleparallel gravity. *Eur. Phys. J. C* **2020**, *80*, 969. [CrossRef]
13. Armendariz-Picon, C.; Damour, T.; Mukhanov, V.F. k-Inflation. *Phys. Lett. B* **1999**, *458*, 209. [CrossRef]
14. Armendariz-Picon, C.; Mukhanov, V.F.; Steinhardt, P.J. Dynamical Solution to the Problem of a Small Cosmological Constant and Late-Time Cosmic Acceleration. *Phys. Rev. Lett.* **2000**, *85*, 4438. [CrossRef] [PubMed]
15. Bronnikov, K.; Clement, G.; Constantinidis, C.; Fabris, J. Structure and stability of cold scalar-tensor black holes. *Phys. Lett. A* **1998**, *243*, 121. [CrossRef]
16. Bronnikov, K.; Clement, G.; Constantinidis, C.; Fabris, J. Cold Scalar-Tensor Black Holes: Causal Structure, Geodesics, Stability. *Grav. Cosmol.* **1998**, *4*, 128.
17. Rastall, P. Generalization of the Einstein Theory. *Phys. Rev. D* **1972**, *6*, 3357. [CrossRef]
18. Fabris, J.C.; Piattella, O.F.; Rodrigues, D.C.; Daouda, M.H. Rastall's cosmology and its observational constraints. *AIP Conf. Proc.* **2015**, *1647*, 50. [CrossRef]
19. Calogero, S. A kinetic theory of diffusion in general relativity with cosmological scalar field. *JCAP* **2011**, *11*, 016. [CrossRef]
20. Nashed, G.G.L. Energy and momentum of a spherically symmetric dilaton frame as regularized by teleparallel gravity. *Annalen Phys.* **2011**, *523*, 450. [CrossRef]
21. Nashed, G.G.L.; Hanafy, W.E. Analytic rotating black-hole solutions in N-dimensional f(T) gravity. *Eur. Phys. J.* **2017**, *77*, 90. [CrossRef]
22. Calogero, S.; Velten, H. Cosmology with matter diffusion. *JCAP* **2013**, *11*, 025. [CrossRef]
23. Velten, H.; Calogero, S. In Proceedings of the 2nd Argentinian-Brazilian Meeting on Gravitation, Relativistic Astrophysics and Cosmology (GRACo II), Buenos Aires, Argentina, 22–25 April 2014; pp.171–176.
24. Koivisto, T. A note on covariant conservation of energy–momentum in modified gravities. *Class. Quant. Grav.* **2006**, *23*, 4289. [CrossRef]
25. Minazzoli, O. Conservation laws in theories with universal gravity/matter coupling. *Phys. Rev. D* **2013**, *88*, 027506. [CrossRef]
26. Almeida, T.S.; Pucheu, M.L.; Romero, C.; Formiga, J.B. From Brans-Dicke gravity to a geometrical scalar-tensor theory. *Phys. Rev. D* **2014**, *89*, 064047. [CrossRef]
27. Chauvineau, B.; Rodrigues, D.C.; Fabris, J.C. Scalar–tensor theories with an external scalar. *Gen. Rel. Grav.* **2016**, *48*, 80. [CrossRef]
28. Nashed, G.G.L.; Hanafy, W.E. Non-trivial class of anisotropic compact stellar model in Rastall gravity. *Eur. Phys. J. C* **2022**, *82*, 679. [CrossRef]
29. Övgün, A.; Sakallı, I.; Saavedra, J.; Leiva, C., Shadow cast of noncommutative black holes in Rastall gravity. *Mod. Phys. Lett. A* **2020**, *35*, 2050163. [CrossRef]
30. Gogoi, D.J.; Goswami, U.D. Quasinormal modes of black holes with non-linear-electrodynamic sources in Rastall gravity. *Phys. Dark Univ.* **2021**, *33*, 100860. [CrossRef]
31. Shao, C.-Y.; Hu, Y.; Tan, Y.-J.; Shao, C.-G.; Lin, K.; Qian, W.-L. Dirac quasinormal modes of power-Maxwell charged black holes in Rastall gravity. *Mod. Phys. Lett. A* **2020**, *35*, 2050193. [CrossRef]
32. Gogoi, D.J.; Karmakar, R.; Goswami, U.D. Quasinormal Modes of Non-Linearly Charged Black Holes surrounded by a Cloud of Strings in Rastall Gravity. *arXiv* **2021**, arXiv:2111.00854v1.
33. Cai, X.-C.; Miao, Y.-G. Quasinormal modes and spectroscopy of a Schwarzschild black hole surrounded by a cloud of strings in Rastall gravity. *Phys. Rev. D* **2020**, *101*, 104023. [CrossRef]
34. Moradpour, H.; Heydarzade, Y.; Ziaie, C.A.H.; Ghaffari, S. Black hole solutions and Euler equation in Rastall and generalized Rastall theories of gravity. *Mod. Phys. Lett. A* **2019**, *34*, 1950304. [CrossRef]
35. Ziaie, A.H.; Moradpour, H.; Ghaffari, S. Gravitational collapse in Rastall gravity. *Phys. Lett. B* **2019**, *793*, 276. [CrossRef]
36. Oliveira, A.M.; Velten, H.E.S.; Fabris, J.C.; Casarini, L. Neutron stars in Rastall gravity. *Phys. Rev. D* **2015**, *92*, 044020. [CrossRef]
37. Batista, C.E.; Daouda, M.H.; Fabris, J.C.; Piattella, O.F.; Rodrigues, D.C., Rastall cosmology and the ΛCDM model. *Phys. Rev. D* **2012**, *85*, 084008. [CrossRef]
38. Fabris, J.C.; Piattella, O.F.; Rodrigues, D.C.; Batista, C.E.M.; Daouda, M.H. Rastall cosmology. *Int. J. Mod. Phys. Conf. Ser.* **2012**, *18*, 67. [CrossRef]
39. Kanzi, S.; Sakallı, I. GUP modified Hawking radiation in bumblebee gravity. *Nucl. Phys. B* **2019**, *946*, 114703. [CrossRef]
40. Bronnikov, K.; Fabris, J.; Piattella, O.; Santos, E. Static, spherically symmetric solutions with a scalar field in Rastall gravity. *Gen. Rel. Grav.* **2016**, *48*, 162. [CrossRef]
41. Heydarzade, Y.; Moradpour, H.; Darabi, F. Black hole solutions in Rastall theory. *Can. J. Phys.* **2017**, *95*, 1253. [CrossRef]
42. Spallucci, E.; Smailagic, A. Gaussian black holes in Rastall gravity. *Int. J. Mod. Phys. D* **2017**, *27*, 1850003. [CrossRef]
43. Ma, M.-S.; Zhao, R., Noncommutative geometry inspired black holes in Rastall gravity. *Eur. Phys. J. C* **2017**, *77*, 629. [CrossRef]
44. Kumar, R.; Ghosh, S.G. Rotating black hole in Rastall theory. *Eur. Phys. J. C* **2018**, *78*, 750. [CrossRef]

45. Xu, Z.; Hou, X.; Gong, X.; Wang, J. Kerr–Newman-AdS black hole surrounded by perfect fluid matter in Rastall gravity. *Eur. Phys. J. C* **2018**, *78*, 513. [CrossRef]
46. de Mello, E.R.B.; Fabris, J.C.; Hartmann, B. Abelian–Higgs strings in Rastall gravity. *Class. Quant. Grav.* **2015**, *32*, 085009. [CrossRef]
47. Santos, A.; Ulhoa, S., On Gödel-type solution in Rastall's gravity. *Mod. Phys. Lett. A* **2015**, *30*, 1550039. [CrossRef]
48. Sadeghi, M., Black brane solution in Rastall AdS massive gravity and viscosity bound. *Mod. Phys. Lett. A* **2018**, *33*, 1850220. [CrossRef]
49. Moradpour, H.; Sadeghnezhad, N.; Hendi, S. Traversable asymptotically flat wormholes in Rastall gravity. *Can. J. Phys.* **2017**, *95*, 1257. [CrossRef]
50. Heydarzade, Y.; Darabi, F. Black hole solutions surrounded by perfect fluid in Rastall theory. *Phys. Lett. B* **2017**, *771*, 365. [CrossRef]
51. Graça, J.M.; Lobo, I.P. Scalar QNMs for higher dimensional black holes surrounded by quintessence in Rastall gravity. *Eur. Phys. J. C* **2018**, *78*, 101. [CrossRef]
52. Lobo, I.P.; Moradpour, H.; Graça, J.P.M.; Salako, I. Thermodynamics of black holes in Rastall gravity. *Int. J. Mod. Phys. D* **2018**, *27*, 1850069. [CrossRef]
53. Licata, I.; Moradpour, H.; Corda, C. The commutator algebra of covariant derivative as general framework for extended gravity. The Rastall theory case and the role of the torsion. *Int. J. Geom. Meth. Mod. Phys.* **2017**, *14*, 1730003. [CrossRef]
54. Darabi, F.; Moradpour, H.; Licata, I.; Heydarzade, Y.; Corda, C. Einstein and Rastall theories of gravitation in comparison. *Eur. Phys. J. C* **2018**, *78*, 25. [CrossRef]
55. Caramês, T.; Fabris, J.C.; Piattella, O.F.; Strokov, V.; Daouda, M.H.; Oliveira, A.M. In Proceedings of the II Argentinian-Brazilian Meeting on Gravitation, Astrophysics and Cosmology, Buenos Aires, Argentina, 22–25 April 2014; pp. 81–86
56. Salako, I.G.; Houndjo, M.; Jawad, A. Generalized Mattig's relation in Brans–Dicke–Rastall gravity. *Int. J. Mod. Phys. D* **2016**, *25*, 1650076. [CrossRef]
57. Visser, M. Rastall gravity is equivalent to Einstein gravity. *Phys. Lett. B* **2018**, *782*, 83. [CrossRef]
58. Rastall, P. A theory of gravity. *Can. J. Phys.* **1976**, *54*, 66. [CrossRef]
59. Capozziello, S.; Gonzalez, P.A.; Saridakis, E.N.; Vasquez, Y. Exact charged black-hole solutions in D-dimensional $f(T)$ gravity: torsion vs curvature analysis. *JHEP* **2013**, *2*, 039. [CrossRef]
60. Ayon-Beato, E. New regular black hole solution from nonlinear electrodynamics. *Phys. Lett. B* **1999**, *464*, 25. [CrossRef]
61. Salazar, I.H.; García, D.A.; Plebański, J. Duality rotations and type D solutions to Einstein equations with nonlinear electromagnetic sources. *J. Math. Phys.* **1987**, *28*, 2171. [CrossRef]
62. Guo, S.; He, K.-J.; Li, G.-R.; Li, G.-P. The shadow and photon sphere of the charged black hole in Rastall gravity. *Class. Quant. Grav.* **2021**, *38*, 165013. [CrossRef]
63. Nashed, G.G.L.; Saridakis, E.N. New rotating black holes in nonlinear Maxwell $f(R)$ gravity. *Phys. Rev. D* **2020**, *102*, 124072. [CrossRef]
64. Prihadi, H.L.; Sakti, M.F.A.R.; Hikmawan, G.; Zen, F.P. Dynamics of charged and rotating NUT black holes in Rastall gravity. *Int. J. Mod. Phys. D* **2020**, *29*, 2050021. [CrossRef]
65. Shahidi, S. osmological implications of Rastall-$f(R)$ theory. *Phys. Rev. D* **2021**, *104*, 084033. [CrossRef]
66. Misner, C.W.; Thorne, K.S.; Wheeler, J.A. *Gravitation*; Freeman, W.H., Ed.; Macmillan: San Francisco, CA, USA, 1973.
67. D'Inverno, R.A. *Internationale Elektronische Rundschau*; Oxford University Press: Oxford, UK, 1992.
68. Nashed, G.G.L. Stability of the vacuum non-singular black hole. *Chaos Solitons Fractals* **2003**, *15*, 841. [CrossRef]
69. Nashed, G.G.L.; Saridakis, E.N. tability of motion and thermodynamics in charged black holes in $f(T)$ gravity. *JCAP* **2022**, *5*, 017. [CrossRef]
70. Nashed, G.G.L.; Nojiri, S. Black holes with Lagrange multiplier and potential in mimetic-like gravitational theory: Multi-horizon black holes. *JCAP* **2022**, *5*, 011. [CrossRef]
71. Hunter, C.J. Action of instantons with a nut charge. *Phys. Rev.* **1999**, *59*, 024009. [CrossRef]
72. Hawking, S.W.; Hunter, C.J.; Page, D.N. NUT charge, anti–de Sitter space, and entropy. *Phys. Rev.* **1999**, *59*, 044033. [CrossRef]
73. Bekenstein, J.D. Black Holes and the Second Law. *Lett. Nuovo Cim.* **1972**, *4*, 737. [CrossRef]
74. Nashed, G.G.L. Stationary axisymmetric solutions and their energy contents in teleparallel equivalent of Einstein theory. *Astrophys. Space Sci.* **2010**, *330*, 173. [CrossRef]
75. Bekenstein, J.D. Black Holes and Entropy. *Phys. Rev.* **1973**, *7*, 2333. [CrossRef]
76. Gibbons, G.W.; Hawking, S.W. Cosmological event horizons, thermodynamics, and particle creation. *Phys. Rev.* **1977**, *15*, 2738. [CrossRef]
77. Myung, Y.S.; Kim, Y.-W.; Park, Y.-J. Quantum cooling evaporation process in regular black holes. *Phys. Lett.* **2007**, *656*, 221. [CrossRef]
78. Kim, W.; Shin, H.; Yoon, M. Anomaly and Hawking Radiation from Regular Black Holes. *J. Korean Phys. Soc.* **2008**, *53*, 1791. [CrossRef]
79. Myung, Y.S.; Kim, Y.-W.; Park, Y.-J. Thermodynamics of regular black hole. *Gen. Rel. Grav.* **2009**, *41*, 1051. [CrossRef]
80. Dymnikova, I.G. De sitter-schwarzschild black hole: Its particlelike core and thermodynamical properties. *Int. J. Mod. Phys. D* **1996**, *5*, 529. [CrossRef]

81. Dymnikova, I. The cosmological term as a source of mass. *Class. Quant. Grav.* **2002**, *19*, 725. [CrossRef]
82. Nashed, G.G.L. Reissner–nordström solutions and energy in teleparallel theory. *Mod. Phys. Lett. A* **2006**, *21*, 2241. [CrossRef]
83. Dymnikova, I. Generic Features of Thermodynamics of Horizons in Regular Spherical Space-Times of the Kerr-Schild Class. *Universe* **2018**, *4*, 63. [CrossRef]
84. Shirafuji, T.; Nashed, G.G.L.; Kobayashi, Y. Equivalence Principle in the New General Relativity. *Prog. Theor. Phys.* **1996**, *96*, 933. [CrossRef]
85. Hayward, S.A. Formation and Evaporation of Nonsingular Black Holes. *Phys. Rev. Lett.* **2006**, *96*, 031103. [CrossRef]
86. Nicolini, P.; Smailagic, A.; Spallucci, E. Noncommutative geometry inspired Schwarzschild black hole. *Phys. Lett.* **2006**, *632*, 547–551. [CrossRef]
87. Sharif, M.; Javed, W. Thermodynamics of a Bardeen black hole in noncommutative space. *Can. J. Phys.* **2011**, *89*, 1027. [CrossRef]
88. Bronnikov, K.A.; Dymnikova, A.I.G. Nonsingular vacuum cosmologies with a variable cosmological term. *Class. Quant. Grav.* **2003**, *20*, 3797. [CrossRef]
89. Bronnikov, K.; Dymnikova, I.; Galaktionov, E. Multihorizon spherically symmetric spacetimes with several scales of vacuum energy. *Class. Quant. Grav.* **2012**, *29*, 095025. [CrossRef]
90. Nouicer, K. Black hole thermodynamics to all orders in the Planck length in extra dimensions. *Class. Quant. Grav.* **2007**, *24*, 6435. [CrossRef]
91. Dymnikova, I.; Korpusik, M. Thermodynamics of Regular Cosmological Black Holes with the de Sitter Interior. *Entropy* **2011**, *13*, 1967. [CrossRef]
92. Chamblin, A.; Emparan, R.; Johnson, C.V.; Myers, R.C. Charged AdS black holes and catastrophic holography. *Phys. Rev.* **1999**, *60*, 064018. [CrossRef]
93. Hawking, S.W. Particle creation by black holes. In *Euclidean Quantum Gravity*; World Scientific: Singapore, 1975; Volume 43, pp. 199–220. [CrossRef]
94. Davies, P.C.W. The thermodynamic theory of black holes. *Proc. Roy. Soc. Lond.* **1977**, *A353*, 499. [CrossRef]
95. Babichev, E.O.; Dokuchaev, V.I.; Eroshenko, Y.N. Black holes in the presence of dark energy. *Phys. Usp.* **2013**, *56*, 1155, [CrossRef]
96. Wang, P.; Wu, H.; Yang, H.; Yao, F. Extended phase space thermodynamics for black holes in a cavity. *JHEP* **2020**, *9*, 154. [CrossRef]
97. Chaloshtary, S.R.; Zangeneh, M.; Hajkhalili, S.; Sheykhi, A.; Zebarjad, S. Thermodynamics and reentrant phase transition for logarithmic nonlinear charged black holes in massive gravity. *Int. J. Mod. Phys. D* **2020**, *29*, 2050081. [CrossRef]
98. Sajadi, S.; Riazi, N.; Hendi, S. Dynamical and thermal stabilities of nonlinearly charged AdS black holes. *Eur. Phys. J. C* **2019**, *79*, 775. [CrossRef]
99. Yu, S.; Gao, C. Exact black hole solutions with nonlinear electrodynamic field. *Int. J. Mod. Phys. D* **2020**, *29*, 2050032. [CrossRef]
100. Ali, M.S.; Ghosh, S.G. Thermodynamics of rotating Bardeen black holes: Phase transitions and thermodynamics volume. *Phys. Rev. D* **2019**, *99*, 024015. [CrossRef]
101. Caldarelli, M.M.; Cognola, G.; Klemm, D. Thermodynamics of Kerr-Newman-AdS black holes and conformal field theories. *Class. Quant. Grav.* **2000**, *17*, 399. [CrossRef]
102. Nashed, G.G.L.; Capozziello, S. Charged spherically symmetric black holes in $f(R)$ gravity and their stability analysis. *Phys. Rev. D* **2019**, *99*, 104018. [CrossRef]

Review

Observational Constraints on Dynamical Dark Energy Models

Olga Avsajanishvili [1,2,*], Gennady Y. Chitov [3,4], Tina Kahniashvili [1,2,5], Sayan Mandal [2,5] and Lado Samushia [1,2,6]

[1]   E.Kharadze Georgian National Astrophysical Observatory, 47/57 Kostava St., Tbilisi 0179, Georgia; tinatin@andrew.cmu.edu (T.K.); lado@phys.ksu.edu (L.S.)
[2]   School of Natural Sciences and Medicine, Ilia State University, 3/5 Cholokashvili Ave., Tbilisi 0162, Georgia; sayanm@andrew.cmu.edu
[3]   Bogoliubov Laboratory of Theoretical Physics, Joint Institute for Nuclear Research, Dubna 141980, Russia; gchitov@theor.jinr.ru
[4]   Département de Physique, Université de Sherbrooke, Sherbrooke, QC J1K 2R1, Canada
[5]   McWilliams Center for Cosmology and Department of Physics, Carnegie Mellon University, Pittsburgh, PA 15213, USA
[6]   Department of Physics, Kansas State University, 116 Cardwell Hall, Manhattan, KS 66506, USA
*   Correspondence: olga.avsajanishvili@iliauni.edu.ge

Abstract: Scalar field ϕCDM models provide an alternative to the standard ΛCDM paradigm, while being physically better motivated. Dynamical scalar field ϕCDM models are divided into two classes: the quintessence (minimally and non-minimally interacting with gravity) and phantom models. These models explain the phenomenology of late-time dark energy. In these models, energy density and pressure are time-dependent functions under the assumption that the scalar field is described by the ideal barotropic fluid model. As a consequence of this, the equation of state parameter of the ϕCDM models is also a time-dependent function. The interaction between dark energy and dark matter, namely their transformation into each other, is considered in the interacting dark energy models. The evolution of the universe from the inflationary epoch to the present dark energy epoch is investigated in quintessential inflation models, in which a single scalar field plays a role of both the inflaton field at the inflationary epoch and of the quintessence scalar field at the present epoch. We start with an overview of the motivation behind these classes of models, the basic mathematical formalism, and the different classes of models. We then present a compilation of recent results of applying different observational probes to constraining ψCDM model parameters. Over the last two decades, the precision of observational data has increased immensely, leading to ever tighter constraints. A combination of the recent measurements favors the spatially flat ΛCDM model but a large class of ϕCDM models is still not ruled out.

Keywords: dark energy; observational data; dynamical dark energy models

check for
updates

Citation: Avsajanishvili, O.; Chitov, G.Y.; Kahniashvili, T.; Mandal, S.; Samushia, L. Observational Constraints on Dynamical Dark Energy Models. *Universe* **2024**, *10*, 122. https://doi.org/10.3390/universe10030122

Academic Editors: Joan Solà Peracaula and Kazuharu Bamba

Received: 26 October 2023
Revised: 26 January 2024
Accepted: 6 February 2024
Published: 4 March 2024

1. Introduction

The accelerated expansion of our universe was first discovered in 1998 on the basis of measurements of the type Ia supernovae (SNe Ia) apparent magnitudes [1–3]. This fact was later confirmed by other cosmological observations, in particular, by measurements of the temperature anisotropy and the polarization in the cosmic microwave background (CMB) radiation [4–13], by studies of the large-scale structure (LSS) of the universe [14–19], by measurements of baryon acoustic oscillations (BAO) peak length scale [20–26], and by measurements of Hubble parameter [27–35].

One possible explanation for this empirical fact is that the energy density of the universe is dominated by dark energy or dark fluid, an energy component with an effective negative pressure (see Refs. [36–43] for reviews). The presence of dark matter in the universe, first discovered through the anomalously high rotation velocity of the outer regions of galaxies [44], is another major mystery of modern cosmology. Different models for dark matter have been proposed including cold dark matter (CDM), consisting of heavy particles with mass $m_{\rm CDM} \geq 100$ KeV, warm dark matter (WDM), composed of particles with a mass of $m_{\rm WDM} \approx 3\text{--}30$ KeV, and hot dark matter (HDM) consisting of ultrarelativistic particles [45]. Assuming general relativity is the correct description of gravity on cosmological scales, about 95% of the energy in the universe has to be in the "dark" form, i.e., in the form of dark energy and dark matter, to explain available observations. According to the last Planck data release (PR4), our universe consists of 4.86% of ordinary matter, 25.95% of dark matter, and 70.39% of dark energy [46].

The true nature and origin of dark energy and dark matter are still unresolved issues of modern cosmology. The simplest description of dark energy is vacuum energy or the cosmological constant Λ (see Refs. [36,39,47] for reviews). The cosmological model based on such a description of dark energy in the spatially flat universe is called the Lambda Cold Dark Matter (ΛCDM) model, which established itself as the standard or concordance model of the universe in the last two decades (see Ref. [36] for a pioneering work and Ref. [48] for recent review and discussion). In this model, dark matter is presented in the form of non-relativistic cold weakly interacting particles which either have never been in equilibrium with the primordial plasma or have been decoupled from it after becoming non-relativistic at an early stage. A good pedagogical overview of the ΛCDM model is available in many recent books [49–53] and reviews [37,54–58].

Despite explaining various observations of our universe to a remarkable degree of accuracy, the ΛCDM model has several unsolved problems and tensions [57,59–65], including the fine-tuning or cosmological constant problem, the coincidence problem, the Hubble and S_8 tensions, and the problem of the shape of the universe. A large number of cosmological models that go beyond the standard ΛCDM scenario with modified dynamics of the expansion of the universe both in early and late times have been considered in order to resolve these tensions. For reviews, see [66–76]. To solve the problems of the ΛCDM model, models with gravity different from general relativity on cosmological scales in the universe, so-called modified gravity (MG) models, have also been proposed [37,77–88]. For reviews, see [43,89,90] and especially a comprehensive analysis by Ishak [91] on a large class of the MG theories leading to the accelerated universe and the observational constraints on those theories.

The value of the energy density of the cosmological constant ρ_Λ following from the quantum field theory estimates is [60] $\rho_\Lambda \sim \hbar M_{\rm pl}^4 \sim 10^{72}$ Gev$^4 \sim 2 \cdot 10^{110}$ erg/cm^3, where $M_{\rm pl} \sim 10^{19}$ GeV is the Planck mass and $\hbar$ is the reduced Planck constant, while cosmological observations of the cosmological constant (like dark energy) show a very different result [60]: $\rho_\Lambda^{\rm obs} \sim 10^{-48}$ Gev$^4 \sim 2 \cdot 10^{-10}$ erg/cm^3. This discrepancy in 120 orders of magnitude between the predicted and observed values of the energy scale of the cosmological constant is called the cosmological constant problem or the fine-tuning problem [54–56,92,93]. An alternative point of view compatible with Einstein's equations of general relativity is to abandon the attempts to explain the minuscule value of the cosmological constant due to some "magic" cancellations of the quantum field theory vacuum terms and to assume its pure geometric origin. The drawback is that in such case the trivial space–time without sources would be the de Sitter universe with an intrinsic curvature [93].

The coincidence is that, based on precise cosmological observations [13,94], the energy density in dark energy (68.7%) is comparable (within an order of magnitude) to that of non-relativistic matter (31.3%) at present. This problem can also be presented as the why now problem, namely: "Why did the acceleration occur in the present epoch of cosmic evolution?" (surely any earlier event would have prevented the formation of structures in the universe) [36,57,59,61]. This fact is an enigma [52,56,93,95–101], because, in the ΛCDM model, the energy density of the cosmological constant does not depend on time, $\rho_\Lambda = \text{const}$, while the energy density of matter varies over time as $\rho_{\rm DM} \sim a^{-3}(t)$ ($a(t)$ and t are the scale factor and cosmic time, respectively), so the ratio of these quantities is time-dependent, $\rho_{\rm DM}/\rho_\Lambda \propto a^{-3}(t)$. Since the vacuum energy does not change over time, it was insignificant during both the radiation domination epoch and the matter domination epoch, but it has become the dominant component recently, at a scale factor $a \approx 0.77$ (or a redshift $z \approx 0.3$), according to Planck 2018 data [13], and it will be the only component in the universe in the future. The energy density of matter and the energy density of the cosmological constant are comparable for a very short period of time, so the following question arises: "Why did it happen that we live in this short (by the cosmological scale) period of time?" After all, this fact is in contradiction to the Copernican principle, since this coincidence implies that the present epoch is a special time, between the matter- and dark-energy-dominated epochs, and may hint at some physical mechanism at play which ensures these energy densities are similar.

The anthropic principle [102,103] can explain the cosmological constant problems and answer the questions: "Why is the energy density of the cosmological constant so small?" and "Why has the accelerated expansion of the universe started recently?" According to the anthropic principle, the energy density of the cosmological constant observed today should be suitable for the evolution of intelligent beings in the universe [92,104–106]. For a more detailed discussion and approaches to solve this problem, see Ref. [107].

The Hubble tension problem is that there is a discrepancy at the level of $\sim 5\sigma$ between the value of the Hubble parameter at the present epoch $H_0 = 100h$ km c^{-1} Mpc^{-1}, where h is a dimensionless normalized Hubble constant, obtained by the local measurements, and CMB temperature, polarization, and lensing anisotropy data [13,108–113]. In particular, supernova measurements give $H_0 = 73.04 \pm 1.04$ km/s/Mpc [114], while CMB measurements (TT,TE,EE + lowE + lensing) lead to $H_0 = 67.36 \pm 0.54$ km/s/Mpc [13].

The S_8 tension problem is that there is a discrepancy at the level of $\sim 2\sigma$–3σ confidence level between the primary CMB temperature anisotropy measurements by the Planck satellite [13] in the strength of matter clustering compared to lower redshift measurements such as the weak gravitational lensing and galaxy clustering [66,68,115–118]. This tension is quantified using the weighted amplitude of the matter fluctuation parameter $S_8 = \sigma_8(\Omega_{\mathrm{m}}/0.3)^{0.5}$, which modulates the amplitude of weak lensing measurements; here, σ_8 is an amplitude of mass fluctuations on scales of $8h^{-1}$ Mpc; $\Omega_{\mathrm{m}} = \Omega_{\mathrm{m}0}a^{-3}/E^2(a)$ is a matter density parameter; $\Omega_{\mathrm{m}0}$ is a matter density parameter at the present epoch; $E(a) = H(a)/H_0$ is a normalized Hubble parameter; $H(a) = \dot{a}/a$ is a Hubble parameter; and $\dot{a}$ is a derivative of the scale factor a with respect to cosmic time.

The problem with the shape of the universe is that the CMB anisotropy power spectra measured by the Planck space telescope show a preference for a spatially closed universe at a more than 3σ confidence level [13,119]. This fact contradicts expectations from the simplest inflationary models [63,66,120] and is interpreted by the cosmological community as a possible crisis of modern cosmology [62,120–124].

One of the alternatives to the ΛCDM model, during the period of time when the accelerated expansion of the universe is governed by the cosmological constant Λ, is dynamical dark energy scalar field ϕCDM models [125–133]. In these models, dark energy is described through the equation of state (EoS) parameter, $w_\phi(t)$, which depends on time: $w_\phi(t) \equiv p_\phi/\rho_\phi$, p_ϕ is the scalar field pressure, ρ_ϕ is the scalar field energy density; whereas, in the ΛCDM model, the EoS parameter is a constant, $w_\Lambda = -1$. At the same time, at the present epoch, the value of the time-dependent EoS parameter in scalar field models becomes approximately equal to minus one, $w_\phi \approx -1$; thus, dynamical dark energy mimics the cosmological constant and becomes almost indistinguishable from it. However, dark energy is a dynamic parameter related to the current value of the scalar field potential, while the universe evolves towards its true vacuum with zero energy, i.e., the zero cosmological term.

Depending on the value of the EoS parameter at the present epoch, ϕCDM scalar field models are divided into quintessence models, with $-1 < w_\phi(t) < -1/3$ [95,98,134–137], see, e.g., Refs. [36,43] (for a review), and phantom models, with $w_\phi(t) < -1$ [138–144]. Quintessence models are divided into two classes: tracker (freezing) models, in which the scalar field evolves slower than the Hubble expansion rate, and thawing models, in which the scalar field evolves faster than the Hubble expansion rate [95,135,141,145,146]. In quintessence tracker models, the energy density of the scalar field first tracks the radiation energy density and then the matter energy density, while it remains subdominant [147]. Only recently does the scalar field become dominant and begins to behave as a component with negative pressure, which leads to the accelerated expansion of the universe [136,148,149]. For certain forms of potentials, the quintessence tracker models have an attractor solution that is insensitive to initial conditions [147].

The interaction between dark energy and dark matter, namely their transformation into each other, is considered in the interacting dark energy (IDE) models [150–154]. In these models, the coincidence problem of the standard ΛCDM model as well as the Hubble constant H_0 tension can be alleviated [67,70,73,155–160].

In the standard ΛCDM cosmological scenario, one assumes the existence of two epochs of accelerated expansion in the universe. The first is inflation [161–169], which happens in the very early universe, and the second is the dark-energy-dominated epoch observed today [50,51,170,171]. Inflation is a theory of the exponential expansion of space in the early universe, which is believed to have lasted approximately from 10^{-36} to 10^{-33}–10^{-32} s after the Big Bang, the exact times being dependent on the microphysics of the model describing inflation. The inflationary models explain the quantum origin of tiny primordial density fluctuations in the universe, which must have been present at very early epochs, as the seeds both for the CMB anisotropies and for the structure formation in the later evolution of the universe. The exponential expansion during inflation comes to an end when a phase transition transforms the vacuum energy into radiation and matter, after which the radiation-dominated epoch begins. This phase transition is called the reheating and its governing dynamics is still debated. A successful inflationary model requires a smooth transition to the decelerated epoch (in which inflation rules the universe as if it were dominated by non-relativistic matter) because, otherwise, the homogeneity of the universe would be violated [172,173]. Inflation resolves several problems in cosmology, namely, the horizon problem, associated with the lack of causal relationship between different regions in the early universe before the recombination epoch (this is an epoch of forming the electrically neutral hydrogen atoms, which began at $t_{\rm rec} \approx 350{,}000$ years after the Big Bang), and the flatness problem, related to the fine tuning of the spatial flatness of the universe in the early epoch so that the spatial flatness of the universe is preserved at the present epoch). The evolution of the universe from the inflationary epoch to the present dark energy epoch is investigated in quintessential inflation models too [174–179]. In these models, a single scalar field plays a role of both the inflaton field at the inflationary epoch and of the quintessence scalar field at the present epoch; thereby, the origin of dark energy at the present epoch is also explained within the same model.

The running vacuum models (RVMs) describe dark energy as a quantum vacuum, the energy density of which slowly evolves with the expansion of the universe [180]. RVM models, like the scalar field ϕCDM models, are associated with scalar fields, but describe dark energy as a quantum vacuum not just the vacuum of a classical scalar field [181]. The EoS parameter of the running vacuum is moderately dynamic in the late universe, $w_{\rm vac} \geqslant -1$, mimicking the quintessence scalar fields [182]. In contrast to classical scalar fields which depend on an arbitrary potential, running cosmic vacuum arises from quantum effects and can be derived from explicit calculations of quantum field theory (QFT) both in the spatially flat and non-flat hypersurfaces (see Ref. [183] for reviews). The latest cosmological data are in good agreement with the RVMs [184,185], while confirming the results of earlier constraints of these models [186–188]. The cosmological constant problem of the ΛCDM model can be resolved in the RVMs [189,190], and the H_0 and σ_8 tensions weaken in these models, as can be seen from the data constraints presented in [184,185].

It has also been suggested that the current accelerated expansion of the universe can be explained by modifications of the general theory of relativity. Several such modifications of general relativity have been proposed, see Refs. [88–90,191], to explain a host of cosmological observations. Although the current observational constraints are still too large to exclude some of the MG theories [91], it seems to be premature at this point to consider such theories as a comprehensive and viable alternative to the minimal model of dark matter and/or dark energy based on Einstein's general relativity, in order to explain the observations. We will therefore not discuss theories of modified gravity in our review.

In this paper we reviewed and analyzed, to the best of our knowledge of current literature, the most relevant studies of the observational constraints on dynamical dark energy models over the past twenty years, in particular, scalar field ϕCDM models, quintessential inflation scalar field ϕCDM models, and IDE models both in the spatially flat and non-flat hypersurfaces. The research effort on the complication, refinement of cosmological data, and the increase in the variety of methods for studying dynamical dark energy models lead to more accurate constraints on the values of the cosmological parameters in these models. Despite the refinement of various observational data and the complication of methods for studying dark energy in the universe, current observational data still favor the standard spatially flat ΛCDM model, while not excluding dynamical dark energy models and spatially closed hyperspaces [192–206]. At the same time, recent studies showed that the currently available observational datasets favor the IDE model at a more than 2σ confidence level [67,158,159,207,208].

This paper is organized as follows: the different cosmological dynamical dark energy models are described in Section 2, observational constraints on dynamical dark energy models by various observational data are presented in Section 3, the main results are summarized in Section 4, the ongoing and upcoming missions are listed in Section 5, and conclusions are presented in Section 6. In this paper, we used the natural system of units: $c = \hbar = k_B = 1$.

2. Cosmological Dark Energy Models

2.1. ΛCDM Model

As highlighted in the Introduction, the Lambda Cold Dark Matter (ΛCDM) model is the standard or concordance model of a spatially flat universe. In this model, dark energy is represented by the cosmological constant Λ, and its energy density is constant

$$\rho_\Lambda = \frac{\Lambda M_{\mathrm{pl}}^2}{8\pi} = \rho_{\mathrm{vac}} = \mathrm{const,} \tag{1}$$

where $\Lambda = 4.33 \cdot 10^{-66}$ eV^2. The pressure and the energy density in the ΛCDM model are related as

$$p_\Lambda = -\rho_\Lambda = \mathrm{const,} \tag{2}$$

leading to the constant EoS parameter

$$w_\Lambda = -1. \tag{3}$$

The action with the cosmological constant Λ is [43]

$$S = -\frac{M_{\mathrm{pl}}^2}{16\pi} \int d^4x \sqrt{-g}(R + 2\Lambda) + S_{\mathrm{M}}, \tag{4}$$

where $g \equiv \det(g_{\mu\nu})$ is the determinant of the metric tensor $g_{\mu\nu}$, R is the Ricci scalar, and S_{M} is the action of the matter. The spatially flat ΛCDM model is typically characterized by six independent parameters [209]: the physical baryon density parameter, $\Omega_b h^2$; the dark matter physical density parameter, $\Omega_c h^2$; the age of the universe, t_0; the scalar spectral index, n_s; the amplitude of the curvature fluctuations, Δ_{R}^2; and the optical depth during the reionization period for $z \in (6, 20)$, τ. In addition to these parameters, the ΛCDM model is described by six extended fixed parameters: the total density parameter, Ω_{tot}; the EoS parameter, w_Λ; the total mass of three types of neutrinos, $\sum m_\nu$; the effective number of the relativistic degrees of freedom, N_{eff}; the tensor/scalar ratio, r; and the scalar spectral index running, a_s.

The extension of the spatially flat ΛCDM model to spatially non-flat hypersurfaces is the oCDM model. The first Friedmann equation describing the evolution of the universe for the spatially non-flat oCDM model (for $\Omega_{k0} = 0$ this equation is applicable for the spatially flat ΛCDM model) has a form

$$E(a) = (\Omega_{r0}a^{-4} + \Omega_{m0}a^{-3} + \Omega_{k0}a^{-2} + \Omega_{\Lambda})^{1/2}, \qquad (5)$$

where $\Omega_{r0} = \rho_{r0}/\rho_{cr}$, $\Omega_{m0} = \rho_{m0}/\rho_{cr}$, and $\Omega_{\Lambda} = \rho_{\Lambda}/\rho_{cr}$ are density parameters at the present epoch for radiation, matter, and vacuum, respectively, where ρ_{r0} and ρ_{m0} are energy densities for radiation and matter at the present epoch, respectively. The value of the critical energy density at the present epoch is equal to $\rho_{cr} = 3M_{pl}^2 H_0^2/8\pi = 1.8791h^2 \cdot 10^{-29}$ g cm^{-3}; $\Omega_{k0} = -k/H_0^2$ is a spatial curvature density parameter at the present epoch, $\rho_{k0} = -3kM_{pl}^2/8\pi$ is a spatial curvature density, k is a curvature parameter, and $E(a) \equiv H(a)/H_0$ is a normalized Hubble parameter.

Observational constraints on the cosmological parameters Ω_{m0} and Ω_{Λ}, obtained from different cosmological datasets for the ΛCDM model and for the oCDM model, are represented in Figure 1.

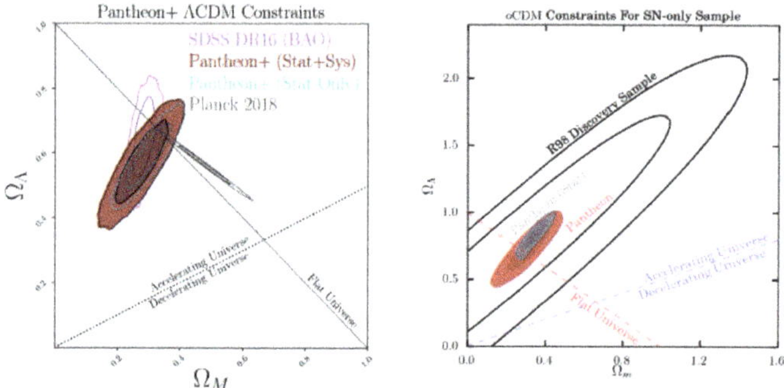

Figure 1. The 1σ and 2σ confidence level contour constraints on Ω_m and Ω_Λ parameters. (Left panel) In the standard spatially flat ΛCDM model from the SNe Ia Pantheon dataset, as well as from the combined BAO peak length scale and Planck datasets. (The figure is adapted from [210]). (Right panel) In the spatially non-flat oCDM model using discovery sample of Riess et al. [211] and the full Pantheon sample of Scolnic et al. [212]. Pantheon constraints with systematic uncertainties are presented in red, while only statistical uncertainties are denoted in gray. (The figure is adapted from [212]).

As we mentioned above, the ΛCDM model is the fiducial model against which all alternative models are compared regarding their fit to observational data. Its predictions agree with the observational data pertaining to the accelerated expansion of the universe, the statistical distribution of LSS, the CMB temperature and polarization anisotropies, and the abundance of light elements in the universe [94,213].

2.2. Dynamical Dark Energy Scalar Field ϕCDM Models

There are numerous physically motivated alternative models for the ΛCDM model [37,43,77–87]. One of the prominent alternatives to the ΛCDM model are the dynamical scalar field ϕCDM models, in which the scalar field can interact with gravity both minimally [36,126,127] and non-minimally via different coupling terms (the so-called extended scalar–tensor models) [214–222]. We will concentrate on the minimally coupled models as the simplest and more natural choice.

In models with minimal interaction with gravity, the role of dark energy is played by a slowly varying uniform self-interacting scalar field ϕ. These ϕCDM models involving a dynamical scalar field do not suffer from the fine-tuning problem of the ΛCDM model, and have a more natural explanation for the observed low-energy scale of dark energy. When the energy density of the scalar field begins to dominate over the energy density of both radiation and matter, the universe begins the stage of the accelerated expansion. At early times during the evolution of the universe, the behavior of the dynamical scalar field is different from that of the cosmological constant Λ, but is almost indistinguishable from that of the cosmological constant during later times.

The dynamical scalar field ϕCDM models are divided into two classes: the quintessence models [147] and phantom models [138,223]. These two classes of models differ from each other by the following attributes:

(i)   The EoS parameter—For quintessence fields, $-1 < w_\phi < -1/3$, while for phantom fields, $w_\phi < -1$.
(ii) The sign of the kinetic term—For quintessence fields, the kinetic term in the Lagrangian has a positive sign, while it is negative for phantom fields.
(iii) The dynamics of the scalar field—The quintessence field rolls gradually to the minimum of its potential, while the phantom field rolls to the maximum of its potential.
(iv) Temporal evolution of dark energy—For quintessence fields, the dark energy density remains almost unchanging with time, while it increases for phantom fields.
(v)   Forecasting the future of the universe—The quintessence models predict either an eternal expansion of the universe or a repeated collapse, depending on the spatial curvature of the universe. On the other hand, the phantom models predict the destruction of any gravitationally related structures in the universe. Depending on the asymptotic behavior of the Hubble parameter $H(t)$, the future scenarios of the universe are divided into a *big rip* for which $H(t) \to \infty$ for a finite future time $t = \text{const}$; a little rip for which $H(t) \to \infty$ at an infinite future time $t \to \infty$, and a pseudo rip for which $H(t) \to \text{const}$ for an infinite future time $t \to \infty$.

The action describing a scalar field in the presence of matter is given by

$$S = \int d^4 x \sqrt{-g}\left(-\frac{M_{\rm pl}^2}{16\pi}R + \mathcal{L}_\phi\right) + S_{\rm M}, \tag{6}$$

where $\mathcal{L}_\phi$ is the Lagrangian density of the scalar field, the form of which depends on the type of the chosen model. We describe the form of $\mathcal{L}_\phi$ for the quintessence and the phantom fields below.

2.2.1. Quintessence Scalar Field

The dynamics of the quintessence scalar field is described by the Lagrangian density

$$\mathcal{L}_\phi = \frac{1}{2}g^{\mu\nu}\partial_\mu\phi\,\partial_\nu\phi - V(\phi), \tag{7}$$

where $V(\phi)$ is a scalar field potential. There are various quintessence potentials discussed in the literature but there is currently no observational constraint to prefer one of these over the others. A list of some of the quintessence potentials is presented in Table 1.

The EoS parameter for the quintessence scalar field is given by

$$w_\phi \equiv \frac{p_\phi}{\rho_\phi} = \frac{\dot{\phi}^2/2 - V(\phi)}{\dot{\phi}^2/2 + V(\phi)}, \tag{8}$$

where $p_\phi = \frac{1}{2}\dot{\phi}^2 - V(\phi)$ and $\rho_\phi = \frac{1}{2}\dot{\phi}^2 + V(\phi)$ are, respectively, the pressure and energy density of the quintessence field. Here, the overdots denote derivatives with respect to the cosmic time t. The equation of motion for the quintessence scalar field can be obtained by varying the action in Equation (6), along with the Lagrangian in Equation (7),

$$\ddot{\phi} + 3H\dot{\phi} + V'(\phi) = 0, \tag{9}$$

with the prime denoting a derivative with respect to the scalar field ϕ. The first Friedmann's equation for a ϕCDM model in a spatially non-flat spacetime has the form

$$E(a) = (\Omega_{r0}a^{-4} + \Omega_{m0}a^{-3} + \Omega_{k0}a^{-2} + \Omega_{\phi}(a))^{1/2}, \tag{10}$$

where $\Omega_{\phi}(a)$ is the dark energy (scalar field) density parameter.

Depending on the shape of potentials, quintessence models are further subdivided into thawing models and freezing (tracking) models [43,135]. In the $w_{\phi} - dw_{\phi}/d\ln a$ plane, thawing and freezing scalar field models are located in strictly designated zones for each of them [135], see Figure 2 (Left panel).

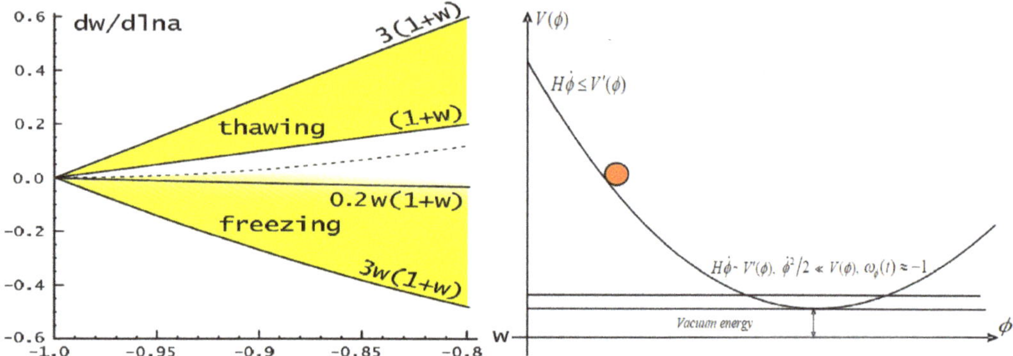

Figure 2. (Left panel) The location of thawing and freezing scalar fields in the $w_{\phi} - dw_{\phi}/d\ln a$ plane. (The figure is adapted from [135]). (Right panel) Regimes of the quick rolling down and slow rolling down of the scalar field to the minimum of its potential.

(a) In the thawing models, the scalar field was too suppressed by the retarding effect of the Hubble expansion, represented by the term $3H\dot{\phi}$ in Equation (9), until recently. This results in a much slower evolution of the scalar field compared to the Hubble expansion and the thawing scalar field manifests itself as the vacuum energy, with the EoS parameter $w_{\phi} \sim -1$. The Hubble expansion rate $H(t)$ decreases with time and, after it falls below $\sqrt{V''(\phi)}$, the scalar field begins to roll to the minimum of its potential, see Figure 2 (Right panel). The value of the EoS parameter for the scalar field thus increases over the time and becomes $w_{\phi} > -1$.

(b) In the freezing models, the scalar field is always suppressed (it is damped), i.e., $H(t) > \sqrt{V''(\phi)}$. Freezing scalar field models have so-called tracking solutions. According to tracking solutions, the quintessence component tracks the background EoS parameter (radiation in the radiation-dominated epoch and matter in the matter-dominated epoch) and eventually only recently grows to dominate the energy density in the universe. This leads to the accelerated expansion of the universe at late times, since the scalar field has a negative effective pressure. The tracker behavior allows the quintessence model to be insensitive to initial conditions. But this requires fine tuning of the potential energy, since $\sqrt{V''(\phi)} \sim H_0 \sim 10^{-33}$ eV.

Table 1. Scalar field quintessence potentials.

Name	Form	Reference
Ratra-Peebles	$V(\phi) \propto \phi^{-\alpha};$ ($\alpha = \text{const} > 0$)	Ratra and Peebles [126]
Exponential	$V(\phi) \propto \exp(-\lambda\phi/M_{\rm pl});$ ($\lambda = \text{const} > 0$)	Wetterich [127], Ratra and Peebles [125], Lucchin and Matarrese [168], Ferreira and Joyce [224]
Zlatev-Wang-Steinhardt	$V(\phi) \propto (\exp(M_{\rm pl}/\phi) - 1)$	Zlatev et al. [147]
Sugra	$V(\phi) \propto \phi^{-\chi} \exp(\gamma\phi^2/M_{\rm pl}^2);$ ($\chi, \gamma = \text{const} > 0$)	Brax and Martin [225]
Sahni-Wang	$V(\phi) \propto (\cosh(\varsigma\phi) - 1)^g;$ ($\varsigma = \text{const} > 0, g = \text{const} < 1/2$)	Sahni and Wang [226]
Barreiro-Copeland-Nunes	$V(\phi) \propto (\exp(\nu\phi) + \exp(\upsilon\phi));$ ($\nu, \upsilon = \text{const} \geq 0$)	Barreiro et al. [227]
Albrecht-Skordis	$V(\phi) \propto ((\phi - B)^2 + A)\exp(-\mu\phi);$ ($A, B = \text{const} \geq 0, \mu = \text{const} > 0$)	Albrecht and Skordis [228]
Urēna-López-Matos	$V(\phi) \propto \sinh^m(nM_{\rm pl}\phi);$ ($n = \text{const} > 0, m = \text{const} < 0$)	Urena-Lopez and Matos [229]
Inverse exponent potential	$V(\phi) \propto \exp(M_{\rm pl}/\phi)$	Caldwell and Linder [135]
Chang-Scherrer	$V(\phi) \propto (\exp(-\tau\phi) + 1);$ ($\tau = \text{const} > 0$)	Chang and Scherrer (2016) [230]

In 1988, Ratra and Peebles introduced a tracker ϕCDM model comprising a scalar field with an inverse power-law potential of the form $V(\phi) = \kappa/2M_{\rm pl}^2\phi^{-\alpha}$, for a model parameter $\alpha > 0$ [125,126]. For $\alpha = 0$, this ϕCDM Ratra–Peebles (RP) model reduces to the ΛCDM model. The positive parameter κ relates to the mass scale of the particles, M_ϕ, as $M_\phi \sim (\kappa M_{\rm pl}^2/2)^{\frac{1}{\alpha+4}}$. The RP ϕCDM model is a typical representative of the behavior of tracker quintessence scalar field ϕCDM models.

2.2.2. Phantom Scalar Field

The Lagrangian density describing a phantom scalar field has the form

$$\mathcal{L}_\phi = -\frac{1}{2}g^{\mu\nu}\partial_\mu\phi\partial_\nu\phi - V(\phi), \tag{11}$$

where the negative sign of the kinetic energy term is required to ensure the dark energy EoS parameter is less than -1, i.e., $w_\phi < -1$, and the energy density increases over time [143]. A phantom or ghost scalar field suffers from quantum instability because its energy density is not limited from below [131]. An incomplete list of phantom potentials is given in Table 2.

Analogous to Equation (8), the EoS parameter for the phantom scalar field is given by

$$w_\phi \equiv \frac{p_\phi}{\rho_\phi} = \frac{-\dot{\phi}^2/2 - V(\phi)}{-\dot{\phi}^2/2 + V(\phi)}, \tag{12}$$

where $p_\phi = -\frac{1}{2}\dot{\phi}^2 - V(\phi)$ and $\rho_\phi = -\frac{1}{2}\dot{\phi}^2 + V(\phi)$ are, respectively, the pressure and the energy density of the phantom field. The equation of motion for the phantom scalar field has the form

$$\ddot{\phi} + 3H\dot{\phi} - V'(\phi) = 0. \tag{13}$$

Table 2. Scalar field phantom potentials.

Name	Form	Reference
Fifth power	$V(\phi) \propto \phi^5$	Scherrer and Sen [140]
Inverse square power	$V(\phi) \propto \phi^{-2}$	Scherrer and Sen [140]
Exponent	$V(\phi) \propto \exp(\beta\phi); (\beta = \text{const} > 0)$	Scherrer and Sen [140]
Quadratic	$V(\phi) \propto \phi^2$	Dutta and Scherrer [141]
Gaussian	$V(\phi) \propto (1 - \exp(\phi^2/\sigma^2));$ $(\sigma = \text{const})$	Dutta and Scherrer [141]
Pseudo-Nambu–Goldstone boson (pNGb)	$V(\phi) \propto (1 - \cos(\phi/\kappa));$ $(\kappa = \text{const} > 0)$	Frieman et al. [231]
Inverse hyperbolic cosine	$V(\phi) \propto \cosh^{-1}(\psi\phi);$ $(\psi = \text{const} > 0)$	Dutta and Scherrer [141]

2.3. Parameterized Dark Energy Models

2.3.1. wCdm Parameterization

In dynamical dark energy models, one can use the wCDM parameterization where the EoS parameter can be expressed as $p = w(a)\rho$. Dark energy models are sometimes characterized only by the EoS parameter and corresponding cosmological models are called wCDM models [232]. This parameterization has no physical motivation, but is commonly used as an ansatz in data analysis to quantify differences and distinguish between dark energy models. The wCDM parameterization in particular makes it possible to differentiate, at the present epoch, the ΛCDM model from other dark energy models.

The time-dependent EoS parameter in the wCDM models is often characterized by the Chevallier–Polarsky–Linder (CPL) $w_0 - w_a$ parameterization [233,234]

$$w(a) = w_0 + w_a(1 - a), \tag{14}$$

where $w_0 = w(a_0)$ and $w_a = 2dw(a)/d\ln(1 + z)|_{z=1} = -2dw(a)/d\ln a|_{a=1/2}$, with z being the cosmological redshift defined as $z = 1/a - 1$ and a_0 being the scale factor at the present time, conventionally normalized as $a_0 = 1$. Although the CPL parameterization is simple and flexible enough to accurately describe EoS parameters in most dark energy models, it cannot describe arbitrary dark energy models with good accuracy (up to a few percent) over a wide redshift range [234]. The dynamical dark energy models where the EoS parameter is expressed through the CPL parameterization are called the $w_0 w_a$CDM models.

The normalized Hubble parameter for the $w_0 w_a$CDM model for the spatially flat universe has the form

$$E(a) = \left[\Omega_{\text{r0}}a^{-4} + \Omega_{\text{m0}}a^{-3} + (1 - \Omega_{\text{m0}})a^{-3(1+w_0+w_a)}e^{-3w_a(1-a)}\right]^{1/2}. \tag{15}$$

The 1σ and 2σ confidence level contour constraints on the cosmological parameters w_0 and w_a in the $w_0 w_a$CDM model from different combinations of datasets—the SNe Ia apparent magnitude (including measurements of the Hubble Space Telescope (HST)), the CMB temperature anisotropy, and the BAO peak length scale—are presented in Figure 3 (left panel).

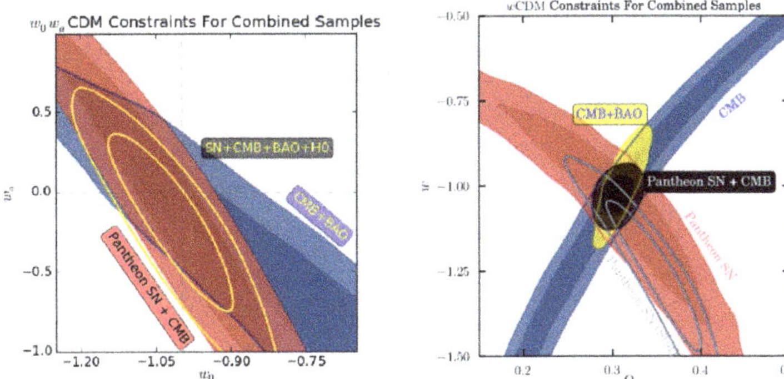

Figure 3. (Left panel) The 1σ and 2σ confidence level contour constraints on cosmological parameters w_0 and w_a in the wCDM model from various datasets: SNe Ia apparent magnitude + CMB temperature anisotropy + BAO peak length scale + HST (yellow), BAO peak length scale + CMB temperature anisotropy (blue), SNe Ia apparent magnitude + CMB temperature anisotropy (red). (Right panel) The 1σ and 2σ confidence level contours constraints on cosmological parameters Ω_{m} and w in the wCDM model from various datasets: SNe Ia apparent magnitude + CMB temperature anisotropy (black), CMB temperature anisotropy (blue), SNe Ia apparent magnitude (red) (with systematic uncertainties), SNe Ia apparent magnitude (gray line) (with only statistical uncertainties). The figure is adapted from [212].

2.3.2. XCDM Models

Cosmological dark energy models with a constant value of the EoS parameter are called XCDM models. These models are defined both in the spatially flat and spatially non-flat hyperspaces. The normalized Hubble parameter expressed through the dark energy EoS parameter w_X has the form

$$E(a) = \left[\Omega_{\mathrm{r}0}a^{-4} + \Omega_{\mathrm{m}0}a^{-3} + \Omega_{\mathrm{k}0}a^{-2} + (1 - \Omega_{\mathrm{m}0})a^{-3(1+w_X)}\right]^{1/2}. \tag{16}$$

The case $w_X = -1$ is equivalent to the standard spatially flat ΛCDM model with the same matter–energy density parameter $\Omega_{\mathrm{m}0}$ and zero spatial curvature, $\Omega_{\mathrm{k}0} = 0$, at the present epoch.

The 1σ and 2σ confidence level contour constraints on the cosmological parameter w_X in the XCDM model from different combinations of datasets—the SNe Ia apparent magnitude, the CMB temperature anisotropy, and the BAO peak length scale—are presented in Figure 3 (right panel).

2.4. Quintessential Inflation Models

Quintessential inflation models describe the evolution of the universe from the inflation epoch till the present dark energy epoch. In these models, a single field ϕ plays the double role of the inflaton field at the inflation epoch and the quintessence scalar field at the present epoch.

The general form of the action for quintessential inflation models reads:

$$S = \int d^4x \sqrt{-g}\left(-\frac{M_{\mathrm{pl}}^2}{16\pi}R + \frac{1}{2}g^{\mu\nu}\partial_\mu\phi\partial_\nu\phi - V(\phi)\right) + S_{\mathrm{M}} + S_{\mathrm{I}}(g_{\mu\nu}, \phi, \psi, \chi, B_\mu), \tag{17}$$

where S_{I} is the action describing the interactions of the inflaton field with the fermion (ψ), scalar (χ), and vector (B_μ) degrees of freedom in the Standard Model and beyond.

To maintain inflation over a long period of time, it is necessary that the acceleration caused by the inflaton field be sufficiently small compared to its velocity over the Hubble time. Under these conditions, the first Friedmann's and Klein–Gordon's equations for the inflaton in the spatially flat universe take the form [168,235,236]

$$\rho = \dot{\phi}^2/2 + V(\phi) \xrightarrow{\dot{\phi}^2/2 \ll V(\phi)} \rho = \frac{3M_{\text{Pl}}^2}{8\pi} H^2 \simeq V(\phi),$$

$$\ddot{\phi} + 3H\dot{\phi} + V'(\phi) = 0 \xrightarrow{|\ddot{\phi}| \ll 3H|\dot{\phi}|} -V'(\phi) \simeq 3H\dot{\phi}. \tag{18}$$

The slow-roll regime of the inflaton field is provided by the potential $V(\phi)$ with certain shapes: exponential [177], power-law [174,175], and plateau-like [178,179]. The slow-roll parameters, which determine the curvature and slope of the potential, should remain small for some period of time to sustain the inflationary behavior:

$$\epsilon = \frac{M_{\text{Pl}}^2}{2}\left(\frac{V'(\phi)}{V(\phi)}\right)^2 \ll 1, \quad \eta = M_{\text{Pl}}^2 \frac{V''(\phi)}{V(\phi)} \ll 1, \quad \theta = M_{\text{Pl}}^2 \frac{V'(\phi)V'''(\phi)}{V(\phi)^2} \ll 1. \tag{19}$$

The scalar spectral index (n_s), tensor spectral index (n_t), scalar spectral index running (a_s), and tensor-to-scalar ratio (r) are defined, respectively, as [177]

$$n_s - 1 = -6\epsilon + 2\eta, \quad n_t = -2\epsilon, \quad a_s \equiv dn_s/d\ln k = 16\epsilon\eta - 24\epsilon^2 - 2\theta, \quad r = 16\epsilon. \tag{20}$$

During the inflationary epoch of the universe, scalar and tensor perturbations are created from quantum vacuum fluctuations and are spatially stretched to superhorizon scales, where they become classical, and the almost scale-invariant tilted primordial power spectrum is formed [237]. The tilted primordial scalar $\mathcal{P}_s(k)$ and tensor $\mathcal{P}_t(k)$ power spectra for spatially flat tilted quintessential inflation models are defined in terms of the wave number k as [168,235,236]

$$\mathcal{P}_s(k) = A_s \left(\frac{k}{k_0}\right)^{n_s - 1}, \quad \mathcal{P}_t(k) = A_t \left(\frac{k}{k_0}\right)^{n_t}, \tag{21}$$

where A_s and A_t are the curvature perturbations amplitude and tensor amplitude at the pivot scale $k_0 = 0.05\ \text{Mpc}^{-1}$ [51].

The untilted primordial power spectrum for untilted spatially non-flat quintessential inflation models is defined as [176,238]

$$P(q) \propto \frac{(q^2 - 4K)^2}{q(q^2 - K)}, \tag{22}$$

where $q = \sqrt{k^2 + K}$ is the wavenumber for scalar perturbations. In the spatially flat limit $K = 0$, $P(q)$ reduces to the $n_s = 1$ spectrum.

2.5. Interacting Dark Energy Models

As mentioned above, one of the major unresolved problems of modern cosmology is the so-called coincidence problem, i.e., the energy densities of dark energy and dark matter are of the same order of magnitude at the present epoch. One way to resolve this problem is to assume that these components somehow interact with each other. IDE models consider the transformation of dark energy and dark matter into each other, with their interaction described by the following modified continuity equations for dark energy and matter, respectively

$$\dot{\rho}_m + 3H\rho_m = \delta_{\text{couple}}, \tag{23}$$

$$\dot{\rho}_\phi + 3H(\rho_\phi + p_\phi) = -\delta_{\text{couple}}, \tag{24}$$

where ρ_m is the matter energy density and δ_{couple} is the interaction function. In IDE models, the following forms of the coupling coefficient δ_{couple} are typically used [152,153]

$$\delta_{\text{couple}} = nQ\rho_m\dot{\phi}, \tag{25}$$

$$\delta_{\text{couple}} = \beta H(\rho_m + \rho_\phi), \tag{26}$$

where $n = \sqrt{8\pi/M_{\text{pl}}^2}$, and β and Q are dimensionless constants. The IDE models are subdivided into two types, as described below [121,152,153].

2.5.1. Coupling of the First Type

The IDE models of the first type are characterized by the exponential potential for the scalar ϕ and the linear interaction determined by the coupling coefficient given by Equation (25), as discussed in [152]. The coupled quintessence scalar field equation is given as

$$\ddot{\phi} + 3H\dot{\phi} + V'(\phi) = -nQ\rho_m\dot{\phi}, \tag{27}$$

where $V(\phi) = V_0 e^{-n\lambda\phi}$ is the scalar field potential and λ is a model parameter. The coupled continuity equation for dark energy is

$$\dot{\rho}_\phi + 3H(\rho_\phi + p_\phi) = -nQ\rho_m\dot{\phi}. \tag{28}$$

The matter energy density evolves according to

$$\dot{\rho}_m + 3H\rho_m = nQ\rho_m, \tag{29}$$

leading to

$$\rho_m = \rho_{m0}a^{-3}e^{nQ\phi}. \tag{30}$$

2.5.2. Coupling of the Second Type

For the second type of IDE models, the scalar potential, and hence the dynamics of the interaction between dark energy and matter is constructed with the requirement that the coincidence parameter $r = \rho_m/\rho_\phi$ takes an analytic expression and for $z \to 0$ becomes a constant, thereby alleviating the coincidence problem of the ΛCDM model [121,153].

The equation of motion for ϕ, Equation (24), can be written as

$$\dot{\phi}\left[\ddot{\phi} + 3H\dot{\phi} + V'(\phi)\right] = -\delta_{\text{couple}}. \tag{31}$$

The coupling coefficient δ_{couple} is constrained by the requirement that the solution to Equation (24) be compatible with a constant relationship between ρ_m and ρ_ϕ energy densities. It is convenient to introduce the quantities Π_m and Π_ϕ by

$$\delta_{\text{couple}} = -3H\Pi_m = 3H\Pi_\phi, \tag{32}$$

by introducing these quantities, the continuity equations for dark energy and matter (Equation (24)) will have the form

$$\dot{\rho}_m + 3H(\rho_m + \Pi_m) = 0, \qquad \dot{\rho}_{\text{DE}} + 3H(\rho_{\text{DE}} + p_{\text{DE}} + \Pi_\phi) = 0. \tag{33}$$

The quantities Π_m and Π_ϕ are related as

$$\Pi_m = -\Pi_\phi = \frac{\rho_m\rho_\phi}{\rho}(\gamma_\phi - 1), \tag{34}$$

$$\gamma_\phi = \frac{p_\phi + \rho_\phi}{\rho_\phi} = \frac{\dot{\phi}^2}{\rho_\phi} \tag{35}$$

where $\rho = \rho_{\mathrm{m}} + \rho_\phi$.

Assuming γ_ϕ is a constant, the value of which $\gamma_\phi \in (0,2)$, it can be found

$$\rho_{\mathrm{m}} \propto \rho_\phi \propto \rho \propto a^{-\nu}, \quad \text{for } \nu = 3\frac{\gamma_\phi + r}{r+1}, \tag{36}$$

where r is a coincidence parameter, which takes an analytic expression for r and becomes a constant, thereby alleviating the coincidence problem. The solution of the second Friedmann's equation $3M_{\mathrm{pl}}H^2 = 8\pi\rho$ for the result obtained in Equation (36), has the form $a \propto t^{2/\nu}$. Thus, the Hubble parameter is defined as

$$H = \frac{2}{\nu t} = \frac{2(r+1)}{3(\gamma_\phi + r)}\frac{1}{t}. \tag{37}$$

The energy density parameter is defined as $\Omega_\phi = \frac{8\pi M_{\mathrm{pl}}^2}{3H^2}\rho_\phi$, as well as $\Omega_\phi = \frac{1}{r+1}$. Equating these equations and inserting Equation (37), we have

$$\rho_\phi = \frac{M_{\mathrm{pl}}^2}{6\pi}\frac{1+r}{(\gamma_\phi + r)^2}\frac{1}{t^2}. \tag{38}$$

The combination with Equation (35) gives

$$\dot{\phi} = \sqrt{\frac{M_{\mathrm{pl}}^2\gamma_\phi(1+r)}{6\pi}}\frac{1}{(\gamma_\phi + r)^2}\frac{1}{t}, \tag{39}$$

thus, the consequence of the condition $\rho_\phi \sim \rho_{\mathrm{m}}$ is the logarithmic evolution of the scalar field ϕ with time.

Applying the equation for the energy density for the scalar field and Equation (35) yields

$$\rho_\phi = \frac{2V(\phi)}{2 - \gamma_\phi} = \frac{\dot{\phi}^2}{\gamma_\phi}, \tag{40}$$

which together with Equations (38) and (39) leads to

$$V(\phi) = \frac{M_{\mathrm{pl}}^2}{6\pi}\left(1 - \frac{\gamma_\phi}{2}\right)\frac{1+r}{(\gamma_\phi + r)^2}\frac{1}{t^2} \;\Rightarrow\; \frac{\partial V(\phi)}{\partial \phi} = -\lambda V(\phi), \tag{41}$$

where $\lambda = \sqrt{\frac{24\pi}{M_{\mathrm{pl}}^2\gamma_\phi(1+r)}}$. Equation (41) implies that the potential has the exponential form

$$V(\phi) = V_0 e^{-\lambda(\phi - \phi_0)}. \tag{42}$$

A significant drawback of this model is the absence of a convincing explanation for the onset of the interaction of dark energy and matter at the epoch of transition from the decelerated to the accelerated expansion of the universe. The thermal quantum-field theory treatment of the quintessence dark energy coupled to the matter field shows that one can consistently recover different expansion regimes of the universe, including the late-time acceleration; however, more work is needed to relate the matter field to a viable dark matter candidate [239,240].

3. Constraints from Observational Data

3.1. Type Ia Supernovae

The observed magnitudes of type Ia supernovas are among the best data for constraining the distance–redshift relationship through the determination of the luminosity distance. In the ϕCDM models, the distances tend to be smaller compared to the ΛCDM predictions at the same redshift. This provides an opportunity for differentiating these models from each other.

One of the first studies in this direction was performed in Podariu and Ratra [241]. They used three datasets of SNe Ia apparent magnitude versus redshift—(i) R98 data [211], both including and excluding the unclassified SNe Ia 1997ck at $z = 0.97$ (with 50 and 49 SNe Ia apparent magnitude data, respectively), (ii) P99 data [2], and (iii) a third set with the corrected/effective stretch factor magnitudes for the 54 Fit C SNe Ia of P99 apparent magnitude data—and obtained constraints on the ϕCDM modelwith RP potential (ϕCDM-RP model) (see Figure 4).

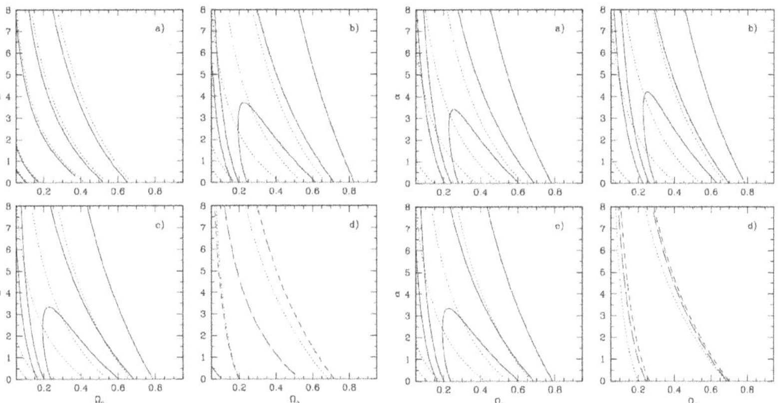

Figure 4. The 1σ, 2σ, and 3σ confidence level contour constraints on parameters of the scalar field ϕCDM model with the RP potential. (Left panel) (**a**) For all R98 SNe Ia apparent magnitude data, (**b**) for R98 SNe Ia apparent magnitude data excluding the $z = 0.97$ measurement, (**c**) for P99 Fit C SNe Ia apparent magnitude dataset, (**d**) for three datasets: all R98 SNe Ia apparent magnitude (long-dashed lines) data, R98 SNe Ia apparent magnitude data excluding the $z = 0.97$ measurement (short-dashed lines), and P99 Fit C SNe Ia apparent magnitude data (dotted lines). (Right panel) (**a**) For all the R98 SNe Ia apparent magnitude data, (**b**) for R98 SNe Ia apparent magnitude data excluding the $z = 0.97$ measurement, (**c**) for P99 Fit C SNe Ia apparent magnitude data, (**d**) for the H_0 and t_0 constraints used in conjunction with all R98 SNe Ia (long-dashed lines) apparent magnitude data, R98 SNe Ia apparent magnitude data excluding the $z = 0.97$ measurement (short-dashed lines), and the P99 Fit C SNe Ia apparent magnitude data (dotted lines). The figure is adapted from [241].

Caresia et al. [214] obtained constraints on the parameters of the ϕCDM with the RP and Sugra potentials [225,242] and also of the extended quintessence models with the inverse power-law RP potential [243,244] from the datasets of apparent magnitude versus redshift measurements of 176 SNe Ia [2,211,245], and the data from the SuperNovae Acceleration Probe (SNAP) satellite [246].[1] The obtained constraints on the model parameters are shown in Figures 5 and 6. No useful constraints on the model parameters were found for the Sugra potential, while 1σ constraints of $\alpha < 0.8$ and $\alpha < 0.6$, for both the extended and ordinary quintessence models using the RP potential, were obtained using the SNe Ia apparent magnitude and SNAP satellite data, respectively.

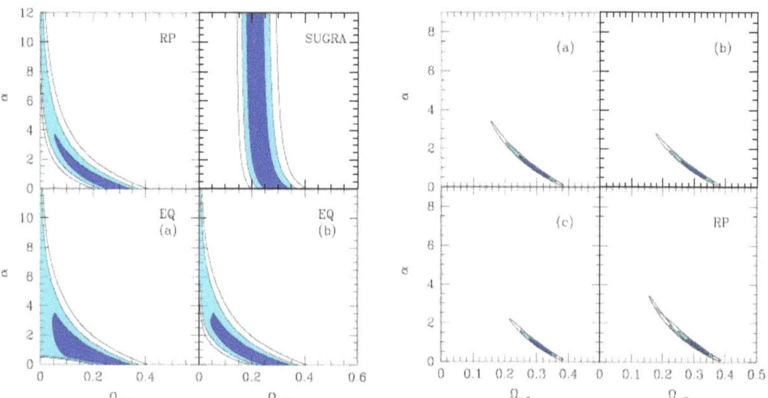

Figure 5. (Left panel) The 1σ, 2σ, and 3σ confidence level contour constraints on Ω_{m0} and α parameters by using a sample of 176 SNe Ia apparent magnitude data. (Upper left sub-panel) For the ordinary quintessence with the inverse power-law RP potential, (upper right sub-panel) for ordinary quintessence with the Sugra potential, (lower left sub-panel) for extended quintessence with the inverse power-law RP potential, (lower right sub-panel) for extended quintessence with the inverse power-law RP potential when upper limits on the time variation of the gravitational constant are satisfied. (Right panel) The 1σ, 2σ, and 3σ confidence level contour constraints on Ω_{m0} and α parameters for the ordinary ϕCDM quintessence model with the inverse power-law RP potential by using SNAP sample data. (Upper left sub-panel) corresponds to constraints obtained by assuming the exact EoS parameter, (upper right sub-panel) corresponds to the linear approximation of the EoS parameter, (lower left sub-panel) corresponds to the constant approximation of the EoS parameter, (lower right sub-panel) corresponds to the superposition of above-mentioned three cases. The figure is adapted from [214].

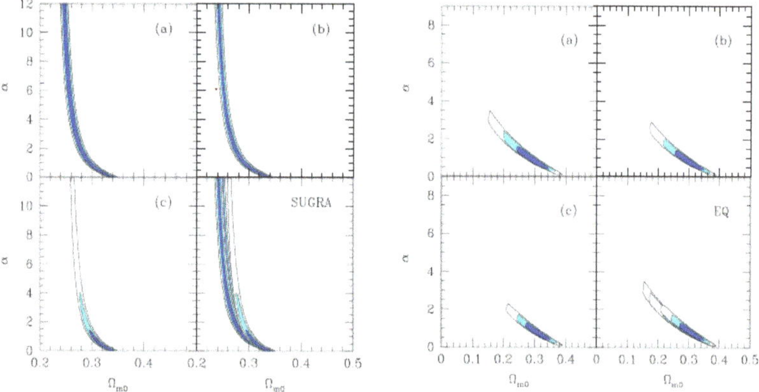

Figure 6. (Left panel) As Figure 5, but for the ordinary ϕCDM quintessence model with the Sugra potential. (Right panel) As Figure 5, but for the extended quintessence model with the inverse power-law RP potential. The results are obtained by imposing the upper bound on the time variation of the gravitational constant. The figure is adapted from [214].

Doran et al. [247] considered a DE model parameterized as [247]

$$\Omega_\phi(a) = \frac{e^{R(a)}}{1 + e^{R(a)}}, \quad w(a) = w_0 \ln a / (1 - b \ln a)^2, \tag{46}$$

where $R(a) = \ln(\Omega_\phi(a)/(1 - \Omega_\phi(a)))$ corresponds to

$$R(a) = R_0 - \frac{2w_0 \ln a}{1 - b \ln a},\tag{47}$$

the constant b is defined by the EoS parameter at the present epoch w_0, the dark energy density parameter at the present epoch $\Omega_{\phi 0}$, and the parameter Ω_e characterizing the amount of dark energy at early times to which it asymptotes for very large redshifts, as

$$b = -3w_0 \left(\ln \frac{1 - \Omega_e}{\Omega_e} + \ln \frac{1 - \Omega_{\phi 0}}{\Omega_{\phi 0}} \right).\tag{48}$$

Using a combination of datasets from SNe Ia [211], Wilkinson Microwave Anisotropy Probe (WMAP) [6], Cosmic Background Imager (CBI) [248], Very Small Array (VSA) [249], SDSS [250], and HST [251], the authors find $w_0 < -0.8$ and $\Omega_e < 0.03$ at the 1σ confidence level; the contours are shown in Figure 7. It should be noted that the SNe Ia apparent magnitude data are most sensitive to w_0, while the CMB temperature anisotropies and the LSS growth rate are the best constraints of Ω_e (see Figure 8).

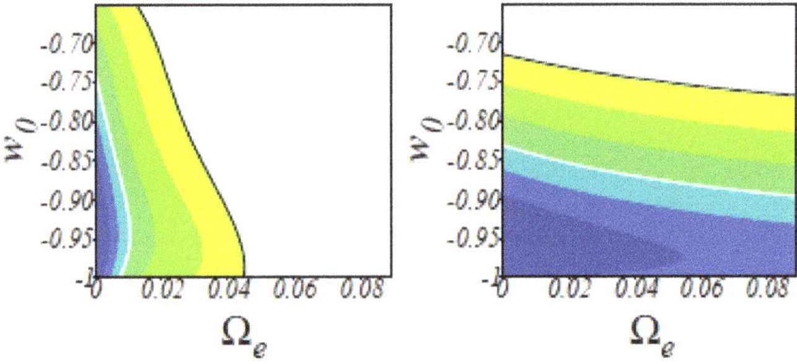

Figure 7. Constraints on parameters Ω_e and w_0. The left picture depicts the distribution from WMAP + CBI + VSA + SDSS + HST data and the right picture is that of SNe Ia apparent magnitude versus redshift data alone. The regions of 1σ (2σ) confidence level are enclosed by a white (black) line. The figure is adapted from [247].

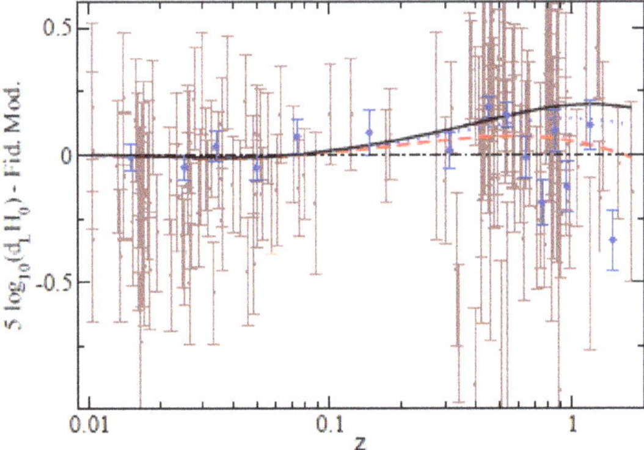

Figure 8. SNe Ia apparent magnitude versus redshift data [211] as data points with thin (brown) error bars. The authors plotted the logarithm of the luminosity distance minus a fiducial model for which $d_L H_0 = (1 + z) \ln(1 + z)$. The solid (black) line is for the spatially flat ΛCDM model, the dotted (blue) line is for $\Omega_e = 10^{-4}$, and the dashed (red) line is for $\Omega_e = 10^{-1}$. For all models, $w_0 = -1$. The figure is adapted from [247].

Pavlov et al. [252] found that, for the ϕCDM-RP model in a spacetime with nonzero spatial curvature, the dynamical scalar field has an attractor solution in the curvature dominated epoch, while the energy density of the scalar field increases relative to that of the spatial curvature. In the left panel of Figure 39, we see that the values $\Omega_{m0} = 0.27$ and $\alpha = 3$ are consistent with these constraints for a range of values Ω_{k0} and for a set of dimensionless time parameter values of $H_0 t_0 = H_0 \int_0^{a_0} da/\dot{a}(t)$, where t_0 is the age of the universe. The right panel of Figure 9 shows a similar analysis for several values of the cosmological test parameter $\triangle(\Omega_{m0}, \Omega_{k0}, \alpha) = \delta(t_0)/(1 + z_i)\delta(t_i)$, where $\delta(t_0)$ and $\delta(t_i)$ are the values of the matter density contrast at, respectively, the present time t_0 and an arbitrary time t_i such that $a(t_i) \ll a(t_0)$, i.e., a time when the universe is well approximated by the Einstein–de Sitter model in the matter-dominated epoch.

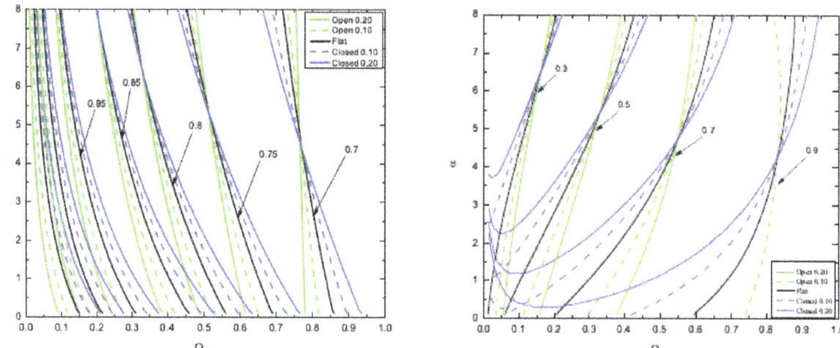

Figure 9. (Left panel) The 2σ contours of the fixed time parameter $H_0 t_0$ as a function of values of the matter density parameter at the present epoch Ω_{m0} and space curvature density parameter at the present epoch Ω_{k0}, as well as the model parameter α in the scalar field ϕCDM model with the RP potential. The results obtained for larger values of free parameters $(\Omega_{m0}, \Omega_{k0}, \alpha)$ and for $H_0 t_0 = [0.7, 0.75, 0.8, 0.85, 0.95, 1.05, 1.15]$. (Right panel) The 2σ contours of the factor by which the growth of matter perturbations falls lower than in the Einstein–de Sitter model. The cosmological test parameter values $\triangle(\Omega_{m0}, \Omega_{k0}, \alpha)$ obtained for the larger values of free parameters $(\Omega_{m0}, \Omega_{k0}, \alpha)$. The figure is adapted from [252].

Fuzfa and Alim [253] studied the ϕCDM model with the RP and Sugra potentials in a spatially closed universe. The estimated values of Ω_{m0} and $\Omega_{\phi0}$, using SNe Ia apparent magnitude data from the SNLS collaboration [254], are quite different from those for the standard spatially flat ΛCDM model (Figure 10). Such a result is expected due to the different cosmic acceleration and dark matter clustering predicted between quintessence models and the standard ΛCDM model, arising from the differences in cosmological parameters, even at $z = 0$. The quintessence scalar field creates more structures outside the filaments, lighter halos with higher internal velocity dispersion, as seen from N-body simulations performed to study the influence of quintessence on the distribution of matter at large scales.

Farooq et al. [255] constrained the ϕCDM-RP model in a spacetime with non-zero spatial curvature, as well as the XCDM model, using the Union2.1 compilation of the 580 SNe Ia apparent magnitude measurements of Suzuki et al. [256], Hubble parameter observations [28,30,257,258], and the $0.1 \leq z \leq 0.75$ BAO peak length scale measurements [21,22,259] (see Figure 11). They constrain the spatial curvature density parameter today to be $|\Omega_{k0}| \leq 0.15$ at a 1σ confidence level and more precise data are required to tighten the bounds on the parameters.

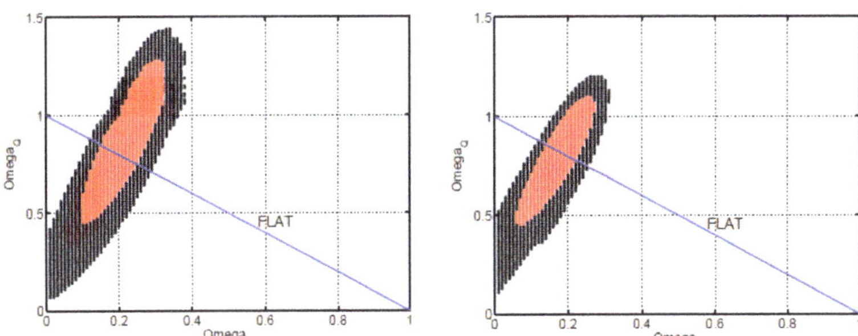

Figure 10. The 1σ and 2σ confidence level contour constraints on the matter density parameter at the present epoch Ω_{m0} and the dark energy density (quintessence) parameter at the present epoch Ω_{Q0} for scalar field ϕCDM models. (Left panel) With the inverse power-law PR potential. (Right panel) With Sugra potential $V(\phi) \propto \phi^{-\alpha} \exp\left(4\pi\phi^2\right)$. The blue lines accord to the flat universe. The figure is adapted from [253].

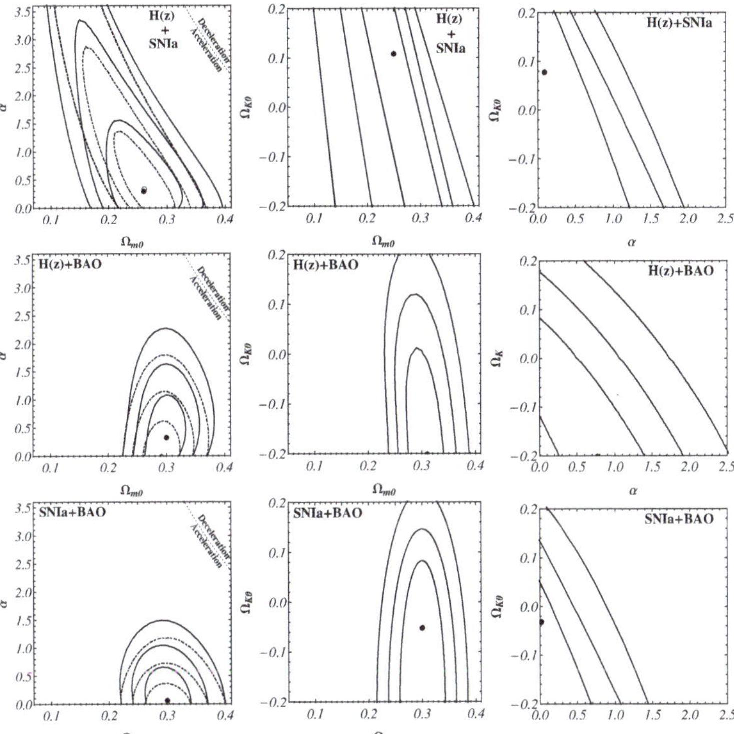

Figure 11. The 1σ, 2σ, and 3σ confidence level contour constraints on parameters of the spatially non-flat scalar field ϕCDM model with the RP potential from compilations of data: $H(z)$ + SNe Ia apparent magnitude (first row), $H(z)$ + BAO peak length scale (second row), and BAO peak length scale + SNe Ia apparent magnitude (third row). Filled circles denote best-fit points. The dot–dashed lines in the first column panels are 1σ, 2σ, and 3σ confidence level contours obtained by Farooq et al. [260] for the spatially flat ϕCDM model (open circles denote best-fit points). Dotted lines separate the accelerating and decelerating models (at zero space curvature). The horizontal axis with $\alpha = 0$ corresponds to the standard spatially flat ΛCDM model. First, second, and third columns correspond to marginalizing over Ω_{k0}, α, and Ω_{m0}, respectively. The figure is adapted from [255].

Assuming that the Hubble constant H_0 tension of the ΛCDM model is actually a tension on the SNe Ia absolute magnitude M_B, Nunes and Di Valentino [158] assessed the M_B tension by comparing the spatially flat ΛCDM model, wCDM, and IDE models using a compilation of two datasets: the SNe Ia Pantheon sample [212] + BAO [22,24,25,261,262] + big bang nucleosynthesis (BBN) [263] and the SNe Ia Pantheon sample + BAO [22,24,25,261,262] + BBN [263] + M_B [264] (see Figure 12). They found that the IDE model can alleviate both the M_B and H_0 tensions with a coupling different from zero at a 2σ confidence level with a preference for a compilation of the Pantheon + BAO + BBN + M_B datasets.

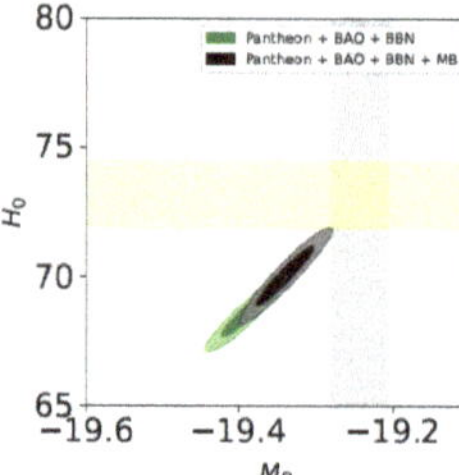
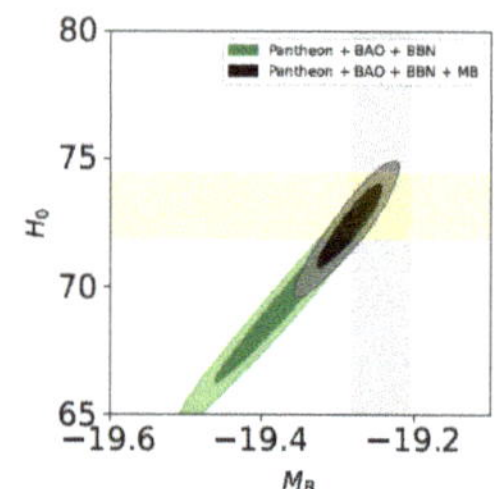
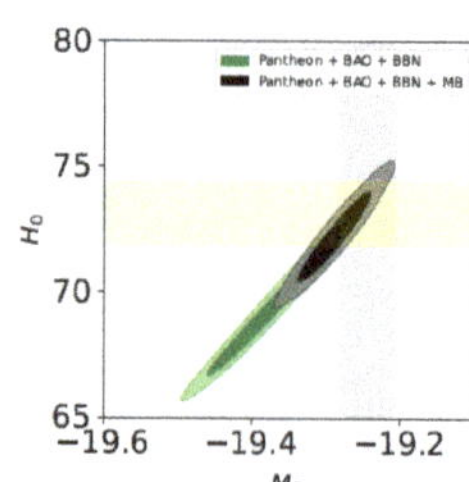

Figure 12. The 1σ and 2σ confidence level contour constraints on M_B and H_0 values for ΛCDM (left panel), wCDM (middle panel), and IDE (right panel) models, obtained from compilations of Pantheon + BAO + BBN and Pantheon + BAO + BBN + M_B datasets. The figure is adapted from [158].

3.2. Cosmic Microwave Background Radiation Data

The CMB provides a very accurate determination of the angular diameter distance d_A at a redshift of $z\sim1000$. This measurement is sensitive to the entire expansion of the universe over this wide range of redshifts. As pointed out before, the ϕCDM models tend to predict smaller distances and can therefore be constrained with the CMB geometric measurements.

In one of the first such studies, Doran et al. [265] used the CMB temperature anisotropy data from the BOOMERANG and MAXIMA experiments [266,267] to distinguish quintessential inflation models with different EoS parameters, described by a kinetic term $k(\phi)$ of the cosmon field this model is described by a Lagrangian of the form $\mathcal{L}_\phi = \frac{1}{2}(\partial_\mu\phi)^2 k^2(\phi) + V(\phi)$: (i) the RP potential with $k(\phi) = 1$, (ii) the leaping kinetic term model where $V(\phi) = M_{\bar{p}l}^4 \exp(-\phi/M_{\bar{p}l})$, $M_{\bar{p}l} = \sqrt{8\pi}M_{pl}$ is the reduced Plank mass, $k(\phi) = k_{\min} + \tanh[(\phi - \phi_1)/M_{\bar{p}l}] + 1$, $\phi_1 \approx 277$ eV, and $k_{\min} = [0.05, 0.1, 0.2, 0.26]$ [268], and (iii) the exponential potential with $V(\phi) = M_{pl}^4 \exp(-\sqrt{2}\alpha\phi/M_{\bar{p}l})$, $\alpha = \sqrt{3/2\Omega_{\phi0}}$, and $k(\phi) = 1$ [150]. The dark energy density parameters today and at the last scattering epoch, $\Omega_{\phi0}$ and $\Omega_{\phi ls}$, respectively, and the averaged EoS parameter of the field ϕ are used to parameterize the separation of peaks in CMB temperature anisotropies, which can be used to measure the value of Ω_ϕ before the last scattering (see Figure 13).

Caldwell et al. [269] investigated how early quintessence dark energy, i.e., a non-negligible quintessence energy density during the recombination and structure formation epochs, affects the baryon–photon fluid and the clustering of dark matter, and thus the CMB temperature anisotropy and the matter power spectra. They showed that early quintessence is characterized by a suppressed ability to cluster at small length scales, as suggested by the compilation of data from WMAP [270,271], CBI [272,273], Arcminute Cosmology Bolometer Array Receiver (ACBAR) [274], the LSS growth rate dataset of Two degree Field Galaxy Redshift Survey (2dFGRS) [275–277], and $L_{y-\alpha}$ forest [278,279]; these are shown in Figure 14.

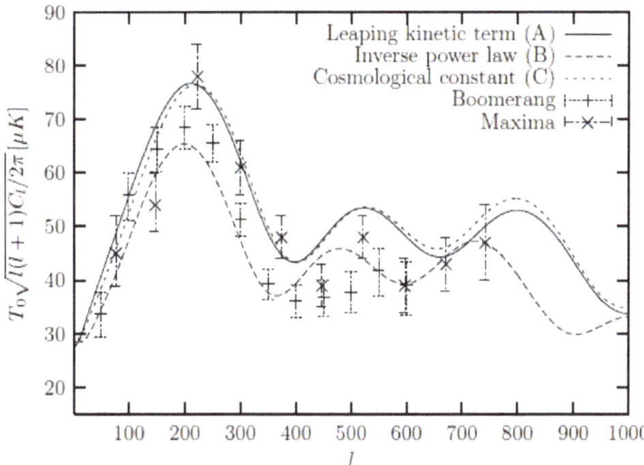

Figure 13. The CMB temperature anisotropy spectrum for different quintessence scalar field ϕCDM models: with the leaping kinetic term (model A), with the inverse power-law RP potential (model B) (here the dark energy density parameter at the present epoch $\Omega_{\phi 0} = 0.6$), and for the ΛCDM model (model C). Data points from the BOOMERANG [266] and MAXIMA [267] experiments are shown for reference. The figure is adapted from Ref. [265].

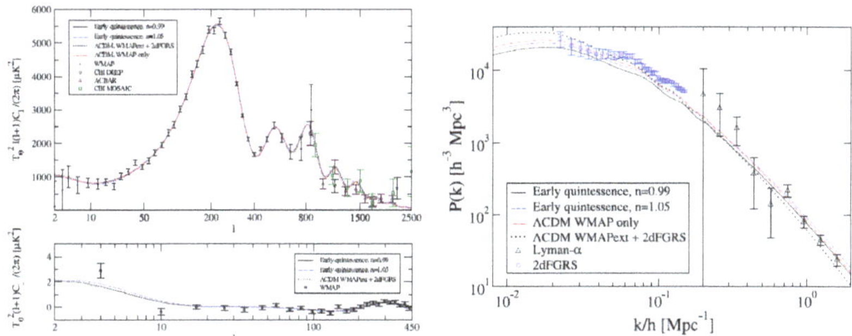

Figure 14. (Left panel) Polarization (TE) and temperature (TT) as functions of the multipole l. Two quintessential inflation models with $n_s = 0.99$ and $n_s = 1.05$ are presented with WMAP data from [270,271]. WMAP-normalized spectra for the best fit for the ΛCDM model (no $L_{y-\alpha}$ data) with the constant spectral index $n = 0.97$ [6] and the ΛCDM model with the running spectral index $n_s = 0.93$, $dn_s/d\ln k = -0.031$ are shown for comparison. For large l, CBI data and ACBAR data are used. (Right panel) The CDM power spectrum at the present epoch as a function of k/h. The linear spectrum for two quintessential inflation models with spectral indices $n_s = 0.99$ and $n_s = 1.05$ are plotted. The best fit for the ΛCDM model with running spectral index $n_s = 0.93$, $dn_s/d\ln k = -0.031$ [6]), normalized to WMAP data (no $L_{y-\alpha}$ data), is shown. 2dFGRS measurements and $L_{y-\alpha}$ data are converted to $z = 0$. The figure is adapted from [269].

Pettorino et al. [215] studied a class of the extended ϕCDM models, where the scalar field is exponentially coupled to the Ricci scalar and is described by the RP potential. The projection of the ISW effect on the CMB temperature anisotropy[2] is found to be considerably larger in the exponential case with respect to a quadratic non-minimal coupling as seen in Figure 15. This reflects the fact that the effective gravitational constant depends exponentially on the dynamics of the scalar field.

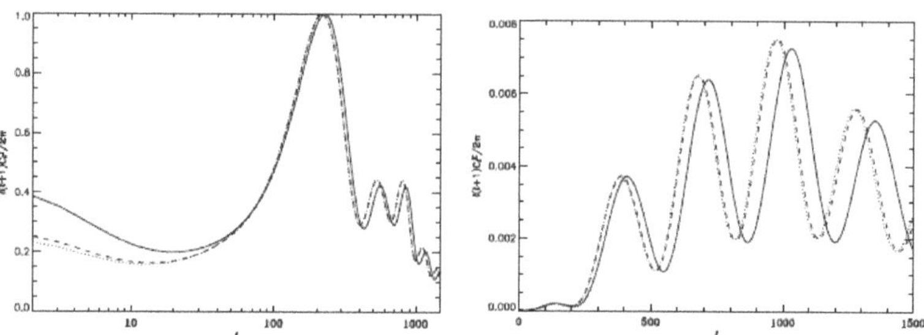

Figure 15. The spectra are in arbitrary units, normalized to unity at the first acoustic peak. (Left panel) CMB angular total intensity power spectra for the scalar field ϕCDM model with the inverse power-law RP potential (dotted), quadratic (dashed) and exponential coupling extended quintessence (solid) with $\omega_{JBD0} = 50$. (Right panel) CMB angular polarization power spectra for the ϕCDM model (dotted), quadratic (dashed), and exponential coupling extended quintessence (solid) with ω_{JBD0}. The figure is adapted from [215].

Mukherjee et al. [280] conducted a likelihood analysis of the Cosmic Background Explorer—Differential Microwave Radiometers (COBE-DMR) sky maps to normalize the ϕCDM-RP model in flat space[3]. As seen from Figure 16, this model remains an observationally viable alternative to the standard spatially flat ΛCDM model.

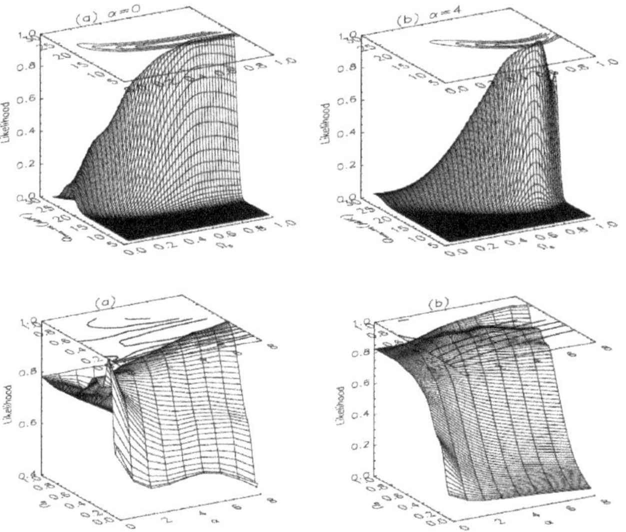

Figure 16. The model with $t_0 = 13$ Gyr and $\Omega_b h^2 = 0.014$. Likelihood functions $L(Q_{rms-PS}, \alpha, \Omega_0)$ (arbitrarily normalized to unity at the highest peak). (Left panel) Derived from a simultaneous analysis of DMR 53 and 90 GHz galactic-frame data. The faint high-latitude foreground galactic emission is corrected and the quadrupole moment in the analysis is included: (a) for $\alpha = 0$ and (b) for $\alpha = 4$. (Right panel) Derived by marginalizing $L(Q_{rms-PS}, \alpha, \Omega_0)$ over Q_{rms-PS} with a uniform prior: (a) the correction for the faint high-latitude foreground galactic emission is ignored and the quadrupole moment from the analysis is excluded, (b) for the faint high latitude foreground galactic emission is corrected and the quadrupole moment in the analysis is included. The figure is adapted from [280].

Samushia and Ratra [283] constrained model parameters of the ΛCDM model, the XCDM model, and the ϕCDM-RP model using galaxy cluster gas mass fraction data [284]; for this, they introduced an auxiliary random variable as opposed to integrating over nuisance parameters of the Markov Chain Monte Carlo (MCMC) method. Two different sets of priors were chosen to study the influence of the type of priors on the obtained results—one set has [7] $h = 0.73 \pm 0.03$, $\Omega_b h^2 = 0.0223 \pm 0.0008$ (1σ errors) and the other set has $h = 0.68 \pm 0.04$ [285,286], and $\Omega_b h^2 = 0.0205 \pm 0.0018$ [287]. The obtained constraints on the ϕCDM model with the RP potential are shown in Figure 17. We see that Ω_m is better constrained than α, whose best-fit value is $\alpha = 0$, corresponding to the standard spatially flat ΛCDM model; however, the scalar field ϕCDM model is not excluded.

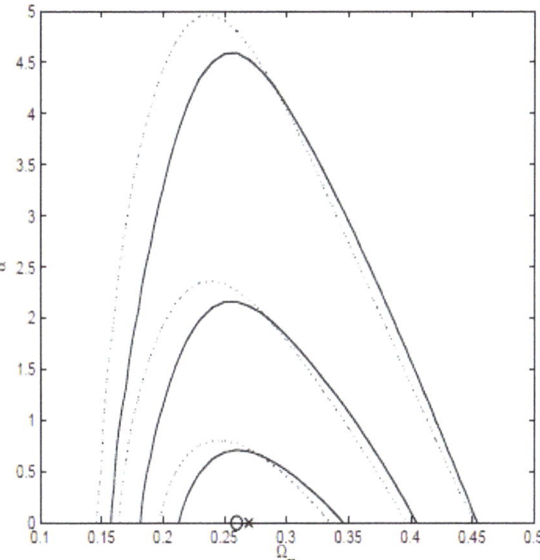

Figure 17. The 1 σ, 2σ, and 3σ confidence level contour constraints on parameters of the scalar field ϕCDM model with the inverse power-law RP potential using cluster gas mass fraction data. Solid lines correspond to WMAP prior while dashed lines correspond to the alternate prior. The cross matches the best fit at $\Omega_{m0} = 0.27$ and $\alpha = 0$. The circle denotes the best fit at $\Omega_{m0} = 0.26$ and $\alpha = 0$. The horizontal axis for which $\alpha = 0$ corresponds to the spatially flat ΛCDM model. The figure is adapted from [283].

Chen et al. [288] constrained the scalar field ϕCDM-RP model and the ΛCDM model with massive neutrinos assuming two different neutrino mass hierarchies in both the spatially flat and non-flat universes, using a joint dataset comprising the CMB temperature anisotropy data [12,289], the BAO peak length scale data from the 6dF Galaxy Survey (6dFGS), SDSS—Main Galaxy Sample (MGS), Baryon Oscillation Spectroscopic Survey (BOSS)-LOWZ (galaxies within the redshift range $0.2 < z < 0.43$), BOSS CMASS-DR11 (galaxies within the redshift range $0.43 < z < 0.7$) [23], the joint light–curve analysis (JLA) compilation from SNe Ia apparent magnitude measurements [290], and the Hubble Space Telescope H_0 prior observations [29]. Assuming three species of degenerate massive neutrinos, they found the 2σ upper bounds of $\sum m_\nu < 0.165$ (0.299) eV and $\sum m_\nu < 0.164$ (0.301) eV, respectively, for the spatially flat (spatially non-flat) ΛCDM model and the spatially flat (spatially non-flat) ϕCDM model (Figure 18). The inclusion of spatial curvature as a free parameter leads to a significant expansion of the confidence regions for $\sum m_\nu$ and other parameters in spatially flat ϕCDM models, but the corresponding differences are larger for both the spatially non-flat ΛCDM and spatially non-flat ϕCDM models.

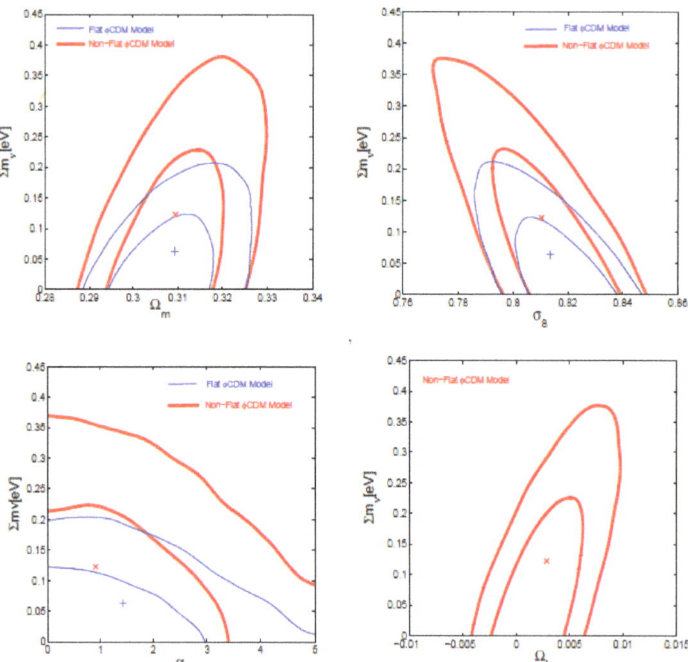

Figure 18. The 1σ and 2σ confidence level contour constraints on parameters of the spatially flat and spatially non-flat scalar field ϕCDM model with the inverse power-law potential from a joint analysis using the HST H_0 prior in the scenario with three species of degenerate massive neutrinos. (Upper left, upper right, and lower left panels) Contours are presented in the $\Omega_m - \sum m_\nu$, $\sigma_8 - \sum m_\nu$, and $\alpha - \sum m_\nu$ planes. The thin blue (thick red) lines correspond to constraints in the spatially flat (spatially non-flat) universe. The "+" ("x") denotes the mean values of the pair in the spatially flat (spatially non-flat) universe. (Lower right panel) Contours are in the $\Omega_k - \sum m_\nu$ plane for the spatially non-flat universe. The "x" denotes the mean values of the $(\Omega_k, \sum m_\nu)$ pair. The figure is adapted from [288].

Park and Ratra [291] constrained the spatially flat tilted and spatially non-flat untilted ϕCDM-RP inflation model by analyzing the CMB temperature anisotropy angular power spectrum data from the Planck 2015 mission [292], the BAO peak length scale measurements [26], a Pantheon collection of 1048 SNe Ia apparent magnitude measurements over the broader redshift range of $0.01 < z < 2.3$ [212], Hubble parameter observation [21,25,28,30–34,257,293], and LSS growth rate measurements [25] (Figures 19 and 20). Constraints on parameters of the spatially non-flat model were improved from a 1.8σ to a more than 3.1σ confidence level by combining CMB temperature anisotropy data with other datasets. Present observations favor a spatially closed universe with the spatial curvature contributing about two-thirds of a percent of the current total cosmological energy budget. The spatially flat tilted ϕCDM model is a 0.4σ better fit to the observational data than the standard spatially flat tilted ΛCDM model, i.e., current observational data allow for the possibility of dynamical dark energy in the universe. The spatially non-flat tilted ϕCDM model better fits the DES bounds on the root mean square (rms) amplitude of mass fluctuations σ_8 as a function of the matter density parameter at the present epoch Ω_{m0}, but it does not provide such good agreement with the larger multipoles of Planck 2015 CMB temperature anisotropy data as the spatially flat tilted ΛCDM model.

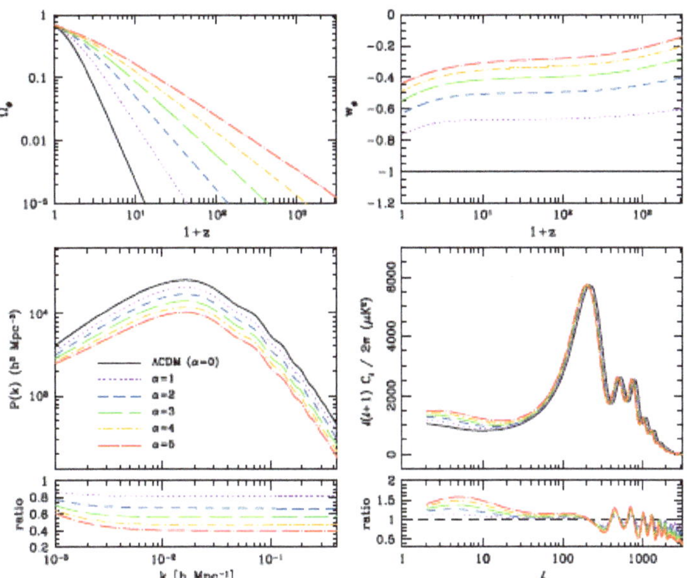

Figure 19. (Upper panels) Evolution of the EoS parameter w_ϕ and dark energy density parameter Ω_ϕ in the tilted spatially flat ϕCDM model for the range of values of α parameter $\alpha \in (1,5)$. The black solid curve accords to the ΛCDM model, which corresponds to reduced ϕCDM model with $\alpha = 0$. (Middle panels) Theoretical predictions for matter density and CMB temperature anisotropy angular power spectra for the ϕCDM model depending on parameter α. (Lower panels) Ratios of the ϕCDM model power spectra relative to the ΛCDM model. The figure is adapted from [291].

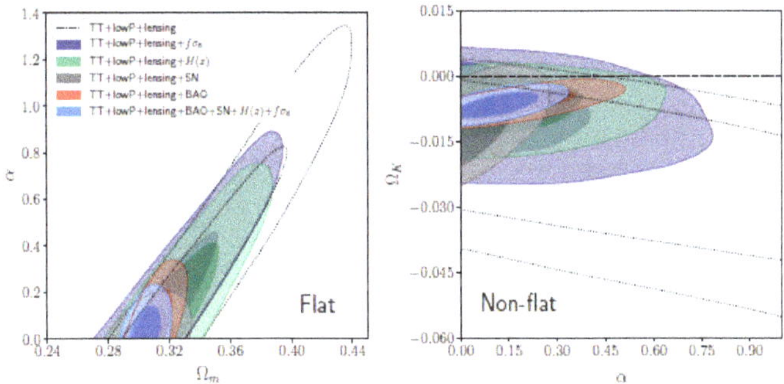

Figure 20. The 1σ and 2σ confidence level contours. (Left panel) In the $\Omega_m - \alpha$ plane for the tilted spatially flat scalar field ϕCDM model. (Right panel) In the $\alpha - \Omega_k$ plane for the untilted spatially non-flat scalar field ϕCDM model. Constraints are derived from Planck CMB TT + lowP + lensing and non-CMB datasets. The horizontal dashed line indicates the spatially flat curvature with $\Omega_k = 0$. For the spatially non-flat ϕCDM model constrained with TT + lowP + lensing, the $h > 0.45$ prior has been used. The figure is adapted from [291].

Constraints on model parameters of the XCDM and ϕCDM-RP (spatially flat tilted) inflation models using the compilation of CMB [292] and BAO data [22–24,294–296] were derived by Ooba et al. [294]. The authors calculated the angular power spectra of the CMB temperature anisotropy using the CLASS code of Blas et al. (2011) [295] and executed the MCMC analysis with Monte Python (Audren et al. [296]). Having one additional parameter compared to the standard spatially flat ΛCDM model, both ϕCDM and XCMB models better fit the TT + lowP + lensing + BAO peak length scale data than does the standard spatially flat ΛCDM model (Figure 21). For the ϕCDM model, $\triangle\chi^2 = -1.60$ and, for the XCDM model, $\triangle\chi^2 = -1.26$ relative to the ΛCDM model. The improvement over the standard spatially flat ΛCDM model in 1.3σ and in 1.1σ for the XCDM model are not significant, but these dynamical dark energy models cannot be ruled out. Both the ϕCDM and XCMB dynamical dark energy models reduce the tension between the Planck 2015 (Aghanim et al. [292]) CMB temperature anisotropy and the weak lensing constraints of the rms amplitude of mass fluctuations σ_8.

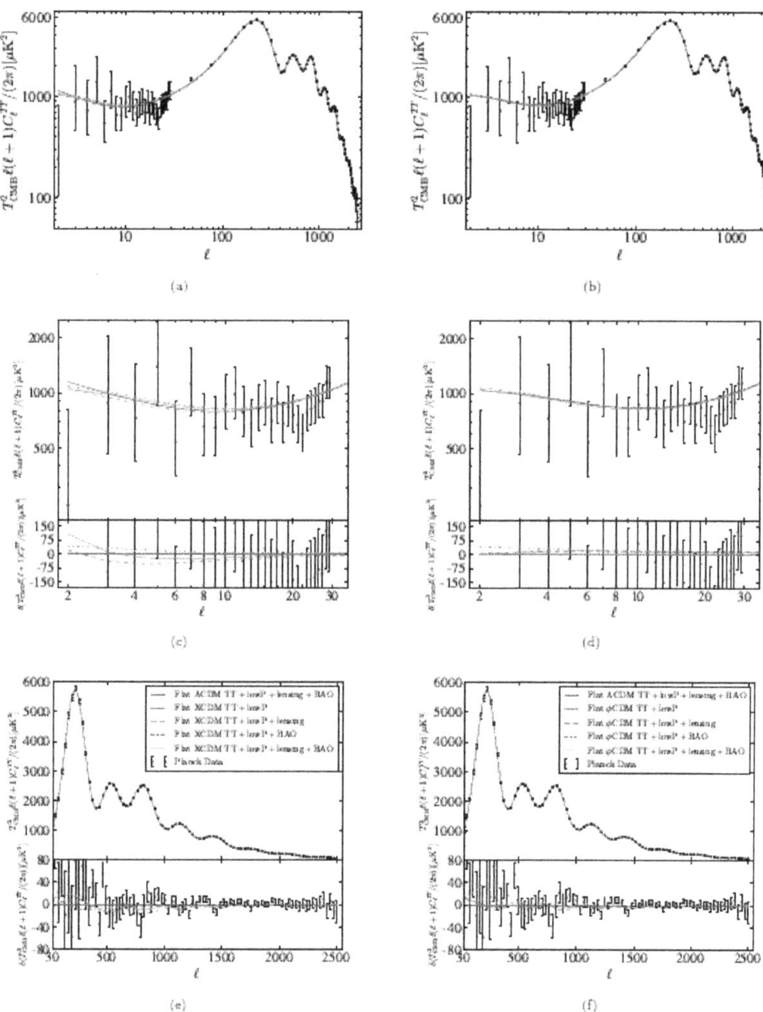

Figure 21. (Left panels (**a**,**c**,**e**)) The comparison of the spatially flat tilted ΛCDM model (gray solid line) with the best-fit C_l's for the XCDM model and the ϕCDM model. (Right panels (**b**,**d**,**f**)) The comparison of the spatially flat tilted ΛCDM model (gray solid line) with the best-fit C_l's for the ϕCDM model. The all-l region is shown on top panels. The low-l region C_l and residuals are represented on middle panels. The high-l region C_l and residuals are demonstrated on bottom panels. The figure is adapted from [294].

Park and Ratra [297] also constrained the Hubble constant H_0 value in the spatially flat and spatially non-flat ΛCDM, XCDM, and ϕCDM-RP models using various combinations of datasets: the BAO peak length scale measurements [26], a Pantheon collection of 1048 SNe Ia apparent magnitude measurements over the broader redshift range of $0.01 < z < 2.3$ [212], and the Hubble parameter observations [21,25,28,30–34,257,293]. According to this analysis, the dataset slightly favors the untilted spatially non-flat dynamical XCDM and ϕCDM quintessential inflation models, as well as smaller Hubble constant H_0 values (Figure 22).

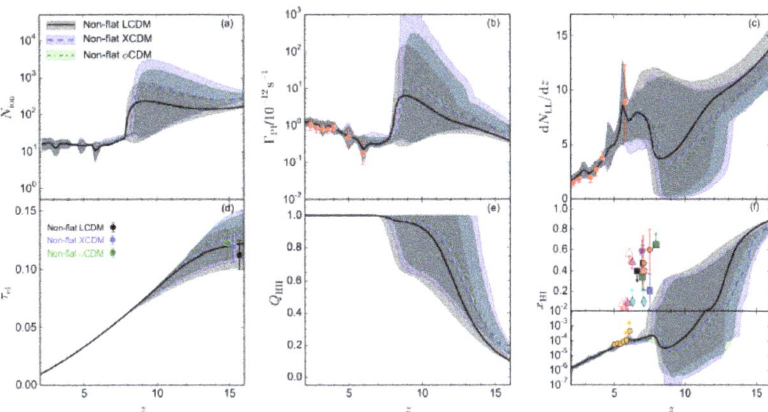

Figure 22. Constraints on various quantities related to reionization obtained from the MCMC analysis for the untilted spatially non-flat ΛCDM, XCDM, and ϕCDM quintessential inflation models that fit best the dataset: Planck 2015 TT + lowP + lensing and SNe Ia apparent magnitude, BAO peak length scale, $H(z)$, and LSS growth rate data. The thick central lines along with surrounding shaded regions correspond to best$-$fit models and their 2σ uncertainty ranges. (Upper panels) (**a**) A number of ionizing photons in the IGM per baryon in stars, (**b**) photoionization rates for hydrogen along with observational data from Wyithe and Bolton [298] and Becker and Bolton [299], (**c**) a specific number of Lyman$-$limit systems with data points from Songaila and Cowie [300] and Prochaska et al. [301]. (Lower panel) (**d**) Electron scattering optical depths along with their values from Park and Ratra [291], (**e**) volume filling factor of ionized regions, (**f**) global neutral hydrogen fraction with different present observational limits. The figure is adapted from [297].

The compilation of the South Pole Telescope polarization (SPTpol) CMB temperature anisotropy data [302], alone and in combination with the Planck 2015 CMB temperature anisotropy data [292] and the non-CMB temperature anisotropy data, consisting of the Pantheon Type SNe Ia apparent magnitude measurements [212], the BAO peak length scale measurements [22,24–26,293], the Hubble parameter $H(z)$ data [21,28,30–34,257], and the LSS growth rate data [25], was used by Park and Ratra [303] to obtain constraints on parameters of the spatially flat and untilted spatially non-flat ΛCDM and XCDM scalar field ϕCDM-RP quintessential inflation models. In each dark energy model, constraints on the cosmological parameters from the SPTpol measurements, the Planck CMB temperature anisotropy, and the non-CMB temperature anisotropy measurements are largely consistent with one another. Smaller angular scale SPTpol measurements (used jointly with only the Planck CMB temperature anisotropy data or with the combination of the Planck CMB temperature anisotropy data and the non-CMB temperature anisotropy data) favor the untilted spatially closed models.

Di Valentino et al. [121] explored the IDE models to find out whether these models can resolve both the Hubble constant H_0 tension problem of the standard spatially flat ΛCDM model and resolve the contradictions between observations of the Hubble constant in high and low redshifts in the spatially non-flat ΛCDM scenario.

The authors constrained parameters of the spatially flat IDE and ΛCDM models as well as the spatially non-flat IDE and ΛCDM models applying the CMB Planck 2018 data [13], BAO [22,24,25] measurements, 1048 data points in the redshift range $z \in (0.01, 2.3)$ of the Pantheon SNe Ia luminosity distance data [212], and a Gaussian prior of the Hubble constant ($H_0 = 74.03 \pm 1.42$ km s^{-1}Mpc^{-1} at 1σ CL), obtained from a reanalysis of the HST data by the SH0ES collaboration [112].

Based on the results of this observational analysis, it was found that the Planck 2018 CMB data favor spatially closed hypersurfaces at more than 99% CL (with a significance of 5σ), while a larger value of the Hubble constant, i.e., alleviation of the Hubble constant tension (with a significance of 3.6σ) has been obtained for the spatially non-flat IDE models. The authors concluded that searches for other forms of the interaction coupling parameter and the EoS for the dark energy component in IDE models are needed, which may further ease the tension of the Hubble constant. The 1σ and 2σ confidence level contours on parameters of the spatially non-flat IDE model are shown in Figure 23.

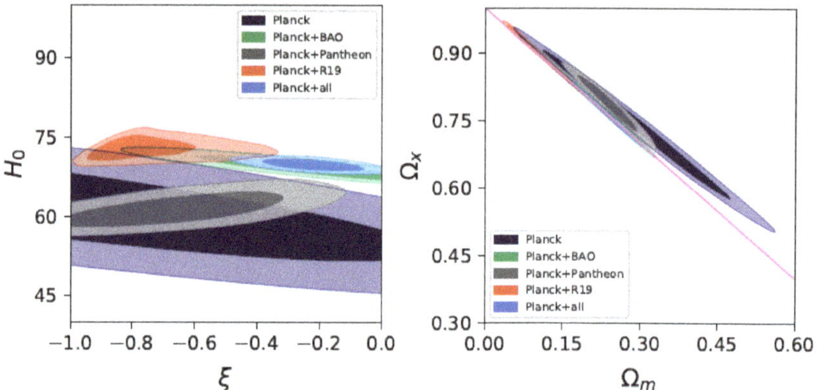

Figure 23. The 1σ and 2σ confidence level contours on parameters of the spatially non-flat IDE model, presented in (H_0, ξ) plane (left panel) and in $(\Omega_\Lambda, \Omega_m)$ plane (right panel). Here ξ is the dimensionless coupling parameter which characterizes the strength of the interaction between the dark sectors. The figure is adapted from [121].

Investigating both the minimally and non-minimally coupled to gravity the spatially-flat scalar field ϕCDM-RP and the extended quintessence models, Davari et al. [221] applied the following dataset: the Pantheon SNe Ia luminosity distance data [212], BAO (6dFGS, SDSSLRG, BOSS-MGS, BOSS-LOWZ, WiggleZ, BOSS-CMASS, BOSS-DR12), CMB [304], $H(z)$ [112], and redshift space distortion (RSD) [305]. According to their results, the ΛCDM model has a strong advantage when local measurements of the Hubble constant H_0 [13] are not taken into account and, conversely, this statement is weakened when local measurements of H_0 are included to the data analysis.

3.3. Large-Scale Structure Growth Rate Data

Another potentially powerful probe of the ϕCDM signatures is the growth rate in the low redshift LSS. The growth rate is expected to be stronger in the ϕCDM models compared to their ΛCDM counterparts.

Pavlov et al. [306] constrained the spatially flat ϕCDM-RP, the XCDM, the wCDM, and the ΛCDM models from future LSS growth rate data, by considering that the full sky space-based survey will observe H_α-emitter galaxies over 15,000 deg^2 of the sky. For the bias and density of observed galaxies, they applied the predictions of Orsi et al. [307] and Geach et al. [308], respectively. They also assumed that half of the galaxies would be detected within the reliable redshift range, which roughly reflects the expected outcomes of proposed space missions, such as the ESA's Euclidean Space Telescope (Euclid) mission and the NASA's Nancy Grace Roman Space Telescope mission. The obtained results are shown in Figure 24, where we see that measurements of the LSS growth rate in the near future will be able to constrain scalar field ϕCDM models with an accuracy of about 10%, considering the fiducial spatially flat ΛCDM model, an improvement of almost an order of magnitude compared to those from currently available datasets [259,309–314]. Constraints on the growth index parameter γ are more restrictive in the ΛCDM model than in other models. For the ϕCDM model, constraints on the growth index parameter γ are about a third tighter than for the wCDM and XCDM models.

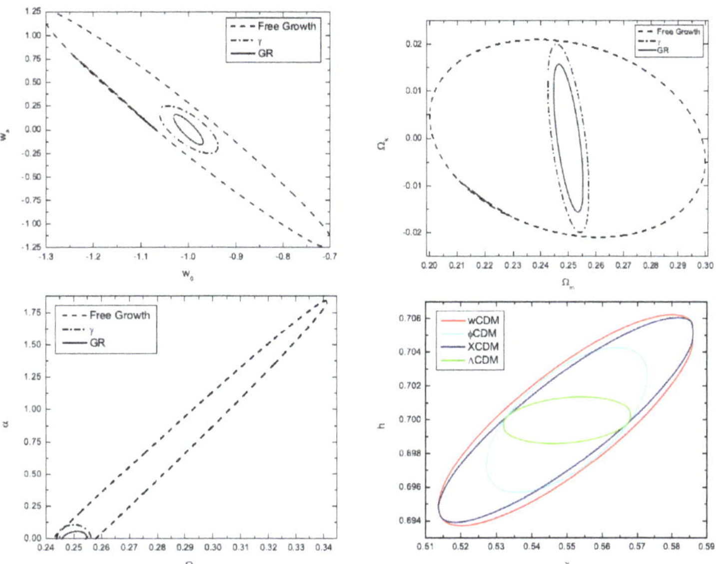

Figure 24. (Left upper panel) The 1σ confidence level contour constraints on parameters w_a and w_0 of the wCDM model. (Right upper panel) The 1σ confidence level contour constraints on parameters Ω_k and Ω_m of the wCDM model. (Left lower panel) The 1σ confidence level contour constraints on parameters α and Ω_m of the scalar field spatially flat ϕCDM model with the inverse power-law RP potential. (Right lower panel) The 1σ confidence level contour constraints on the normalized Hubble constant h and the parameter γ describing deviations from general relativity for various dark energy models. The figure is adapted from [306].

Pavlov et al. [315] also obtained constraints on the above DE models from the Hubble parameter $H(z)$ observations [28,30,257,258], the Union2.1 compilation of 580 SNe Ia apparent magnitude measurements [256], and a compilation of 14 independent LSS growth rate measurements within the redshift range $0.067 \leq z \leq 0.8$ [21,22,259,316]. The authors performed two joint analyses, first for the combination of the $H(z)$ and SNe Ia apparent magnitude data, and second for measurements of the LSS growth rate, the Hubble parameter $H(z)$, and the SNe Ia apparent magnitude. The results of these analyses are presented in Figure 25. Constraints on cosmological parameters of the spatially flat ϕCDM model from LSS growth rate data are quite restrictive. In combination with the SNe Ia apparent magnitude versus the redshift data and the Hubble parameter measurements, the LSS growth rate data are consistent with the standard spatially flat ΛCDM model, as well as with the spatially flat ϕCDM model.

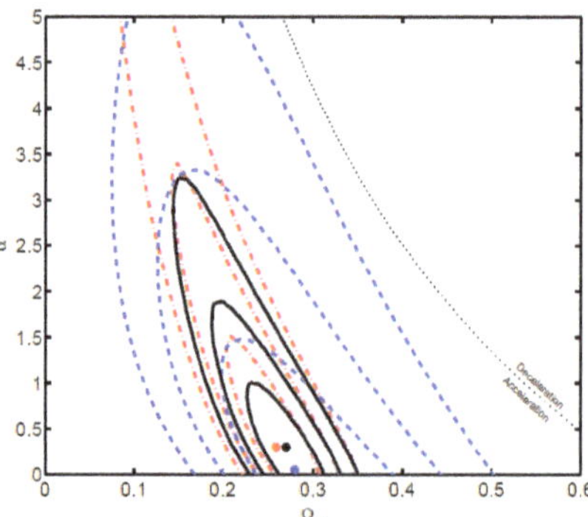

Figure 25. The 1σ, 2σ, and 3σ confidence level contour constraints on parameters of the spatially flat scalar field ϕCDM model with the inverse power-law RP potential from LSS growth rate measurements (blue dashed lines with blue filled circle at best fit $(\Omega_m, \alpha) = (0.28, 0.052)$, $\chi^2_{\min}/\text{dof} = 8.62/12$); SNe Ia apparent magnitude + $H(z)$ data (red dot–dashed lines with red filled circle at best fit $(\Omega_m, \alpha) = (0.26, 0.302)$, $\chi^2_{\min}/\text{dof} = 562/598$); and a combination of all datasets (black solid lines and black filled circle at the best fit $(\Omega_m, \alpha) = (0.27, 0.300)$, $\chi^2_{\min}/\text{dof} = 570/612$). The horizontal axis with $\alpha = 0$ corresponds to the standard spatially flat ΛCDM model and the curved dotted line denotes zero-acceleration models. The figure is adapted from [315].

Avsajanishvili et al. [317] constrained the parameters Ω_m and α of the spatially flat ϕCDM-RP model. Applying only measurements of the LSS growth rate [318], the authors obtained a strong degeneracy between the model parameters Ω_m and α, Figure 26 (left panel). This was followed by obtaining constraints from a compilation of data from the LSS growth rate measurements [318], and the distance–redshift ratio of the BAO peak length scale observations, and prior distance from the CMB temperature anisotropy [319], which eliminated the degeneracy between Ω_m and α, giving $\Omega_m = 0.30 \pm 0.04$ and $0 \leq \alpha \leq 1.30$ at a 1σ confidence level (the best-fit value for the model parameter α is $\alpha = 0$). Constraints on Ω_m and α from the data compilation of Gupta et al. (2012) [318] and Giostri et al. (2012) [319] are presented in Figure 41 (right panel).

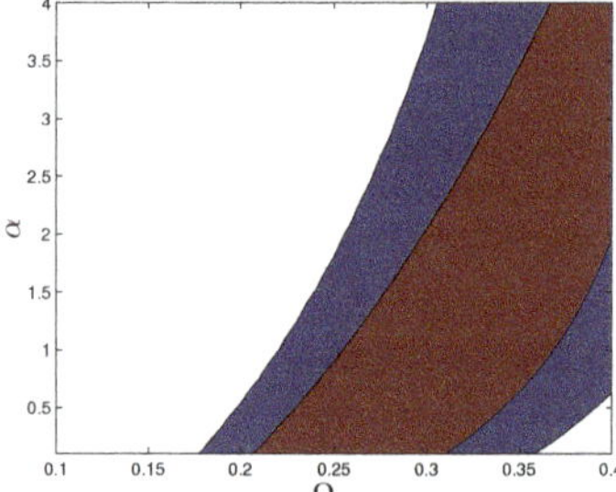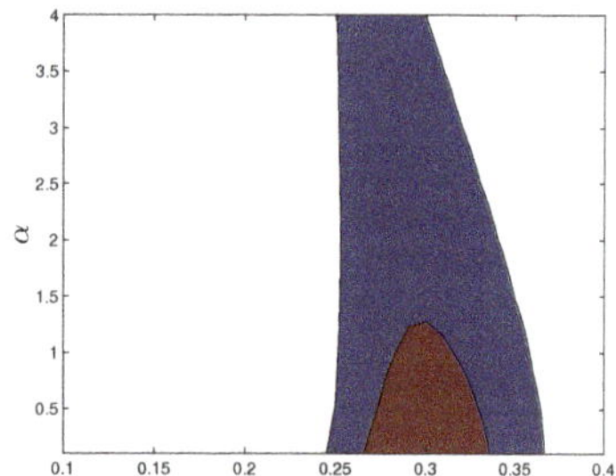

Figure 26. The 1σ and 2σ confidence level contour constraints on parameters Ω_m and α in the scalar field ϕCDM model with the inverse power-law RP potential. (Left panel) Constraints are obtained from the LSS growth rate data [318]. (Right panel) Constraints are obtained from the data compilation of Gupta et al. (2012) [318] and Giostri et al. (2012) [319]. The figure is adapted from [317].

Avsajanishvili et al. [320] also constrained various quintessence and phantom scalar field ϕCDM models, presented in Tables 1 and 2, using observational data predicted for the Dark Energy Spectroscopic Instrument (DESI) [293]. The parameters of these models were constrained using the MCMC methods by comparing measurements of the expansion rate of the universe $H(z)$, the angular diameter distance d_A, and the LSS growth rate, predicted for the standard spatially flat ΛCDM model with corresponding values calculated for the ϕCDM models. Results of constraints for the Zlatev–Wang–Steinhardt potential, the phantom pNGb potential, and the inverse power-law RP potential are shown in Figures 27–29. To compare quintessence and phantom models, Bayesian statistical tests were conducted, namely, the Bayes factor, and the *AIC* and *BIC* information criteria were calculated. The ϕCDM scalar field models could not be unambiguously preferred, from the DESI predictive data, over the standard ΛCDM spatially flat model, the latter still being the most preferred dark energy model. The authors also investigated how the ϕCDM models can be approximated by the CPL parameterization, by plotting the CPL-ΛCDM 3σ confidence level contours, using MCMC techniques, and displayed on them the largest ranges of the current EoS parameters for each ϕCDM model. These ranges were obtained for different values of model parameters or initial conditions from the prior ranges. The authors classified the scalar field models based on whether they can or cannot be distinguished from the standard spatially flat ΛCDM model at the present epoch, as seen in Figure 30. They found that all studied models can be divided into two classes: models that have attractor solutions and models whose evolution depends on initial conditions.

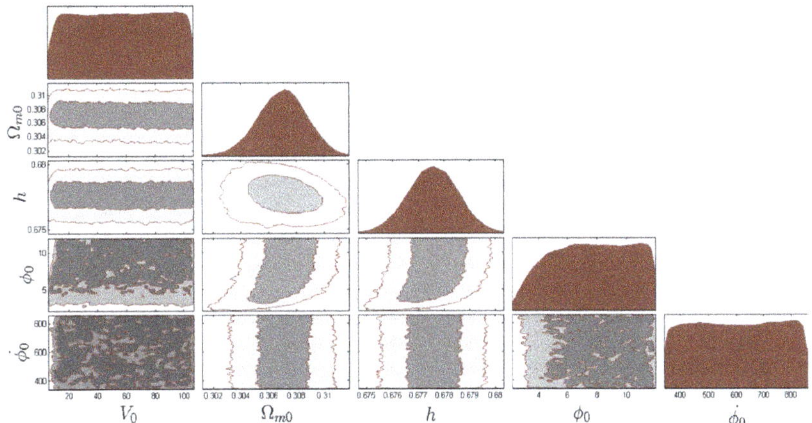

Figure 27. The 1σ and 2σ confidence level contour plots for various pairs of free parameters (V_0, Ω_{m0}, h, ϕ_0, and $\dot{\phi}_0$), for which the spatially flat ϕCDM model with the Zlatev–Wang–Steinhardt potential is in the best fit with the standard spatially flat ΛCDM model. The figure is adapted from [320].

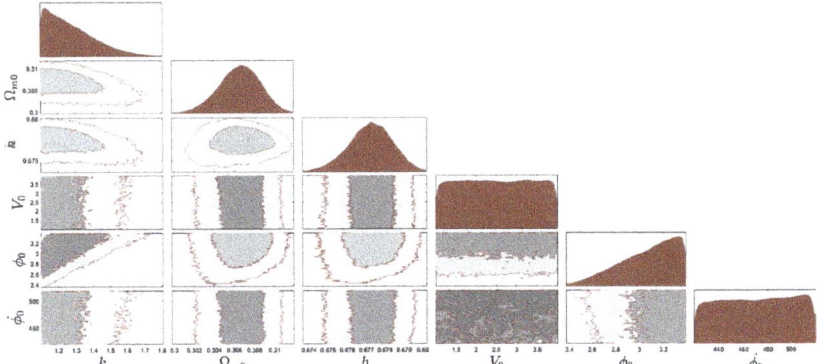

Figure 28. The 1σ and 2σ confidence level contour plots for various pairs of free parameters (k, Ω_{m0}, h, V_0, ϕ_0, and $\dot{\phi}_0$), for which the spatially flat ϕCDM model with the phantom PNGB potential is in the best fit with the standard spatially flat ΛCDM model. The figure is adapted from [320].

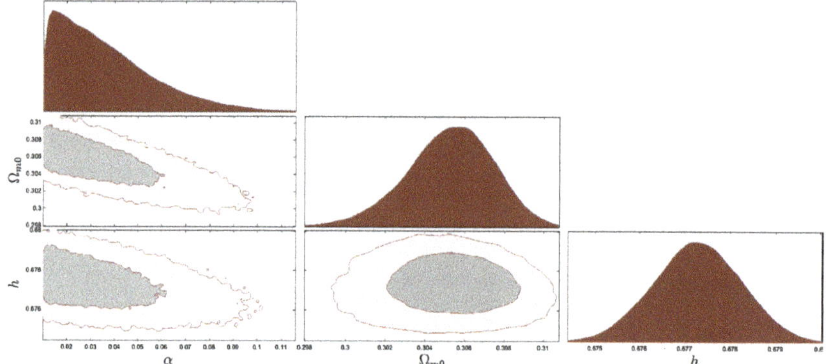

Figure 29. The 1σ and 2σ confidence level contour plots for various pairs of free parameters (α, Ω_{m0}, h), for which the spatially flat ϕCDM model with the RP potential is the best fit with the standard spatially flat ΛCDM model. The figure is adapted from [320].

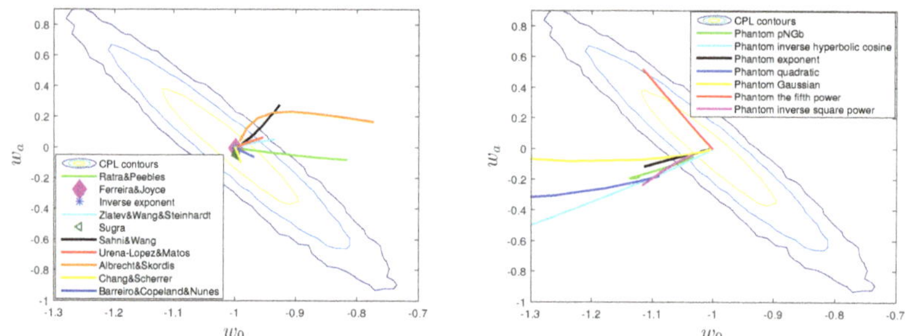

Figure 30. (Left panel) The comparison of the possible w_0 and w_a values for quintessence dark energy potentials in the spatially flat scalar field ϕCDM models with the CPL-ΛCDM 1σ, 2σ, and 3σ confidence level contours. (Right panel) The comparison of possible w_0 and w_a values for phantom dark energy potentials in the spatially flat scalar field ϕCDM models with the CPL-ΛCDM 1σ, 2σ, and 3σ confidence level contours. The figure is adapted from [320].

Peracaula et al. [321] constrained the spatially flat ΛCDM, XCDM, and ϕCDM-RP models by constructing three datasets: DS1/SP consisting of SNe Ia apparent magnitude + $H(z)$ + BAO peak length scale + LSS growth rate + CMB temperature anisotropy data with matter power spectrum SP; DS1/BSP consisting of SNe Ia apparent magnitude + $H(z)$ + BAO peak length scale + LSS growth rate + CMB temperature anisotropy data with both matter power spectrum and bispectrum; and DS2/BSP, which involves BAO peak length scale + LSS growth rate + CMB temperature anisotropy data with both matter power spectrum and bispectrum. These datasets include 1063 SNe Ia apparent magnitude data [110,212], 31 measurements of $H(z)$ from cosmic chronometers [35,257], 16 BAO peak length scale data [322,323], LSS growth rate data, specifically 18 points from data [21,323,324], one point from the weak lensing observable S_8 [325], and full CMB likelihood from Planck 2015 TT + lowP + lensing [12]. The obtained constraints are shown in Figures 31 and 32. The authors tested the effect of separating the expansion history data (SNe Ia apparent magnitude + $H(z)$) from the CMB temperature anisotropy characteristics and the LSS formation data (BAO peak length scale + LSS), where LSS includes RSD and weak lensing measurements, and found that the expansion history data are not particularly sensitive to the dynamic effects of dark energy, while the data compilation of BAO peak length scale + LSS + CMB temperature anisotropy is more sensitive. Also, the influence of the bispectral component of the matter correlation function on the dynamics of dark energy is studied. For this the

BAO peak length scale + LSS data were considered, including both the conventional power spectrum and the bispectrum. As a result, when the bispectral component is excluded, the results obtained are consistent with previous studies by other authors, which means that no clear signs of dynamical dark energy have been found in this case. On the contrary, when the bispectrum component was included in the BAO peak length scale + LSS growth rate dataset for the ϕCDM model, a significant dynamical dark energy signal was achieved at a $2.5 - 3\sigma$ confidence level. The bispectrum can therefore be a very useful tool for tracking and examining the possible dynamical features of dark energy and their influence on the LSS formation in the linear regime.

Park and Ratra [326] constrained the tilted spatially flat and untilted spatially non-flat XCDM models by applying the Planck 2015 CMB temperature anisotropy data [292], BAO peak length scale measurements [26], a Pantheon collection of 1048 SNe Ia apparent magnitude measurements over the broader redshift range $0.01 < z < 2.3$ [212], Hubble parameter observations [21,25,28,30–34,257,293], and LSS growth rate measurements [25], and obtained results as shown in Figures 33 and 34. These data slightly favor the spatially closed XCDM model over the spatially flat ΛCDM model at a 1.2σ confidence level, while also being in better agreement with the untilted spatially flat XCDM model than with the spatially flat ΛCDM model at the 0.3σ confidence level. Current observational data are unable to rule out dynamical dark energy models. The dynamical untilted spatially non-flat XCDM model is compatible with the Dark Energy Survey (DES) limits on the current value of the rms mass fluctuation amplitude σ_8 as a function of the matter density parameter at the present epoch Ω_{m0} but it does not give such a good agreement with higher multipoles of CMB temperature anisotropy data as the standard spatially flat ΛCDM model.

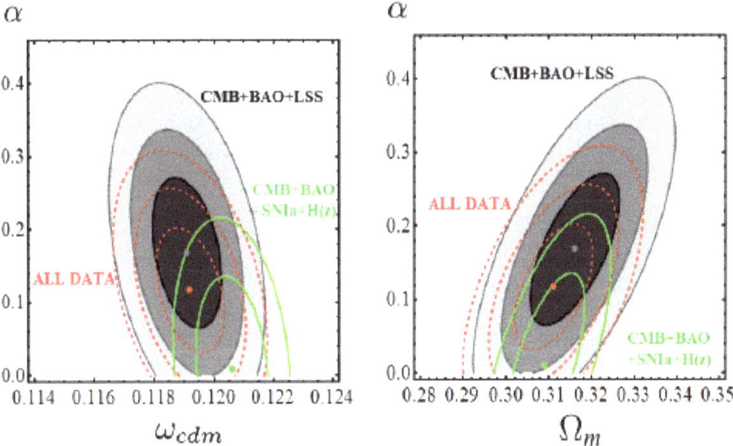

Figure 31. The 1σ, 2σ, and 3σ confidence level contour constraints on parameters of the spatially flat scalar field ϕCDM model with the inverse power-law RP potential using different combinations of datasets and the compressed Planck 2018 data [13]. The results obtained from DS2/BSP dataset CMB temperature anisotropy + BAO peak length scale+LSS growth rate (gray contours), DS1/BSP dataset: SNe Ia apparent magnitude + $H(z)$ + BAO peak length scale + LSS growth rate + CMB temperature anisotropy (dashed red contours), and DS1/BSP dataset without LSS growth rate data CMB temperature anisotropy + BAO peak length scale + SNe Ia apparent magnitude + $H(z)$ (solid green contours). The results are presented in the $w_{cdm} - \alpha$ plane (left panel) and in the $w_{cdm} - \Omega_m$ plane (right panel), where $w_{cdm} = \Omega_m h^2$ is a physical matter density parameter. The figure is adapted from [321].

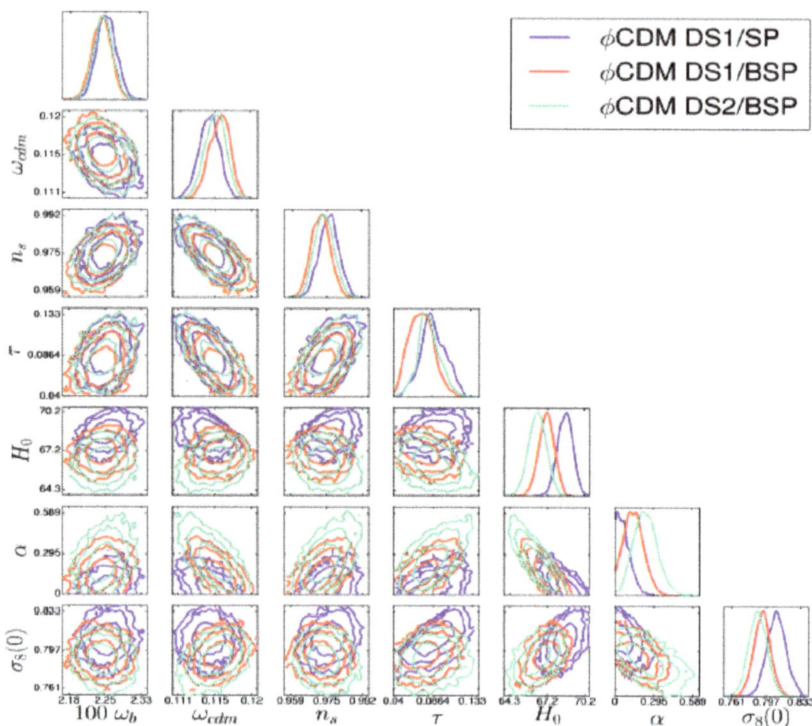

Figure 32. The 1σ, 2σ, and 3σ confidence level contour constraints on parameters of the spatially flat scalar field ϕCDM model with the inverse power-law RP potential from DS1/SP, DS1/BSP, and DS2/BSP datasets. The figure is adapted from [321].

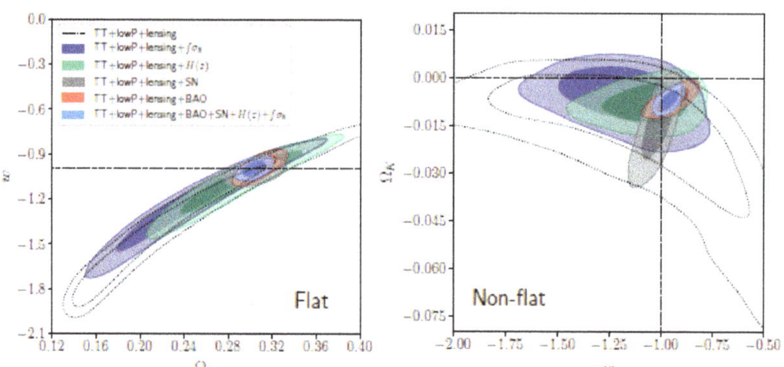

Figure 33. The 1σ and 2σ confidence level contour for the tilted spatially flat XCDM model (left panel) and for the untilted spatially non-flat XCDM model (right panel), constrained by Planck CMB TT + lowP + lensing and non-CMB datasets. The horizontal and vertical dashed lines indicate the standard spatially flat ΛCDM model (with $w = -1$ and $\Omega_k = 0$). The figure is adapted from [326].

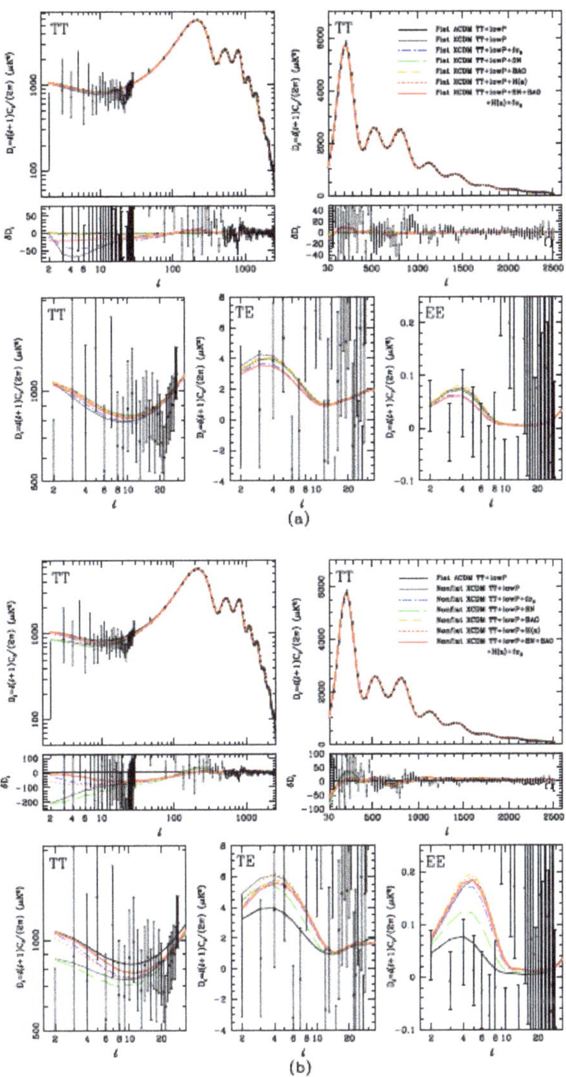

Figure 34. Best-fit CMB temperature anisotropy power spectra of (**a**) the tilted spatially flat XCDM model (top five panels) and (**b**) the untilted spatially non-flat XCDM model (bottom five panels) constrained by Plank CMB TT + lowP data (excluding lensing data) together with data on SNe Ia apparent magnitude, BAO peak length scale, $H(z)$, and LSS growth rate. The best-fit power spectra of the tilted spatially flat ΛCDM model are shown as black curves. The residual δD_l of the TT power spectra are shown with respect to the spatially flat ΛCDM power spectrum that best fits the TT + lowP data. The high-l region C_l and residuals are shown on the bottom panels. The figure is adapted from [326].

3.4. Baryon Acoustic Oscillations Data

Samushia and Ratra [327] constrained the standard spatially flat ΛCDM, the XCDM, and the ϕCDM-RP models from BAO peak length scale measurements [17,20], in conjunction with WMAP measurements of the apparent acoustic horizon angle and galaxy cluster gas mass fraction measurements [284]. These constraints are presented in Figure 35. It is seen that the measurements of Percival et al. (2007) [17] constrain the ϕCDM model less effectively (left panel of Figure 35), while measurements of the joint BAO peak length scale and the galaxy cluster gas mass give consistent and more accurate constraints on the parameters of the ϕCDM model than those derived from other data, i.e., $\alpha < 3.5$ (right panel of Figure 35).

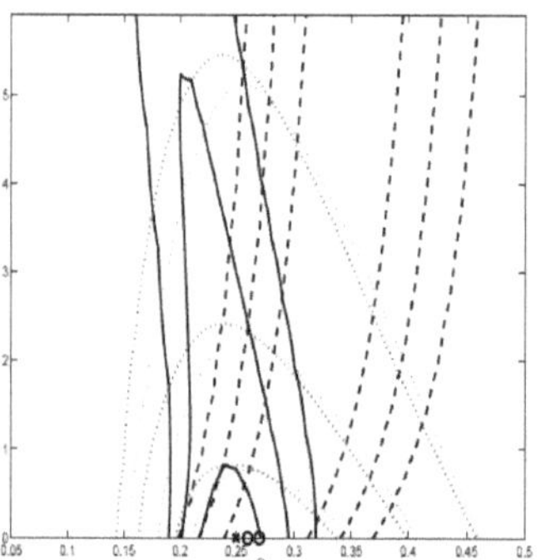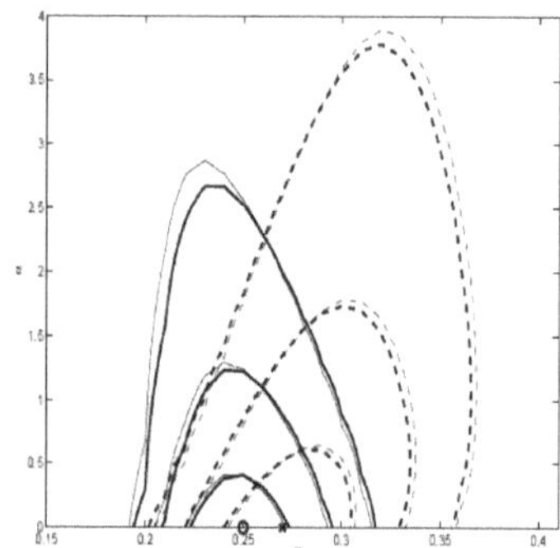

Figure 35. The 1σ, 2σ, and 3σ confidence level contour constraints on parameters of the scalar field ϕCDM model with the inverse power-law RP scalar field potential. The $\alpha = 0$ axis corresponds to the standard spatially flat ΛCDM model. (Left panel) Solid lines are constraints derived by Percival et al. (2007) [17] using BAO peak length scale data in conjunction with WMAP data on acoustic horizon angle. Dashed lines are constraints obtained by Eisenstein et al. (2005) [20] from BAO peak length scale data. The circle denotes the best-fit value. Two sets of dotted lines are constraints obtained from galaxy cluster gas mass fraction measurements of Samushia and Ratra (2008) [327]; thick dotted lines are derived using WMAP priors for h and $\Omega_b h^2$ while thin dotted lines are obtained for alternate priors. The cross denotes the best-fit value. (Right panel) Solid lines are joint constraints obtained by Percival et al. (2007) from BAO peak length scale data in conjunction with WMAP data on acoustic horizon angle and galaxy cluster gas mass fraction measurements. The circle denotes the best-fit value with a suitable $\chi^2 \simeq 58$ for 42 degrees of freedom; dashed lines are joint constraints derived by Eisenstein et al. (2005) [20] using BAO peak length scale data. The cross denotes the best-fit value with a suitable $\chi^2 \simeq 52$ for 41 degrees of freedom. Thick lines are derived using the WMAP priors for h and $\Omega_b h^2$, and thin lines are for alternate priors. Joint best-fit values for two prior sets overlap. Here, Ω_m and α ranges are smaller than those shown on the left panel. The figure is adapted from [327].

The above models were also constrained by Samushia et al. [328] using the lookback time versus redshift data [329], the passively evolving galaxies data [257], the current BAO peak length scale data, and the SNe Ia apparent magnitude measurements. Applying a Bayesian prior on the total age of the universe based on WMAP data, the authors obtained constraints on the ϕCDM model as shown in Figure 36. Constraints on the ϕCDM model by joint datasets consisting of measurements of the age of the universe, SNe Ia Union apparent magnitude, and BAO peak length scale are tighter than those obtained from datasets consisting of data on the lookback time and the age of the universe.

The quintessential inflation model with the generalized exponential potential $V(\phi) \propto \exp(-\lambda \phi^n / M_{\mathrm{pl}}^n)$, $n > 1$ was studied by Geng et al. [177]. The authors extended this model including massive neutrinos that are non-minimally coupled to a scalar field, obtaining observational constraints on parameters from combinations of data: the CMB temperature anisotropy [289], the BAO peak length scale from BOSS [23,314], and the 11 SNe Ia apparent magnitudes from Supernova Legacy Survey (SNLS) [254]. It was found that the upper bound on possible values of the sum of neutrino masses $\sum m_\nu < 2.5$ eV is significantly larger than in the spatially flat ΛCDM model (Figure 37). The authors concluded that the model under consideration is in good agreement with observations and represents a successful scheme for the unification of the primordial inflaton field causing inflation in the very early universe and dark energy causing the accelerated expansion of the universe at the present epoch.

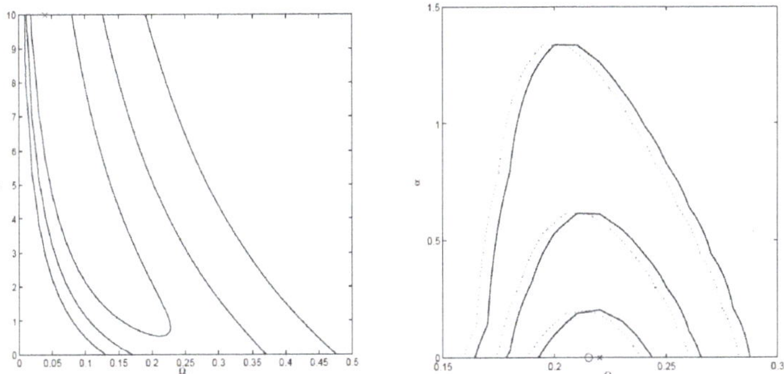

Figure 36. The 1σ, 2σ, and 3σ confidence level contour constraints on parameters of the scalar field ϕCDM model with the inverse power-law RP potential. The horizontal axis with $\alpha = 0$ corresponds to the standard spatially flat ΛCDM model. (Left panel) Dotted lines obtained from the lookback time data and measurements of the age of the universe. The cross denotes the best-fit parameters $\Omega_m = 0.04$ and $\alpha = 10$ with $\chi^2 = 22$, for $\alpha = 0$ with $\chi^2 = 359$ for 346 degrees of freedom derived using measurements of the lookback time, the age of the universe, SNe Ia apparent magnitude, and BAO peak length scale, while solid lines are derived using only SNe Ia apparent magnitude measurements and BAO peak length scale data. The cross denotes the best-fit point at $\Omega_m = 0.22$ and $\alpha = 0$ with $\chi^2 = 329$ for 307 degrees of freedom. The figure is adapted from [328].

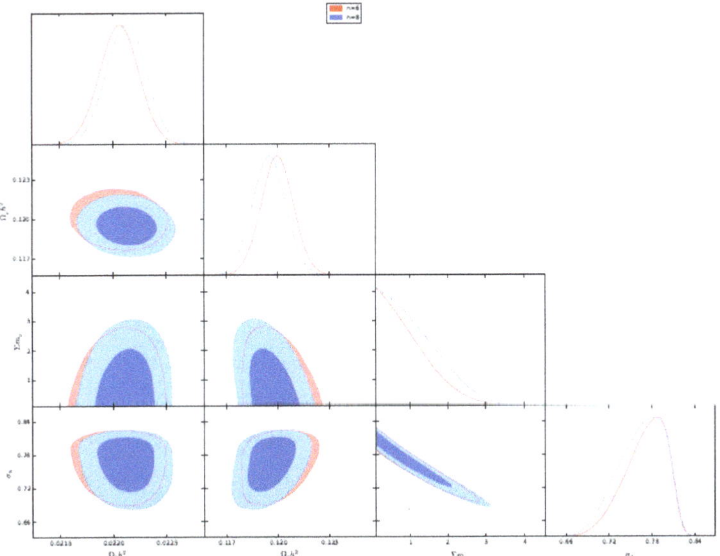

Figure 37. The 1σ and 2σ confidence level contours of one- and two-dimensional distributions of $\Omega_b h^2$, $\Omega_m h^2$, $\sum m_\nu$, and σ_8 for the quintessential inflation model with the exponential potential $V(\phi) \propto \exp(-\lambda \phi^n / M_{\mathrm{pl}}^n)$, $n = 6$ (orange line) and $n = 8$ (blue line). The figure is adapted from [177].

The compilation of CMB angular power spectrum data from the Planck 2015 mission [292] and BAO peak length scale measurements from the matter power spectra obtained by missions 6dFGS [22], BOSS, LOWZ and CMASS [23], and SDSS-MGS [24] were applied by Ooba et al. [330] to obtain constraints on the spatially non-flat quintessential inflation ϕCDM-RP model. The theoretical angular power spectra of the CMB temperature anisotropy were calculated using the Cosmic Linear Anisotropy Solving System (CLASS) code of Blas et al. [295] and the MCMC analysis was performed with Monte Python from Audren et al. [296]. The authors also used a physically consistent power spectrum for energy density inhomogeneities in the spatially non-flat (spatially closed) quintessential inflation ϕCDM model and found that the spatially closed ϕCDM model provides a better fit to the lower multipole region of CMB temperature anisotropy data compared to that provided by the tilted spatially flat ΛCDM model. The former reduces the tension between the Planck and the weak lensing σ_8 constraints, while the higher multipole region of the CMB temperature anisotropy data is in better agreement with the tilted spatially flat ΛCDM model than with the spatially closed ϕCDM model (Figure 38).

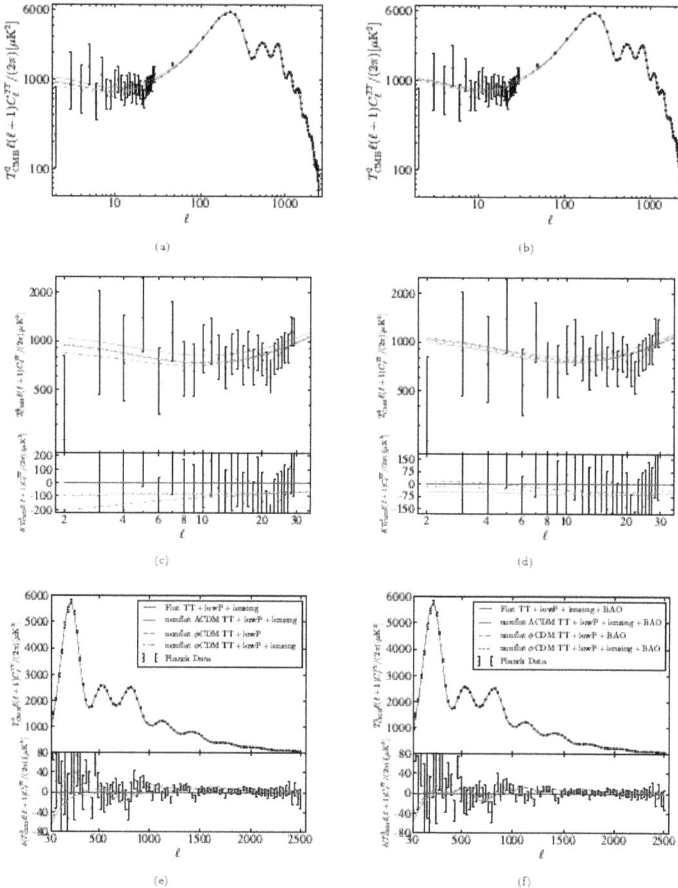

Figure 38. The C_l for the best-fit spatially non-flat ϕCDM, spatially non$-$flat ΛCDM, and spatially flat tilted ΛCDM (gray solid line) models. (Left panels) (**a,c,e**) Results obtained from only CMB temperature anisotropy data. (Right panels) (**b,d,f**) results obtained only from CMB temperature anisotropy + BAO peak length scale data. All-l regions are demonstrated in top panels. The low-l region C_l and residuals are shown in the middle panels. The high-l region C_l and residuals are presented in the bottom panels. The figure is adapted from [330].

Ryan et al. [331] constrained the parameters of the ϕCDM-RP, the XCDM, and the ΛCDM models from BAO peak length scale measurements [22,24–26,293] and the Hubble parameter $H(z)$ data [21,28,30–34,257]. The results obtained for the ϕCDM model are presented in Figure 39, which shows that this dataset is consistent with the standard spatially flat ΛCDM model. Depending on the value of the Hubble constant H_0 as a prior and the cosmological model under consideration, the data provides evidence in favor of the spatially non-flat scalar field ϕCDM model.

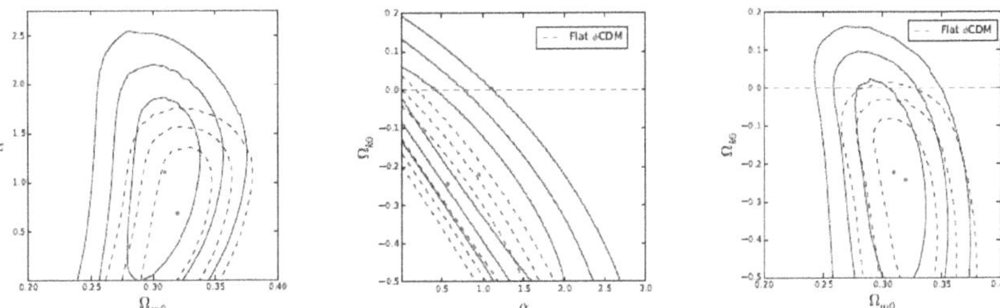

Figure 39. The 1σ, 2σ, and 3σ confidence level contour constraints on parameters of the spatially non-flat ϕCDM model with the RP potential. Solid (dashed) contours correspond to $H_0 = 68 \pm 2.8$ (73.24 ± 1.74) km s^{-1}Mpc^{-1} prior; the red dots indicate the location of the best-fit point in each prior case. The horizontal axis with $\alpha = 0$ denotes the spatially flat ΛCDM model. (Left panel) The results obtained for the Ω_{k0} marginalization. (Center panel) The results obtained for the Ω_{m0} marginalization. (Right panel) The results obtained for the parameter α marginalization. The figure is adapted from [331].

Chudaykin et al. [332] obtained constraints on the parameters of the oCDM, XCDM (here w_0CDM), and wCDM models by using the joint analysis from data on the BAO peak length scale, BBN, and SNe Ia apparent magnitude. The resulting constraints are completely independent of the CMB temperature anisotropy data but compete with the CMB temperature anisotropy constraints in terms of parameter error bars. The authors consequently obtained the value of the spatial curvature density parameter at the present epoch $\Omega_{k0} = -0.043^{+0.036}_{-0.036}$ at a 1σ confidence level, which is consistent with the spatially flat universe; in the spatially flat XCDM model, the value of the dark energy EoS parameter at the present epoch $w_0 = -1.031^{+0.052}_{-0.048}$ at a 1σ confidence level, which approximately equals the value of the EoS parameter for the ΛCDM model; values of the w_0 and w_a in the CPL parameterization of the EoS parameter of the wCDM model $w_0 = -0.98^{+0.099}_{-0.11}$ and $w_a = -0.33^{+0.63}_{-0.48}$ at a 1σ confidence level. The authors also found that the exclusion of the SNe Ia apparent magnitude data from the joint data analysis does not significantly weaken the resulting constraints. This means that, when using a single external BBN prior, full-shape and BAO peak length scale data can provide reliable constraints independent of CMB temperature anisotropy constraints. The authors also tightened the observational constraints on cosmological parameters with the inclusion of the hexadecapole ($l = 4$) moment of the redshift-space power spectrum.

Bernui et al. [67] investigated the effect of the BAO measurements on the IDE models that have significantly different dynamic behavior compared to the prediction of the standard ΛCDM model. The authors used the compilation of 15 transversal 2D BAO measurements [333,334] and CMB data [119] to constrain the IDE models. It was found that the transversal 2D BAO and traditional 3D BAO measurements can generate completely different observational constraints on the coupling parameter in the IDE models. Moreover, in contrast to the joint Planck + BAO analysis, where it is not possible to solve the Hubble constant H_0 tension, the joint Planck + BAO (transversal) analysis agrees well with the measurements made by the SH0ES team, and when applied to the IDE models, solves the Hubble constant H_0 tension. The 1σ and 2σ confidence level contour constraints on the coupling parameter ξ in the IDE model using the 2D transversal 2D BAO are shown in Figure 40.

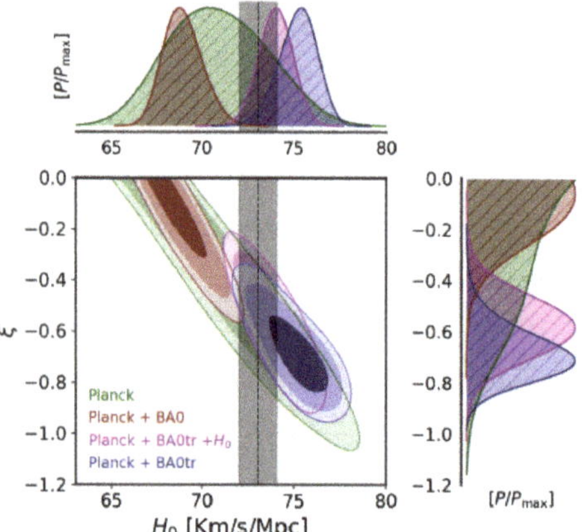

Figure 40. The 1σ and 2σ confidence level 2D contour constraints on the coupling parameter ξ in the IDE model and 1D posteriors for the cases without lensing. The grey vertical stripe refers to the value of H_0 measured by the SH0ES team ($H_0 = 73.04 \pm 1.04 \ \mathrm{km\ s^{-1}\ Mpc^{-1}}$ at 1σ confidence level). The figure is adapted from [67].

3.5. Hubble Parameter Data

Samushia and Ratra [335] used the Simon, Verde, and Jimenez (SVJ) [257] definition of the redshift dependence of the Hubble parameter $H(z)$ (so-called SVJ $H(z)$ data) to constrain cosmological parameters in the scalar field ϕCDM-RP model. According to the results obtained (Figure 41), using the $H(z)$ data, the constraints on the matter density parameter Ω_{m} are more stringent than those on the model parameter α. Constraints on the matter density Ω_{m} are approximately as tight as the ones derived from the galaxy cluster gas mass fraction data [336] and from the SNe Ia apparent magnitude data [337].

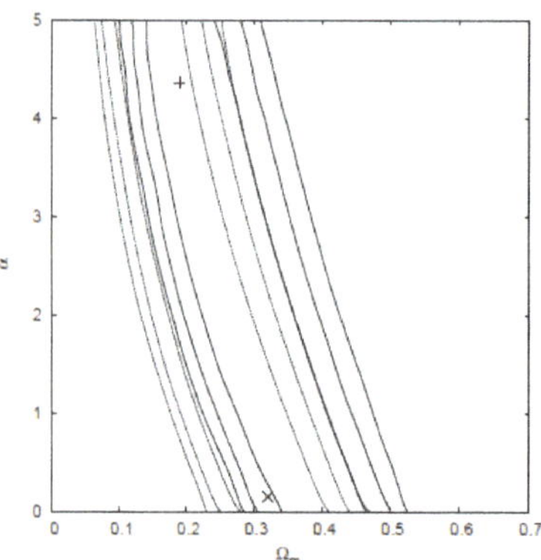

Figure 41. The 1σ, 2σ, and 3σ confidence level contour constraints on parameters of the ϕCDM model with the RP potential. Solid lines correspond to $H_0 = 73 \pm 3 \ \mathrm{km\ s^{-1}\ Mpc^{-1}}$, while dashed lines correspond to $H_0 = 68 \pm 4 \ \mathrm{km\ s^{-1}\ Mpc^{-1}}$. The plus sign denotes the maximum likelihood at $\Omega_{\mathrm{m0}} = 0.32$ and $\alpha = 0.15$ with reduced $\chi^2 = 1.8$. The cross denotes the maximum likelihood at $\Omega_{\mathrm{m0}} = 0.19$ and $\alpha = 4.37$ with reduced $\chi^2 = 1.89$. The horizontal axis for which $\alpha = 0$ corresponds to the spatially flat ΛCDM model. The figure is adapted from [335].

Chen and Ratra [338] analyzed constraints on the model parameters of the ϕCDM-RP, the XCDM, and the ΛCDM models, using 13 Hubble parameter $H(z)$ data versus redshift [28,258]. The authors showed (see Figure 42) that the Hubble parameter $H(z)$ data yield quite strong constraints on the parameters of the ϕCDM model. The constraints derived from the $H(z)$ measurements are almost as restrictive as those implied by the currently available lookback time observations and the GRB luminosity data, but more stringent than those based on the currently available galaxy cluster angular size data. However, they are less restrictive than those following from the joint analysis of SNe Ia apparent magnitude and BAO peak length scale data. The joint analysis of the Hubble parameter $H(z)$ data with SNe Ia apparent magnitude and BAO peak length scale data favor the standard spatially flat ΛCDM model but do not exclude the dynamical scalar field ϕCDM model.

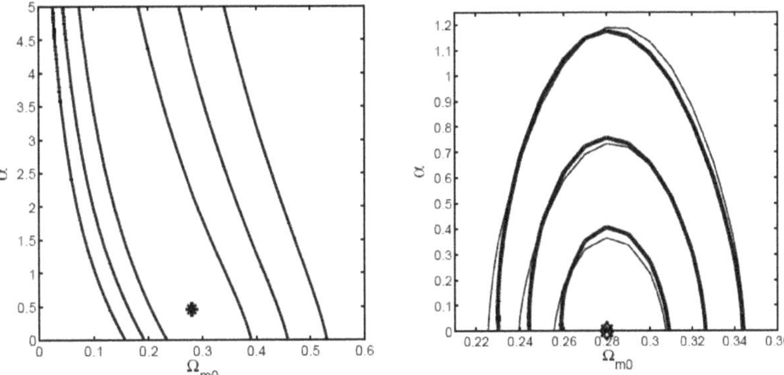

Figure 42. The 1σ, 2σ, and 3σ confidence level contour constraints on parameters of the ϕCDM model with the RP potential. The horizontal axis with $\alpha = 0$ corresponds to the standard spatially flat ΛCDM model. (Left panel) Contours obtained from $H(z)$ data. The star denotes the best-fit pair $(\Omega_{m0}, \alpha) = (0.28, 0.46)$, $\chi^2_{min} = 10.1$. (Right panel) Contours were obtained from a joint analysis of the BAO peak length scale and SNe Ia apparent magnitude data (with systematic errors), with (and without) $H(z)$ data. The cross denotes the best-fit point determined from the joint sample with $H(z)$ data at $\Omega_{m0} = 0.28$ and $\alpha = 0$, with $\chi^2_{min} = 531$. The diamond denotes the best-fit point obtained from the joint sample with $H(z)$ data at $\Omega_{m0} = 0.28$ and $\alpha = 0$, $\chi^2_{min} = 541$. The figure is adapted from [338].

In [339], Farooq et al. obtained constraints on the parameters of the ϕCDM-RP, the XCDM, the wCDM, and the ΛCDM models from analysis of measurements of the BAO peak length scale, SNe Ia apparent magnitude [256], and 21 Hubble parameter $H(z)$ [28,30,257,258]. The results of this analysis are shown in Figure 43. Constraints are more restrictive with the inclusion of eight new $H(z)$ measurements [30] than those derived by Chen and Ratra [338]. This analysis favors the standard spatially flat ΛCDM model but does not exclude the scalar field ϕCDM model.

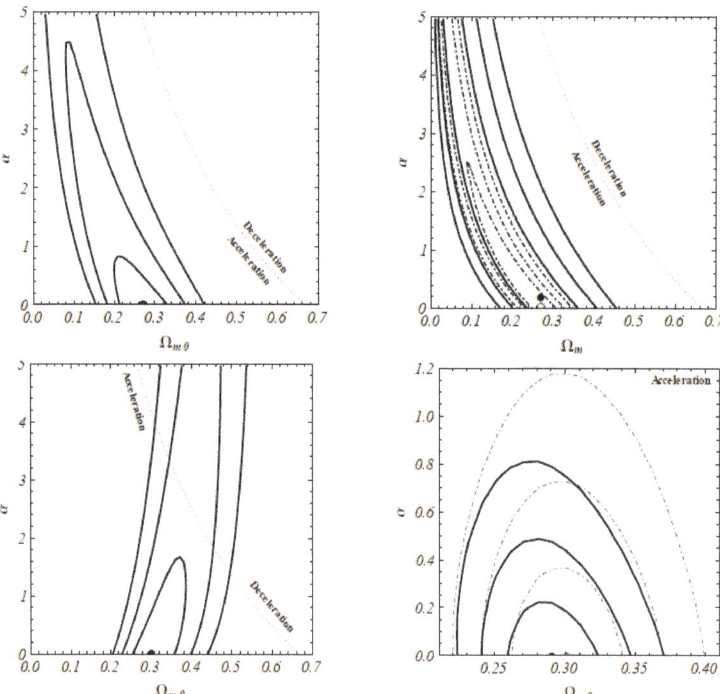

Figure 43. Thick solid lines are 1σ, 2σ, and 3σ confidence level contour constraints on the parameters of the spatially flat ϕCDM model with the RP potential, for the prior $H_0 = 73.8 \pm 2.4 \, \mathrm{km \, s^{-1} Mpc^{-1}}$. The horizontal axis with $\alpha = 0$ corresponds to the standard spatially flat ΛCDM model. (Left upper panel) Contours obtained from $H(z)$ data. Thin dot–dashed lines are 1σ, 2σ, and 3σ confidence level contours reproduced from [338], where the prior is $H_0 = 68 \pm 3.5 \, \mathrm{km \, s^{-1} Mpc^{-1}}$; the empty circle corresponds to the best-fit point. The curved dotted lines denote zero-acceleration models. The filled black circles correspond to best-fit points. (Right upper panel) Contours obtained from only SNe Ia apparent magnitude data with (without) systematic errors. Filled (open) circles denote likelihood maxima for the case of data with (without) systematic errors. (Left lower panel) Contours were obtained from only the BAO peak length scale data. Filled circles denote likelihood maxima. (Right lower panel) Contours obtained from data on the BAO peak length scale and SNe Ia apparent magnitude (with systematic errors), with (without) $H(z)$ data. The full (empty) circle denotes the best-fit point determined from a joint analysis with (without) $H(z)$ data. The figure is adapted from [339].

Farooq and Ratra [340] worked out constraints on the parameters of the ϕCDM-RP, the XCDM, and the ΛCDM models from measurements of the Hubble parameter $H(z)$ at redshift $z = 2.3$ [341] and 21 lower redshift measurements [28,30,257,258]. Constraints with the inclusion of the new $H(z)$ measurement of Busca et al. are more restrictive than those derived by Farooq et al. (Figure 44). As seen in this figure, the $H(z)$ constraints depend on the Hubble constant prior to H_0 used in the analysis. The resulting constraints are more stringent than those which follow from measurements of the SNe Ia apparent magnitude of Suzuki et al. (2012) [256]. This joint analysis consisting of measurements of $H(z)$, SNe Ia apparent magnitude, and BAO peak length scale favors the standard spatially flat ΛCDM model, but the dynamical scalar field ϕCDM model is not excluded either.

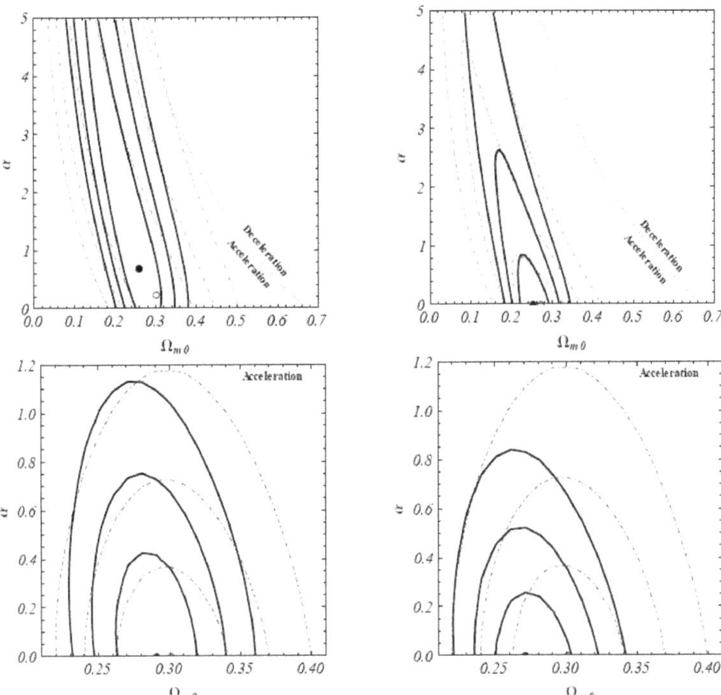

Figure 44. Thick solid (thin dot–dashed) lines are 1σ, 2σ, and 3σ confidence level contour constraints on the parameters of the spatially flat ϕCDM model with the RP potential from the new $H(z)$ data (old $H(z)$ data were used in [339]). The filled (empty) circle is the best-fit point from new (old) $H(z)$ measurements. The horizontal axis with $\alpha = 0$ corresponds to the standard spatially flat ΛCDM model. The curved dotted lines denote zero-acceleration models. (Left upper panel) Contours obtained for the $H_0 = 68 \pm 3.5$ kms^{-1}Mpc^{-1} prior. The filled circles correspond to the best-fit pair $(\Omega_{m0}, \alpha) = (0.36, 0.70)$, $\chi^2_{\min} = 15.2$. The empty circles correspond to the best-fit pair $(\Omega_{m0}, \alpha) = (0.30, 0.25)$, $\chi^2_{\min} = 14.6$. (Right upper panel) Contours obtained for the $H_0 = 73.8 \pm 2.4$ kms^{-1}Mpc^{-1} prior. The filled circles correspond to the best-fit pair $(\Omega_{m0}, \alpha) = (0.25, 0)$, $\chi^2_{\min} = 16.1$. Empty circles correspond to the best-fit pair $(\Omega_{m0}, \alpha) = (0.27, 0)$, $\chi^2_{\min} = 15.6$. (Left lower panel) Contours obtained from joint analysis with SNe Ia apparent magnitude data (with systematic errors) and BAO peak length scale data, with (without) $H(z)$ data. The full (empty) circle marks the best-fit point determined from a joint analysis with (without) $H(z)$ data. Contours obtained for $H_0 = 68 \pm 3.5$ kms^{-1}Mpc^{-1} prior. The full circle indicates the best-fit pair $(\Omega_{m0}, \alpha) = (0.29, 0)$, $\chi^2_{\min} = 567$ while the empty circle corresponds to the best-fit pair $(\Omega_{m0}, \alpha) = (0.30, 0)$, $\chi^2_{\min} = 551$. (Right lower panel) Contours obtained for the $H_0 = 73.8 \pm 2.4$ kms^{-1}Mpc^{-1} prior. The empty circle denotes the best-fit pair $(\Omega_{m0}, \alpha) = (0.30, 0)$, $\chi^2_{\min} = 551$ while the full circle denotes the best-fit pair $(\Omega_{m0}, \alpha) = (0.27, 0)$, $\chi^2_{\min} = 569$. The figure is adapted from [340].

Farooq and Ratra [342] found constraints on the parameters of the ϕCDM-RP model from the compilation of 28 independent measurements of the Hubble parameter $H(z)$ within the range of redshift $0.07 \leq z \leq 2.3$. Measurements of $H(z)$ require a currently accelerating cosmological expansion at a 3σ confidence level. The authors determined the deceleration–acceleration transition redshift $z_{\mathrm{da}} = 0.74 \pm 0.05$. This result is in good agreement with the result obtained by Busca et al. [341], which is $z_{\mathrm{da}} = 0.82 \pm 0.08$ based on 11 measurements of $H(z)$ from BAO peak length scale data within the range of redshift $0.2 \leq z \leq 2.3$. The resulting constraints with different priors of H_0 are demonstrated in Figure 45.

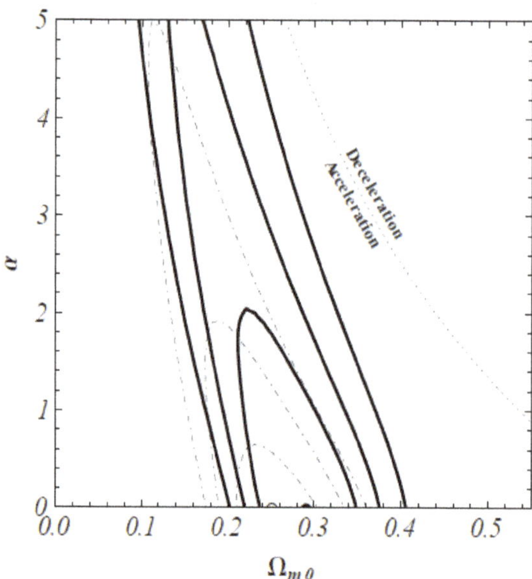

Figure 45. Thick solid and thin dot–dashed lines are 1σ, 2σ, and 3σ confidence level contour constraints on the parameters of the scalar field ϕCDM model with the RP potential from the compilation of $H(z)$ data for $H_0 = 68 \pm 3.5$ km s^{-1}Mpc^{-1} and $H_0 = 73.8 \pm 2.4$ km s^{-1}Mpc^{-1} priors, respectively. The horizontal axis with $\alpha = 0$ corresponds to the standard spatially flat ΛCDM model and the curved dotted line denotes zero-acceleration models. Filled and empty circles are best-fit points for which $(\Omega_{m0}, \alpha) = (0.29, 0)$, $\chi^2_{\min} = 18.24$ and $(\Omega_{m0}, \alpha) = (0.25, 0)$, $\chi^2_{\min} = 20.64$, respectively. The figure is adapted from [342].

Farooq et al. [260] analyzed constraints on the parameters of the spatially flat ϕCDM-RP, the XCDM, and the ΛCDM models from a compilation of measurements of the Hubble parameter $H(z)$. To obtain this compilation, the authors used weighted mean and median statistics techniques to combine 23 independent lower redshifts, $z < 1.04$, and Hubble parameter $H(z)$ measurements, and define binned forms of them. Then, this compilation was combined with 5 $H(z)$ measurements at the higher redshifts $1.3 \leq z \leq 2.3$. The resulting constraints are shown in Figure 46. As seen from the figure, the weighted mean binned data are almost identical to those derived from analysis using 28 independent measurements of $H(z)$. Binned weighted-mean values of $H(z)/(1 + z)$ versus redshift data are presented in Figure 47. These results are consistent with a moment of the deceleration–acceleration transition at redshift $z_{da} = 0.74 \pm 0.05$ derived by Farooq and Ratra [342], which corresponds to the standard spatially flat ΛCDM model.

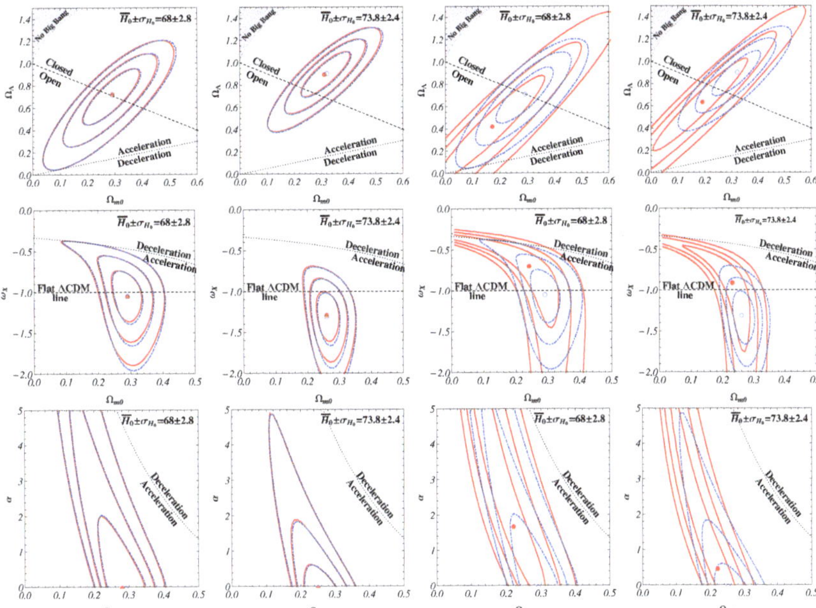

Figure 46. Thick solid and thin dot–dashed lines are 1σ, 2σ, and 3σ confidence level contour constraints on the parameters of the ϕCDM model with the RP potential, the XCDM model, and the ΛCDM model from 7 or 9 measurements per bin data. In these three rows, the first two plots include red weighted-mean constraints while the second two include red median statistics. Filled red and empty blue circles correspond to the best-fit points. Dashed diagonal lines denote spatially flat models and dotted lines show zero-acceleration models. The figure is adapted from [260].

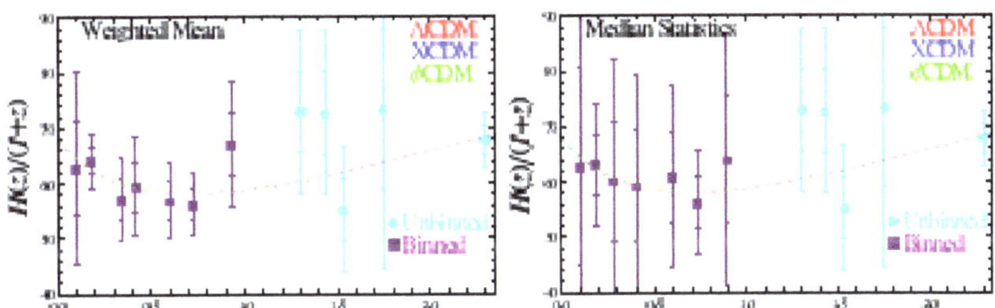

Figure 47. The $H(z)/(1+z)$ data binned with 7 or 9 measurements per bin, as well as 5 higher measurements of redshift, and Farooq and Ratra [260] best-fit model predictions. Dashed and dotted lines correspond to $H_0 = 68 \pm 3.5$ km s^{-1}Mpc^{-1} and $H_0 = 73.8 \pm 2.4$ km s^{-1}Mpc^{-1} priors, respectively. The figure is adapted from [342].

Chen et al. [343] used 28 measurements of the Hubble parameter $H(z)$ within the redshift range $0.07 \leq z \leq 2.3$ [21,28,30,31,257,341,344] to determine the value of the Hubble constant H_0 in the ϕCDM-RP, wCDM, and the spatially flat and spatially non-flat ΛCDM models. The result obtained for the ϕCDM-RP model is shown in Figure 48. The value of the Hubble constant H_0 is found as follows: for the spatially flat and spatially non-flat ΛCDM model, $H_0 = 68.3^{+2.7}_{-3.3}$ km s^{-1}Mpc^{-1} and $H_0 = 68.4^{+2.9}_{-3.3}$ km s^{-1}Mpc^{-1}; for the wCDM model, $H_0 = 65.0^{+6.5}_{-6.6}$ km s^{-1}Mpc^{-1}; for the ϕCDM model, $H_0 = 67.9^{+2.4}_{-2.4}$ km s^{-1}Mpc^{-1} (at a 1σ confidence level). The obtained H_0 values are more consistent with the smaller values determined from the recent CMB temperature anisotropy and BAO peak length scale data, and with the values derived from the median statistics analysis of Huchra's compilation of H_0 data.

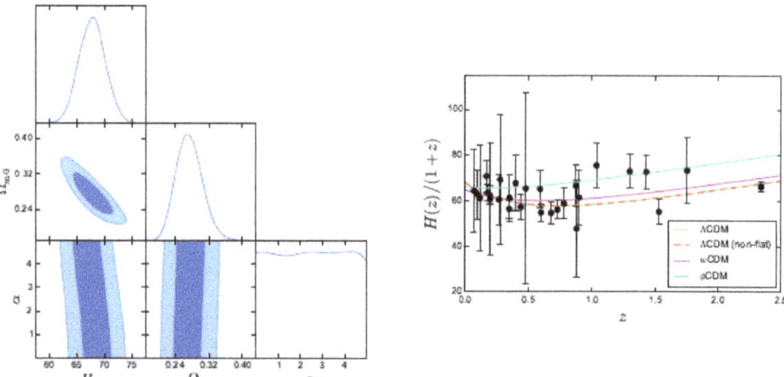

Figure 48. (Left panel) The 1σ and 2σ confidence level contour constraints on the parameters of the ϕCDM model with the RP potential. (Right panel) Best-fit model curves from the 28 $H(z)$ data points for the spatially flat ϕCDM model, wCDM model, and the spatially flat and spatially non-flat ΛCDM model. The figure is adapted from [343].

Farooq et al. [192] determined constraints on the parameters of the ϕCDM-RP, XCDM, wCDM, and the ΛCDM models in the spatially flat and spatially non-flat universe. The authors used the updated compilation of 38 measurements of the Hubble parameter $H(z)$ within the redshift range $0.07 \leq z \leq 2.36$ [21,25,28,30–34,257,293]. The result for these constraints is shown in Figure 49. The authors determined the redshift of the cosmological deceleration–acceleration transition, z_{da}, and the value of the Hubble constant H_0 from the $H(z)$ measurements. The determined values of z_{da} are insensitive to the chosen model and depend only on the assumed value of the Hubble constant H_0. The weighted mean of these measurements is $z_{da} = 0.72 \pm 0.05$ (0.84 ± 0.03) for $H_0 = 68 \pm 2.8$ (73.24 ± 1.74) km s^{-1}Mpc^{-1}. The authors proposed a model-independent method to determine the value of the Hubble constant H_0. The $H(z)$ data are consistent with the standard spatially flat ΛCDM model while they do not rule out the spatially non-flat XCDM and spatially non-flat ϕCDM models.

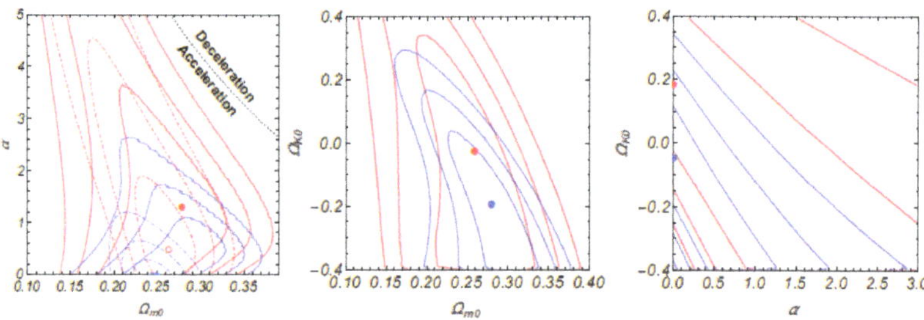

Figure 49. The 1σ, 2σ, and 3σ confidence level contours constraints on the parameters of the spatially non-flat ϕCDM model with the RP potential. Red (blue) solid lines are for the lower (higher) H_0 prior. (Left, center, and right panels) The results obtained correspond to the marginalization over Ω_{k0}, α, and Ω_{m0}, respectively. Red (blue) solid circles are the best-fit points for the lower (higher) H_0 prior. Red (blue) dot–dashed lines in the left panel are 1σ, 2σ, and 3σ for the lower (higher) H_0 prior in the spatially flat ϕCDM model. The figure is adapted from [192].

3.6. Quasar Angular Size Data

Ryan et al. [193] determined constraints on the parameters of the spatially flat and spatially non-flat ΛCDM, XCDM, and ϕCDM-RP models using BAO peak length scale measurements [22,24–26,293], the Hubble parameter $H(z)$ data [21,30–34,257], and quasar (QSO) angular size data [345,346]. The 1σ, 2σ, and 3σ confidence level contour constraints on the parameters of the spatially non-flat ϕCDM model with the RP potential from $H(z)$, QSO, and BAO peak length scale datasets are presented in Figure 50. Depending on the chosen model and dataset, the observational data slightly favor both the spatially closed hypersurfaces with $\Omega_{k0} < 0$ at 1.7σ confidence level and the dynamical dark energy models over the standard spatially flat ΛCDM model at a slightly higher than 2σ confidence level. Furthermore, depending on the dataset and the model, the observational data favor a lower Hubble constant value over the one measured by the local data at a 1.8σ confidence level to 3.4σ confidence level.

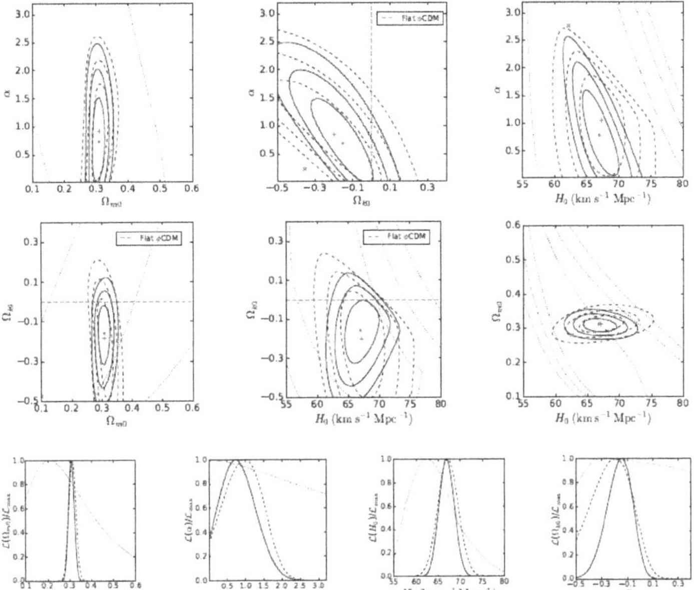

Figure 50. The 1σ, 2σ, and 3σ confidence level contour constraints on the parameters of the spatially non-flat ϕCDM model with the RP potential from data on $H(z)$, QSO, and BAO peak length scale. (Upper and middle panels) The vertical green dashed line in the upper center panel, and the horizontal green dashed lines in the middle left and middle center panels separate spatially closed models (with $\Omega_{k0} < 0$) from spatially open models (with $\Omega_{k0} > 0$). The horizontal line with $\alpha = 0$ in the upper panels corresponds to the spatially non-flat ΛCDM model. (Lower panel) One-dimensional likelihoods for $\Omega_{m0}, \alpha, H_0, \Omega_{k0}$. The figure is adapted from [193].

Cao et al. [347] found constraints on the parameters of the spatially flat and non-flat ΛCDM, XCDM, and ϕCDM-RP models using H_{II} starburst galaxy apparent magnitude measurements [348,349], the compilation of 1598 X-ray and UV flux measurements of QSO 2015 data within the redshift range $0.036 \leq z \leq 5.1003$ and 2019 QSO data [350,351] alone and in conjunction with BAO peak length scale measurements [22,24–26,293], and Hubble parameter $H(z)$ data [21,28,30–34,257]. The constraints on the parameters of the spatially flat and spatially non-flat ϕCDM model with the RP potential obtained from the datasets mentioned above are shown in Figure 51. A combined analysis of all datasets leads to the relatively model-independent and restrictive estimates for the values of the matter density parameter at the present epoch Ω_{m0} and the Hubble constant H_0. Depending on the cosmological model, these estimates are consistent with a lower value of H_0 in the range of a 2.0σ to 3.4σ confidence level. Combined datasets favor the spatially flat ΛCDM, while at the same time they do not rule out dynamical dark energy models.

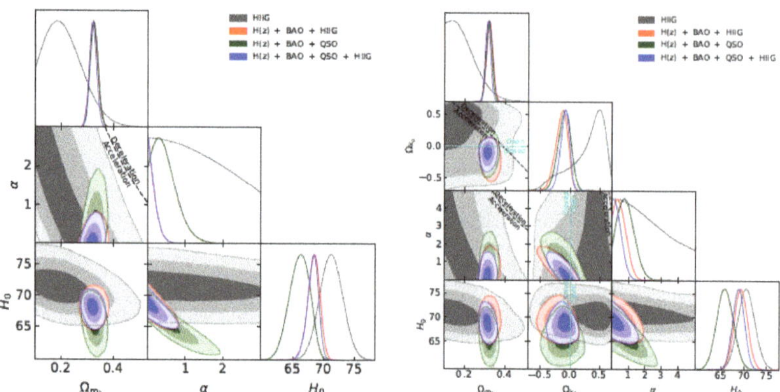

Figure 51. The 1σ, 2σ, and 3σ confidence level contour constraints on the parameters of the spatially flat ϕCDM model with the RP potential (left panel). The black dotted line splits the parameter space into accelerated and decelerated regions. The axis with $\alpha = 0$ denotes the spatially flat ΛCDM model. Constraints for the spatially non-flat ϕCDM model with the RP potential are depicted in the right panel. The figure is adapted from [347].

The compilation of 1598 X-ray and UV flux measurements of QSO 2015 data within the redshift range $0.036 \leq z \leq 5.1003$, and 2019 QSO data [350,351] alone and in conjunction with BAO peak length scale measurements [22,24–26,293], and Hubble parameter $H(z)$ data [21,28,30–34,257] was applied by Khadka and Ratra [195] to impose constraints on the parameters of the tilted spatially flat and untilted spatially non-flat ΛCDM, XCDM, and ϕCDM-RP quintessential inflation models. The obtained constraints for the untilted spatially non-flat ϕCDM-RP model from the combination of various datasets and extended QSO data only are presented in Figure 52. In most of the models, the QSO data favor the values of the matter density parameter at the present epoch $\Omega_{m0} \sim 0.5$–0.6, while, in a combined analysis of QSO data with $H(z)$ + BAO peak length scale dataset, the values of the matter density parameter at the present epoch Ω_{m0} are shifted slightly towards larger values. A combined set of data, QSO + BAO peak length scale + $H(z)$, is consistent with the standard spatially flat ΛCDM model, but favors slightly both the spatially closed hypersurfaces and the dynamical dark energy models.

Khadka and Ratra [196] obtained constraints on the parameters of the tilted spatially flat and untilted spatially non-flat ΛCDM, XCDM, and ϕCDM-RP quintessential inflation models from a compilation of 808 X-ray and UV flux measurements of QSOs (quasi-stellar objects) within the redshift range $0.061 \leq z \leq 6.28$ alone [350] and in combination with the BAO peak length scale measurements [22,24–26,293], and the Hubble parameter $H(z)$ data [21,28,30–34,257]. The 1σ, 2σ, and 3σ confidence level contours constraints on the parameters of the untilted spatially non-flat ϕCDM model with the RP potential from the combination of various datasets are presented in Figure 53. The constraints using only the QSO data are significantly weaker but consistent with those from the combination of the $H(z)$ + BAO peak length scale data. Combined analysis from QSO + $H(z)$ + BAO peak length scale data is consistent with the standard spatially flat ΛCDM model but slightly favors both closed spatial hypersurfaces and the untilted spatially non-flat ϕCDM model.

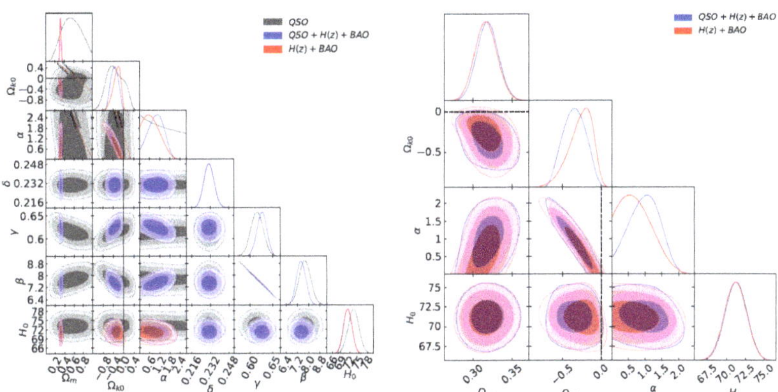

Figure 52. The 1σ, 2σ, and 3σ confidence level contour constraints on the parameters of the untilted spatially non-flat ϕCDM model with RP potential using the combination of datasets: QSO (gray line), $H(z)$ + BAO peak length scale (red line), and QSO + $H(z)$ + BAO peak length scale (blue line). (Left panel) Contours and one-dimensional likelihoods for all free parameters. The red dotted curved lines denote zero-acceleration lines. (Right panel) Plots for $\Omega_{\mathrm{m0}}, \Omega_{\mathrm{k0}}, \alpha, H_0$ cosmological parameters, without constraints from QSO data. These plots are for $H_0 = 73.24 \pm 1.74$ km s^{-1}Mpc^{-1} as a prior. The black dashed straight lines denote the flat hypersurface with $\Omega_{\mathrm{k0}} = 0$. The figure is adapted from [195].

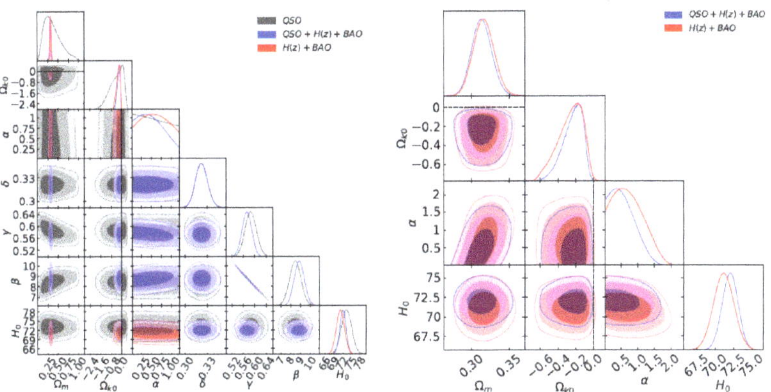

Figure 53. The 1σ, 2σ, and 3σ confidence level contours constraints on the parameters of the untilted spatially non-flat ϕCDM model with the RP potential using the combination of datasets: QSO (gray line), $H(z)$ + BAO peak length scale (red line), and QSO + $H(z)$ + BAO peak length scale (blue line). (Left panel) Contours and one-dimensional likelihoods for all free parameters. (Right panel) Plots for only $\Omega_{\mathrm{m0}}, \Omega_{\mathrm{k0}}, \alpha, H_0$ cosmological parameters, without constraints only from QSO data. These plots are for $H_0 = 73.24 \pm 1.74$) km s^{-1}Mpc^{-1} as a prior. Black dashed straight lines denote the spatially flat hypersurface with $\Omega_{\mathrm{k0}} = 0$. The figure is adapted from [196].

Cao et al. [352] found constraints on the parameters of the spatially flat and non-flat ΛCDM, XCDM, and ϕCDM-RP models using the higher-redshift GRB data [353,354], the starburst galaxy (H_{II}G) measurements [348,349,355], and the QSO angular size (QSO-AS) data [345,346]. Constraints from the combined analysis of cosmological parameters of the spatially flat and non-flat ϕCDM-RP model are presented in Figure 54. The constraints from the combined analysis of these datasets are consistent with the currently accelerating cosmological expansion, as well as with the constraints obtained from analysis of the Hubble parameter $H(z)$ data and the measurements of the BAO peak length scale. From the analysis of the $H(z)$ + BAO peak length scale + QSO-AS + H_{II}G + GRB dataset, the model-independent values of the matter density parameter at the present epoch $\Omega_{m0} = 0.313 \pm 0.013$ and the Hubble constant $H_0 = 69.3 \pm 1.2$ km s^{-1}Mpc^{-1} are obtained. This analysis favors the spatially flat ΛCDM model but also does not rule out dynamical dark energy models.

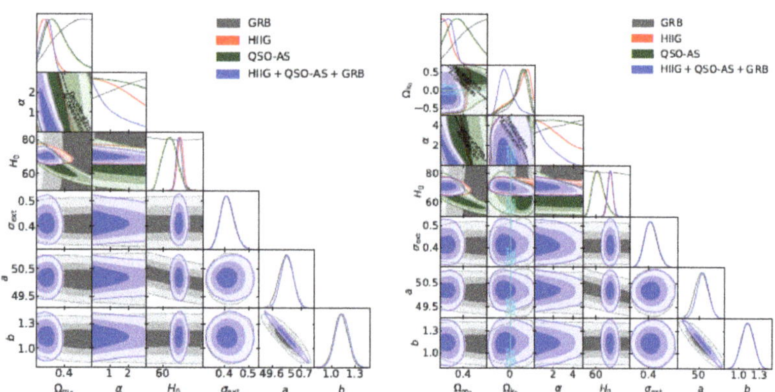

Figure 54. The 1σ, 2σ, and 3σ confidence level contours constraints on the parameters of the spatially flat (left panel) and spatially non-flat (right panel) ϕCDM models with the RP potential from various datasets. The black dotted line splits the parameter space into the regions of the currently decelerating and accelerating cosmological expansion. The axis with $\alpha = 0$ denotes the spatially flat ΛCDM model. The figure is adapted from [352].

Khadka and Ratra [197] determined constraints on the parameters of the spatially flat and non-flat ΛCDM, XCDM, and ϕCDM-RP models from compilation of the X-ray and UV flux measurements of 2038 QSOs which span the redshift range $0.009 \leq z \leq 7.5413$ [351,356]. The authors found that, for the full QSO dataset, parameters of the X-ray and UV luminosity $L_X - L_{UV}$ relation used to standardize these QSO data depends on the cosmological model, and therefore cannot be used to constrain the cosmological parameters in these models. Subsets of these QSOs, limited by redshift $z \leq 1.5 - 1.7$, obey the $L_X - L_{UV}$ relation in a way that is independent of the cosmological model and can therefore be used to constrain the cosmological parameters. Constraints from these smaller subsets of lower redshift QSO data are generally consistent, but much weaker than those inferred from the Hubble parameter $H(z)$ and the BAO peak length scale measurements (Figure 55).

Cao et al. [201] determined constraints on the parameters of the spatially flat and non-flat ΛCDM, XCDM, and ϕCDM-RP models by analyzing a total of 1383 measurements consisting of 1048 Pantheon SNe Ia apparent magnitude measurements from Scolnic et al. (2018) [212], and 20 binned SNe Ia apparent magnitude measurements of DES Collaboration [357,358], 120 QSO measurements [350,351,356], 153 H_{II}G data [348,349,355], 11 BAO peak length scale measurements [22,24–26,293], and 31 Hubble parameter $H(z)$ data [21,28,30–34,257]. Constraints on the parameters of the spatially non-flat ϕCDM model with the RP potential from that analysis of the data are shown in Figure 56. From the analysis of those datasets, the model-independent estimates of the Hubble constant, $H_0 = 68.8 \pm 1.8$ km s^{-1}Mpc^{-1}, as well as the matter density parameter at the present epoch, $\Omega_{m0} = 0.294 \pm 0.020$, are obtained. While the constraints favor dynamical dark energy and slightly spatially closed hypersurfaces, they do not preclude dark energy from being a cosmological constant and spatially flat hypersurfaces.

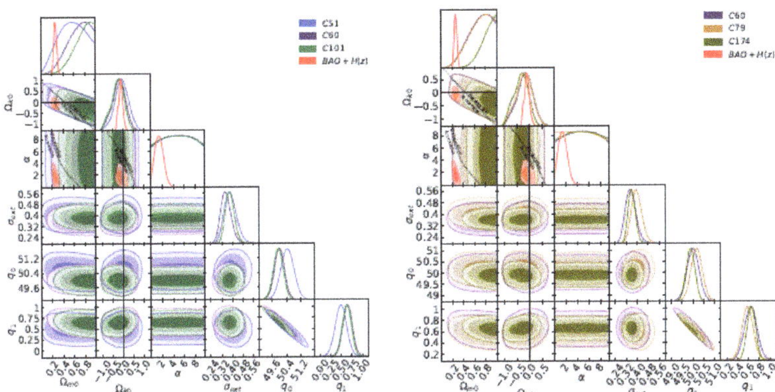

Figure 55. The 1σ, 2σ, and 3σ confidence level contour constraints on the parameters of the spatially flat (left panel) and non-flat (right panel) ϕCDM model with the RP potential, using QSO (blue) and the BAO peak length scale + $H(z)$ (red) datasets. The black dotted line in the $\alpha - \Omega_{m0}$ sub-panels is the line of zero acceleration, under which the accelerated cosmological expansion occurs. The axis with $\alpha = 0$ denotes the spatially flat ΛCDM model. The figure is adapted from [197].

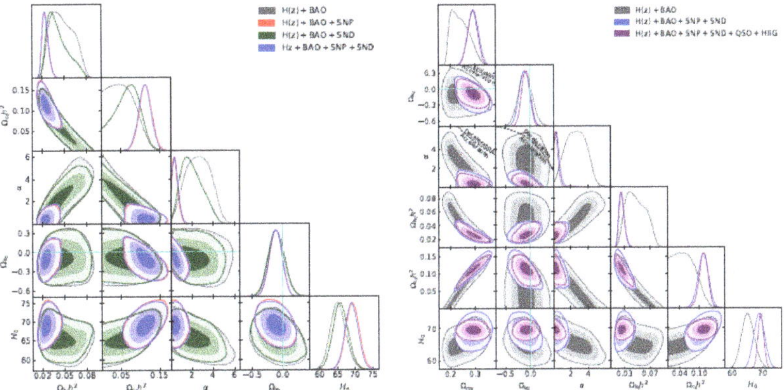

Figure 56. The 1σ, 2σ, and 3σ confidence level contour constraints on the parameters of the spatially non-flat ϕCDM model with the RP potential. The zero-acceleration line splits the parameter space into regions of the currently decelerating and accelerating cosmological expansion. The cyan dash–dot lines show the spatially flat ϕCDM model; regions with spatially closed geometry are located either below or to the left. The axis with $\alpha = 0$ denotes the spatially flat ΛCDM model. The figure is adapted from [201].

Khadka and Ratra [198] found constraints on the parameters of the spatially flat and non-flat ΛCDM, XCDM, and ϕCDM-RP models from 78 reverberation-measured Mg_{II} time-lag QSOs within the redshift range $0.0033 \leq z \leq 1.89$ [359,360]. The authors applied the radius–luminosity or $R - L$ relation to standardized 78 Mg_{II} QSOs data. In each model, the authors simultaneously determined the $R - L$ relation and parameters in these models, thus avoiding the problem of circularity. It was found that values of the $R - L$ relation parameter are independent of the model used in the analysis, which makes it possible to establish that current Mg_{II} QSOs data are standardizable candles. Constraints derived from the QSO data only are significantly weaker than those derived from the combined set of the BAO peak length scale and the Hubble parameter $H(z)$ measurements, but are consistent with both of them. The constraints obtained from the Mg_{II} QSOs data in conjunction with the BAO peak length scale + $H(z)$ measurements agree with the spatially flat ΛCDM model as well as with spatially non-flat dynamical dark energy models.

Khadka and Ratra [199] found that the recent compilation of the QSO X-ray and UV flux measurements [356] includes the QSO data that appear not to be standardized via the X-ray luminosity and the UV luminosity $L_X - L_{UV}$ relation parameters that are dependent on both the cosmological model and the redshift, so it should not be used to constrain the model parameters. These data include a compilation of seven different subsamples. The authors analyzed these subgroups and some combinations of subgroups to define which QSO subgroups are responsible for questions specified in the paper by Khadka and Ratra [197]. They considered that the largest of the seven subsamples in this compilation, SDSS-4XMM QSOs, which contributes about two-thirds of all QSOs, has the $L_X - L_{UV}$ ratios that depend on both the accepted cosmological model and the redshift, and thus is the source of a similar problem found earlier when collecting the full QSO data.

The second and third largest subsamples, SDSS-Chandra and XXL QSOs, which together account for about 30% of total QSO data, appear to be standardized. Constraints on the cosmological parameters from these subsamples are weak and consistent with the standard spatially flat ΛCDM model or with the constraints from the better-established cosmological probes. Constraints on the cosmological parameters of the spatially flat and spatially non-flat ϕCDM models with the RP potential, using SDSS-Chandra and XXL QSO data as well as $H(z)$ data and BAO peak length scale data, are presented in Figure 57.

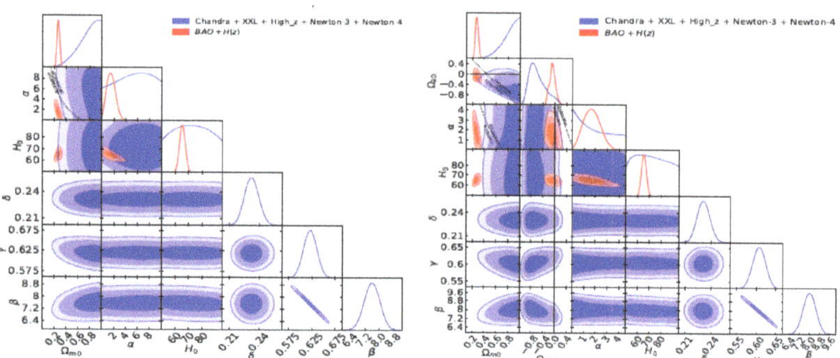

Figure 57. The 1σ, 2σ, and 3σ confidence level contour constraints on the parameters of the spatially flat (left panel) and spatially non-flat (right panel) ϕCDM models with the RP potential, using Chandra + XXL + High-z + Newton-3 + Newton-4 (blue) and BAO peak length scale + $H(z)$ (red) datasets. In all plots, black dotted lines are zero-acceleration lines, which split the parameter space into the regions of current acceleration and deceleration. Black dashed line corresponds to $\Omega_{k0} = 0$. The axis with $\alpha = 0$ denotes the spatially flat ΛCDM model. The figure is adapted from [199].

Khadka et al. [200] used 118 $H\beta$ QSO measurements [361] within the redshift range $0.0023 \leq z \leq 0.89$ to simultaneously constrain cosmological model parameters and QSO two-parameter radius–luminosity $R - L$ relation parameters of the spatially flat and non-flat ΛCDM, XCDM, and ϕCDM-RP models. The authors found that the $R - L$ relation parameters for $H\beta$ QSO data are independent in models under investigation, therefore QSO data seem to be standardizable through $R - L$ relation parameters. The constraints derived using $H\beta$ QSO data are weak, slightly favoring the currently accelerating cosmological expansion, and are generally in the 2σ tension with the constraints derived from analysis of the measurements of the BAO peak length scale and the Hubble parameter $H(z)$. Constraints on the cosmological parameters of the spatially flat and non-flat ϕCDM-RP model, from the $H\beta$ QSO measurements, and the $H(z)$ and BAO peak length scale data are presented in Figure 58.

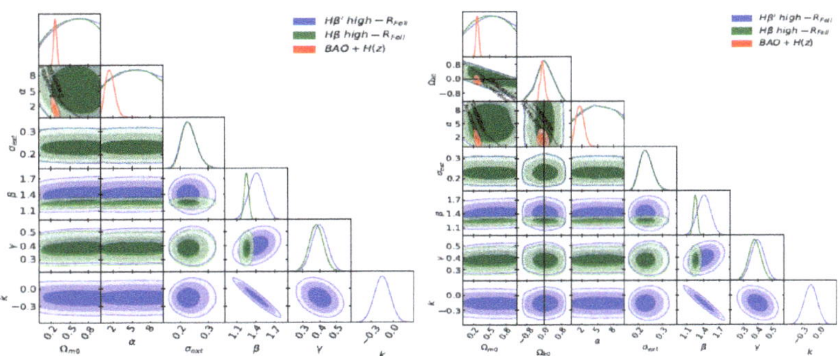

Figure 58. The 1σ, 2σ, and 3σ confidence level contour constraints on the parameters of the spatially flat (left panel) and spatially non-flat (right panel) ϕCDM model with the RP potential from 3-parameter $H\beta'$ high-$\Re_{FeII}$ (blue), 2-parameter $H\beta'$ high-$\Re_{FeII}$ (green), and $H(z)$ + BAO peak length scale (red) measurements. Black dotted lines correspond to zero-acceleration lines. Black dashed lines represent $\Omega_{k0} = 0$. The figure is adapted from [200].

Khadka et al. [362] determined constraints on the parameters of the spatially flat and spatially non-flat ΛCDM, XCDM, and ϕCDM-RP models using the observations of 66 reverberation-measured Mg_{II} QSOs within the redshift range $0.36 \leq z \leq 1.686$. Constraints on the cosmological parameters of the spatially flat and spatially non-flat ϕCDM models with the RP potential from various QSO datasets are shown in Figure 59. The authors also studied the two- and three-parameter radius–luminosity $R - L$ relations [363,364] for Mg_{II} QSO sources, and found that these relations do not depend on the assumed cosmological model; therefore, they can be used to standardize QSO data. The authors found for the two-parameter $R - L$ relation that the data subsets with low-$\Re_{FeII}$ and high-$\Re_{FeII}$ obey the same $R - L$ relation within the error bars. Extending the two-parameter $R - L$ relation to three parameters does not lead to the expected decrease in the intrinsic variance in the $R - L$ relation. None of the three-parameter $R - L$ relations provides a significantly better measurement fit than the two-parameter $R - L$ relation. The results obtained differ significantly from those found by Khadka et al. [200] from analysis of reverberation-measured $H\beta$ QSOs.

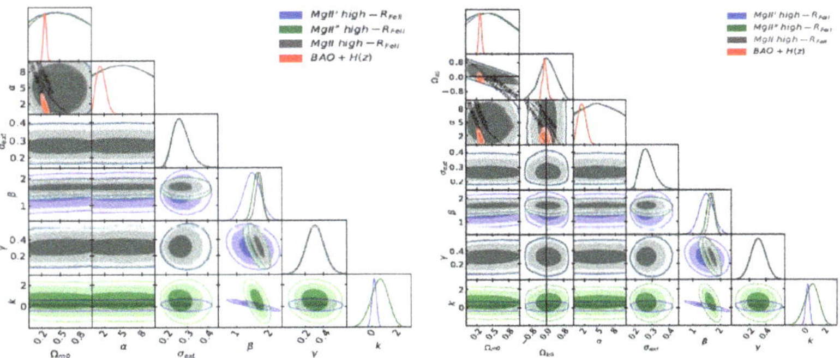

Figure 59. The 1σ, 2σ, and 3σ confidence level contour constraints on the parameters of the spatially flat (left panel) and spatially non-flat (right panel) ϕCDM models with the RP potential from the measurements of $Mg_{II'}$ high-$\Re_{FeII}$ (blue), $Mg_{II''}$ high-$\Re_{FeII}$ (green), Mg_{II} high-$\Re_{FeII}$ (gray), and BAO peak length scale + $H(z)$. Black dotted lines correspond to zero-acceleration lines. Black dashed lines represent $\Omega_{k0} = 0$. The figure is adapted from [362].

Cao et al. [203] determined constraints on the parameters of the spatially flat and non-flat ΛCDM, XCDM, ϕCDM-RP models, as well as on the QSO radius–luminosity $R - L$ relation parameters from the 38 C_{IV} QSO reverberation-measured data in the redshift range $0.001064 \leq z \leq 3.368$. An improved method is used that takes into account more accurately the asymmetric error bars for the time-delay measurements. The authors found that the parameters of the $R - L$ relation do not depend on the cosmological models considered and, therefore, the $R - L$ relation can be used to standardize the C_{IV} QSO data. Mutually consistent constraints on the cosmological parameters from C_{IV}, Mg_{II}, and $H(z)$ + BAO peak length scale data allow conducting the analysis from C_{IV} + Mg_{II} dataset as well as from the $H(z)$ + BAO peak length scale + C_{IV} + Mg_{II} datasets. Although the C_{IV} + Mg_{II} cosmological constraints are weak, they slightly (at a $\sim 0.1\sigma$ confidence level) change the constraints from the $H(z)$ + BAO peak length scale + C_{IV} + Mg_{II} datasets. The constraints on the cosmological parameters of the spatially flat and non-flat ϕCDM-RP from various QSO datasets are shown in Figure 60.

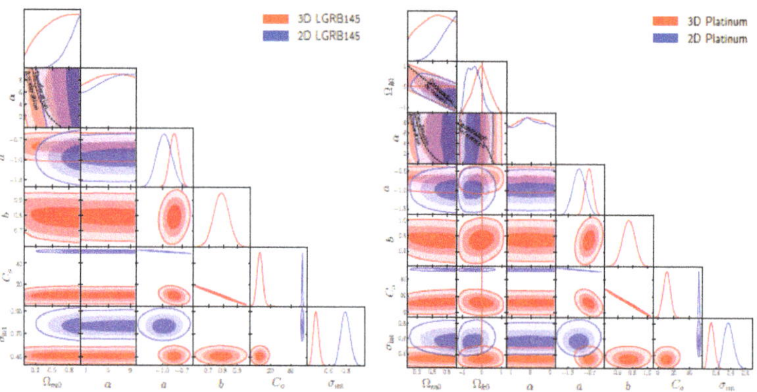

Figure 60. The 1σ, 2σ, and 3σ confidence level contour constraints on the parameters of the spatially flat (left panel) and spatially non-flat (right panel) ϕCDM models with the RP potential from various combinations of datasets. The axis with $\alpha = 0$ denotes the spatially flat ΛCDM model. The black dash–dotted lines denote spatially flat hypersurfaces $\Omega_{k0} = 0$; closed spatial hypersurfaces are located either below or to the left. The black dotted lines correspond to the lines of zero acceleration and split the parameter space into currently accelerating (bottom left) and decelerating (top right) regions. The figure is adapted from [203].

3.7. Gamma Ray Burst Distance Data

Samushia and Ratra [365] derived constraints on the parameters of the spatially flat ΛCDM, XCDM, and ϕCDM-RP models using the observational datasets of SNe Ia Union apparent magnitude data [366] and BAO peak length scale data [17], and measurements of gamma-ray burst (GRB) distances [367,368]. The authors applied two methods for analyzing the GRB data-fitting luminosity relation of GRB, Wang's method [368] and Schaefer's method [367]. The constraints on the cosmological parameters of the ϕCDM model from analysis of the SNe Ia Union apparent magnitude data and the BAO peak length scale measurements, with and without the GRB measurements, are presented in Figure 61. The constraints from the GRB data obtained by two different methods disagree with each other at a more than 2σ confidence level. The cosmological parameters of the ϕCDM model could not be tightly constrained only by the current GRB data.

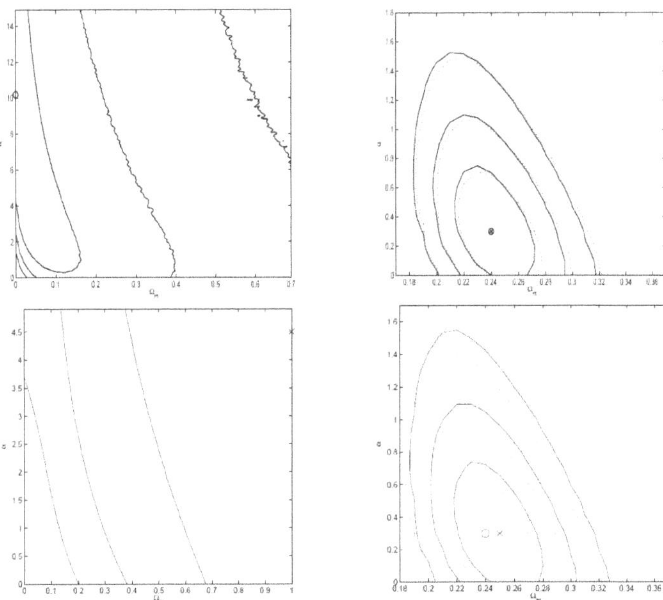

Figure 61. The 1σ, 2σ, and 3σ confidence level contours constraints on parameters of the ϕCDM model with the RP potential. The horizontal axis with $\alpha = 0$ corresponds to the standard spatially flat ΛCDM model. (Left upper panel) Contours are obtained by Wang's (2008) [368] method. The circle indicates the best-fit parameter values $\Omega_m = 0$, $\alpha = 10.2$ with $\chi^2 = 1.39$ for 4 degrees of freedom. (Right upper panel) Contours are derived using the GRB data by Wang's (2008) [368] method, the SNe Ia Union apparent magnitude data, and the BAO peak length scale measurements, while dotted lines (here the cross denotes the best-fit point) are derived using only the SNe Ia apparent magnitude data and the BAO peak length scale measurements. The best-fit parameters in both cases are $\Omega_m = 0.24$, $\alpha = 0.3$ with $\chi^2 = 326$ for 313 degrees of freedom (solid lines) and $\chi^2 = 321$ for 307 degrees of freedom. (Left lower panel) Contours are obtained using GRB data by Schaefer's (2007) method (here the cross indicates the best-fit parameter values): $\Omega_m = 1$, $\alpha = 4.5$ with $\chi^2 = 77.8$ for 67 degrees of freedom. (Right lower panel) Contours are obtained using Schaefer's (2007) [367] method, the SNe Ia Union apparent magnitude data, and the BAO peak length scale measurements, while dotted lines are obtained using the SNe Ia apparent magnitude data and the BAO peak length scale measurements only. Solid lines (circle denotes best-fit point) are derived using GRB data, here the cross denotes the best-fit point. The best-fit matter density parameters are $\Omega_m = 0.24$, $\alpha = 0.30$ with $\chi^2 = 401$ for 376 degrees of freedom (solid lines), and $\Omega_m = 0.25$, $\alpha = 0.3$ with $\chi^2 = 321$ for 307 degrees of freedom (dotted lines). The figure is adapted from [365].

Khadka and Ratra [369] performed an analysis of the constraints on the parameters of the spatially flat and non-flat ΛCDM, XCDM, and ϕCDM-RP models from measurements of the peak photon energy and bolometric fluence of 119 GRBs extending over the redshift range of $0.34 \leq z \leq 8.2$ [353,354], and Amati relation parameters [370], BAO peak length scale measurements [22,24–26,293], and Hubble parameter $H(z)$ data [21,28,30–34,257]. Resulting constraints on the parameters of the spatially flat and spatially non-flat ϕCDM models with the RP potential are presented in Figure 62.

The Amati relation between the peak photon energy of the GRB in the cosmological rest frame, E_p, and E_{iso} is given as

$$\log(E_{\mathrm{iso}}) = a + b \log(E_p), \tag{50}$$

where a and b are free parameters defined from data, representing points of intersection and slope in the Amati relation, respectively. E_p and E_{iso} are specified as

$$E_{iso} = \frac{4\pi D_L^2(z, p) S_{bolo}}{(1+z)}, \quad E_p = E_{p,obs}(1+z),\tag{51}$$

where $D_L(z, p)$ is the luminosity distance, p is a cosmological parameter, S_{bolo} is the measured bolometric fluence, and $E_{p,obs}$ is the measured peak energy of the GRB.

The resulting Amati relation parameters are almost identical in all considered cosmological models, which confirms the use of the Amati relation parameters to standardize these GRB data. The constraints on the cosmological parameters of the models under consideration from the GRB data are consistent with the constraints obtained from the analysis of the BAO peak length scale and the measurements of the Hubble parameter $H(z)$, but are less restrictive.

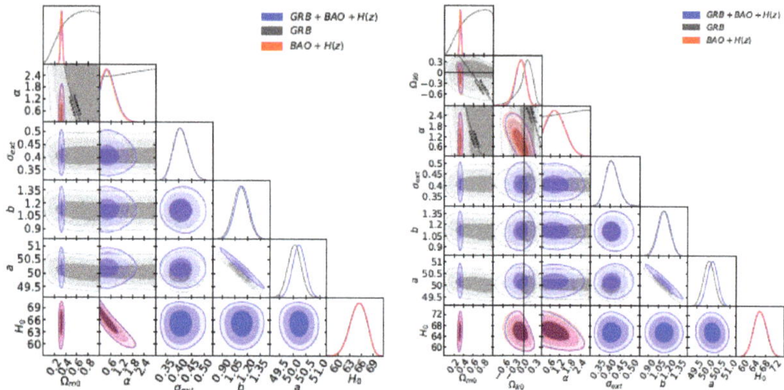

Figure 62. The 1σ, 2σ, and 3σ confidence level contour constraints on the parameters of the spatially flat (left panel) and spatially non-flat (right panel) ϕCDM models with the RP potential, using the combination of datasets: GRB (gray line), $H(z)$ + BAO peak length scale (red line), and GRB + $H(z)$ + BAO peak length scale (blue line). The black dotted line splits the parameter space into accelerating and decelerating regions. The axis with $\alpha = 0$ denotes the spatially flat ΛCDM model. The figure is adapted from [369].

Khadka et al. [371] analyzed constraints on the parameters of the spatially flat and non-flat ΛCDM, XCDM, and ϕCDM-RP models from the GRB data. The authors considered eight different GRB datasets to test whether the current GRB measurements, which probe a largely unexplored range of cosmological redshifts, can be used to reliably constrain the parameters of these models. The authors applied the MCMC analysis implemented in Monte Python to find the most appropriate correlations and cosmological parameters for the eight GRB samples, with and without the BAO peak length scale and the $H(z)$ data.

They applied three Amati correlation samples [370] and five Combo correlation samples [372] to obtain correlations and constraints on the model parameters. Constraints on the parameters of the spatially non-flat ϕCDM-RP model, using various datasets of GRB, as well as the BAO peak length scale + $H(z)$ data. The authors found that the cosmological constraints, determined from the A118 sample consisting of 118 bursts, agree but are much weaker than those following from the BAO peak length scale and the $H(z)$ data. These constraints are consistent with the spatially flat ΛCDM as well as with the spatially non-flat dynamical dark energy models.

Cao et al. [202] applied the spatially flat and non-flat ΛCDM, XCDM, and ϕCDM-RP models in the analysis of the three (ML, MS, and GL) $(L_0 - t_b)$ Dainotti-correlated sets of GRB measurements collected by Wang et al. [373] and Hu et al. [374] that together explore the redshift range $0.35 \leq z \leq 5.91$. The authors found that each dataset, as well as the combinations of MS + GL, ML + GL, and ML + MS, obey the cosmological model-independent Dainotti correlations [375,376]) and therefore are standardized. The luminosity of the plateau phase for GRBs that obey the Dainotti correlation is defined as

$$L_0 = \frac{4\pi D_L^2 F_0}{(1+z)}^{1-\beta'}, \tag{52}$$

where F_0 is the GRB X-ray flux, β' is the spectral index in the plateau phase, and D_L is the luminosity distance.

The authors applied these GRB data in combination with the best currently available Amati-correlated GRB data of Amati [370] that explore the redshift range $0.3399 \leq z \leq 8.2$ to constrain the cosmological model parameters. As a result, constraints are weak, providing lower bounds on the matter density parameter at the present epoch Ω_{m0}, moderately favoring the non-zero spatial curvature, and largely consistent with both the currently accelerated cosmological expansion and with constraints determined on the basis of more reliable data. Constraints of cosmological parameters of the spatially flat and non-flat ϕCDM-RP model, using the Dainotti-correlated sets of the GRB measurements as well as the $H(z)$ and BAO peak length scale data are presented in Figure 63.

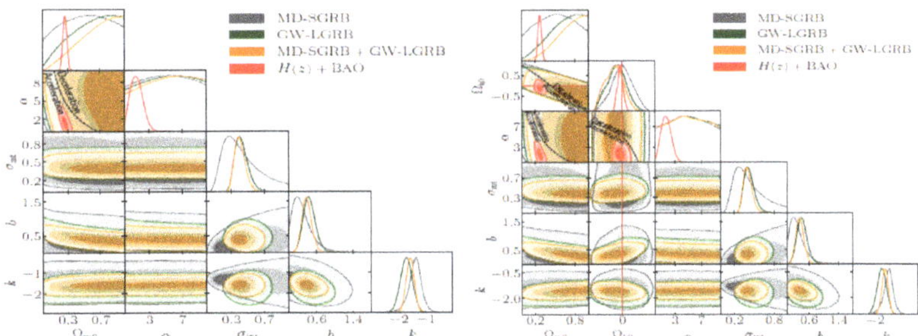

Figure 63. The 1σ, 2σ, and 3σ confidence level contour constraints on the parameters of the spatially flat (left panel) and spatially non-flat (right panel) ϕCDM models with the RP potential from MD-SGRB (gray), GWLGRB (green), MD-SGRB + GW-LGRB (orange), and $H(z)$ + BAO peak length scale (red) data. Black dashed lines denote zero-acceleration lines, which split the parameter space into two regions of current acceleration and deceleration. Dash–dotted crimson lines correspond to spatially flat hypersurfaces with spatially closed hypersurfaces either below or to the left. The magenta lines correspond to the ϕCDM model; the closed spatial geometry are either below or to the left. The axis with $\alpha = 0$ denotes the spatially flat ΛCDM model. The figure is adapted from [202].

Cao et al. [204] used the spatially flat and non-flat ΛCDM, XCDM, and ϕCDM-RP models to analyze the compilation of data from 50 Platinum GRBs within the redshift range $0.553 \leq z \leq 5.0$. The authors found that these data obey the three-parameter fundamental plane or Dainotti correlation, independent of the cosmological model, and therefore they are amenable to standardization and can be used to constrain cosmological parameters. To improve the accuracy of the constraints for the GRB data only, the authors excluded ordinary GRB data from the larger Amati-correlated A118 dataset of 118 GRBs and analyzed the remaining 101 Amati-correlated GRBs with 50 Platinum GRB datasets. This joint dataset of 151 GRBs is being investigated within the little-studied redshift range $z \in (2.3, 8.2)$. Due to the consistency of cosmological constraints from the platinum GRB data with the $H(z)$ + BAO peak length scale dataset, the authors combined platinum GRB and the $H(z)$ + BAO peak length scale data to carry out the analysis and found small changes in the cosmological parameter constraints compared to the constraints from the $H(z)$ + BAO peak length scale data. The resulting constraints from the GRBs only are more stringent than those from the $H(z)$ + BAO peak length scale dataset but are less precise.

Cao et al. [205] proposed the constraints on the parameters of the spatially flat and non-flat ΛCDM, XCDM, and ϕCDM-RP models, using the extended set of the GRB data including the 50 platinum GRBs within the redshift range $0.553 \leq z \leq 5$ by Dainotti et al. [377], the LGRB95 data that contains 95 long GRBs measurements within the redshift range $0.297 \leq z \leq 9.4$ by Dainotti et al. [377]. The compilation of the 145 GRB datasets was also used. The constraints on the cosmological parameters of the spatially flat and spatially non-flat ϕCDM model with the RP potential, using various GRB datasets, are shown in Figure 64. The authors also examined which of two correlations, the two-dimensional Dainotti correlation [378] or the three-dimensional Dainotti correlation [379,380], fits better the GRB datasets. Based on the results of AIC, BIC, and Deviation Information Criterion (DIC) analysis, the authors found that the three-dimensional Dainotti correlation is much preferable to the two-dimensional one for the GRB datasets.

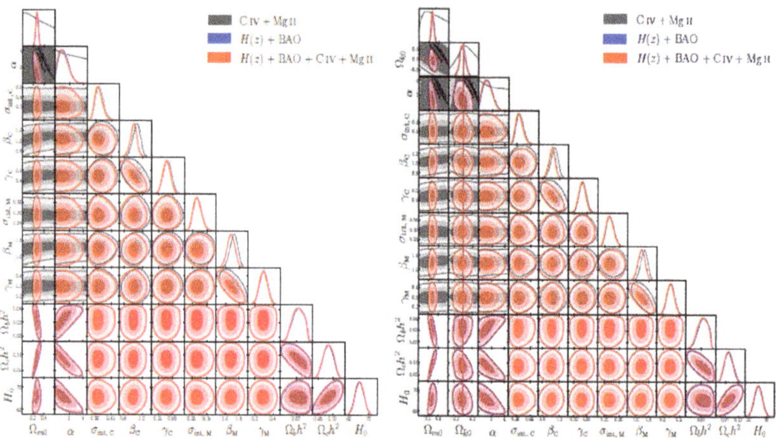

Figure 64. The 1σ, 2σ, and 3σ confidence level contours on parameters of the spatially flat (left panel) and spatially non-flat (right panel) ϕCDM models with the RP potential, using various combinations of GRB datasets. The axis with $\alpha = 0$ denotes the spatially flat ΛCDM model. The black dotted lines correspond to lines of zero acceleration and split the parameter space into currently accelerating (bottom left) and currently decelerating (top right) regions. The crimson dash–dot lines denote spatially flat hypersurfaces $\Omega_{k0} = 0$; closed spatial hypersurfaces are located either below or to the left. The figure is adapted from [205].

3.8. Starburst Galaxy Data

Mania and Ratra [381] analyzed constraints on the parameters of the ϕCDM-RP, the XCDM, and the ΛCDM models from the H_{II} starburst galaxy apparent magnitude versus redshift data of Siegel et al. [382]. The authors followed the Percival et al. [259] procedure to obtain these constraints. The results are demonstrated in Figure 65. These constraints are largely consistent but not as restrictive as those derived from the measurements of the BAO peak length scale, the SNe Ia apparent magnitude, and the CMB temperature anisotropy.

Cao et al. [383] derived constraints on the parameters of the spatially flat and non-flat ΛCDM, XCDM, and **ϕCDM-RP models** from the compilation of the H_{II} starburst galaxy (H_{II}G) data of González-Morán et al. [355] and the H_{II}G data of González-Morán et al. [384]. The authors tested the model independence of the QSO angular size measurements. They found that the new compilation of 2019 H_{II}G data provides tighter constraints and favors lower values of the cosmological parameters than those from the 2019 H_{II}G data. The use of the QSO measurements gives model-independent constraints on the characteristic linear size l_m of QSO within a sample. Analysis of the data on the $H(z)$, BAO peak length scale, the SNe Ia apparent magnitude-Pantheon, the SNe Ia apparent magnitude-DES, QSO, and the latest compilation of the H_{II}G data provides almost model-independent estimates of the Hubble constant, the matter density parameter at the present epoch, and the characteristic linear size, respectively, as $H_0 = 69.7 \pm 1.2$ km s^{-1}Mpc^{-1}, $\Omega_{\mathrm{m0}} = 0.295 \pm 0.021$, and $l_m = 10.93 \pm 0.25$ pc.

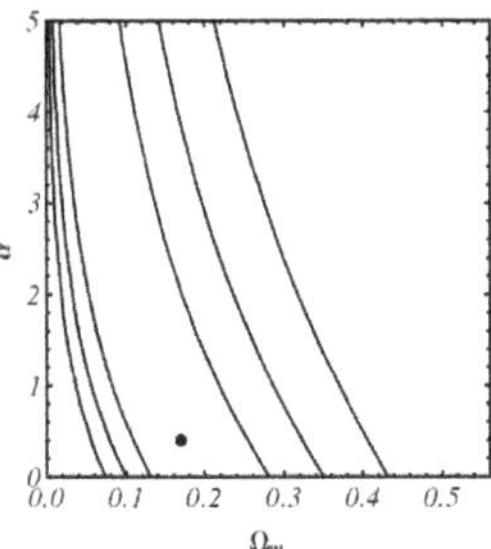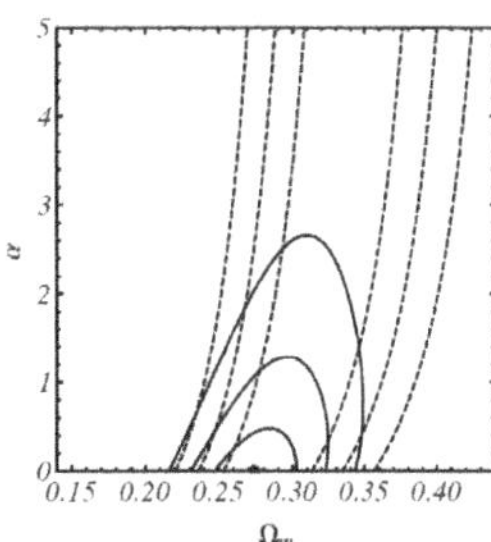

Figure 65. The 1σ, 2σ, and 3σ confidence level contour constraints on the parameters of the spatially flat ϕCDM model with the RP potential. The horizontal axis with $\alpha = 0$ corresponds to the standard spatially flat ΛCDM model. (Left panel) Contours obtained from H_{II} galaxy data. The best-fit point with $\chi^2_{\mathrm{min}} = 53.3$ is indicated by the solid black circle at $\Omega_{\mathrm{m0}} = 0.17$ and $\alpha = 0.39$. (Right panel) Contours obtained from joint H_{II} galaxy and BAO peak length scale data (solid lines) and BAO peak length scale data only (dashed lines). The best-fit point with $-2\log(L_{\mathrm{max}}) = 55.6$ is indicated by the solid black circle at $\Omega_{\mathrm{m0}} = 0.27$ and $\alpha = 0$. The figure is adapted from [381].

Cao and Ratra [206] performed analysis of constraints on the parameters of the spatially flat and non-flat ΛCDM, XCDM, and ϕCDM-RP models from joint datasets consisting of data on the updated 32 $H(z)$ Hubble parameter, 12 BAO peak length scale, 1048 Pantheon SNe Ia apparent magnitudes, 20 binned DES-3yr SNe Ia apparent magnitudes, 120 QSO-AS and 78 Mg_{II} reverberation-measured QSO, 181 H_{II} starburst galaxy, and 50 Platinum Amati-correlated GRB. As a result, the authors found that constraints from each dataset are mutually consistent. There is a slight difference between constraints determined from the QSO-AS + H_{II}G + Mg_{II} QSO + A118 dataset and those from QSO-AS + H_{II}G + Mg_{II} QSO + Platinum + A101 dataset, so the authors considered only the cosmological constraints from the joint dataset $H(z)$ + BAO peak length scale + SNe Ia apparent magnitudes + QSO-AS + H_{II}G + Mg_{II} QSO + A118 (HzBSNQHMA). The model-independent value of the Hubble constant, $H_0 = 69.7 \pm 1.2$ km s^{-1}Mpc^{-1}, and the matter density parameter at the present epoch, $\Omega_{\mathrm{m0}} = 0.295 \pm 0.017$, were obtained by using the HzBSNQHMA dataset. The obtained value of the constraint for H_0 lies in the middle of the spatially flat ΛCDM model result of Planck Collaboration 2018 of Aghanim et al. [13] and the local expansion rate $H(z)$ result of Riess et al. [113], a bit closer to the former. Based on the DIC analysis, the spatially flat ΛCDM model is the most preferable, but both dynamic dark energy models and space curvature are not ruled out.

3.9. X-ray Gas Mass Fraction of Cluster Data

Using Chandra measurements of X-ray gas mass fraction of 26 rich clusters obtained by Allen et al. [385], Chen and Ratra [336] constrained the parameters of the ϕCDM-RP, ΛCDM, and the XCDM models. Resulting constraints are consistent with those derived from other cosmological tests but favor the spatially flat ΛCDM model more, Figure 14. Constraints on the parameters of the ϕCDM model are tighter than those derived from the SNe Ia apparent magnitude data of Podariu and Ratra [241], Waga and Frieman [386], redshift–angular size data of Chen and Ratra [387], Podariu et al. [388], and gravitational lensing statistics of Chae et al. [389], Figure 66 (left panel).

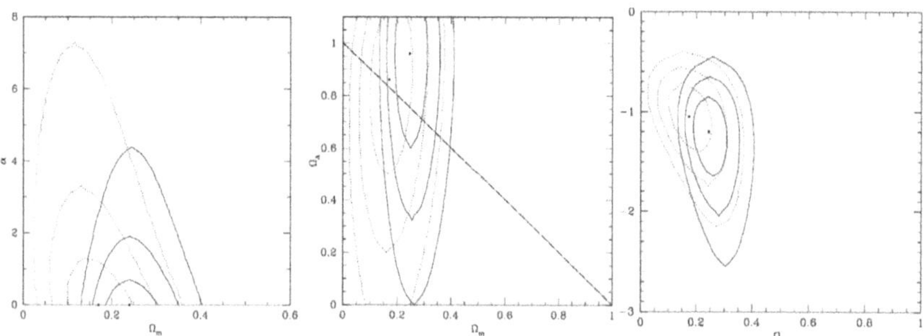

Figure 66. The 1σ, 2σ, and 3σ confidence level contour constraints on parameters. (Left panel) For the spatially flat ϕCDM model with the RP potential and non-relativistic CDM. Continuous lines are obtained for $h = 0.72 \pm 0.08$ and $\Omega_b h^2 = 0.0214 \pm 0.002$ while dotted lines match $h = 0.68 \pm 0.04$ and $\Omega_b h^2 = 0.014 \pm 0.004$. (Middle panel) For the spatially flat ΛCDM model. Continuous lines are obtained for $h = 0.72 \pm 0.08$ and $\Omega_b h^2 = 0.0214 \pm 0.002$ while dotted lines obtained for $h = 0.68 \pm 0.04$ and $\Omega_b h^2 = 0.014 \pm 0.004$. The diagonal dash–dotted line delimits spatially flat models. (Right panel) For the XCDM model. Continuous lines are obtained for $h = 0.72 \pm 0.08$ and $\Omega_b h^2 = 0.0214 \pm 0.002$ while dotted lines are derived for $h = 0.68 \pm 0.04$ and $\Omega_b h^2 = 0.014 \pm 0.004$. In all pictures, two dots indicate the place of maximum probability. The figure is adapted from [336].

Wilson et al. [337] used the R04 gold SNe Ia apparent magnitude versus the redshift data of Riess et al. [211] and X-ray gas mass fraction of cluster data from Allen et al. [385] to constrain the ϕCDM-RP model; the results are given in Figure 67. According to these results, the standard spatially flat ΛCDM model is preferable, but the ϕCDM model is not ruled out either. The contours obtained from joint R04 gold SNe Ia apparent magnitude data and galaxy cluster gas mass fraction data are tighter constrained than those obtained by Podariu and Ratra [241] from earlier SNe Ia apparent magnitude versus redshift data.

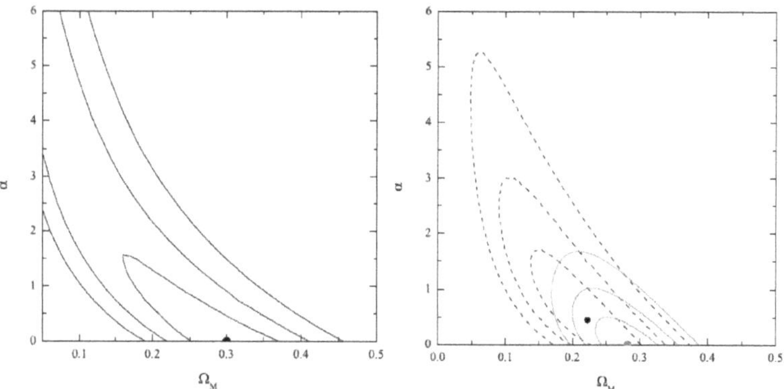

Figure 67. The 1σ, 2σ, and 3σ confidence level contour constraints on the parameters of the ϕCDM model with the RP potential. (Left panel) For the R04 gold SNe Ia apparent magnitude sample. The dot indicates the maximum likelihood for which $\Omega_{m0} = 0.30$ and $\alpha = 0$. (Right panel) For the joint R04 gold SNe Ia apparent magnitude sample and galaxy cluster gas mass fraction data. The solid gray lines are computed for $h = 0.72 \pm 0.08$ and $\Omega_b h^2 = 0.0214 \pm 0.002$ with maximum likelihood at $\Omega_{m0} = 0.28$ and $\alpha = 0$. The black dotted lines are computed for $h = 0.68 \pm 0.04$ and $\Omega_b h^2 = 0.014 \pm 0.004$, with maximum likelihood at $\Omega_{m0} = 0.22$ and $\alpha = 0.45$. The figure is derived from constraints on the parameters of the spatially flat ΛCDM, the XCDM, and the ϕCDM models with the RP potential adapted from [337].

Constraints on the model parameters w_0 and w_a of the wCDM model using the X-ray temperature data on massive galaxy clusters within the redshift range $0.05 \leq z \leq 0.83$ with massive galaxy clusters ($M_{\text{cluster}} > 8 \times 10^{14}\, h^{-1} M_\odot$ within a comoving radius of $R_{\text{cluster}} = 1.5\, h^{-1}\text{Mpc}$), were determined by Campanelli et al. [390]. The results are presented in Figure 68. Current data on massive clusters weakly constrain w_0 and w_a parameters around the $(w_0, w_a) = (-1, 0)$ values corresponding to the ΛCDM model. In the analysis including data from the galaxy cluster number count, Hubble parameter $H(z)$, CMB temperature anisotropy, BAO peak length scale, and the SNe Ia apparent magnitude, the values of $w_0 = -1.14^{+0.14}_{-0.16}$ and $w_a = 0.85^{+0.42}_{0.60}$ were obtained at a 1σ confidence level.

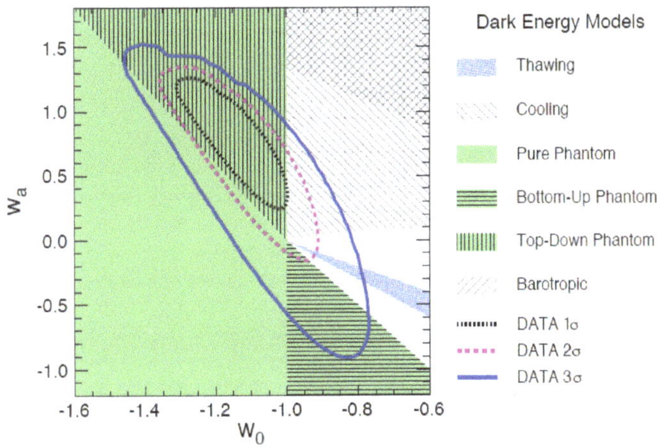

Figure 68. The 1σ, 2σ, and 3σ confidence level contours in the w_0-w_a plane from the data analysis of the galaxy cluster number count, Hubble parameter $H(z)$, CMB temperature anisotropy, BAO peak length scale, and the SNe Ia apparent magnitude. The shaded areas represent various types of dynamical dark energy models. The figure is adapted from [390].

Chen and Ratra [391] applied angular size versus redshift measurements for galaxy clusters from Bonamente et al. [392] to constraint parameters of the ϕCDM-RP, the XCDM, and the ΛCDM models. X-ray observations of the intracluster medium in combination with radio observations of the Sunyaev–Zel'dovich effect of galaxy clusters make it possible to estimate the distance from the angular diameter d_A of galaxy clusters. The authors applied the 38 angular diameter distance d_A from Bonamente et al. (2006) [392] to constrain the cosmological parameters of the models presented above. The results are demonstrated in Figure 69. The analysis of the angular size measurements along with the more restrictive BAO peak length scale data and the SNe Ia apparent magnitude measurements favors the spatially flat ΛCDM model but does not exclude the ϕCDM model.

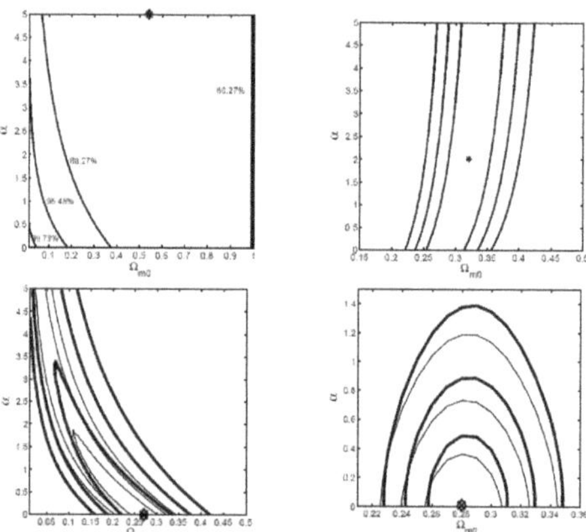

Figure 69. The 1σ, 2σ, and 3σ confidence level contour constraints on the parameters of the ϕCDM model with the RP potential. The horizontal axis with $\alpha = 0$ corresponds to the standard spatially flat ΛCDM model. (Left upper panel) Contours obtained from angular diameter distance d_A data. The star denotes the best-fit point $(\Omega_{m0}, \alpha) = (0.54, 5)$, $\chi^2_{\min} = 37.3$. (Right upper panel) Contours obtained using BAO peak length scale data. The star marks the best-fit point $(\Omega_{m0}, \alpha) = (0.32, 2.01)$, $\chi^2_{\min} = 0.169$. (Left lower panel) Contours obtained from SNe Ia apparent magnitude data. Thin solid lines (best-fit pair at $(\Omega_{m0}, \alpha) = (0.27, 0.00)$, $\chi^2_{\min} = 543$, marked by cross "x") exclude systematic errors, while thick solid lines (best-fit pair at $(\Omega_{m0}, \alpha) = (0.27, 0.00)$, $\chi^2_{\min} = 531$, marked by diamond "$\diamond$") calculated for systematics. (Right lower panel) Thick (thin) solid lines are contours obtained from a joint analysis of BAO peak length scale and SNe Ia apparent magnitude (with systematic errors) data, with (and without) angular diameter distance d_A. The cross "x" means the best-fit point defined from the joint sample without the d_A data at $(\Omega_{m0}, \alpha) = (0.28, 0.00)$ with $\chi^2_{\min} = 531$. The diamond "$\diamond$" denotes the best-fit point determined from the joint sample with the d_A data at $(\Omega_{m0}, \alpha) = (0.28, 0.01)$ with $\chi^2_{\min} = 572$. The figure is adapted from [391].

3.10. Reionization Data

Mitra et al. [194] studied the influence of dynamical dark energy and spatial curvature on cosmic reionization. With this aim, the authors examined reionization in the tilted spatially flat and untilted spatially non-flat XCDM, and ϕCDM-RP quintessential inflation models. Statistical analysis was performed based on a principal component analysis and the MCMC analysis using a compilation of the lower-redshift reionization data by Wyithe and Bolton [298] and Becker and Bolton [299] to estimate uncertainties in the model reionization histories. The authors found that, regardless of the nature of dark energy, there are significant differences between the reionization histories of the spatially flat and spatially non-flat cosmological models. Although both the flat and non-flat models fit well the low-redshift $z \leq 6$ reionization observations, there is a discrepancy between high-redshift $z \geq 7$ Lyman-α emitter (LAE) data from Songaila and Cowie [300] and Prochaska et al. [301], and the predictions from spatially non-flat models. Unlike spatially flat models, the spatially non-flat ones have a much earlier and more extended reionization scenario that is completed around $z \approx 7$. The authors found that the higher the electron scattering optical depths τ_{el} the more extended is the reionization process. Moreover, such models predict a much lower fraction of neutral hydrogen at higher redshifts, $7 \lesssim z \lesssim 13$ (from 2σ limits for $Q_{HII} \sim 1$), namely, $\tau_{el} < 1$ at $z \sim 8$, is clearly contradictory to most current observation limits from distant QSO, GRB, and LAE data. Also, a serious disadvantage of spatially non-flat models can be seen from the results obtained on the evolution of the photon escape fraction $f_{esc}(z)$: in spatially non-flat models $f_{esc}(z) \gtrsim 1$ even at $z \gtrsim 8$, given its 2σ limits. Such non-physical $f_{esc}(z)$ values indicate the possibility of excluding these models. However, the Planck 2018 [13] reduction in the value of the τ_{el} in the six-parameter tilted flat ΛCDM inflation model by about 0.9σ reconciles the predictions of the non-flat model with observations.

3.11. Gravitational Lensing Data

Constraints on the parameters of two ϕCDM models, with the RP and the pNGb potentials, were analyzed by Waga and Frieman [386]. These models predict radically different futures for our universe. In the model with the RP potential, the expansion of the universe will continue to accelerate. In the model with the pNGb potential, the present epoch of the expansion of the universe with acceleration will be followed by a return to the matter-dominated epoch. For these observational tests, the authors used the compilation of measurements: gravitational lensing statistics [393–397] and the high-z SNe Ia apparent magnitudes [211,398,399]. The results of these studies are presented in Figure 70, where it is shown that a large region of parameter space for the considered models is consistent with the SNe Ia apparent magnitude data if $\Omega_{m0} > 0.15$. The authors obtained the constraint on the model parameter α of the RP potential, $\alpha < 5$. The ϕCDM model with the pNGb potential is constrained by the SNe Ia apparent magnitude and lensing measurements at a 2σ confidence level.

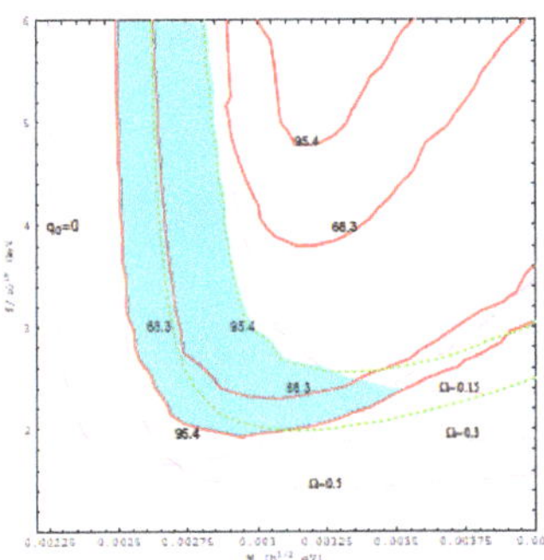
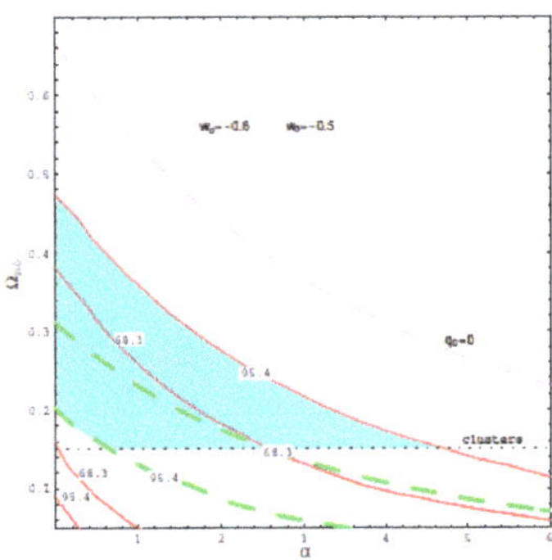

Figure 70. The 1σ and 2σ confidence level contours arising from lensing statistics and SNe Ia apparent magnitude versus redshift data (solid curves). (Left panel) The ϕCDM model with the pNGb potential. Solid curves correspond to constraints from SNe Ia apparent magnitude data. Contours of the constant matter energy parameter at the present epoch Ω_{m0} and the limit for the acceleration parameter at the present epoch $q_0 = 0$ are depicted. (Right panel) The ϕCDM model with the RP potential. The lower bound of $\Omega_{m0} = 0.15$ from clusters and curves for the constant value of the EoS parameter at the present epoch w_0 are shown. The figure is adapted from [386].

Statistics on strong gravitational lensing based on Cosmic Lens All-Sky Survey data from Myers et al. [400] and Browne et al. [401] was applied by Chae et al. [389] to constrain parameters of the ϕCDM-RP model. The results are presented in Figure 71. The maximum of the likelihood accords to the values of the matter density parameter at the present epoch $\Omega_{m0} = 0.34$ and the model parameter $\alpha = 0$, i.e., to the standard spatially flat ΛCDM model. For the 68% confidence level, $0.18 < \Omega_{m0} < 0.62$ and $\alpha < 2.7$, while, for the 95% confidence level, $\Omega_{m0} = 1$ and $\alpha = 8$. Strong gravitational lensing constraints are favorable for the standard spatially flat ΛCDM model, and consistent with Chen and Ratra [387] constraints from the SNe Ia apparent magnitude data, but are weaker.

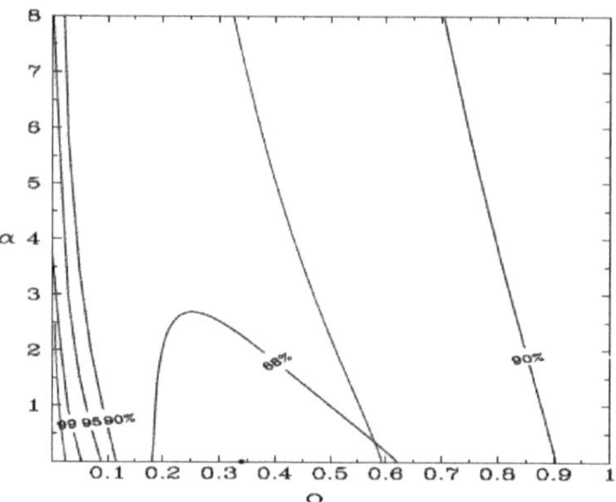

Figure 71. The 68%, 90%, 95%, and 99% confidence level contour constraints on the parameters of the ϕCDM model with the RP potential from the strong gravitational lensing data. The thin line represents the 68% confidence level derived from the SNe Ia apparent magnitude versus redshift data by Chen and Ratra [387]. The horizontal axis for which $\alpha = 0$ corresponds to the spatially flat ΛCDM model. The figure is adapted from [389].

3.12. Compact Radio Source Data

The compact radio source angular size versus redshift data of Gurvits et al. [402] were used by Chen and Ratra [387] to derive constraints on the parameters of the ϕCDM-RP model. These constraints are consistent with the results obtained from the SNe Ia apparent magnitude data of [2,211] but they are less restrictive, Figure 72.

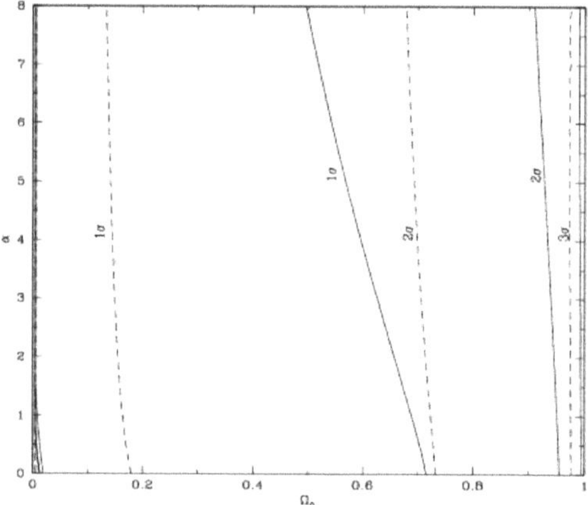

Figure 72. The 1σ, 2σ, and 3σ confidence level contours constraints on the parameters of the spatially flat ϕCDM model with the RP potential. Solid lines are contours computed for the uniform prior $p(\Omega_0) = 1$. Short dashed lines are obtained for the logarithmic prior $p(\Omega_0) = 1/\Omega_0$. The figure is adapted from [387].

Podariu et al. [388] used the redshift–angular size data from double radio galaxies called FRIIb sources to constrain the parameters of the ϕCDM-RP model in the spatially flat universe. These constraints are consistent both with the results obtained from the SNe Ia apparent magnitude data of [2,211] and with the results obtained from the compact radio source redshift–angular size data of Chen and Ratra in [387], but they are less restrictive, Figure 73.

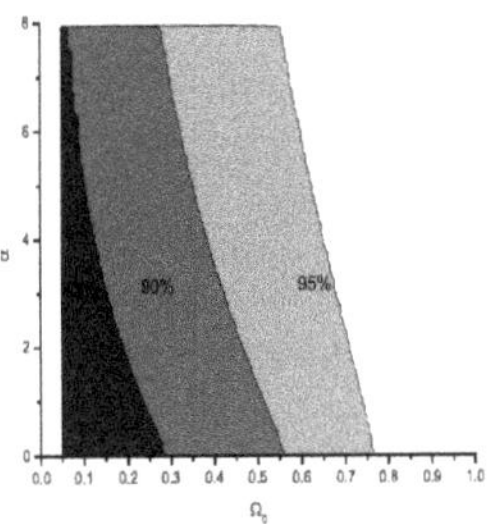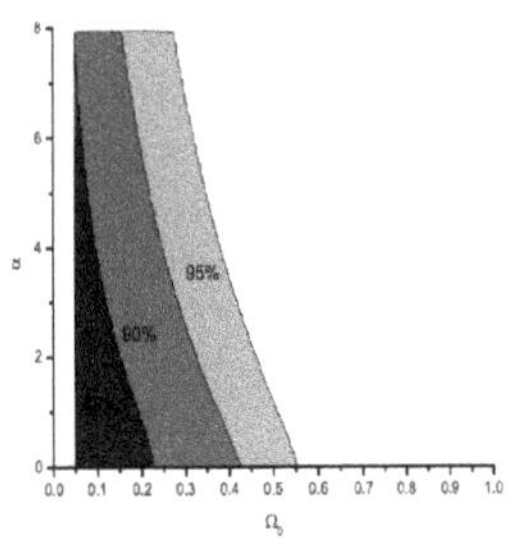

Figure 73. The 1σ, 2σ, and 3σ confidence level contour constraints on the parameters of the spatially flat ϕCDM model with the RP potential. (Left panel) Constraints were obtained using all twenty radio galaxies (including 3C 427.1). (Right panel) Constraints were obtained using only nineteen radio galaxies (excluding 3C 427.1). The figure is adapted from [388].

4. Summary and Results

The results of research based on the papers considered in the review are summarized and grouped into the following topics: results of constraints on the parameters of dynamical dark energy models, alleviation and resolving of the ΛCDM model tensions, data preferences, disadvantages of models to data, failure and incompatibility of data, sensibility of various data, consistency of constraints results by various data, comparing constraints with various data, model-independent estimate of the Hubble constant H_0 and matter density parameter at the present epoch Ω_{m0}, and problems with QSO available data. The main research results are presented in more detail below.

- Results of constraints on the parameters of dynamical dark energy models

    1. For both the extended and ordinary quintessence ϕCDM-RP models, 1σ constraints were obtained of $\alpha < 0.8$ and $\alpha < 0.6$, using the SNe Ia+SNAP data, Caresia et al. [214].

    2. Constraints on the spatial curvature density parameter today to be $|\Omega_{k0}| \leq 0.15$ at a 1σ confidence level in the spatially non-flat ϕCDM-RP model as well as the XCDM model, from SNe Ia $+H(z)+$ BAO data. More precise data are required to tighten the bounds on the parameters, Farooq et al. [255].

    3. In constraints on the model parameters of the ΛCDM model, the XCDM model, and the ϕCDM-RP model using galaxy cluster gas mass fraction data, Ω_m is better constrained than α, whose best-fit value is $\alpha = 0$, corresponding to the standard spatially flat ΛCDM model; however, the scalar field ϕCDM model is not excluded [283].

    4. The deceleration–acceleration transition redshift $z_{da} = 0.74 \pm 0.05$ was obtained as a result of the constraints on the parameters of the ϕCDM-RP model from $H(z)$ data [342].

    5. A likelihood analysis of the COBE-DMR sky maps to normalize the spatially flat ϕCDM-RP model shows that this model remains an observationally viable alternative to the standard spatially flat ΛCDM model [280].

    6. The 2σ upper bounds of $\sum m_\nu < 0.165$ (0.299) eV and $\sum m_\nu < 0.164$ (0.301) eV, respectively, for the spatially flat (spatially non-flat) ΛCDM model and the spatially flat (spatially non-flat) ϕCDM model were defined using CMB + BAO + SNe Ia and the Hubble Space Telescope H_0 prior observations. The inclusion of spatial curvature as a free parameter leads to a significant expansion of the confidence regions for $\sum m_\nu$ and other parameters in spatially flat ϕCDM models, but the corresponding differences are larger for both the spatially non-flat ΛCDM and spatially non-flat ϕCDM models [288].

    7. When the bispectrum component is included in the BAO + LSS data for the ϕCDM model, a significant dynamical dark energy signal was achieved at a $2.5 - 3\sigma$ confidence level. Thus, the bispectrum can be a very useful tool for

tracking and examining the possible dynamical features of dark energy and their influence on the LSS formation in the linear regime [321]. (The bispectrum component has been used by Solà et al. [187] before to study the running cosmic vacuum in the RVMs!)

8. As a result of constraints on the parameters of the oCDM, XCDM (here w_0CDM), and wCDM models by using the BAO+BBN +SNe Ia data the value of epoch $\Omega_{k0} = -0.043^{+0.036}_{-0.036}$ at a 1σ confidence level, which is consistent with the spatially flat universe; in the spatially flat XCDM model, the value of the dark energy EoS parameter at the present epoch $w_0 = -1.031^{+0.052}_{-0.048}$ at a 1σ confidence level, which approximately equals the value of the EoS parameter for the ΛCDM model; and values of the w_0 and w_a in the CPL parameterization of the EoS parameter of the wCDM model $w_0 = -0.98^{+0.099}_{-0.11}$ and $w_a = -0.33^{+0.63}_{-0.48}$ at 1σ confidence level were obtained. The exclusion of the SNe Ia data from the joint data analysis does not significantly weaken the resulting constraints. This means that, when using a single external BBN prior, full-shape and BAO peak length scale data can provide reliable constraints independent of CMB temperature anisotropy constraints [332].

9. Current X-ray temperature data on massive galaxies weakly constrain the w_0 and w_a parameters of the wCDM model around the $(w_0, w_a) = (-1, 0)$ values of the wCDM model corresponding to the ΛCDM model. In the analysis including data from the galaxy cluster number count+ $H(z)$ + CMB temperature anisotropy + BAO + SNe Ia, the values of $w_0 = -1.14^{+0.14}_{-0.16}$ and $w_a = 0.85^{+0.42}_{0.60}$ were obtained at a 1σ confidence level [390].

10. In constraints on parameters in the spatially flat ΛCDM model, the spatially closed ϕCDM models with the RP and Sugra potentials using SNe Ia data, values of Ω_{m0} and $\Omega_{\phi0}$, are quite different from those for the ΛCDM. The quintessence scalar field creates more structures outside the filaments, lighter halos with higher internal velocity dispersion, as seen from N-body simulations performed by the authors to study the influence of quintessence on the distribution of matter on large scales [253].

11. In the ϕCDM-RP model in a spacetime with non-zero spatial curvature, the dynamical scalar field has an attractor solution in the curvature dominated epoch, while the energy density of the scalar field increases relative to that of the spatial curvature [252].

12. In constraints on H_0 in the ϕCDM-RP, the wCDM, and the spatially flat and spatially non-flat ΛCDM models from measurements of $H(z)$, the value of the H_0 is found as follows: for the spatially flat and spatially non-flat ΛCDM model, $H_0 = 68.3^{+2.7}_{-3.3}$ km s^{-1}Mpc^{-1} and $H_0 = 68.4^{+2.9}_{-3.3}$ km s^{-1}Mpc^{-1}; for the wCDM model, $H_0 = 65.0^{+6.5}_{-6.6}$ km s^{-1}Mpc^{-1}; for the ϕCDM model, $H_0 = 67.9^{+2.4}_{-2.4}$ km s^{-1}Mpc^{-1} (at a 1σ confidence level) [343].

13. In constraints on the parameters of the spatially flat and non-flat ΛCDM, XCDM, and ϕCDM-RP models, as well as on the QSO radius–luminosity $R - L$ relation parameters from QSO reverberation measured, the parameters of the $R - L$ relation do not depend on the cosmological models considered and, therefore, the $R - L$ relation can be used to standardize the C_{IV} QSO data. Mutually consistent constraints on the cosmological parameters from C_{IV}, Mg_{II}, and $H(z)$ + BAO peak length scale data allow conducting the analysis from the C_{IV} + Mg_{II} dataset as well as from the $H(z)$ + BAO peak length scale + C_{IV} + Mg_{II} datasets. Although the C_{IV} + Mg_{II} cosmological constraints are weak, they slightly (at a $\sim0.1\sigma$ confidence level) change the constraints from the $H(z)$ + BAO peak length scale + C_{IV} + Mg_{II} datasets [203].

14. The quintessential inflation model with the generalized exponential potential including massive neutrinos that are non-minimally coupled with a scalar field obtains observational constraints on parameters using combinations of data:

CMB + BAO (BOSS) + SNe Ia (SNLS). The upper bound on possible values of the sum of neutrino masses $\sum m_\nu < 2.5$ eV is significantly larger than in the spatially flat ΛCDM model [177].

- Alleviation and resolving of the ΛCDM model tensions

  1. The joint Planck + BAO (transversal) analysis agrees well with the measurements made by the SH0ES team and, applied to the IDE models, solves the Hubble constant H_0 tension [67].
  2. A larger value of the Hubble constant, i.e., alleviation of the Hubble constant tension (with a significance of 3.6σ), has been obtained for the spatially non-flat IDE models. Searches for other forms of the interaction function and the EoS for the dark energy component in IDE models are needed, which may further ease the tension of the Hubble constant [121].
  3. The lower multipole region of CMB + BAO (6dFGS, SDSS-MGS) in the spatially closed quintessential inflation ϕCDM model reduces the tension between the Planck and the weak lensing σ_8 constraints [330].
  4. The maximum of the likelihood in the constraint parameters in the ϕCDM-RP model from the strong gravitational lensing data accords to the values of the matter density parameter at the present epoch $\Omega_{m0} = 0.34$ and the model parameter $\alpha = 0$, i.e., to the standard spatially flat ΛCDM model. For the 68% confidence level, $0.18 < \Omega_{m0} < 0.62$ and $\alpha < 2.7$, while, for 95% confidence level, $\Omega_{m0} = 1$ and $\alpha = 8$ [389].
  5. In extended ϕCDM- RP models with exponential coupling to the Ricci scalar, the projection of the ISW effect on the CMB temperature anisotropy is found to be considerably larger in the exponential case with respect to a quadratic non-minimal coupling. This reflects the fact that the effective gravitational constant depends exponentially on the dynamics of the scalar field [215].
  6. The value of the cosmological deceleration–acceleration transition z_{da} is insensitive to the chosen model from the spatially flat and spatially non-flat ϕCDM-RP, the XCDM, and the wCDM using $H(z)$ data, and depend only on the assumed value of the Hubble constant H_0. The weighted mean of these measurements is $z_{da} = 0.72 \pm 0.05 \, (0.84 \pm 0.03)$ for $H_0 = 68 \pm 2.8 \, (73.24 \pm 1.74) \, \mathrm{km\,s^{-1}Mpc^{-1}}$ [192].
  7. In contrast to the joint Planck + BAO analysis, where it is not possible to solve the Hubble constant H_0 tension, the joint Planck + BAO (transversal) analysis agrees well with the measurements made by the SH0ES team and, applied to the IDE models, solves the Hubble constant H_0 tension [67].

- Data preferences

  1. Planck 2018 CMB data favor spatially closed hypersurfaces in spatially non-flat IDE models at more than 99% CL (with a significance of 5σ) [121].
  2. The higher multipole region of the CMB temperature anisotropy data is in better agreement with the tilted spatially flat ΛCDM model than with the spatially closed ϕCDM model [330].
  3. Depending on the value of the Hubble constant H_0 as a prior and the cosmological model under consideration, the data provides evidence in favor of the spatially non-flat scalar field ϕCDM model [331].
  4. The spatially closed quintessential inflation ϕCDM model provides a better fit to the lower multipole region of CMB temperature anisotropy data +BAO (6dFGS, SDSS-MGS) data compared to that provided by the tilted spatially flat ΛCDM model [330].
  5. In most of the tilted spatially flat and untilted spatially non-flat ΛCDM, XCDM, and ϕCDM-RP quintessential inflation models, the QSO data favor $\Omega_{m0} \sim 0.5$–0.6, while, in a combined analysis of QSO + $H(z)$ + BAO, the values of the Ω_{m0} are shifted slightly towards larger values. A combined QSO + BAO peak length scale + $H(z)$ dataset is consistent with the standard spatially flat ΛCDM model,

but favors slightly both the spatially closed hypersurfaces and the dynamical dark energy models [195].

6. Depending on the chosen model (from spatially flat and spatially non-flat ΛCDM, XCDM, and ϕCDM-RP models) and dataset (from BAO + $H(z)$ + QSO), the data slightly favor both the spatially closed hypersurfaces with $\Omega_{k0} < 0$ at a 1.7σ confidence level and the dynamical dark energy models over the standard spatially flat ΛCDM model at a slightly higher than 2σ confidence level. Furthermore, depending on the dataset and the model, the observational data favor a lower Hubble constant value than the one measured by the local data at a 1.8σ confidence level to 3.4σ confidence level [193].

7. The analysis with the $H(z)$ + BAO + QSO-AS + H_{II}G + GRB dataset favors the spatially flat ΛCDM model but also does not rule out dynamical dark energy models [352].

8. The Hubble constant H_0 value is constrained in the spatially flat and spatially non-flat ΛCDM, XCDM, and ϕCDM-RP models using various combinations of datasets: BAO +SNe Ia $H(z)$. The BAO + SNe Ia $H(z)$ dataset slightly favors the untilted spatially non-flat dynamical XCDM and ϕCDM quintessential inflation models, as well as smaller Hubble constant H_0 values [297].

9. Smaller angular scale SPTpol measurements (used jointly with only Planck CMB temperature anisotropy data or with the combination of Planck CMB temperature anisotropy data and non-CMB temperature anisotropy data) favor the untilted spatially closed models [303].

10. The spatially flat ϕCDM scalar field models could not be unambiguously preferred, from the DESI predictive data ($H(z)$ + $H(z)$ + angular diameter distance d_A), over the standard ΛCDM spatially flat model, the latter still being the most preferred dark energy model [320].

11. CMB (Planck 2015) + BAO + SNe Ia +$H(z)$ + LSS growth data slightly favor the spatially closed XCDM model over the spatially flat ΛCDM model at a 1.2σ confidence level, while also being in better agreement with the untilted spatially flat XCDM model than with the spatially flat ΛCDM model at the 0.3σ confidence level [326].

12. The analysis of the BAO + SNe Ia+ angular diameter distance d_A (using X-ray observations of the intracluster medium + radio observations of the Sunyaev–Zel'dovich effect of galaxy clusters) data favors the spatially flat ΛCDM model but does not exclude the spatially flat ϕCDM-RP model [391].

13. SNe Ia + X-ray gas mass fraction of cluster data is preferable to the standard spatially flat ΛCDM model, but the ϕCDM model is not ruled out either [337].

14. The spatially flat ΛCDM model is the most preferable, but both dynamic dark energy models and space curvature are not ruled out [206].

15. Combined analysis from QSO + $H(z)$ + BAO data is consistent with the standard spatially flat ΛCDM model, but slightly favors both closed spatial hypersurfaces and the untilted spatially non-flat ϕCDM model [196].

16. Constraints on the parameters in the spatially flat and non-flat ΛCDM, XCDM, and ϕCDM-RP models from three (ML, MS, and GL) ($L_0 - t_b$) Dainotti-correlated sets of GRB measurements are weak, providing lower bounds on parameter Ω_{m0}, moderately favoring the non-zero spatial curvature, and largely consistent with both the currently accelerated cosmological expansion and with constraints determined on the basis of more reliable data [202].

17. In constraints on the parameters in the spatially flat and non-flat ΛCDM, XCDM, and ϕCDM-RP models by GRB data, the three-dimensional Dainotti correlation [379,380] is much preferable to the two-dimensional [378] one for the GRB datasets [205].

18. The $R - L$ relation parameters for $H\beta$ QSO data are independent in models under investigation, from the spatially flat and non-flat ΛCDM, XCDM, and ϕCDM-RP

models; therefore, QSO data seem to be standardizable through $R - L$ relation parameters. The constraints derived using $H\beta$ QSO data are weak, slightly favoring the currently accelerating cosmological expansion, and are generally in the 2σ tension with the constraints derived from analysis of the measurements of the BAO peak length scale and the Hubble parameter $H(z)$ [200].

19. Constraints on the parameters of the spatially non-flat untilted ϕCDM-RP inflation model were improved from a 1.8σ to a more than 3.1σ confidence level by combining by CMB (Planck 2015) + BAO + SNe Ia + $H(z)$ + LSS data. CMB (Planck 2015) + BAO + SNe Ia + $H(z)$ + LSS data favor a spatially closed universe with the spatial curvature contributing about two-thirds of a percent of the current total cosmological energy budget. The spatially flat tilted ϕCDM inflation model is a 0.4σ better fit to the observational data than is the standard spatially flat tilted ΛCDM model, i.e., current observational data allow for the possibility of dynamical dark energy in the universe. The spatially non-flat tilted ϕCDM model better fits the DES bounds on the rms amplitude of mass fluctuations σ_8 as a function of the parameter Ω_{m0} [291].

20. The ΛCDM model has a strong advantage, investigating both the minimally coupled with gravity scalar field spatially flat ϕCDM-RP model and non-minimally coupled scalar field extended quintessence model with gravity (with the Ricci scalar), applying the dataset: the Pantheon SNe + BAO (6dFGS, SDSSLRG, BOSS-MGS, BOSS-LOWZ, WiggleZ, BOSS-CMASS, BOSS-DR12) + CMB + $H(z)$ + RSD, when local measurements of the Hubble constant H_0 [13] are not taken into account and, conversely, this statement is weakened when local measurements of H_0 are included in the data analysis [221].

- Disadvantages of models to data

    1. Spatially non-flat ϕCDM-RP quintessential inflation models predict a much lower fraction of neutral hydrogen at higher redshifts $7 \lesssim z \lesssim 13$ (from 2σ limits for $Q_{HII} \sim 1$), namely, $\tau_{el} < 1$ at $z \sim 8$, are clearly contradictory to most current observation limits from distant QSO + GRB + LAE data [194].

    2. A serious disadvantage of spatially non-flat ϕCDM-RP quintessential inflation models can be seen from the results obtained from the evolution of the photon escape fraction $f_{esc}(z)$: in spatially non-flat models $f_{esc}(z) \gtrsim 1$ even at $z \gtrsim 8$, given its 2σ limits. Such non-physical $f_{esc}(z)$ values indicate the possibility of excluding these models. (However, the Planck 2018 [13] reduction in the value of the τ_{el} in the six-parameter tilted flat ΛCDM inflation model by about 0.9σ reconciles the predictions of the non-flat model with observations) [194].

    3. H_{II} starburst galaxy apparent magnitude + QSO only(or) + BAO datasets favor the spatially flat ΛCDM model, while at the same time do not rule out dynamical spatially flat and non-flat ϕCDM-RP models [347].

    4. Constraints on the parameters of the spatially flat and non-flat ΛCDM, XCDM, and ϕCDM-RP models using SNe Ia (Pantheon + DES) + QSO + H_{II}G data + BAO + $H(z)$ favor dynamical dark energy and slightly spatially closed hypersurfaces; they do not preclude dark energy from being a cosmological constant and spatially flat hypersurfaces [201].

    5. Constraints on the parameters of the spatially non-flat untilted ϕCDM-RP inflation model by CMB (Planck 2015) + BAO + SNe Ia + $H(z)$ + LSS data do not provide such good agreement with the larger multipoles of CMB (Planck 2015) data as the spatially flat tilted ΛCDM model [291].

- Failure and incompatibility of data

    1. CMB (Planck 2015) + BAO + SNe Ia + $H(z)$ + LSS growth data are unable to rule out dynamical scalar field spatially flat ϕCDM models [326].

2. The dynamical untilted spatially non-flat XCDM model is not compatible with with higher multipoles of CMB temperature anisotropy data, as is the standard spatially flat ΛCDM model [326].

3. The parameters of the spatially flat ϕCDM model could not be tightly constrained only by the current GRB data [365].

4. There is a strong degeneracy between the model parameters Ω_m and α in the spatially flat ϕCDM-RP model applying only LSS data. According to constraints from LSS growth rate + BAO + CMB data, $\Omega_m = 0.30 \pm 0.04$ and $0 \leq \alpha \leq 1.30$ at a 1σ confidence level (the best-fit value for the model parameter α is $\alpha = 0$) [317].

- Sensibility of various data

  1. Studying dark energy in the early universe using SNe Ia + WMAP + CBI + VSA + SDSS + HST data, the values $w_0 < -0.8$ and density parameter in the early universe $\Omega_e < 0.03$ at the 1σ confidence level are found. SNe Ia data are most sensitive to w_0, while CMB temperature anisotropies and LSS growth rate are the best constraints of Ω_e, Doran et al. [247].

  2. Expansion history data are not particularly sensitive to the dynamic effects of dark energy, while the data compilation BAO + LSS + CMB anisotropy is more sensitive [321].

- Consistency of constraint results with various data

  1. Constraints on the parameters of the ϕCDM-RP model from compact radio source angular size versus redshift data are consistent with the results obtained from the SNe Ia apparent magnitude data of [2,211], but they are less restrictive [387].

  2. Constraints of the spatially flat ϕCDM-RP model from radio galaxies FRIIb sources+redshift–angular size data are consistent both with the results obtained from the SNe Ia apparent magnitude data of [2,211] and with the results obtained from the compact radio source redshift–angular size [387], but they are less restrictive [388].

  3. Constraints on the parameters of the ϕCDM-RP, the XCDM, and the ΛCDM models from BAO + $H(z)$ data. The BAO + $H(z)$ data dataset is consistent with the standard spatially flat ΛCDM model [331].

  4. Constraints on the parameters of the spatially flat ϕCDM-RP, the XCDM, and the ΛCDM models from measurements of $H(z)$ are consistent with a moment of the deceleration–acceleration transition at redshift $z_{da} = 0.74 \pm 0.05$ derived by Farooq and Ratra [342], which corresponds to the standard spatially flat ΛCDM model [260].

  5. Constraints on the parameters of the spatially flat and non-flat ΛCDM, XCDM, and ϕCDM-RP models using the higher-redshift GRB H_{II}G + QSO are consistent with the currently accelerating cosmological expansion, as well as with the constraints obtained from the analysis of the $H(z)$ + BAO peak length scale. From the analysis of the $H(z)$ + BAO + QSO-AS + H_{II}G + GRB dataset, the model-independent values of epoch $\Omega_{m0} = 0.313 \pm 0.013$ and $H_0 = 69.3 \pm 1.2$ km s^{-1}Mpc^{-1} are obtained [352].

  6. In each dark energy model (from the spatially flat and untilted spatially non-flat ΛCDM, XCDM, and scalar field ϕCDM-RP quintessential inflation models), constraints on cosmological parameters from SPTpol measurements+ CMB temperature anisotropy and non-CMB temperature anisotropy measurements are largely consistent with one another [303].

  7. The dynamical untilted spatially non-flat XCDM model is compatible with the DES limits on the current value of the rms mass fluctuation amplitude σ_8 as a Ω_{m0} [326].

  8. A large region of parameter space for the ϕCDM models, with the RP and the pNGb potential models, is consistent with the SNe Ia data if $\Omega_{m0} > 0.15$, wherein the constraints on the model parameter α of the RP potential is $\alpha < 5$. The ϕCDM

model with the pNGb potential is constrained by the SNe Ia apparent magnitude + lensing measurements at a 2σ confidence level [386].

9. The constraints obtained from the Mg_{II} QSOs + BAO + $H(z)$ agree with the spatially flat ΛCDM model as well as with spatially non-flat dynamical dark energy models [198].

10. The $H(z)$ data are consistent with the standard spatially flat ΛCDM model while they do not rule out the spatially non-flat XCDM and spatially non-flat ϕCDM models [192].

11. The obtained H_0 values, as a result of constraints in ϕCDM-RP, the wCDM, and the spatially flat and spatially non-flat ΛCDM models by $H(z)$ measurements, are more consistent with the smaller values determined from the recent CMB temperature anisotropy and BAO peak length scale data and with the values derived from the median statistics analysis of Huchra's compilation of H_0 data [343].

12. Constraints on spatially flat and non-flat ΛCDM, XCDM, and ϕCDM-RP models from GRB are consistent with the spatially flat ΛCDM as well as with the spatially non-flat dynamical dark energy models [371].

13. The second and third largest subsamples, SDSS-Chandra and XXL QSOs, which together account for about 30% of total QSO data, appear to be standardized. Constraints on the cosmological parameters from these subsamples are weak and consistent with the standard spatially flat ΛCDM model or with the constraints from the better-established cosmological probes [199].

14. The quintessential inflation model with the generalized exponential potential is in good agreement with observations and represents a successful scheme for the unification of the primordial inflaton field causing inflation in the very early universe and dark energy causing the accelerated expansion of the universe at the present epoch [177].

15. In quintessential inflation models, the early quintessence is characterized by a suppressed ability to cluster at small scales, as suggested by the compilation of data from WMAP + CBI + ACBAR + 2dFGRS + $L_{y-\alpha}$. Quintessential inflation models are compatible with these data for a constant spectral index of primordial density perturbations [269].

- Comparing constraints with various data

1. Constraints on cosmological parameters in the spatially flat ϕCDM model by joint datasets consisting of measurements of the age of the universe+SNe Ia + BAO are tighter than those obtained from datasets consisting of data on the lookback time + age of the universe [328].

2. Constraints on cosmological parameters in the scalar field ϕCDM-RP model from SVJ $H(z)$ data [335]. Using the $H(z)$ data, the constraints on the Ω_m are more stringent than those on the model parameter α. Constraints on the matter density Ω_m are approximately as tight as the ones derived from the galaxy cluster gas mass fraction data [336] and from the SNe Ia apparent magnitude data [337].

3. Constraints on the parameters of the ϕCDM-RP, the XCDM, the wCDM, and the ΛCDM models using BAO + SNe Ia data are more restrictive with the inclusion of eight new $H(z)$ measurements than those derived by Chen and Ratra [338]. This analysis favors the standard spatially flat ΛCDM model but does not exclude the scalar field ϕCDM model [339].

4. Constraints on the parameters of the ϕCDM-RP, the XCDM, and the ΛCDM models using $H(z)$ data. $H(z)$ data yield quite strong constraints on the parameters of the ϕCDM model. The constraints derived from the $H(z)$ measurements are almost as restrictive as those implied by the currently available lookback time observations and the GRB luminosity data, but more stringent than those based on the currently available galaxy cluster angular size data. However, they are less restrictive than those following from the joint analysis of SNe Ia + BAO. The joint

analysis of the $H(z)$ + SNe Ia + BAO favors the standard spatially flat ΛCDM model but does not exclude the dynamical scalar field ϕCDM model [338].

5. Constraints on the parameters of the ϕCDM-RP, the XCDM, and the ΛCDM models with the inclusion of new $H(z)$ measurement of Busca et al. are more restrictive than those derived by Farooq et al. The $H(z)$ constraints depend on the Hubble constant prior to H_0 used in the analysis. The resulting constraints are more stringent than those which follow from measurements of the SNe Ia apparent magnitude of Suzuki et al. (2012). This joint analysis consisting of measurements of $H(z)$+SNe Ia + BAO favors the standard spatially flat ΛCDM model but the dynamical scalar field ϕCDM model is not excluded either [340].

6. SNe Ia + $H(z)$ + LSS growth rate data are consistent with the standard spatially flat ΛCDM model, as well as with the spatially flat ϕCDM-RP model [315].

7. Strong gravitational lensing constraints are favorable for the standard spatially flat ΛCDM model and consistent with [387] constraints from the SNe Ia apparent magnitude data, but are weaker [389].

8. Constraints on the parameters of the spatially flat ϕCDM model obtained from joint R04 gold SNe Ia apparent magnitude data and galaxy cluster gas mass fraction data are tighter than those obtained by Podariu and Ratra [241] from earlier SNe Ia apparent magnitude versus redshift data [337].

9. Constraints obtained in [336] on the parameters of the ϕCDM model using Chandra measurements of X-ray gas mass fraction of the clusters are tighter than those derived from the SNe Ia apparent magnitude data [241,386], redshift–angular size data of [387,388], and gravitational lensing statistics [389]

10. Constraints on the parameters of the spatially flat and non-flat ΛCDM, XCDM, and ϕCDM-RP models derived from the QSO data only are significantly weaker than those derived from the combined set of the BAO + $H(z)$, but are consistent with both of them [198].

11. QSO data are significantly weaker but consistent with those from the combination of the $H(z)$ + BAO data in tilted spatially flat and untilted spatially non-flat ΛCDM, XCDM, and ϕCDM-RP quintessential inflation models [196].

12. The constraints in spatially flat ΛCDM, XCDM, and ϕCDM-RP models from the GRB data obtained by Wang's method [368] and Schaefer's method [367] disagree with each other at a more than 2σ confidence level [365].

13. Constraints on spatially flat and non-flat ΛCDM, XCDM, and ϕCDM-RP models from GRB data agree but are much weaker than those following from the BA + $H(z)$ data [371].

14. Constraints on the parameters of the spatially flat and non-flat ΛCDM, XCDM, and ϕCDM-RP models from the GRB data are consistent with the constraints obtained from the analysis of the BAO + $H(z)$ but are less restrictive [369].

15. Constraints on parameters in spatially flat and non-flat ΛCDM, XCDM, and ϕCDM-RP models from GRB + $H(z)$ + BAO data take small changes in parameter constraints compared to the constraints from the $H(z)$ + BAO data. The constraints from the GRBs only are more stringent than those from the $H(z)$ + BAO dataset but are less precise [204].

16. Constraints on the parameters of the ϕCDM-RP, the XCDM, and the ΛCDM models from the H_{II}G are largely consistent but not as restrictive as those derived from the measurements of the BAO + SNe Ia + CMB temperature anisotropy [381].

17. Subsets of full QSO data, limited by redshift $z \leq 1.5$–1.7, obey the $L_X - L_{UV}$ relation in a way that is independent of the cosmological model (from the spatially flat and non-flat ΛCDM, XCDM, and ϕCDM-RP models) and can therefore be used to constrain the cosmological parameters. Constraints from these smaller subsets of lower redshift QSO data are generally consistent but much weaker than those inferred from the Hubble parameter $H(z)$ + BAO measurements [197].

18. WMAP + BAO + galaxy cluster gas mass fraction measurements give consistent and more accurate constraints on the parameters of the spatially flat ϕCDM model than those derived from other data, wherein, constraints on the parameter α, $\alpha < 3.5$ [327].

19. Future measurements of the LSS growth rate in the near future will be able to constrain the spatially flat ϕCDM-RP models with an accuracy of about 10%, considering the fiducial spatially flat ΛCDM model, an improvement of almost an order of magnitude compared to those from currently available datasets. Constraints on the growth index parameter γ are more restrictive in the ΛCDM model than in other models. In the ϕCDM model, constraints on the growth index parameter γ are about a third tighter than in the wCDM and XCDM models [306].

- Model-independent estimate of the Hubble constant H_0 and matter density parameter at the present epoch Ω_{m0}

  1. Constraints on the parameters of the spatially flat and non-flat ΛCDM, XCDM, and ϕCDM-RP models from the $H(z)$ + BAO + SNe Ia (Pantheon, DES, QSO) + H_{II}G data provides almost model-independent estimates of the Hubble constant, the matter density parameter at the present epoch, and the characteristic linear size, respectively, as $H_0 = 69.7 \pm 1.2$ km s^{-1}Mpc^{-1}, $\Omega_{m0} = 0.295 \pm 0.021$, and $l_m = 10.93 \pm 0.25$ pc. [383].

  2. The model-independent value of the Hubble constant $H_0 = 69.7 \pm 1.2$ km s^{-1}Mpc^{-1} and the parameter $\Omega_{m0} = 0.295 \pm 0.017$ were obtained by using the $H(z)$ + BAO + SNe Ia + QSO-AS + H_{II}G + Mg_{II} QSO + A118 (HzBSNQHMA) data in the spatially flat and non-flat ΛCDM, XCDM, and ϕCDM-RP models [206].

  3. An analysis of all H_{II} starburst galaxy apparent magnitude + QSO only(or) + BAO datasets in the spatially flat and non-flat ΛCDM, XCDM, and ϕCDM-RP models leads to the relatively model-independent and restrictive estimates for the values of the parameter Ω_{m0} and the Hubble constant H_0. Depending on the cosmological model, these estimates are consistent with a lower value of H_0 in the range of a 2.0σ to 3.4σ confidence level [347].

- Problems with QSO available data

  1. In constraints on the parameters of the spatially flat and spatially non-flat ΛCDM, XCDM, and ϕCDM-RP models using Mg_{II} QSOs data, two- and three-parameter radius–luminosity $R - L$ relations do not depend on the assumed cosmological model; therefore, they can be used to standardize QSO data. The authors found for the two-parameter $R - L$ relation that the data subsets with low-$\Re_{FeII}$ and high-$\Re_{FeII}$ obey the same $R - L$ relation within the error bars. Extending the two-parameter $R - L$ relation to three parameters does not lead to the expected decrease in the intrinsic variance of the $R - L$ relation. None of the three-parameter $R - L$ relations provides a significantly better measurement fit than the two-parameter $R - L$ relation. The results obtained differ significantly from those found by Khadka et al. [200] from analysis of reverberation-measured $H\beta$ QSOs [362].

  2. Constraints on the parameters of the spatially flat and non-flat ΛCDM, XCDM, and ϕCDM-RP models for the full QSO dataset; parameters of the X-ray and UV luminosity $L_X - L_{UV}$ relation used to standardize these QSO data depends on the cosmological model and therefore cannot be used to constrain the cosmological parameters in these models [197].

  3. A compilation of the QSO X-ray and UV flux measurements [356] includes the QSO data that appear not to be standardized via the X-ray luminosity and the UV luminosity $L_X - L_{UV}$ relation parameters that are dependent on both the cosmological model and the redshift, so it should not be used to constrain the model parameters [199].

5. Ongoing and Upcoming Cosmological Missions

We look forward to more precise various cosmological data which will be obtained by the ongoing and upcoming cosmological missions (a list of which are presented below): GRB, X-ray temperature of massive galaxy clusters, the expansion rate of the universe $H(z)$, the BAO peak length scale, the CMB radiation, the angular diameter distance, and the LSS growth rate of matter density fluctuations. We hope that these observational data will allow cosmologists to obtain tighter constraints on the bounds of cosmological parameters in the dynamical dark energy models than have been obtained to date.

To study GRB and X-ray for investigating the early universe, the following are scheduled to launch: the Space-based multiband astronomical Variable Objects Monitor (SVOM) mission [403] (will be access on 14 January 2024), the Transient High-Energy Sky and Early Universe Surveyor (THESEUS) mission [404] (will be access on 2037), the Hydrogen Intensity and Real-time Analysis Experiment (HIRAX) (https://hirax.ukzn.ac.za) (will be accessed on 2024). To investigate the weak lensing survey BAO peak length scale and RSD, it is planned to launch the following: the 4-metre Multi-Object Spectroscopic Telescope (4MOST) (https://www.eso.org/sci/facilities/develop/instruments/4MOST.html) (accessed on 2023) and the BAO from the Integrated Neutral Gas Observations (BINGO) radio telescope (https://bingotelescope.org) (currently under construction). To investigate the acceleration expansion of the universe, the nature of the dark universe, the dynamics and evolution of the universe, and the growth of LSS in the universe, the Euclidean Space Telescope (Euclid) (https://www.euclid-ec.org/) (will be accessed on 1 July 2023); the following are scheduled to launch: the Spectro-Photometer for the History of the universe (SPHEREx) (https://spherex) (will be accessed on April 2025), the Nancy Grace Roman Space Telescope (https://www.jpl.nasa.gov/missions/the-nancy-grace-roman-space-telescope) (will be accessed on May 2027), the ArmazoNes high Dispersion Echelle Spectrograph (ANDES) (https://elt.eso.org/instrument/ANDES/) (will be accessed on 2030), the Extremely Large Telescope (ELT) (https://elt.eso.org) (will be accessed on 2028), the Rubin Observatory Legacy Survey of Space and Time (Rubin/LSST) (https://lsst.org) (will be access on 2025), and the Simons Observatory (SO) (https://simonsobservatory.org) (will be accessed on 2024). To explore CMB radiation, it is planned to launch the following: the Cosmic Microwave Background—Stage IV (CMB-S4) experiment (https://cmb-s4.org (will be accessed on 2029)) and the Lite Satellite for the studies of B-mode polarization and Inflation from Cosmic Background Radiation Detection (LiteBIRD) (https://litebird.html (will be accessed on 2032); to study the fingerprint of primordial gravitational waves in CMB, the SPIDER experiment (https://spider.princeton.edu (will be accessed on December 2024)).

6. Conclusions

In this review we analyzed and summarized the current research effort to constrain the parameters of the dynamical dark energy models through cosmological observations. Our review does not claim to be a complete presentation of all research conducted by scientists in this field. We did our best to account for the results of the papers that seem to us the most relevant, among numerous ones, where the authors applied different types of methods and observational constraints with various observational datasets to investigate the dynamical dark energy models. The main results of the papers considered in this review are summarized and grouped in Section 4 into the following topics: results of constraints on the parameters of the dynamical dark energy models, alleviation and resolving of the ΛCDM model tensions, data preferences, disadvantages of models to data, failure and incompatibility of data, sensibility of various data, consistency and comparisons of the constraint results with various data, model-independent estimate of the Hubble constant H_0 and the matter density parameter at the present epoch Ω_{m0}, and problems with the QSO available data.

In most of papers presented in this review the quintessence scalar field ϕCDM model with the inverse power-law RP potential has been studied. This model is the simplest one, and it is a typical representative of the large family of the tracker quintessence scalar field models. It grasps the general properties of the tracker models. Due to its simplicity, this model has attracted and will continue to attract the community in its effort to study it.

This review is a kind of historical cross-section of the study of dynamical dark energy models, in which it is clearly seen that the complication, refinement, and increase in the diversity of cosmological data and methods for studying dynamical dark energy models lead to more precise constraints on the values of cosmological parameters. According to the results of the papers considered in this review, the accuracy of the available cosmological data is insufficient to tighten the bounds on the parameters of the dynamical dark energy models and more precise data are required (the main ongoing and upcoming cosmological experiments are presented in Section 5). At the same time, despite the refinement of observational data and complications of methods for studying dark energy in the universe, current observational data favor the standard spatially flat ΛCDM model, while not excluding dynamical dark energy models and spatially closed hyperspaces.

Since the current observational constraints seem to leave room for ϕCDM as a viable alternative to the true minimalist ΛCDM model, it is appropriate to point out some salient features which can naturally complement the big picture, if the idea that a scalar field is responsible for the observed dark energy is taken seriously.

One of the important problems to be addressed is the *coincidence problem*—at the present epoch, the energy scales of dark matter and dark energy are of comparable magnitudes, and these are again comparable to that of the cosmological neutrinos.

A proposal for the explanation of the neutrino mass due to coupling through the dark energy field (i.e., beyond the Higgs mechanism of the Standard Model) was put forward by Fardon, Nelson, and Weiner [405], followed by Peccei [406]. These works predict a time-dependent mass for dark matter particles and neutrinos. It has been shown in the following work [239,240] that it is possible to simultaneously obtain the values of the late-time evolution of the universe ($w \to -1$, $V(\phi) \sim M^4 \sim \Lambda$), asymptotically close to that of the ΛCDM model, as well as the dark-energy-generated neutrino masses consistent with the observations for several dark energy potentials coupled to the fermionic field.

The models analyzed in [239,240] are toy models since only a single fermionic species coupled to the scalar field is considered. An important and promising direction of the future work, aiming to reveal the origins of the dark sector of the universe and the neutrino mass generation, appears to be an approach unifying the scalar field(s) coupled to the matter fields of the Standard Model. Various proposals to incorporate the dark matter component of the universe along with dark energy and neutrinos within a common framework were put forward [407–409]. More detailed quantitative analysis needs to be done to fully explore the observable effects predicted by these models with multiple fermionic flavors and different DE potentials on the expansion history of the universe.

It has been pointed out many times in the literature that too many scalar fields are not very natural. There have been efforts to relate, e.g., the Higgs field to inflation [410,411]. A quite plausible conjecture is that the inflation and quintessence represent the same physical field analyzed at different regimes of the universe evolution [412]. Thus, the objective is to advance such unifying theories back in time to incorporate the inflation.

Funding: O.A. acknowledges partial support from the Shota Rustaveli Georgian NSF grant YS-22-998. T.K. acknowledges partial support from the NASA ATP award 80NSSC22K0825.

Data Availability Statement: No new data were created or anlyzed in this study. Date sharing is not applicable to this article.

Acknowledgments: We greatly appreciate helpful comments and discussions of some of the issues and research findings with Bharat Ratra and Chan-Gyung Park.

Conflicts of Interest: The authors declare no conflict of interest.

Abbreviations

Abbreviation	Full Form
ACBAR	Arcminute Cosmology Bolometer Array Receiver
AIC	Akaike Information Criterion
BIC	Bayesian Information Criterion
BOSS	Baryon Oscillation Spectroscopic Survey
BAO	Baryon Acoustic Oscillations
BBN	Big Bang Nucleosynthesis
GRB	Gamma-ray Bursts
CBI	Cosmic Background Imager
CDM	Cold Dark Matter
CLASS	Cosmic Linear Anisotropy Solving System
COBE-DMR	Cosmic Background Explorer—Differential Microwave Radiometers
CPL	Chevallier–Polarsky–Linder
CMB	Cosmic Microwave Background Radiation
DES	Dark Energy Survey
DESI	Dark Energy Spectroscopic Instrument
DIC	Deviation Information Criterion
EoS	Equation of State
Euclid	Euclidean Space Telescope
FRII	Fanaroff–Riley Type II
FLRW	Friedmann–Lemaître–Robertson–Walker
HDM	Hot Dark Matter
HST	Hubble Space Telescope
HzBSNQHMA	$H(z)$ + BAO + SNe Ia + QSO-AS + H_{II}G + Mg_{II} QSO + A118
H_{II}S	H_{II} Starburst Galaxy
IDE	Interacting Dark Energy
ISW	Integrated Sachs–Wolfe
JBD	Jordan–Brans–Dicke
JLA	Joint Light-Curve Analysis
LAE	Lyman-α Emitter
LSS	Large-Scale Structure
MCMC	Markov Chain Monte Carlo
MGS	Main Galaxy Sample
pNGb	Pseudo-Nambu–Goldstone Boson
QFT	Quantum Field Theory
QSO	Quasar
PR4	Last Planck Data Release
RP	Ratra–Peebles
rms	root mean square
RSD	Redshift Space Distortion
RVM	Running Vacuum Model
SDSS	Sloan Digital Sky Survey
SNe Ia	Supernovae Ia
SNLS	Supernova Legacy Survey
SPTpol	South Pole Telescope Polarization
SVJ$H(z)$ data	Simon, Verde, and Jimenez $H(z)$ Data
UV	Ultraviolet
VSA	Very Small Array
wCDM	w Cold Dark Matter
WMAP	Wilkinson Microwave Anisotropy Probe
WDM	Warm Dark Matter
WFIRST	Wide-Field Infrared Survey Telescope
XCDM	X Cold Dark Matter
ΛCDM	Lambda Cold Dark Matter

ϕCDM	Phi Cold Dark Matter
oCDM	ΛCDM Extension to Non-Flat Hypersurfaces
2dFGRS	Two-Degree Field Galaxy Redshift Survey
6dFGS	Six-Degree Field Galaxy Survey

Notes

[1] For the latter model, the first Friedmann's equation and the Klein–Gordon scalar field equation for these models are defined, respectively, as

$$H^2(a) = \frac{1}{3F}\left(a^2 \rho_{\text{fluid}} + \frac{1}{2}\dot{\phi}^2 + a^2 V(\phi) - 3H(a)\dot{F}(\phi)\right), \tag{43}$$

$$\ddot{\phi} + 2H\dot{\phi} = \frac{a^2}{2}F'(\phi)R - a^2 V'(a), \tag{44}$$

where the dot now denotes the derivative with respect to the conformal time and the prime denotes a derivative with respect to the scalar field ϕ. The quantity ρ_{fluid} is the energy density associated with all components of the universe except for the quintessential scalar field, and the function $F(\phi)$ defines the non-minimal coupling between gravity and the scalar field ϕ, with the form [243]

$$F(\phi) = 1/8\pi G + \tilde{F}(\phi) - \tilde{F}(\phi_0), \tag{45}$$

where $\tilde{F}(\phi) = \vartheta \phi^2$, ϑ is a dimensionless constant, and ϕ_0 is a value of the scalar field at the present epoch.

[2] The observational constraints on a projection of the Integrated Sachs–Wolfe (ISW) effect on the CMB temperature anisotropy was obtained for a fixed value of the Jordan–Brans–Dicke (JBD) parameter at the present epoch ω_{JBD0}, the latter being defined as

$$\omega_{\text{JBD}} = F\left(\frac{dF}{d\phi}\right)^{-2} = \frac{8\pi}{\vartheta^2}\exp\left[-\frac{\vartheta(\phi - \phi_0)}{M_{\text{pl}}}\right], \quad \omega_{\text{JBD0}} = 8\pi/\vartheta^2, \tag{49}$$

where ζ is a dimensionless coupling, ϕ_0 is the present value for the scalar field, and $F = \frac{1}{16\pi G}\exp\left(\frac{\vartheta}{M_{\text{pl}}}(\phi - \phi_0)\right)$ is a generalized function of the gravity term $R/16\pi G$.

[3] This is performed by fixing at the present epoch the amplitude of the initial energy density fluctuations generated in the early inflation epoch for this model and comparing the model predictions of the large angular scale spatial anisotropy in the CMB radiation with observational data. The authors computed model predictions as a function of the model parameter α, as well as other cosmological parameters, following Brax et al. [281], and then determined the normalization amplitude by comparing these predictions with the COBE-DMR CMB temperature anisotropy measurements of Bennett [5] and Gorski et al. [282].

References

1. Riess, A.G.; Filippenko, A.V.; Challis, P.; Clocchiatti, A.; Diercks, A.; Garnavich, P.M.; Gilliland, R.L.; Hogan, C.J.; Jha, S.; Kirshner, R.P.; et al. Observational Evidence from Supernovae for an Accelerating Universe and a Cosmological Constant. *Astron. J.* **1998**, *116*, 1009–1038. [CrossRef]
2. Perlmutter, S.; Aldering, G.; Goldhaber, G.; Knop, R.A.; Nugent, P.; Castro, P.G.; Deustua, S.; Fabbro, S.; Goobar, A.; Groom, D.E.; et al. Measurements of Ω and Λ from 42 High-Redshift Supernovae. *Astron. Astrophys.* **1999**, *517*, 565–586. [CrossRef]
3. Riess, A.G.; Strolger, L.G.; Casertano, S.; Ferguson, H.C.; Mobasher, B.; Gold, B.; Challis, P.J.; Filippenko, A.V.; Jha, S.; Li, W.; et al. New Hubble Space Telescope Discoveries of Type Ia Supernovae at z >= 1: Narrowing Constraints on the Early Behavior of Dark Energy. *Astron. Astrophys.* **2007**, *659*, 98–121.
4. Smoot, G.F.; Bennett, C.L.; Kogut, A.; Wright, E.; Aymon, J.; Boggess, N.; Cheng, E.; De Amici, G.; Gulkis, S.; Hauser, M.; et al. Structure in the COBE differential microwave radiometer first-year maps. *Astron. Astrophys.* **1992**, *396*, L1–L5. [CrossRef]
5. Bennett, C.L.; Banday, A.; Gorski, K.M.; Hinshaw, G.; Jackson, P.; Keegstra, P.; Kogut, A.; Smoot, G.F.; Wilkinson, D.T.; Wright, E.L. Four year COBE DMR cosmic microwave background observations: Maps and basic results. *Astrophys. J. Lett.* **1996**, *464*, L1–L4. [CrossRef]
6. Spergel, D.N.; Verde, L.; Peiris, H.V.; Komatsu, E.; Nolta, M.R.; Bennett, C.L.; Halpern, M.; Hinshaw, G.; Jarosik, N.; Kogut, A.; et al. First-Year Wilkinson Microwave Anisotropy Probe (WMAP) Observations: Determination of Cosmological Parameters. *Astrophys. J. Suppl.* **2003**, *148*, 175–194. [CrossRef]
7. Spergel, D.N.; Bean, R.; Doré, O.; Nolta, M.R.; Bennett, C.L.; Dunkley, J.; Hinshaw, G.; Jarosik, N.; Komatsu, E.; Page, L.; et al. Three-Year Wilkinson Microwave Anisotropy Probe (WMAP) Observations: Implications for Cosmology. *Astrophys. J. Suppl.* **2007**, *170*, 377–408. [CrossRef]
8. Hinshaw, G.; Weiland, J.L.; Hill, R.S.; Odegard, N.; Larson, D.; Bennett, C.L.; Dunkley, J.; Gold, B.; Greason, M.R.; Jarosik, N.; et al. Five-Year Wilkinson Microwave Anisotropy Probe Observations: Data Processing, Sky Maps, and Basic Results. *Astrophys. J. Suppl.* **2009**, *180*, 225–245. [CrossRef]

9. Nolta, M.R.; Dunkley, J.; Hill, R.S.; Hinshaw, G.; Komatsu, E.; Larson, D.; Page, L.; Spergel, D.N.; Bennett, C.L.; Gold, B.; et al. Five-Year Wilkinson Microwave Anisotropy Probe Observations: Angular Power Spectra. *Astrophys. J. Suppl.* **2009**, *1*, 296–305. [CrossRef]

10. Komatsu, E.; Smith, K.M.; Dunkley, J.; Bennett, C.L.; Gold, B.; Hinshaw, G.; Jarosik, N.; Larson, D.; Nolta, M.R.; Page, L.; et al. Seven-year Wilkinson Microwave Anisotropy Probe (WMAP) Observations: Cosmological Interpretation. *Astrophys. J. Suppl.* **2011**, *192*, 18. [CrossRef]

11. Planck Collaboration; Ade, P.A.R.; Aghanim, N.; Armitage-Caplan, C.; Arnaud, M.; Ashdown, M.; Atrio-Barandela, F.; Aumont, J.; Baccigalupi, C.; Banday, A.J.; et al. Planck 2013 results. XVI. Cosmological parameters. *Astron. Astrophys.* **2014**, *571*, A16.

12. Planck Collaboration; Ade, P.A.R.; Aghanim, N.; Arnaud, M.; Ashdown, M.; Aumont, J.; Baccigalupi, C.; Banday, A.J.; Barreiro, R.B.; Bartlett, J.G.; et al. Planck 2015 results. XIII. Cosmological parameters. *Astron. Astrophys.* **2016**, *594*, A13.

13. Planck Collaboration; Aghanim, N.; Akrami, Y.; Ashdown, M.; Aumont, J.; Baccigalupi, C.; Ballardini, M.; Banday, A.J.; Barreiro, R.B.; Bartolo, N.; et al. Planck 2018 results. VI. Cosmological parameters. *Astron. Astrophys.* **2020**, *641*, A6.

14. SDSS. 2001. Available online: http://www.sdss.org/ (accessed on 5 February 2024).

15. Dodelson, S.; Narayanan, V.K.; Tegmark, M.; Scranton, R.; Budavári, T.; Connolly, A.; Csabai, I.; Eisenstein, D.; Frieman, J.A.; Gunn, J.E.; et al. The Three-dimensional Power Spectrum from Angular Clustering of Galaxies in Early Sloan Digital Sky Survey Data. *Astron. Astrophys.* **2002**, *572*, 140–156. [CrossRef]

16. 2dFGRS. 2002. Available online: http://www.mso.anu.edu.au/2dFGRS/ (accessed on 5 February 2024).

17. Percival, W.J.; Cole, S.; Eisenstein, D.J.; Nichol, R.C.; Peacock, J.A.; Pope, A.C.; Szalay, A.S. Measuring the Baryon Acoustic Oscillation scale using the SDSS and 2dFGRS. *Mon. Not. R. Astron. Soc.* **2007**, *381*, 1053–1066. [CrossRef]

18. DES. 2013. Available online: https://www.darkenergysurvey.org/ (accessed on 5 February 2024).

19. Kwan, J.; Sánchez, C.; Clampitt, J.; Blazek, J.; Crocce, M.; Jain, B.; Zuntz, J.; Amara, A.; Becker, M.R.; Bernstein, G.M.; et al. Cosmology from large-scale galaxy clustering and galaxy-galaxy lensing with Dark Energy Survey Science Verification data. *Mon. Not. R. Astron. Soc.* **2017**, *464*, 4045–4062. [CrossRef]

20. Eisenstein, D.J.; Zehavi, I.; Hogg, D.W.; Scoccimarro, R.; Blanton, M.R.; Nichol, R.C.; Scranton, R.; Seo, H.J.; Tegmark, M.; Zheng, Z.; et al. Detection of the Baryon Acoustic Peak in the Large-Scale Correlation Function of SDSS Luminous Red Galaxies. *Astron. Astrophys.* **2005**, *633*, 560–574. [CrossRef]

21. Blake, C.; Kazin, E.A.; Beutler, F.; Davis, T.M.; Parkinson, D.; Brough, S.; Colless, M.; Contreras, C.; Couch, W.; Croom, S.; et al. The WiggleZ Dark Energy Survey: Mapping the distance-redshift relation with baryon acoustic oscillations. *Mon. Not. R. Astron. Soc.* **2011**, *418*, 1707–1724. [CrossRef]

22. Beutler, F.; Blake, C.; Colless, M.; Jones, D.H.; Staveley-Smith, L.; Campbell, L.; Parker, Q.; Saunders, W.; Watson, F. The 6dF Galaxy Survey: Baryon acoustic oscillations and the local Hubble constant. *Mon. Not. R. Astron. Soc.* **2011**, *416*, 3017–3032. [CrossRef]

23. Anderson, L.; Aubourg, É.; Bailey, S.; Beutler, F.; Bhardwaj, V.; Blanton, M.; Bolton, A.S.; Brinkmann, J.; Brownstein, J.R.; Burden, A.; et al. The clustering of galaxies in the SDSS-III Baryon Oscillation Spectroscopic Survey: Baryon acoustic oscillations in the Data Releases 10 and 11 Galaxy samples. *Mon. Not. R. Astron. Soc.* **2014**, *441*, 24–62. [CrossRef]

24. Ross, A.J.; Samushia, L.; Howlett, C.; Percival, W.J.; Burden, A.; Manera, M. The clustering of the SDSS DR7 main Galaxy sample – I. A 4 per cent distance measure at $z = 0.15$. *Mon. Not. R. Astron. Soc.* **2015**, *449*, 835–847. [CrossRef]

25. Alam, S.; Ata, M.; Bailey, S.; Beutler, F.; Bizyaev, D.; Blazek, J.A.; Bolton, A.S.; Brownstein, J.R.; Burden, A.; Chuang, C.H.; et al. The clustering of galaxies in the completed SDSS-III Baryon Oscillation Spectroscopic Survey: Cosmological analysis of the DR12 galaxy sample. *Mon. Not. R. Astron. Soc.* **2017**, *470*, 2617–2652. [CrossRef]

26. Ata, M.; Baumgarten, F.; Bautista, J.; Beutler, F.; Bizyaev, D.; Blanton, M.R.; Blazek, J.A.; Bolton, A.S.; Brinkmann, J.; Brownstein, J.R.; et al. The clustering of the SDSS-IV extended Baryon Oscillation Spectroscopic Survey DR14 quasar sample: First measurement of baryon acoustic oscillations between redshift 0.8 and 2.2. *Mon. Not. R. Astron. Soc.* **2018**, *473*, 4773–4794. [CrossRef]

27. HST. 1990. Available online: https://www.stsci.edu/hst (accessed on 5 February 2024).

28. Stern, D.; Jimenez, R.; Verde, L.; Kamionkowski, M.; Stanford, S.A. Cosmic Chronometers: Constraining the Equation of State of Dark Energy. I: H(z) Measurements. *JCAP* **2010**, *2*, 8. [CrossRef]

29. Riess, A.G.; Macri, L.; Casertano, S.; Lampeitl, H.; Ferguson, H.C.; Filippenko, A.V.; Jha, S.W.; Li, W.; Chornock, R. A 3% Solution: Determination of the Hubble Constant with the Hubble Space Telescope and Wide Field Camera 3. *Astron. Astrophys.* **2011**, *730*, 119. [CrossRef]

30. Moresco, M.; Verde, L.; Pozzetti, L.; Jimenez, R.; Cimatti, A. New constraints on cosmological parameters and neutrino properties using the expansion rate of the Universe to z~1.75. *JCAP* **2012**, *7*, 53. [CrossRef]

31. Zhang, C.; Zhang, H.; Yuan, S.; Liu, S.; Zhang, T.J.; Sun, Y.C. Four new observational H(z) data from luminous red galaxies in the Sloan Digital Sky Survey data release seven. *Res. Astron. Astrophys.* **2014**, *14*, 1221. [CrossRef]

32. Delubac, T.; Bautista, J.E.; Busca, N.G.; Rich, J.; Kirkby, D.; Bailey, S.; Font-Ribera, A.; Slosar, A.; Lee, K.G.; Pieri, M.M.; et al. Baryon acoustic oscillations in the Lyα forest of BOSS DR11 quasars. *Astron. Astrophys.* **2015**, *574*, A59. [CrossRef]

33. Moresco, M. Raising the bar: New constraints on the Hubble parameter with cosmic chronometers at $z \sim 2$. *Mon. Not. R. Astron. Soc.* **2015**, *450*, L16–L20. [CrossRef]

34. Moresco, M.; Pozzetti, L.; Cimatti, A.; Jimenez, R.; Maraston, C.; Verde, L.; Thomas, D.; Citro, A.; Tojeiro, R.; Wilkinson, D. A 6% measurement of the Hubble parameter at $z \sim 0.45$: Direct evidence of the epoch of cosmic re-acceleration. *JCAP* **2016**, *5*, 14. [CrossRef]

35. Ratsimbazafy, A.L.; Loubser, S.I.; Crawford, S.M.; Cress, C.M.; Bassett, B.A.; Nichol, R.C.; Väisänen, P. Age-dating Luminous Red Galaxies observed with the Southern African Large Telescope. *Mon. Not. R. Astron. Soc.* **2017**, *467*, 3239–3254. [CrossRef]

36. Peebles, P.J.E.; Ratra, B. The Cosmological constant and dark energy. *Rev. Mod. Phys.* **2003**, *75*, 559–606. [CrossRef]

37. Copeland, E.J.; Sami, M.; Tsujikawa, S. Dynamics of dark energy. *Int. J. Mod. Phys.* **2006**, *D15*, 1753–1936. [CrossRef]

38. Frieman, J.; Turner, M.; Huterer, D. Dark Energy and the Accelerating Universe. *Ann. Rev. Astron. Astrophys.* **2008**, *46*, 385–432. [CrossRef]

39. Caldwell, R.R.; Kamionkowski, M. The Physics of Cosmic Acceleration. *Ann. Rev. Nucl. Part. Sci.* **2009**, *59*, 397–429. [CrossRef]

40. Tsujikawa, S. Modified gravity models of dark energy. *Lect. Notes Phys.* **2010**, *800*, 99–145.

41. Tsujikawa, S. Dark energy: Investigation and modeling. *arXiv* **2010**, arXiv:1004.1493.

42. Weinberg, D.H.; Mortonson, M.J.; Eisenstein, D.J.; Hirata, C.; Riess, A.G.; Rozo, E. Observational probes of cosmic acceleration. *Phys. Rept.* **2013**, *530*, 87–255. [CrossRef]

43. Yoo, J.; Watanabe, Y. Theoretical Models of Dark Energy. *Int. J. Mod. Phys.* **2012**, *D21*, 1230002. [CrossRef]

44. Rubin, V.C.; Thonnard, N.; Ford, W.K., Jr. Rotational properties of 21 SC galaxies with a large range of luminosities and radii, from NGC 4605 (R = 4kpc) to UGC 2885 (R = 122 kpc). *Astrophys. J.* **1980**, *238*, 471. [CrossRef]

45. Rubakov, V.A. Cosmology. In Proceedings of the 2011 European School of High-Energy Physics (ESHEP 2011), Cheile Gradistei, Romania, 7–20 September 2011; pp. 151–195.

46. Tristram, M.; Banday, A.J.; Douspis, M.; Garrido, X.; Górski, K.M.; Henrot-Versillé, S.; Ilić, S.; Keskitalo, R.; Lagache, G.; Lawrence, C.R.; et al. Cosmological parameters derived from the final (PR4) Planck data release. *arXiv* **2023**, arXiv:2309.10034.

47. Silvestri, A.; Trodden, M. Approaches to Understanding Cosmic Acceleration. *Rept. Prog. Phys.* **2009**, *72*, 096901. [CrossRef]

48. Lopez-Corredoira, M. History and Problems of the Standard Model in Cosmology. *arXiv* **2023**, arXiv:2307.10606.

49. Peebles, P.J.E. *Principles of Physical Cosmology*; Princeton University Press: Princeton, NJ, USA, 1994.

50. Weinberg, S. *Cosmology*; Oxford University Press: Oxford, UK, 2008.

51. Dodelson, S.; Schmidt, F. *Modern Cosmology*; Academic Press: London, UK, 2021.

52. Baumann, D. *Cosmology*; Cambridge University Press: Cambridge, UK, 2022.

53. Huterer, D. *A Course in Cosmology*; Cambridge University Press: Cambridge, UK, 2023.

54. Carroll, S.M.; Press, W.H.; Turner, E.L. The cosmological constant. *Annu. Rev. Astro. Astrophys.* **1992**, *30*, 499–542. [CrossRef]

55. Carroll, S.M. The Cosmological constant. *Living Rev. Rel.* **2001**, *4*, 1. [CrossRef]

56. Martin, J. Everything You Always Wanted to Know about the Cosmological Constant Problem (But Were Afraid to Ask). *C. R. Phys.* **2012**, *13*, 566–665. [CrossRef]

57. Padilla, A. Lectures on the Cosmological Constant Problem. *arXiv* **2015**, arXiv:1502.05296.

58. Deruelle, N.; Uzan, J.P. *Relativity in Modern Physics*; Oxford Graduate Texts; Oxford University Press: Oxford, UK, 2018.

59. Weinberg, S. The Cosmological Constant Problem. *Rev. Mod. Phys.* **1989**, *61*, 1–23. [CrossRef]

60. Weinberg, S. The Cosmological constant problems. In *Sources and Detection of Dark Matter and Dark Energy in the Universe, Proceedings of the 4th International Symposium, DM 2000, Marina del Rey, CA, USA, 23–25 February 2000*; Springer: Berlin, Germany, 2000; pp. 18–26.

61. Padmanabhan, T. Cosmological constant: The Weight of the vacuum. *Phys. Rept.* **2003**, *380*, 235–320. [CrossRef]

62. Di Valentino, E. Challenges of the Standard Cosmological Model. *Universe* **2022**, *8*, 399. [CrossRef]

63. Abdalla, E.; Abellán, G.F.; Aboubrahim, A.; Agnello, A.; Akarsu, Ö.; Akrami, Y.; Alestas, G.; Aloni, D.; Amendola, L.; Anchordoqui, L.A.; et al. Cosmology intertwined: A review of the particle physics, astrophysics, and cosmology associated with the cosmological tensions and anomalies. *J. High Energy Astrophys.* **2022**, *34*, 49–211. [CrossRef]

64. Peebles, P.J.E. Anomalies in physical cosmology. *Ann. Phys.* **2022**, *447*, 169159. [CrossRef]

65. Perivolaropoulos, L.; Skara, F. Challenges for ΛCDM: An update. *New Astron. Rev.* **2022**, *95*, 101659. [CrossRef]

66. Di Valentino, E.; Mena, O.; Pan, S.; Visinelli, L.; Yang, W.; Melchiorri, A.; Mota, D.F.; Riess, A.G.; Silk, J. In the realm of the Hubble tension—A review of solutions. *Class. Quant. Grav.* **2021**, *38*, 153001. [CrossRef]

67. Bernui, A.; Di Valentino, E.; Giarè, W.; Kumar, S.; Nunes, R.C. Exploring the H0 tension and the evidence for dark sector interactions from 2D BAO measurements. *Phys. Rev. D* **2023**, *107*, 103531. [CrossRef]

68. Niedermann, F.; Sloth, M.S. New Early Dark Energy as a solution to the H_0 and S_8 tensions. *arXiv* **2023**, arXiv:2307.03481.

69. Benaoum, H.B.; García, L.Á.; Castañeda, L. Early dark energy induced by non-linear electrodynamics. *arXiv* **2023**, arXiv:2307.05917.

70. Hoerning, G.A.; Landim, R.G.; Ponte, L.O.; Rolim, R.P.; Abdalla, F.B.; Abdalla, E. Constraints on interacting dark energy revisited: Implications for the Hubble tension. *arXiv* **2023**, arXiv:2308.05807.

71. Wei, J.J.; Melia, F. Investigating Cosmological Models and the Hubble Tension Using Localized Fast Radio Bursts. *Astron. Astrophys.* **2023**, *955*, 101. [CrossRef]

72. Smith, T.L.; Poulin, V. Current small-scale CMB constraints to axion-like early dark energy. *arXiv* **2023**, arXiv:2309.03265.

73. Teixeira, E.M.; Daniel, R.; Frusciante, N.; van de Bruck, C. Forecasts on interacting dark energy with standard sirens. *arXiv* **2023**, arXiv:2309.06544.

74. Raveri, M. Resolving the Hubble tension at late times with Dark Energy. *arXiv* **2023**, arXiv:2309.06795.

75. Kodama, T.; Shinohara, T.; Takahashi, T. Generalized early dark energy and its cosmological consequences. *arXiv* **2023**, arXiv:2309.11272.

76. Vagnozzi, S. Seven Hints That Early-Time New Physics Alone Is Not Sufficient to Solve the Hubble Tension. *Universe* **2023**, *9*, 393. [CrossRef]

77. Fierz, M.; Pauli, W. On relativistic wave equations for particles of arbitrary spin in an electromagnetic field. *Proc. R. Soc. Lond.* **1939**, *A173*, 211–232.

78. Polchinski, J. *String Theory. Vol. 1: An Introduction to the Bosonic String*; Cambridge Monographs on Mathematical Physics; Cambridge University Press: Cambridge, UK, 2007.

79. Arkani-Hamed, N.; Dimopoulos, S.; Dvali, G.R. The Hierarchy problem and new dimensions at a millimeter. *Phys. Lett.* **1998**, *B429*, 263–272. [CrossRef]

80. Dvali, G.R.; Gabadadze, G.; Porrati, M. 4-D gravity on a brane in 5-D Minkowski space. *Phys. Lett.* **2000**, *B485*, 208–214. [CrossRef]

81. Kamenshchik, A.Y.; Moschella, U.; Pasquier, V. An Alternative to quintessence. *Phys. Lett.* **2001**, *B511*, 265–268. [CrossRef]

82. Capozziello, S.; Carloni, S.; Troisi, A. Quintessence without scalar fields. *Recent Res. Dev. Astron. Astrophys.* **2003**, *1*, 625.

83. Scherrer, R.J. Purely kinetic k-essence as unified dark matter. *Phys. Rev. Lett.* **2004**, *93*, 011301. [CrossRef]

84. Nicolis, A.; Rattazzi, R.; Trincherini, E. The Galileon as a local modification of gravity. *Phys. Rev.* **2009**, *D79*, 064036. [CrossRef]

85. Shifman, M. Large Extra Dimensions: Becoming acquainted with an alternative paradigm. *Int. J. Mod. Phys.* **2010**, *A25*, 199–225. [CrossRef]

86. Khoury, J.; Wyman, M. N-Body Simulations of DGP and Degravitation Theories. *Phys. Rev.* **2009**, *D80*, 064023. [CrossRef]

87. Wang, S.; Wang, Y.; Li, M. Holographic Dark Energy. *Phys. Rept.* **2017**, *696*, 1–57. [CrossRef]

88. Martens, P. Challenging Mysteries of the Universe with Gravity beyond General Relativity. Ph.D. Thesis, Kyoto University, Kyoto, Japan, 2023.

89. Joyce, A.; Lombriser, L.; Schmidt, F. Dark Energy Versus Modified Gravity. *Annu. Rev. Nucl. Part. Sci.* **2016**, *66*, 95–122. [CrossRef]

90. Slosar, A.; Davis, T.; Eisenstein, D.; Hložek, R.; Ishak-Boushaki, M.; Mandelbaum, R.; Marshall, P.; Sakstein, J.; White, M. Dark Energy and Modified Gravity. *arXiv* **2019**, arXiv:1903.12016.

91. Ishak, M. Testing General Relativity in Cosmology. *Living Rev. Rel.* **2019**, *22*, 1. [CrossRef]

92. Sakharov, A. Cosmological transitions with changes in the signature of the metric. *Sov. Phys. JETP* **1984**, *60*, 214–218.

93. Ishak, M. Remarks on the formulation of the cosmological constant/dark energy problems. *Found. Phys.* **2007**, *37*, 1470–1498. [CrossRef]

94. Aiola, S.; Calabrese, E.; Maurin, L.; Naess, S.; Schmitt, B.L.; Abitbol, M.H.; Addison, G.E.; Ade, P.A.R.; Alonso, D.; Amiri, M.; et al. The Atacama Cosmology Telescope: DR4 maps and cosmological parameters. *JCAP* **2020**, *2020*, 47. [CrossRef]

95. Steinhardt, P.J.; Wang, L.M.; Zlatev, I. Cosmological tracking solutions. *Phys. Rev.* **1999**, *D59*, 123504. [CrossRef]

96. Garriga, J.; Vilenkin, A. Solutions to the cosmological constant problems. *Phys. Rev. D* **2001**, *64*, 023517. [CrossRef]

97. Dalal, N.; Abazajian, K.; Jenkins, E.; Manohar, A.V. Testing the Cosmic Coincidence Problem and the Nature of Dark Energy. *Phys. Rev. Lett.* **2001**, *87*, 141302. [CrossRef]

98. Steinhardt, P.J. A quintessential introduction to dark energy. *Philos. Trans. R. Soc. Lond. Ser. A Math. Phys. Eng. Sci.* **2003**, *361*, 2497–2513. [CrossRef]

99. Starkman, G.D.; Trotta, R. Why Anthropic Reasoning Cannot Predict Λ. *Phys. Rev. Lett.* **2006**, *97*, 201301. [CrossRef]

100. Maor, I.; Krauss, L.; Starkman, G. Anthropic Arguments and the Cosmological Constant, with and without the Assumption of Typicality. *Phys. Rev. Lett.* **2008**, *100*, 041301. [CrossRef]

101. Zheng, J.; Chen, Y.; Xu, T.; Zhu, Z.H. Diagnosing the cosmic coincidence problem and its evolution with recent observations. *arXiv* **2021**, arXiv:2107.08916.

102. Weinberg, S. Anthropic Bound on the Cosmological Constant. *Phys. Rev. Lett.* **1987**, *59*, 2607. [CrossRef]

103. Barrow, J.D.; Tipler, F.J. *The Anthropic Cosmological Principle*; Oxford University Press: Oxford, UK, 1988.

104. Vilenkin, A. Anthropic Approach to the Cosmological Constant Problems. *Int. J. Theor. Phys.* **2003**, *42*, 1193–1209. [CrossRef]

105. Garriga, J.; Vilenkin, A. Anthropic Prediction for Λ and the Q Catastrophe. *Prog. Theor. Phys. Suppl.* **2006**, *163*, 245–257. [CrossRef]

106. Vilenkin, A. Anthropic predictions: The case of the cosmological constant. In *Universe or Multiverse?* Carr, B., Ed.; Cambridge University Press: Cambridge, UK, 2007; pp. 163–180.

107. Velten, H.E.S.; vom Marttens, R.F.; Zimdahl, W. Aspects of the cosmological "coincidence problem". *Eur. Phys. J. C* **2014**, *74*, 3160. [CrossRef]

108. Riess, A.G.; Macri, L.M.; Hoffmann, S.L.; Scolnic, D.; Casertano, S.; Filippenko, A.V.; Tucker, B.E.; Reid, M.J.; Jones, D.O.; Silverman, J.M.; et al. A 2.4% Determination of the Local Value of the Hubble Constant. *Astron. Astrophys.* **2016**, *826*, 56. [CrossRef]

109. Bonvin, V.; Courbin, F.; Suyu, S.H.; Marshall, P.J.; Rusu, C.E.; Sluse, D.; Tewes, M.; Wong, K.C.; Collett, T.; Fassnacht, C.D.; et al. New COSMOGRAIL time delays of HE 0435-1223: H_0 to 3.8 per cent precision from strong lensing in a flat ΛCDM model. *Mon. Not. R. Astron. Soc.* **2017**, *465*, 4914–4930. [CrossRef]

110. Riess, A.G.; Rodney, S.A.; Scolnic, D.M.; Shafer, D.L.; Strolger, L.G.; Ferguson, H.C.; Postman, M.; Graur, O.; Maoz, D.; Jha, S.W.; et al. Type Ia Supernova Distances at Redshift >1.5 from the Hubble Space Telescope Multi-cycle Treasury Programs: The Early Expansion Rate. *Astron. Astrophys.* **2018**, *853*, 126. [CrossRef]

111. Birrer, S.; Treu, T.; Rusu, C.E.; Bonvin, V.; Fassnacht, C.D.; Chan, J.H.H.; Agnello, A.; Shajib, A.J.; Chen, G.C.F.; Auger, M.; et al. H0LiCOW - IX. Cosmographic analysis of the doubly imaged quasar SDSS 1206+4332 and a new measurement of the Hubble constant. *Mon. Not. R. Astron. Soc.* **2019**, *484*, 4726–4753. [CrossRef]
112. Riess, A.G.; Casertano, S.; Yuan, W.; Macri, L.M.; Scolnic, D. Large Magellanic Cloud Cepheid Standards Provide a 1% Foundation for the Determination of the Hubble Constant and Stronger Evidence for Physics beyond ΛCDM. *Astrophys. J.* **2019**, *876*, 85. [CrossRef]
113. Riess, A.G.; Casertano, S.; Yuan, W.; Bowers, J.B.; Macri, L.; Zinn, J.C.; Scolnic, D. Cosmic Distances Calibrated to 1% Precision with Gaia EDR3 Parallaxes and Hubble Space Telescope Photometry of 75 Milky Way Cepheids Confirm Tension with ΛCDM. *Astrophys. J. Lett.* **2021**, *908*, L6. [CrossRef]
114. Riess, A.G.; Yuan, W.; Macri, L.M.; Scolnic, D.; Brout, D.; Casertano, S.; Jones, D.O.; Murakami, Y.; Anand, G.S.; Breuval, L.; et al. A Comprehensive Measurement of the Local Value of the Hubble Constant with 1 km s^{-1} Mpc^{-1} Uncertainty from the Hubble Space Telescope and the SH0ES Team. *Astrophys. J. Lett.* **2022**, *934*, L7. [CrossRef]
115. Joudaki, S.; Blake, C.; Heymans, C.; Choi, A.; Harnois-Deraps, J.; Hildebrandt, H.; Joachimi, B.; Johnson, A.; Mead, A.; Parkinson, D.; et al. CFHTLenS revisited: Assessing concordance with Planck including astrophysical systematics. *Mon. Not. R. Astron. Soc.* **2017**, *465*, 2033–2052. [CrossRef]
116. Abbott, T.M.C.; Aguena, M.; Alarcon, A.; Allam, S.; Alves, O.; Amon, A.; Andrade-Oliveira, F.; Annis, J.; Avila, S.; Bacon, D.; et al. Dark Energy Survey Year 3 results: Cosmological constraints from galaxy clustering and weak lensing. *Phys. Rev. D* **2022**, *105*, 023520. [CrossRef]
117. Philcox, O.H.E.; Ivanov, M.M. BOSS DR12 full-shape cosmology: ΛCDM constraints from the large-scale galaxy power spectrum and bispectrum monopole. *Phys. Rev. D* **2022**, *105*, 043517. [CrossRef]
118. Nunes, R.C.; Vagnozzi, S. Arbitrating the S8 discrepancy with growth rate measurements from redshift-space distortions. *Mon. Not. R. Astron. Soc.* **2021**, *505*, 5427–5437. [CrossRef]
119. Planck Collaboration; Aghanim, N.; Akrami, Y.; Ashdown, M.; Aumont, J.; Baccigalupi, C.; Ballardini, M.; Banday, A.J.; Barreiro, R.B.; Bartolo, N.; et al. Planck 2018 results. V. CMB power spectra and likelihoods. *Astron. Astrophys.* **2020**, *641*, A5.
120. Di Valentino, E.; Melchiorri, A.; Silk, J. Planck evidence for a closed Universe and a possible crisis for cosmology. *Nature Astron.* **2019**, *4*, 196–203. [CrossRef]
121. Di Valentino, E.; Melchiorri, A.; Mena, O.; Pan, S.; Yang, W. Interacting Dark Energy in a closed universe. *Mon. Not. R. Astron. Soc.* **2021**, *502*, L23–L28. [CrossRef]
122. Vagnozzi, S.; Di Valentino, E.; Gariazzo, S.; Melchiorri, A.; Mena, O.; Silk, J. The galaxy power spectrum take on spatial curvature and cosmic concordance. *Phys. Dark Univ.* **2021**, *33*, 10085. [CrossRef]
123. Vagnozzi, S.; Loeb, A.; Moresco, M. Eppur è piatto? The Cosmic Chronometers Take on Spatial Curvature and Cosmic Concordance. *Astrophys. J.* **2021**, *908*, 84. [CrossRef]
124. Dhawan, S.; Alsing, J.; Vagnozzi, S. Non-parametric spatial curvature inference using late-Universe cosmological probes. *Mon. Not. R. Astron. Soc.* **2021**, *506*, L1–L5. [CrossRef]
125. Ratra, B.; Peebles, P.J.E. Cosmological Consequences of a Rolling Homogeneous Scalar Field. *Phys. Rev.* **1988**, *D37*, 3406. [CrossRef]
126. Ratra, B.; Peebles, P.J.E. Cosmology with a time-variable cosmological constant. *Astrophys. J.* **1988**, *325*, L17–L20.
127. Wetterich, C. Cosmologies with Variable Newton's 'Constant'. *Nucl. Phys.* **1988**, *B302*, 645–667. [CrossRef]
128. Brax, P.; Martin, J. Quintessence and the accelerating universe. *arXiv* **2002**, arXiv:astro-ph/0210533.
129. Linder, E.V. The Dynamics of Quintessence, The Quintessence of Dynamics. *Gen. Rel. Grav.* **2008**, *40*, 329–356. [CrossRef]
130. Cai, Y.F.; Saridakis, E.N.; Setare, M.R.; Xia, J.Q. Quintom Cosmology: Theoretical implications and observations. *Phys. Rept.* **2010**, *493*, 1–60. [CrossRef]
131. Amendola, L.; Tsujikawa, S. *Dark Energy*; Cambridge University Press: Cambridge, UK, 2015.
132. Bahamonde, S.; Böhmer, C.G.; Carloni, S.; Copeland, E.J.; Fang, W.; Tamanini, N. Dynamical systems applied to cosmology: Dark energy and modified gravity. *Phys. Rept.* **2018**, *775–777*, 1–122. [CrossRef]
133. Piras, D.; Lombriser, L. A representation learning approach to probe for dynamical dark energy in matter power spectra. *arXiv* **2023**, arXiv:2310.10717.
134. Frampton, P.H.; Ludwick, K.J.; Scherrer, R.J. The Little Rip. *Phys. Rev.* **2011**, *D84*, 063003. [CrossRef]
135. Caldwell, R.R.; Linder, E.V. The Limits of quintessence. *Phys. Rev. Lett.* **2005**, *95*, 141301. [CrossRef]
136. Schimd, C.; Tereno, I.; Uzan, J.P.; Mellier, Y.; van Waerbeke, L.; Semboloni, E.; Hoekstra, H.; Fu, L.; Riazuelo, A. Tracking quintessence by cosmic shear—Constraints from virmos-descart and cfhtls and future prospects. *Astron. Astrophys.* **2007**, *463*, 405–421. [CrossRef]
137. Vagnozzi, S.; Dhawan, S.; Gerbino, M.; Freese, K.; Goobar, A.; Mena, O. Constraints on the sum of the neutrino masses in dynamical dark energy models with $w(z) \geq -1$ are tighter than those obtained in ΛCDM. *Phys. Rev. D* **2018**, *98*, 083501. [CrossRef]
138. Caldwell, R.R. A Phantom menace? *Phys. Lett.* **2002**, *B545*, 23–29. [CrossRef]
139. Elizalde, E.; Nojiri, S.; Odintsov, S.D. Late-time cosmology in (phantom) scalar-tensor theory: Dark energy and the cosmic speed-up. *Phys. Rev.* **2004**, *D70*, 043539. [CrossRef]
140. Scherrer, R.J.; Sen, A.A. Phantom Dark Energy Models with a Nearly Flat Potential. *Phys. Rev.* **2008**, *D78*, 067303. [CrossRef]

141. Dutta, S.; Scherrer, R.J. Dark Energy from a Phantom Field Near a Local Potential Minimum. *Phys. Lett.* **2009**, *B676*, 12–15. [CrossRef]
142. Frampton, P.H.; Ludwick, K.J.; Scherrer, R.J. Pseudo-rip: Cosmological models intermediate between the cosmological constant and the little rip. *Phys. Rev.* **2012**, *D85*, 083001. [CrossRef]
143. Ludwick, K.J. The Viability of Phantom Dark Energy: A Brief Review. *Mod. Phys. Lett.* **2017**, *A32*, 1730025. [CrossRef]
144. Escamilla, L.A.; Giarè, W.; Di Valentino, E.; Nunes, R.C.; Vagnozzi, S. The state of the dark energy equation of state circa 2023. *arXiv* **2023**, arXiv:2307.14802.
145. Chiba, T.; De Felice, A.; Tsujikawa, S. Observational constraints on quintessence: Thawing, tracker, and scaling models. *Phys. Rev.* **2013**, *D87*, 083505. [CrossRef]
146. Lima, N.A.; Liddle, A.R.; Sahlén, M.; Parkinson, D. Reconstructing thawing quintessence with multiple datasets. *Phys. Rev. D* **2016**, *93*, 063506. [CrossRef]
147. Zlatev, I.; Wang, L.M.; Steinhardt, P.J. Quintessence, cosmic coincidence, and the cosmological constant. *Phys. Rev. Lett.* **1999**, *82*, 896–899. [CrossRef]
148. Linder, E.V. Quintessence's last stand? *Phys. Rev.* **2015**, *D91*, 063006. [CrossRef]
149. Bag, S.; Mishra, S.S.; Sahni, V. New tracker models of dark energy. *JCAP* **2018**, *8*, 9. [CrossRef]
150. Wetterich, C. The Cosmon model for an asymptotically vanishing time dependent cosmological 'constant'. *Astron. Astrophys.* **1995**, *301*, 321–328.
151. Amendola, L. Perturbations in a coupled scalar field cosmology. *Mon. Not. R. Astron. Soc.* **2000**, *312*, 521. [CrossRef]
152. Amendola, L. Coupled quintessence. *Phys. Rev.* **2000**, *D62*, 043511. [CrossRef]
153. Zimdahl, W.; Pavon, D. Interacting quintessence. *Phys. Lett.* **2001**, *B521*, 133–138. [CrossRef]
154. Liu, X.; Tsujikawa, S.; Ichiki, K. Observational constraints on interactions between dark energy and dark matter with momentum and energy transfers. *arXiv* **2023**, arXiv:2309.13946.
155. Di Valentino, E.; Melchiorri, A.; Mena, O.; Vagnozzi, S. Nonminimal dark sector physics and cosmological tensions. *Phys. Rev. D* **2020**, *101*, 063502. [CrossRef]
156. Yang, W.; Pan, S.; Di Valentino, E.; Nunes, R.C.; Vagnozzi, S.; Mota, D.F. Tale of stable interacting dark energy, observational signatures, and the H_0 tension. *JCAP* **2018**, *9*, 19. [CrossRef]
157. Di Valentino, E.; Melchiorri, A.; Mena, O.; Vagnozzi, S. Interacting dark energy in the early 2020s: A promising solution to the H_0 and cosmic shear tensions. *Phys. Dark Univ.* **2020**, *30*, 100666. [CrossRef]
158. Nunes, R.C.; Di Valentino, E. Dark sector interaction and the supernova absolute magnitude tension. *Phys. Rev. D* **2021**, *104*, 063529. [CrossRef]
159. Zhai, Y.; Giarè, W.; van de Bruck, C.; Di Valentino, E.; Mena, O.; Nunes, R.C. A consistent view of interacting dark energy from multiple CMB probes. *JCAP* **2023**, *7*, 32. [CrossRef]
160. Pan, S.; Yang, W. On the interacting dark energy scenarios—The case for Hubble constant tension. *arXiv* **2023**, arXiv:2310.07260.
161. Starobinsky, A.A. A New Type of Isotropic Cosmological Models without Singularity. *Phys. Lett. B* **1980**, *91*, 99–102. [CrossRef]
162. Guth, A.H. Inflationary universe: A possible solution to the horizon and flatness problems. *Phys. Rev. D* **1981**, *23*, 347–356. [CrossRef]
163. Mukhanov, V.F.; Chibisov, G.V. Quantum Fluctuations and a Nonsingular Universe. *JETP Lett.* **1981**, *33*, 532–535.
164. Starobinsky, A.A. Dynamics of Phase Transition in the New Inflationary Universe Scenario and Generation of Perturbations. *Phys. Lett. B* **1982**, *117*, 175–178. [CrossRef]
165. Linde, A.D. A New Inflationary Universe Scenario: A Possible Solution of the Horizon, Flatness, Homogeneity, Isotropy and Primordial Monopole Problems. *Phys. Lett. B* **1982**, *108*, 389–393. [CrossRef]
166. Linde, A.D. Chaotic Inflation. *Phys. Lett. B* **1983**, *129*, 177–181. [CrossRef]
167. Lucchin, F.; Matarrese, S.; Pollock, M.D. Inflation with a Nonminimally Coupled Scalar Field. *Phys. Lett. B* **1986**, *167*, 163–168. [CrossRef]
168. Lucchin, F.; Matarrese, S. Power-law inflation. *Phys. Rev.* **1985**, *D32*, 1316–1322. [CrossRef]
169. Lucchin, F.; Matarrese, S. Kinematical properties of generalized inflation. *Phys. Lett. B* **1985**, *164*, 282–286. [CrossRef]
170. Dodelson, S. *Modern Cosmology*; Academic Press: Amsterdam, The Netherlands, 2003.
171. Gorbunov, D.S.; Rubakov, V.A. *Introduction to the Theory of the Early Universe: Cosmological Perturbations and Inflationary Theory*; World Scientific: Hackensack, NJ, USA, 2011.
172. Mukhanov, V. *Physical Foundations of Cosmology*; Cambridge University Press: Cambridge, UK, 2005.
173. Senatore, L. Lectures on Inflation. In Proceedings of the Theoretical Advanced Study Institute in Elementary Particle Physics: New Frontiers in Fields and Strings, Boulder, CO, USA, 1–26 June 2015.
174. Peebles, P.J.E.; Vilenkin, A. Quintessential inflation. *Phys. Rev. D* **1999**, *59*, 063505. [CrossRef]
175. Rubio, J.; Wetterich, C. Emergent scale symmetry: Connecting inflation and dark energy. *Phys. Rev. D* **2017**, *96*, 063509. [CrossRef]
176. Ratra, B. Inflation in a closed universe. *Phys. Rev.* **2017**, *D96*, 103534. [CrossRef]
177. Geng, C.Q.; Lee, C.C.; Sami, M.; Saridakis, E.N.; Starobinsky, A.A. Observational constraints on successful model of quintessential Inflation. *JCAP* **2017**, *6*, 11. [CrossRef]
178. Akrami, Y.; Kallosh, R.; Linde, A.; Vardanyan, V. Dark energy, α-attractors, and large-scale structure surveys. *JCAP* **2018**, *6*, 41. [CrossRef]

179. Dimopoulos, K.; Owen, C. Quintessential Inflation with α-attractors. *JCAP* **2017**, *6*, 27. [CrossRef]
180. Shapiro, I.L.; Sola, J.; Espana-Bonet, C.; Ruiz-Lapuente, P. Variable cosmological constant as a Planck scale effect. *Phys. Lett. B* **2003**, *574*, 149–155. [CrossRef]
181. Sola, J.; Gomez-Valent, A.; de Cruz Pérez, J. Dynamical dark energy: Scalar fields and running vacuum. *Mod. Phys. Lett. A* **2017**, *32*, 1750054. [CrossRef]
182. Moreno-Pulido, C.; Peracaula, J.S. Equation of state of the running vacuum. *Eur. Phys. J. C* **2022**, *82*, 1137. [CrossRef]
183. Sola Peracaula, J. The cosmological constant problem and running vacuum in the expanding universe. *Philos. Trans. R. Soc. A* **2022**, *380*, 20210182. [CrossRef]
184. Solà Peracaula, J.; Gómez-Valent, A.; de Cruz Perez, J.; Moreno-Pulido, C. Running vacuum against the H_0 and σ_8 tensions. *EPL* **2021**, *134*, 19001. [CrossRef]
185. Sola Peracaula, J.; Gomez-Valent, A.; de Cruz Perez, J.; Moreno-Pulido, C. Running Vacuum in the Universe: Phenomenological Status in Light of the Latest Observations, and Its Impact on the σ_8 and H_0 Tensions. *Universe* **2023**, *9*, 262. [CrossRef]
186. Sola, J.; Gomez-Valent, A.; de Cruz Pérez, J. Hints of dynamical vacuum energy in the expanding Universe. *Astrophys. J. Lett.* **2015**, *811*, L14. [CrossRef]
187. Solà, J.; Gómez-Valent, A.; de Cruz Pérez, J. First evidence of running cosmic vacuum: Challenging the concordance model. *Astrophys. J.* **2017**, *836*, 43. [CrossRef]
188. Solà Peracaula, J.; de Cruz Pérez, J.; Gomez-Valent, A. Possible signals of vacuum dynamics in the Universe. *Mon. Not. R. Astron. Soc.* **2018**, *478*, 4357–4373. [CrossRef]
189. Moreno-Pulido, C.; Sola, J. Running vacuum in quantum field theory in curved spacetime: Renormalizing ρ_{vac} without $\sim m^4$ terms. *Eur. Phys. J. C* **2020**, *80*, 692. [CrossRef]
190. Moreno-Pulido, C.; Sola Peracaula, J. Renormalizing the vacuum energy in cosmological spacetime: Implications for the cosmological constant problem. *Eur. Phys. J. C* **2022**, *82*, 551. [CrossRef]
191. Nojiri, S.; Odintsov, S.; Oikonomou, V. Modified gravity theories on a nutshell: Inflation, bounce and late-time evolution. *Phys. Rep.* **2017**, *692*, 1–104. [CrossRef]
192. Farooq, O.; Madiyar, F.R.; Crandall, S.; Ratra, B. Hubble Parameter Measurement Constraints on the Redshift of the Deceleration–acceleration Transition, Dynamical Dark Energy, and Space Curvature. *Astrophys. J.* **2017**, *835*, 26. [CrossRef]
193. Ryan, J.; Chen, Y.; Ratra, B. Baryon acoustic oscillation, Hubble parameter, and angular size measurement constraints on the Hubble constant, dark energy dynamics, and spatial curvature. *Mon. Not. R. Astron. Soc.* **2019**, *488*, 3844–3856. [CrossRef]
194. Mitra, S.; Park, C.G.; Choudhury, T.R.; Ratra, B. First study of reionization in tilted flat and untilted non-flat dynamical dark energy inflation models. *Mon. Not. R. Astron. Soc.* **2019**, *487*, 5118–5128. [CrossRef]
195. Khadka, N.; Ratra, B. Using quasar X-ray and UV flux measurements to constrain cosmological model parameters. *Mon. Not. R. Astron. Soc.* **2020**, *497*, 263–278. [CrossRef]
196. Khadka, N.; Ratra, B. Quasar X-ray and UV flux, baryon acoustic oscillation, and Hubble parameter measurement constraints on cosmological model parameters. *Mon. Not. R. Astron. Soc.* **2020**, *492*, 4456–4468. [CrossRef]
197. Khadka, N.; Ratra, B. Determining the range of validity of quasar X-ray and UV flux measurements for constraining cosmological model parameters. *Mon. Not. R. Astron. Soc.* **2021**, *502*, 6140–6156. [CrossRef]
198. Khadka, N.; Yu, Z.; Zajaček, M.; Martinez-Aldama, M.L.; Czerny, B.; Ratra, B. Standardizing reverberation-measured Mg II time-lag quasars, by using the radius–luminosity relation, and constraining cosmological model parameters. *Mon. Not. R. Astron. Soc.* **2021**, *508*, 4722–4737. [CrossRef]
199. Khadka, N.; Ratra, B. Do quasar X-ray and UV flux measurements provide a useful test of cosmological models? *Mon. Not. R. Astron. Soc.* **2022**, *510*, 2753–2772. [CrossRef]
200. Khadka, N.; Martínez-Aldama, M.L.; Zajaček, M.; Czerny, B.; Ratra, B. Do reverberation-measured $H\beta$ quasars provide a useful test of cosmology? *Mon. Not. R. Astron. Soc.* **2022**, *513*, 1985. [CrossRef]
201. Cao, S.; Ryan, J.; Ratra, B. Using Pantheon and DES supernova, baryon acoustic oscillation, and Hubble parameter data to constrain the Hubble constant, dark energy dynamics, and spatial curvature. *Mon. Not. R. Astron. Soc.* **2021**, *504*, 300–310. [CrossRef]
202. Cao, S.; Khadka, N.; Ratra, B. Standardizing Dainotti-correlated gamma-ray bursts, and using them with standardized Amati-correlated gamma-ray bursts to constrain cosmological model parameters. *Mon. Not. R. Astron. Soc.* **2022**, *510*, 2928–2947. [CrossRef]
203. Cao, S.; Zajaček, M.; Panda, S.; Martínez-Aldama, M.L.; Czerny, B.; Ratra, B. Standardizing reverberation-measured C iv time-lag quasars, and using them with standardized Mg ii quasars to constrain cosmological parameters. *Mon. Not. R. Astron. Soc.* **2022**, *516*, 1721–1740. [CrossRef]
204. Cao, S.; Dainotti, M.; Ratra, B. Standardizing Platinum Dainotti-correlated gamma-ray bursts, and using them with standardized Amati-correlated gamma-ray bursts to constrain cosmological model parameters. *Mon. Not. R. Astron. Soc.* **2022**, *512*, 439–454. [CrossRef]
205. Cao, S.; Dainotti, M.; Ratra, B. Gamma-ray burst data strongly favour the three-parameter fundamental plane (Dainotti) correlation over the two-parameter one. *Mon. Not. R. Astron. Soc.* **2022**, *516*, 1386–1405. [CrossRef]
206. Cao, S.; Ratra, B. Using lower-redshift, non-CMB, data to constrain the Hubble constant and other cosmological parameters. *Mon. Not. R. Astron. Soc.* **2022**, *513*, 5686–5700. [CrossRef]

207. Gariazzo, S.; Di Valentino, E.; Mena, O.; Nunes, R.C. Late-time interacting cosmologies and the Hubble constant tension. *Phys. Rev. D* **2022**, *106*, 023530. [CrossRef]
208. Yang, W.; Pan, S.; Mena, O.; Di Valentino, E. On the dynamics of a dark sector coupling. *JHEAP* **2023**, *40*, 19–40. [CrossRef]
209. Spergel, D.N. The dark side of cosmology: Dark matter and dark energy. *Science* **2015**, *347*, 1100–1102. [CrossRef]
210. Brout, D.; Scolnic, D.; Popovic, B.; Riess, A.G.; Carr, A.; Zuntz, J.; Kessler, R.; Davis, T.M.; Hinton, S.; Jones, D.; et al. The Pantheon+ Analysis: Cosmological Constraints. *Astron. Astrophys.* **2022**, *938*, 110. [CrossRef]
211. Riess, A.G.; Strolger, L.G.; Tonry, J.; Casertano, S.; Ferguson, H.C.; Mobasher, B.; Challis, P.; Filippenko, A.V.; Jha, S.; Li, W.; et al. Type Ia Supernova Discoveries at z > 1 from the Hubble Space Telescope: Evidence for Past Deceleration and Constraints on Dark Energy Evolution. *Astron. Astrophys.* **2004**, *607*, 665–687.
212. Scolnic, D.M.; Jones, D.O.; Rest, A.; Pan, Y.C.; Chornock, R.; Foley, R.J.; Huber, M.E.; Kessler, R.; Narayan, G.; Riess, A.G.; et al. The Complete Light-curve Sample of Spectroscopically Confirmed SNe Ia from Pan-STARRS1 and Cosmological Constraints from the Combined Pantheon Sample. *Astron. Astrophys.* **2018**, *859*, 101. [CrossRef]
213. Madhavacheril, M.S.; Qu, F.J.; Sherwin, B.D.; MacCrann, N.; Li, Y.; Abril-Cabezas, I.; Ade, P.A.R.; Aiola, S.; Alford, T.; Amiri, M.; et al. The Atacama Cosmology Telescope: DR6 Gravitational Lensing Map and Cosmological Parameters. *arXiv* **2023**, arXiv:2304.05203.
214. Caresia, P.; Matarrese, S.; Moscardini, L. Constraints on extended quintessence from high-redshift supernovae. *Astrophys. J.* **2004**, *605*, 21–29. [CrossRef]
215. Pettorino, V.; Baccigalupi, C.; Mangano, G. Extended quintessence with an exponential coupling. *JCAP* **2005**, *1*, 14. [CrossRef]
216. Nesseris, S.; Perivolaropoulos, L. The Limits of Extended Quintessence. *Phys. Rev. D* **2007**, *75*, 023517. [CrossRef]
217. Hrycyna, O.; Szydlowski, M. Non-minimally coupled scalar field cosmology on the phase plane. *JCAP* **2009**, *4*, 26. [CrossRef]
218. Li, B.; Mota, D.F.; Barrow, J.D. N-body simulations for extended quintessence models. *Astrophys. J.* **2011**, *728*, 109. [CrossRef]
219. Fan, Y.; Wu, P.; Yu, H. Spherical collapse in the extended quintessence cosmological models. *Phys. Rev. D* **2015**, *92*, 083529. [CrossRef]
220. Fan, Y.; Wu, P.; Yu, H. Cosmological perturbations of non-minimally coupled quintessence in the metric and Palatini formalisms. *Phys. Lett. B* **2015**, *746*, 230–236. [CrossRef]
221. Davari, Z.; Marra, V.; Malekjani, M. Cosmological constrains on minimally and non-minimally coupled scalar field models. *Mon. Not. R. Astron. Soc.* **2020**, *491*, 1920–1933. [CrossRef]
222. Figueroa, D.G.; Florio, A.; Opferkuch, T.; Stefanek, B.A. Lattice simulations of non-minimally coupled scalar fields in the Jordan frame. *SciPost Phys.* **2023**, *15*, 77. [CrossRef]
223. Caldwell, R.R.; Kamionkowski, M.; Weinberg, N.N. Phantom energy and cosmic doomsday. *Phys. Rev. Lett.* **2003**, *91*, 071301. [CrossRef]
224. Ferreira, P.G.; Joyce, M. Cosmology with a primordial scaling field. *Phys. Rev.* **1998**, *D58*, 023503. [CrossRef]
225. Brax, P.; Martin, J. Quintessence and supergravity. *Phys. Lett. B* **1999**, *468*, 40–45. [CrossRef]
226. Sahni, V.; Wang, L.M. A New cosmological model of quintessence and dark matter. *Phys. Rev.* **2000**, *D62*, 103517. [CrossRef]
227. Barreiro, T.; Copeland, E.J.; Nunes, N.J. Quintessence arising from exponential potentials. *Phys. Rev.* **2000**, *D61*, 127301. [CrossRef]
228. Albrecht, A.; Skordis, C. Phenomenology of a realistic accelerating universe using only Planck scale physics. *Phys. Rev. Lett.* **2000**, *84*, 2076–2079. [CrossRef]
229. Urena-Lopez, L.A.; Matos, T. A New cosmological tracker solution for quintessence. *Phys. Rev.* **2000**, *D62*, 081302.
230. Chang, H.Y.; Scherrer, R.J. Reviving Quintessence with an Exponential Potential. *arXiv* **2016**, arXiv:1608.03291.
231. Frieman, J.A.; Hill, C.T.; Stebbins, A.; Waga, I. Cosmology with ultralight pseudo Nambu-Goldstone bosons. *Phys. Rev. Lett.* **1995**, *75*, 2077–2080. [CrossRef] [PubMed]
232. Barger, V.; Gao, Y.; Marfatia, D. Accelerating cosmologies tested by distance measures. *Phys. Lett.* **2007**, *B648*, 127–132. [CrossRef]
233. Chevallier, M.; Polarski, D. Accelerating universes with scaling dark matter. *Int. J. Mod. Phys.* **2001**, *D10*, 213–224. [CrossRef]
234. Linder, E.V. Exploring the expansion history of the universe. *Phys. Rev. Lett.* **2003**, *90*, 091301. [CrossRef]
235. Ratra, B. Quantum Mechanics of Exponential Potential Inflation. *Phys. Rev. D* **1989**, *40*, 3939. [CrossRef]
236. Ratra, B. Restoration of Spontaneously Broken Continuous Symmetries in de Sitter Space-Time. *Phys. Rev. D* **1985**, *31*, 1931–1955. [CrossRef]
237. Bettoni, D.; Rubio, J. Quintessential Inflation: A Tale of Emergent and Broken Symmetries. *Galaxies* **2022**, *10*, 22. [CrossRef]
238. Ratra, B.; Peebles, P.J.E. Inflation in an open universe. *Phys. Rev.* **1995**, *D52*, 1837–1894. [CrossRef]
239. Chitov, G.Y.; August, T.; Natarajan, A.; Kahniashvili, T. Mass Varying Neutrinos, Quintessence, and the Accelerating Expansion of the Universe. *Phys. Rev. D* **2011**, *83*, 045033. [CrossRef]
240. Mandal, S.; Chitov, G.Y.; Avsajanishvili, O.; Singha, B.; Kahniashvili, T. Mass varying neutrinos with different quintessence potentials. *JCAP* **2021**, *5*, 18. [CrossRef]
241. Podariu, S.; Ratra, B. Supernovae Ia constraints on a time variable cosmological 'constant'. *Astrophys. J.* **2000**, *532*, 109–117. [CrossRef]
242. Binetruy, P. Models of dynamical supersymmetry breaking and quintessence. *Phys. Rev. D* **1999**, *60*, 063502. [CrossRef]
243. Perrotta, F.; Baccigalupi, C.; Matarrese, S. Extended quintessence. *Phys. Rev. D* **1999**, *61*, 023507. [CrossRef]
244. Baccigalupi, C.; Matarrese, S.; Perrotta, F. Tracking extended quintessence. *Phys. Rev. D* **2000**, *62*, 123510. [CrossRef]

245. Tonry, J.L.; Schmidt, B.P.; Barris, B.; Candia, P.; Challis, P.; Clocchiatti, A.; Coil, A.L.; Filippenko, A.V.; Garnavich, P.; Hogan, C.; et al. Cosmological Results from High-z Supernovae. *Astron. Astrophys.* **2003**, *594*, 1–24.

246. Aldering, G.; Akerlof, C.W.; Amanullah, R.; Astier, P.; Barrelet, E.; Bebek, C.; Bergstrom, L.; Bercovitz, J.; Bernstein, G.M.; Bester, M.; et al. Overview of the Supernova/Acceleration Probe (SNAP). *Proc. SPIE Int. Soc. Opt. Eng.* **2002**, *4835*, 146.

247. Doran, M.; Karwan, K.; Wetterich, C. Observational constraints on the dark energy density evolution. *JCAP* **2005**, *11*, 7. [CrossRef]

248. Readhead, A.C.S.; Mason, B.S.; Contaldi, C.R.; Pearson, T.J.; Bond, J.R.; Myers, S.T.; Padin, S.; Sievers, J.L.; Cartwright, J.K.; Shepherd, M.C.; et al. Extended Mosaic Observations with the Cosmic Background Imager. *Astron. Astrophys.* **2004**, *609*, 498–512. [CrossRef]

249. Rebolo, R.; Battye, R.A.; Carreira, P.; Cleary, K.; Davies, R.D.; Davis, R.J.; Dickinson, C.; Genova-Santos, R.; Grainge, K.; Gutiérrez, C.M.; et al. Cosmological parameter estimation using Very Small Array data out to l = 1500. *Mon. Not. R. Astron. Soc.* **2004**, *353*, 747–759. [CrossRef]

250. Tegmark, M.; Strauss, M.A.; Blanton, M.R.; Abazajian, K.; Dodelson, S.; Sandvik, H.; Wang, X.; Weinberg, D.H.; Zehavi, I.; Bahcall, N.A.; et al. Cosmological parameters from SDSS and WMAP. *Phys. Rev. D* **2004**, *69*, 103501. [CrossRef]

251. Freedman, W.L.; Madore, B.F.; Gibson, B.K.; Ferrarese, L.; Kelson, D.D.; Sakai, S.; Mould, J.R.; Kennicutt, R.C., Jr.; Ford, H.C.; Graham, J.A.; et al. Final Results from the Hubble Space Telescope Key Project to Measure the Hubble Constant. *Astron. Astrophys.* **2001**, *553*, 47–72.

252. Pavlov, A.; Westmoreland, S.; Saaidi, K.; Ratra, B. Nonflat time-variable dark energy cosmology. *Phys. Rev. D* **2013**, *88*, 123513. [CrossRef]

253. Fuzfa, A.; Alimi, J.M. Some impacts of quintessence models on cosmic structure formation. *AIP Conf. Proc.* **2006**, *861*, 858–866.

254. Astier, P.; Guy, J.; Regnault, N.; Pain, R.; Aubourg, E.; Balam, D.; Basa, S.; Carlberg, R.G.; Fabbro, S.; Fouchez, D.; et al. The Supernova Legacy Survey: Measurement of Ω_M, Ω_Λ and w from the first year data set. *Astron. Astrophys.* **2006**, *447*, 31–48. [CrossRef]

255. Farooq, O.; Mania, D.; Ratra, B. Observational constraints on non-flat dynamical dark energy cosmological models. *Astrophys. Space Sci.* **2015**, *357*, 11. [CrossRef]

256. Suzuki, N.; Rubin, D.; Lidman, C.; Aldering, G.; Amanullah, R.; Barbary, K.; Barrientos, L.F.; Botyanszki, J.; Brodwin, M.; Connolly, N.; et al. The Hubble Space Telescope Cluster Supernova Survey. V. Improving the Dark-energy Constraints above z > 1 and Building an Early-type-hosted Supernova Sample. *Astron. Astrophys.* **2012**, *746*, 85.

257. Simon, J.; Verde, L.; Jimenez, R. Constraints on the redshift dependence of the dark energy potential. *Phys. Rev. D* **2005**, *71*, 123001. [CrossRef]

258. Gaztanaga, E.; Cabre, A.; Hui, L. Clustering of Luminous Red Galaxies IV: Baryon Acoustic Peak in the Line-of-Sight Direction and a Direct Measurement of H(z). *Mon. Not. R. Astron. Soc.* **2009**, *399*, 1663–1680. [CrossRef]

259. Percival, W.J.; Reid, B.A.; Eisenstein, D.J.; Bahcall, N.A.; Budavari, T.; Frieman, J.A.; Fukugita, M.; Gunn, J.E.; Ivezić, Ž.; Knapp, G.R.; et al. Baryon acoustic oscillations in the Sloan Digital Sky Survey Data Release 7 galaxy sample. *Mon. Not. R. Astron. Soc.* **2010**, *401*, 2148–2168. [CrossRef]

260. Farooq, O.; Crandall, S.; Ratra, B. Binned Hubble parameter measurements and the cosmological deceleration-acceleration transition. *Phys. Lett. B* **2013**, *726*, 72–82. [CrossRef]

261. Blomqvist, M.; du Mas des Bourboux, H.; Busca, N.G.; de Sainte Agathe, V.; Rich, J.; Balland, C.; Bautista, J.E.; Dawson, K.; Font-Ribera, A.; Guy, J.; et al. Baryon acoustic oscillations from the cross-correlation of Lyα absorption and quasars in eBOSS DR14. *Astron. Astrophys* **2019**, *629*, A86. [CrossRef]

262. Halder, A.; Pandey, M. Probing the effects of primordial black holes on 21-cm EDGES signal along with interacting dark energy and dark matter–baryon scattering. *Mon. Not. R. Astron. Soc.* **2021**, *508*, 3446–3454. [CrossRef]

263. Cooke, R.J.; Pettini, M.; Steidel, C.C. One Percent Determination of the Primordial Deuterium Abundance. *Astrophys. J.* **2018**, *855*, 102. [CrossRef]

264. Camarena, D.; Marra, V. On the use of the local prior on the absolute magnitude of Type Ia supernovae in cosmological inference. *Mon. Not. R. Astron. Soc.* **2021**, *504*, 5164–5171. [CrossRef]

265. Doran, M.; Lilley, M.J.; Schwindt, J.; Wetterich, C. Quintessence and the separation of CMB peaks. *Astrophys. J.* **2001**, *559*, 501–506. [CrossRef]

266. de Bernardis, P.; Ade, P.A.R.; Bock, J.J.; Bond, J.R.; Borrill, J.; Boscaleri, A.; Coble, K.; Crill, B.P.; de Gasperis, G.; de Troia, G.; et al. First results from the BOOMERanG experiment. *AIP Conf. Proc.* **2001**, *555*, 85–94.

267. Hanany, S.; Ade, P.; Balbi, A.; Bock, J.; Borrill, J.; Boscaleri, A.; de Bernardis, P.; Ferreira, P.G.; Hristov, V.V.; Jaffe, A.H.; et al. MAXIMA-1: A Measurement of the Cosmic Microwave Background Anisotropy on Angular Scales of 10′-5°. *Astrophys. J. Lett* **2000**, *545*, L5–L9. [CrossRef]

268. Hebecker, A.; Wetterich, C. Natural quintessence? *Phys. Lett. B* **2001**, *497*, 281–288. [CrossRef]

269. Caldwell, R.R.; Doran, M.; Mueller, C.M.; Schafer, G.; Wetterich, C. Early quintessence in light of WMAP. *Astrophys. J. Lett.* **2003**, *591*, L75–L78. [CrossRef]

270. Hinshaw, G.; Spergel, D.N.; Verde, L.; Hill, R.S.; Meyer, S.S.; Barnes, C.; Bennett, C.L.; Halpern, M.; Jarosik, N.; Kogut, A.; et al. First-Year Wilkinson Microwave Anisotropy Probe (WMAP) Observations: The Angular Power Spectrum. *Astrophys. J. Suppl.* **2003**, *148*, 135–159. [CrossRef]

271. Kogut, A.; Spergel, D.N.; Barnes, C.; Bennett, C.L.; Halpern, M.; Hinshaw, G.; Jarosik, N.; Limon, M.; Meyer, S.S.; Page, L.; et al. First-Year Wilkinson Microwave Anisotropy Probe (WMAP) Observations: Temperature-Polarization Correlation. *Astrophys. J. Suppl.* **2003**, *148*, 161–173. [CrossRef]

272. Pearson, T.J.; Mason, B.S.; Readhead, A.C.S.; Shepherd, M.C.; Sievers, J.L.; Udomprasert, P.S.; Cartwright, J.K.; Farmer, A.J.; Padin, S.; Myers, S.T.; et al. The Anisotropy of the Microwave Background to l = 3500: Mosaic Observations with the Cosmic Background Imager. *Astron. Astrophys.* **2003**, *591*, 556–574. [CrossRef]

273. Mason, B.S.; Pearson, T.J.; Readhead, A.C.S.; Shepherd, M.C.; Sievers, J.; Udomprasert, P.S.; Cartwright, J.K.; Farmer, A.J.; Padin, S.; Myers, S.T.; et al. The Anisotropy of the Microwave Background to l = 3500: Deep Field Observations with the Cosmic Background Imager. *Astron. Astrophys.* **2003**, *591*, 540–555. [CrossRef]

274. Kuo, C.L.; Ade, P.A.R.; Bock, J.J.; Cantalupo, C.; Daub, M.D.; Goldstein, J.; Holzapfel, W.L.; Lange, A.E.; Lueker, M.; Newcomb, M.; et al. High-Resolution Observations of the Cosmic Microwave Background Power Spectrum with ACBAR. *Astron. Astrophys.* **2004**, *600*, 32–51. [CrossRef]

275. Percival, W.J.; Baugh, C.M.; Bland-Hawthorn, J.; Bridges, T.; Cannon, R.; Cole, S.; Colless, M.; Collins, C.; Couch, W.; Dalton, G.; et al. The 2dF Galaxy Redshift Survey: The power spectrum and the matter content of the Universe. *Mon. Not. R. Astron. Soc.* **2001**, *327*, 1297–1306. [CrossRef]

276. Peacock, J.A.; Cole, S.; Norberg, P.; Baugh, C.M.; Bland-Hawthorn, J.; Bridges, T.; Cannon, R.D.; Colless, M.; Collins, C.; Couch, W.; et al. A measurement of the cosmological mass density from clustering in the 2dF Galaxy Redshift Survey. *Nature* **2001**, *410*, 169–173. [CrossRef]

277. Verde, L.; Heavens, A.F.; Percival, W.J.; Matarrese, S.; Baugh, C.M.; Bland-Hawthorn, J.; Bridges, T.; Cannon, R.; Cole, S.; Colless, M.; et al. The 2dF Galaxy Redshift Survey: The bias of galaxies and the density of the Universe. *Mon. Not. R. Astron. Soc.* **2002**, *335*, 432–440. [CrossRef]

278. Gnedin, N.Y.; Hamilton, A.J.S. Matter power spectrum from the Lyman-alpha forest: Myth or reality? *Mon. Not. R. Astron. Soc.* **2002**, *334*, 107–116. [CrossRef]

279. Croft, R.A.C.; Weinberg, D.H.; Bolte, M.; Burles, S.; Hernquist, L.; Katz, N.; Kirkman, D.; Tytler, D. Towards a precise measurement of matter clustering: Lyman alpha forest data at redshifts 2-4. *Astrophys. J.* **2002**, *581*, 20–52. [CrossRef]

280. Mukherjee, P.; Banday, A.J.; Riazuelo, A.; Gorski, K.M.; Ratra, B. COBE-DMR-Normalized dark energy cosmogony. *Astrophys. J.* **2003**, *598*, 767–778. [CrossRef]

281. Brax, P.; Martin, J.; Riazuelo, A. Quintessence model building. In Proceedings of the 6th Workshop on Non-Perturbative Quantum Chromodynamics, Paris, France, 5–9 June 2001; pp. 315–325.

282. Gorski, K.M.; Ratra, B.; Stompor, R.; Sugiyama, N.; Banday, A.J. COBE—DMR normalized open CDM cosmogonies. *Astrophys. J. Suppl.* **1998**, *114*, 1–30. [CrossRef]

283. Samushia, L.; Ratra, B. Constraints on Dark Energy from Galaxy Cluster Gas Mass Fraction versus Redshift data. *Astrophys. J. Lett.* **2008**, *680*, L1–L4. [CrossRef]

284. Allen, S.W.; Rapetti, D.A.; Schmidt, R.W.; Ebeling, H.; Morris, G.; Fabian, A.C. Improved constraints on dark energy from Chandra X-ray observations of the largest relaxed galaxy clusters. *Mon. Not. R. Astron. Soc.* **2008**, *383*, 879–896. [CrossRef]

285. Gott, J.R., III; Vogeley, M.S.; Podariu, S.; Ratra, B. Median statistics, H(0), and the accelerating universe. *Astrophys. J.* **2001**, *549*, 1–17. [CrossRef]

286. Chen, G.; Gott, J.R., III; Ratra, B. Non-Gaussian error distribution of Hubble constant measurements. *Publ. Astron. Soc. Pac.* **2003**, *115*, 1269–1279. [CrossRef]

287. Fields, B.; Sarkar, S. Big-Bang nucleosynthesis (2006 Particle Data Group mini-review). *J. Phys. G.* **2006**, *33*, 220.

288. Chen, Y.; Ratra, B.; Biesiada, M.; Li, S.; Zhu, Z.H. Constraints on non-flat cosmologies with massive neutrinos after Planck 2015. *Astrophys. J.* **2016**, *829*, 61. [CrossRef]

289. Planck Collaboration; Adam, R.; Ade, P.A.R.; Aghanim, N.; Alves, M.I.R.; Arnaud, M.; Ashdown, M.; Aumont, J.; Baccigalupi, C.; Banday, A.J.; et al. Planck 2015 results. X. Diffuse component separation: Foreground maps. *Astron. Astrophys.* **2016**, *594*, A10. [CrossRef]

290. Betoule, M.; Kessler, R.; Guy, J.; Mosher, J.; Hardin, D.; Biswas, R.; Astier, P.; El-Hage, P.; Konig, M.; Kuhlmann, S.; et al. Improved cosmological constraints from a joint analysis of the SDSS-II and SNLS supernova samples. *Astron.Astrophys.* **2014**, *568*, A22. [CrossRef]

291. Park, C.G.; Ratra, B. Observational constraints on the tilted spatially-flat and the untilted nonflat ϕCDM dynamical dark energy inflation models. *Astrophys. J.* **2018**, *868*, 83. [CrossRef]

292. Planck Collaboration; Aghanim, N.; Akrami, Y.; Ashdown, M.; Aumont, J.; Baccigalupi, C.; Ballardini, M.; Banday, A.J.; Barreiro, R.B.; Bartolo, N.; et al. Planck intermediate results. LI. Features in the cosmic microwave background temperature power spectrum and shifts in cosmological parameters. *Astron. Astrophys.* **2017**, *607*, A95.

293. Font-Ribera, A.; McDonald, P.; Mostek, N.; Reid, B.A.; Seo, H.J.; Slosar, A. DESI and other dark energy experiments in the era of neutrino mass measurements. *JCAP* **2014**, *1405*, 23. [CrossRef]

294. Ooba, J.; Ratra, B.; Sugiyama, N. Planck 2015 constraints on spatially-flat dynamical dark energy models. *Astrophys. Space Sci.* **2019**, *364*, 176. [CrossRef]

295. Blas, D.; Lesgourgues, J.; Tram, T. The Cosmic Linear Anisotropy Solving System (CLASS). Part II: Approximation schemes. *JCAP* **2011**, *2011*, 34. [CrossRef]

296. Audren, B.; Lesgourgues, J.; Benabed, K.; Prunet, S. Conservative constraints on early cosmology with MONTE PYTHON. *JCAP* **2013**, *2013*, 1. [CrossRef]
297. Park, C.G.; Ratra, B. Measuring the Hubble constant and spatial curvature from supernova apparent magnitude, baryon acoustic oscillation, and Hubble parameter data. *Astrophys. Space Sci.* **2019**, *364*, 134. [CrossRef]
298. Wyithe, J.S.B.; Bolton, J.S. Near-zone sizes and the rest-frame extreme ultraviolet spectral index of the highest redshift quasars. *Mon. Not. R. Astron. Soc.* **2011**, *412*, 1926. [CrossRef]
299. Becker, G.D.; Bolton, J.S. New Measurements of the Ionizing Ultraviolet Background over $2 < z < 5$ and Implications for Hydrogen Reionization. *Mon. Not. R. Astron. Soc.* **2013**, *436*, 1023.
300. Songaila, A.; Cowie, L.L. The Evolution of Lyman Limit Absorption Systems to Redshift Six. *Astrophys. J.* **2010**, *721*, 1448–1466. [CrossRef]
301. Prochaska, J.X.; O'Meara, J.M.; Worseck, G. A Definitive Survey for Lyman Limit Systems at z ~ 3.5 with the Sloan Digital Sky Survey. *Astron. Astrophys.* **2010**, *718*, 392–416. [CrossRef]
302. Henning, J.W.; Sayre, J.T.; Reichardt, C.L.; Ade, P.A.R.; Anderson, A.J.; Austermann, J.E.; Beall, J.A.; Bender, A.N.; Benson, B.A.; Bleem, L.E.; et al. Measurements of the Temperature and E-mode Polarization of the CMB from 500 Square Degrees of SPTpol Data. *Astron. Astrophys.* **2018**, *852*, 97. [CrossRef]
303. Park, C.G.; Ratra, B. Using SPT polarization, *Planck* 2015, and non-CMB data to constrain tilted spatially-flat and untilted nonflat ΛCDM, XCDM, and ϕCDM dark energy inflation cosmologies. *Phys. Rev. D* **2020**, *101*, 083508. [CrossRef]
304. Chen, L.; Huang, Q.G.; Wang, K. Distance priors from Planck final release. *J. Cosmol. Astropart. Phys.* **2019**, *2019*, 28. [CrossRef]
305. Nesseris, S.; Pantazis, G.; Perivolaropoulos, L. Tension and constraints on modified gravity parametrizations of $G_{\mathrm{eff}}(z)$ from growth rate and Planck data. *Phys. Rev. D* **2017**, *96*, 023542. [CrossRef]
306. Pavlov, A.; Samushia, L.; Ratra, B. Forecasting cosmological parameter constraints from near-future space-based galaxy surveys. *Astrophys. J.* **2012**, *760*, 19. [CrossRef]
307. Orsi, A.; Baugh, C.; Lacey, C.; Cimatti, A.; Wang, Y.; Zamorani, G. Probing dark energy with future redshift surveys: A comparison of emission line and broad-band selection in the near-infrared. *Mon. Not. R. Astron. Soc.* **2010**, *405*, 1006–1024. [CrossRef]
308. Geach, J.E.; Cimatti, A.; Percival, W.; Wang, Y.; Guzzo, L.; Zamorani, G.; Rosati, P.; Pozzetti, L.; Orsi, A.; Baugh, C.M.; et al. Empirical Hα emitter count predictions for dark energy surveys. *Mon. Not. R. Astron. Soc.* **2010**, *402*, 1330–1338. [CrossRef]
309. Percival, W.J.; Schaefer, B.M. Galaxy peculiar velocities and evolution-bias. *Mon. Not. R. Astron. Soc.* **2008**, *385*, 78. [CrossRef]
310. Guzzo, L.; Pierleoni, M.; Meneux, B.; Branchini, E.; Le Fèvre, O.; Marinoni, C.; Garilli, B.; Blaizot, J.; De Lucia, G.; Pollo, A.; et al. A test of the nature of cosmic acceleration using galaxy redshift distortions. *Nature* **2008**, *451*, 541–544. [CrossRef]
311. Blake, C.; Davis, T.; Poole, G.B.; Parkinson, D.; Brough, S.; Colless, M.; Contreras, C.; Couch, W.; Croom, S.; Drinkwater, M.J.; et al. The WiggleZ Dark Energy Survey: Testing the cosmological model with baryon acoustic oscillations at z = 0.6. *Mon. Not. R. Astron. Soc.* **2011**, *415*, 2892–2909. [CrossRef]
312. Blake, C.; Brough, S.; Colless, M.; Contreras, C.; Couch, W.; Croom, S.; Davis, T.; Drinkwater, M.J.; Forster, K.; Gilbank, D.; et al. The WiggleZ Dark Energy Survey: The growth rate of cosmic structure since redshift z = 0.9. *Mon. Not. R. Astron. Soc.* **2011**, *415*, 2876–2891. [CrossRef]
313. Reid, B.A.; Samushia, L.; White, M.; Percival, W.J.; Manera, M.; Padmanabhan, N.; Ross, A.J.; Sánchez, A.G.; Bailey, S.; Bizyaev, D.; et al. The clustering of galaxies in the SDSS-III Baryon Oscillation Spectroscopic Survey: Measurements of the growth of structure and expansion rate at z = 0.57 from anisotropic clustering. *Mon. Not. R. Astron. Soc.* **2012**, *426*, 2719–2737. [CrossRef]
314. Anderson, L.; Aubourg, E.; Bailey, S.; Bizyaev, D.; Blanton, M.; Bolton, A.S.; Brinkmann, J.; Brownstein, J.R.; Burden, A.; Cuesta, A.J.; et al. The clustering of galaxies in the SDSS-III Baryon Oscillation Spectroscopic Survey: Baryon acoustic oscillations in the Data Release 9 spectroscopic galaxy sample. *Mon. Not. R. Astron. Soc.* **2012**, *427*, 3435–3467. [CrossRef]
315. Pavlov, A.; Farooq, O.; Ratra, B. Cosmological constraints from large-scale structure growth rate measurements. *Phys. Rev. D* **2014**, *90*, 023006. [CrossRef]
316. Samushia, L.; Percival, W.J.; Raccanelli, A. Interpreting large-scale redshift-space distortion measurements. *Mon. Not. R. Astron. Soc.* **2012**, *420*, 2102–2119. [CrossRef]
317. Avsajanishvili, O.; Arkhipova, N.A.; Samushia, L.; Kahniashvili, T. Growth Rate in the Dynamical Dark Energy Models. *Eur. Phys. J.* **2014**, *C74*, 3127. [CrossRef]
318. Gupta, G.; Sen, S.; Sen, A.A. GCG Parametrization for Growth Function and Current Constraints. *JCAP* **2012**, *1204*, 28. [CrossRef]
319. Giostri, R.; Vargas dos Santos, M.; Waga, I.; Reis, R.R.R.; Calvão, M.O.; Lago, B.L. From cosmic deceleration to acceleration: New constraints from SN Ia and BAO/CMB. *JCAP* **2012**, *3*, 27. [CrossRef]
320. Avsajanishvili, O.; Huang, Y.; Samushia, L.; Kahniashvili, T. The observational constraints on the flat ϕCDM models. *Eur. Phys. J.* **2018**, *C78*, 773. [CrossRef]
321. Sola Peracaula, J.; Gomez-Valent, A.; de Cruz Pérez, J. Signs of Dynamical Dark Energy in Current Observations. *Phys. Dark Univ.* **2019**, *25*, 100311. [CrossRef]
322. Gil-Marín, H.; Percival, W.J.; Verde, L.; Brownstein, J.R.; Chuang, C.H.; Kitaura, F.S.; Rodríguez-Torres, S.A.; Olmstead, M.D. The clustering of galaxies in the SDSS-III Baryon Oscillation Spectroscopic Survey: RSD measurement from the power spectrum and bispectrum of the DR12 BOSS galaxies. *Mon. Not. R. Astron. Soc.* **2017**, *465*, 1757–1788. [CrossRef]

323. Gil-Marín, H.; Guy, J.; Zarrouk, P.; Burtin, E.; Chuang, C.H.; Percival, W.J.; Ross, A.J.; Ruggeri, R.; Tojerio, R.; Zhao, G.B.; et al. The clustering of the SDSS-IV extended Baryon Oscillation Spectroscopic Survey DR14 quasar sample: Structure growth rate measurement from the anisotropic quasar power spectrum in the redshift range $0.8 < z < 2.2$. *Mon. Not. R. Astron. Soc.* **2018**, *477*, 1604–1638.

324. Mohammad, F.G.; Bianchi, D.; Percival, W.J.; de la Torre, S.; Guzzo, L.; Granett, B.R.; Branchini, E.; Bolzonella, M.; Garilli, B.; Scodeggio, M.; et al. The VIMOS Public Extragalactic Redshift Survey (VIPERS). Unbiased clustering estimate with VIPERS slit assignment. *Astron. Astrophys.* **2018**, *619*, A17. [CrossRef]

325. Hildebrandt, H.; Viola, M.; Heymans, C.; Joudaki, S.; Kuijken, K.; Blake, C.; Erben, T.; Joachimi, B.; Klaes, D.; Miller, L.; et al. KiDS-450: Cosmological parameter constraints from tomographic weak gravitational lensing. *Mon. Not. R. Astron. Soc.* **2017**, *465*, 1454–1498. [CrossRef]

326. Park, C.G.; Ratra, B. Observational constraints on the tilted flat-XCDM and the untilted nonflat XCDM dynamical dark energy inflation parameterizations. *Astrophys. Space Sci.* **2019**, *364*, 82. [CrossRef]

327. Samushia, L.; Ratra, B. Constraints on dark energy from baryon acoustic peak and galaxy cluster gas mass measurements. *Astrophys. J.* **2009**, *703*, 1904–1910. [CrossRef]

328. Samushia, L.; Dev, A.; Jain, D.; Ratra, B. Constraints on dark energy from the lookback time versus redshift test. *Phys. Lett. B* **2010**, *693*, 509–514. [CrossRef]

329. Capozziello, S.; Cardone, V.F.; Funaro, M.; Andreon, S. Constraining dark energy models using the lookback time to galaxy clusters and the age of the Universe. *Phys. Rev. D* **2004**, *70*, 123501. [CrossRef]

330. Ooba, J.; Ratra, B.; Sugiyama, N. Planck 2015 Constraints on the Nonflat ϕCDM Inflation Model. *Astrophys. J.* **2018**, *866*, 68. [CrossRef]

331. Ryan, J.; Doshi, S.; Ratra, B. Constraints on dark energy dynamics and spatial curvature from Hubble parameter and baryon acoustic oscillation data. *Mon. Not. R. Astron. Soc.* **2018**, *480*, 759–767. [CrossRef]

332. Chudaykin, A.; Ivanov, M.M.; Simonović, M. Optimizing large-scale structure data analysis with the theoretical error likelihood. *Phys. Rev. D* **2021**, *103*, 043525. [CrossRef]

333. Nunes, R.C.; Yadav, S.K.; Jesus, J.F.; Bernui, A. Cosmological parameter analyses using transversal BAO data. *Mon. Not. R. Astron. Soc.* **2020**, *497*, 2133–2141. [CrossRef]

334. de Carvalho, E.; Bernui, A.; Avila, F.; Novaes, C.P.; Nogueira-Cavalcante, J.P. BAO angular scale at zeff = 0.11 with the SDSS blue galaxies. *Astron. Astrophys.* **2021**, *649*, A20. [CrossRef]

335. Samushia, L.; Ratra, B. Cosmological Constraints from Hubble Parameter versus Redshift Data. *Astrophys. J. Lett.* **2006**, *650*, L5–L8. [CrossRef]

336. Chen, G.; Ratra, B. Constraints on scalar-field dark energy from galaxy cluster gas mass fraction versus redshift. *Astrophys. J. Lett.* **2004**, *612*, L1–L4. [CrossRef]

337. Wilson, K.M.; Chen, G.; Ratra, B. Supernova ia and galaxy cluster gas mass fraction constraints on dark energy. *Mod. Phys. Lett. A* **2006**, *21*, 2197–2204. [CrossRef]

338. Chen, G.; Ratra, B. Median statistics and the Hubble constant. *Publ. Astron. Soc. Pac.* **2011**, *123*, 1127–1132. [CrossRef]

339. Farooq, O.; Mania, D.; Ratra, B. Hubble parameter measurement constraints on dark energy. *Astrophys. J.* **2013**, *764*, 138. [CrossRef]

340. Farooq, O.; Ratra, B. Constraints on dark energy from the Lyα forest baryon acoustic oscillations measurement of the redshift 2.3 Hubble parameter. *Phys. Lett. B* **2013**, *723*, 1–6. [CrossRef]

341. Busca, N.G.; Delubac, T.; Rich, J.; Bailey, S.; Font-Ribera, A.; Kirkby, D.; Le Goff, J.M.; Pieri, M.M.; Slosar, A.; Aubourg, É.; et al. Baryon acoustic oscillations in the Lyα forest of BOSS quasars. *Astron. Astrophys.* **2013**, *552*, A96. [CrossRef]

342. Farooq, O.; Ratra, B. Hubble parameter measurement constraints on the cosmological deceleration-acceleration transition redshift. *Astrophys. J. Lett.* **2013**, *766*, L7. [CrossRef]

343. Chen, Y.; Kumar, S.; Ratra, B. Determining the Hubble constant from Hubble parameter measurements. *Astrophys. J.* **2017**, *835*, 86. [CrossRef]

344. Chuang, C.H.; Wang, Y. Modelling the anisotropic two-point galaxy correlation function on small scales and single-probe measurements of H(z), D_A(z) and f(z)σ_8(z) from the Sloan Digital Sky Survey DR7 luminous red galaxies. *Mon. Not. R. Astron. Soc.* **2013**, *435*, 255–262. [CrossRef]

345. Cao, S.; Biesiada, M.; Jackson, J.; Zheng, X.; Zhao, Y.; Zhu, Z.H. Measuring the speed of light with ultra-compact radio quasars. *JCAP* **2017**, *2*, 12. [CrossRef]

346. Cao, S.; Zheng, X.; Biesiada, M.; Qi, J.; Chen, Y.; Zhu, Z.H. Ultra-compact structure in intermediate-luminosity radio quasars: Building a sample of standard cosmological rulers and improving the dark energy constraints up to z ~ 3. *Astron. Astrophys.* **2017**, *606*, A15. [CrossRef]

347. Cao, S.; Ryan, J.; Ratra, B. Cosmological constraints from H ii starburst galaxy apparent magnitude and other cosmological measurements. *Mon. Not. R. Astron. Soc.* **2020**, *497*, 3191–3203. [CrossRef]

348. Chávez, R.; Terlevich, R.; Terlevich, E.; Bresolin, F.; Melnick, J.; Plionis, M.; Basilakos, S. The L-σ relation for massive bursts of star formation. *Mon. Not. R. Astron. Soc.* **2014**, *442*, 3565–3597. [CrossRef]

349. Chávez, R.; Plionis, M.; Basilakos, S.; Terlevich, R.; Terlevich, E.; Melnick, J.; Bresolin, F.; González-Morán, A.L. Constraining the dark energy equation of state with H II galaxies. *Mon. Not. R. Astron. Soc.* **2016**, *462*, 2431–2439. [CrossRef]

350. Risaliti, G.; Lusso, E. A Hubble Diagram for Quasars. *Astrophys. J.* **2015**, *815*, 33. [CrossRef]
351. Risaliti, G.; Lusso, E. Cosmological constraints from the Hubble diagram of quasars at high redshifts. *Nature Astron.* **2019**, *3*, 272–277. [CrossRef]
352. Cao, S.; Ryan, J.; Khadka, N.; Ratra, B. Cosmological constraints from higher redshift gamma-ray burst, H ii starburst galaxy, and quasar (and other) data. *Mon. Not. R. Astron. Soc.* **2021**, *501*, 1520–1538. [CrossRef]
353. Wang, J.S.; Wang, F.Y.; Cheng, K.S.; Dai, Z.G. Measuring dark energy with the $E_{iso} - E_p$ correlation of gamma-ray bursts using model-independent methods. *Astron. Astrophys.* **2016**, *585*, A68. [CrossRef]
354. Dirirsa, F.F.; Razzaque, S.; Piron, F.; Arimoto, M.; Axelsson, M.; Kocevski, D.; Longo, F.; Ohno, M.; Zhu, S. Spectral analysis of Fermi-LAT gamma-ray bursts with known redshift and their potential use as cosmological standard candles. *Astrophys. J.* **2019**, *887*, 13. [CrossRef]
355. González-Morán, A.L.; Chávez, R.; Terlevich, R.; Terlevich, E.; Bresolin, F.; Fernández-Arenas, D.; Plionis, M.; Basilakos, S.; Melnick, J.; Telles, E. Independent cosmological constraints from high-z H II galaxies. *Mon. Not. R. Astron. Soc.* **2019**, *487*, 4669–4694. [CrossRef]
356. Lusso, E.; Risaliti, G.; Nardini, E.; Bargiacchi, G.; Benetti, M.; Bisogni, S.; Capozziello, S.; Civano, F.; Eggleston, L.; Elvis, M.; et al. Quasars as standard candles. III. Validation of a new sample for cosmological studies. *Astron. Astrophys.* **2020**, *642*, A150. [CrossRef]
357. Abbott, T.M.C.; Allam, S.; Andersen, P.; Angus, C.; Asorey, J.; Avelino, A.; Avila, S.; Bassett, B.A.; Bechtol, K.; Bernstein, G.M.; et al. First Cosmology Results using Type Ia Supernovae from the Dark Energy Survey: Constraints on Cosmological Parameters. *Astrophys. J.* **2019**, *872*, L30. [CrossRef]
358. Brout, D.; Scolnic, D.; Kessler, R.; D'Andrea, C.B.; Davis, T.M.; Gupta, R.R.; Hinton, S.R.; Kim, A.G.; Lasker, J.; Lidman, C.; et al. First Cosmology Results Using SNe Ia from the Dark Energy Survey: Analysis, Systematic Uncertainties, and Validation. *Astron. Astrophys.* **2019**, *874*, 150. [CrossRef]
359. Homayouni, Y.; Trump, J.R.; Grier, C.J.; Horne, K.; Shen, Y.; Brandt, W.N.; Dawson, K.S.; Alvarez, G.F.; Green, P.J.; Hall, P.B.; et al. The Sloan Digital Sky Survey Reverberation Mapping Project: Mg II Lag Results from Four Years of Monitoring. *Astron. Astrophys.* **2020**, *901*, 55. [CrossRef]
360. Yu, Z.; Martini, P.; Penton, A.; Davis, T.M.; Malik, U.; Lidman, C.; Tucker, B.E.; Sharp, R.; Kochanek, C.S.; Peterson, B.M.; et al. OzDES Reverberation Mapping Programme: The first Mg II lags from 5 yr of monitoring. *Mon. Not. R. Astron. Soc.* **2021**, *507*, 3771–3788. [CrossRef]
361. Martínez-Aldama, M.L.; Czerny, B.; Kawka, D.; Karas, V.; Panda, S.; Zajaček, M.; Życki, P.T. Can Reverberation-measured Quasars Be Used for Cosmology? *Astron. Astrophys.* **2019**, *883*, 170. [CrossRef]
362. Khadka, N.; Zajaček, M.; Panda, S.; Martínez-Aldama, M.L.; Ratra, B. Consistency study of high- and low-accreting Mg ii quasars: No significant effect of the Fe ii to Mg ii flux ratio on the radius–luminosity relation dispersion. *Mon. Not. R. Astron. Soc.* **2022**, *515*, 3729–3748. [CrossRef]
363. Martínez-Aldama, M.L.; Zajaček, M.; Czerny, B.; Panda, S. Scatter Analysis along the Multidimensional Radius-Luminosity Relations for Reverberation-mapped Mg II Sources. *Astron. Astrophys.* **2020**, *903*, 86. [CrossRef]
364. Zajaček, M.; Czerny, B.; Martinez-Aldama, M.L.; Rałowski, M.; Olejak, A.; Przyłuski, R.; Panda, S.; Hryniewicz, K.; Śniegowska, M.; Naddaf, M.H.; et al. Time Delay of Mg II Emission Response for the Luminous Quasar HE 0435-4312: Toward Application of the High-accretor Radius-Luminosity Relation in Cosmology. *Astron. Astrophys.* **2021**, *912*, 10. [CrossRef]
365. Samushia, L.; Ratra, B. Constraining dark energy with gamma-ray bursts. *Astrophys. J.* **2010**, *714*, 1347–1354. [CrossRef]
366. Kowalski, M.; Rubin, D.; Aldering, G.; Agostinho, R.J.; Amadon, A.; Amanullah, R.; Balland, C.; Barbary, K.; Blanc, G.; Challis, P.J.; et al. Improved Cosmological Constraints from New, Old, and Combined Supernova Data Sets. *Astron. Astrophys.* **2008**, *686*, 749–778. [CrossRef]
367. Schaefer, B.E. The Hubble Diagram to Redshift >6 from 69 Gamma-Ray Bursts. *Astrophys. J.* **2007**, *660*, 16–46. [CrossRef]
368. Wang, Y. Model-Independent Distance Measurements from Gamma-Ray Bursts and Constraints on Dark Energy. *Phys. Rev. D* **2008**, *78*, 123532. [CrossRef]
369. Khadka, N.; Ratra, B. Constraints on cosmological parameters from gamma-ray burst peak photon energy and bolometric fluence measurements and other data. *Mon. Not. R. Astron. Soc.* **2020**, *499*, 391–403. [CrossRef]
370. Amati, L.; Frontera, F.; Tavani, M.; in't Zand, J.J.M.; Antonelli, A.; Costa, E.; Feroci, M.; Guidorzi, C.; Heise, J.; Masetti, N.; et al. Intrinsic spectra and energetics of BeppoSAX Gamma-Ray Bursts with known redshifts. *Astron. Astrophys.* **2002**, *390*, 81–89. [CrossRef]
371. Khadka, N.; Luongo, O.; Muccino, M.; Ratra, B. Do gamma-ray burst measurements provide a useful test of cosmological models? *JCAP* **2021**, *9*, 42. [CrossRef]
372. Fenimore, E.E.; Ramirez-Ruiz, E. Redshifts for 220 BATSE gamma-ray bursts determined by variability and the cosmological consequences. *arXiv* **2017**, arXiv:astro-ph/0004176.
373. Hu, J.P.; Wang, F.Y.; Dai, Z.G. Measuring cosmological parameters with a luminosity-time correlation of gamma-ray bursts. *Mon. Not. R. Astron. Soc.* **2021**, *507*, 730–742. [CrossRef]
374. Wang, F.Y.; Hu, J.P.; Zhang, G.Q.; Dai, Z.G. Standardized Long Gamma-Ray Bursts as a Cosmic Distance Indicator. *Astrophys. J.* **2022**, *924*, 97. [CrossRef]

375. Dainotti, M.G.; Cardone, V.F.; Capozziello, S. A time-luminosity correlation for Gamma Ray Bursts in the X-rays. *Mon. Not. R. Astron. Soc.* **2008**, *391*, 79. [CrossRef]
376. Dainotti, M.G.; Willingale, R.; Capozziello, S.; Fabrizio Cardone, V.; Ostrowski, M. Discovery of a Tight Correlation for Gamma-ray Burst Afterglows with "Canonical" Light Curves. *Astrophys. J.* **2010**, *722*, L215–L219. [CrossRef]
377. Dainotti, M.G.; Lenart, A.; Sarracino, G.; Nagataki, S.; Capozziello, S.; Fraija, N. The X-ray fundamental plane of the Platinum Sample, the Kilonovae and the SNe Ib/c associated with GRBs. *Astrophys. J.* **2020**, *904*, 97. [CrossRef]
378. Dainotti, M.G.; Nagataki, S.; Maeda, K.; Postnikov, S.; Pian, E. A study of gamma ray bursts with afterglow plateau phases associated with supernovae. *Astron. Astrophys.* **2017**, *600*, A98. [CrossRef]
379. Dainotti, M.G.; Lenart, A.L.; Fraija, N.; Nagataki, S.; Warren, D.C.; De Simone, B.; Srinivasaragavan, G.; Mata, A. Closure relations during the plateau emission of Swift GRBs and the fundamental plane. *Publ. Astron. Soc. Jap.* **2021**, *73*, 970–1000. [CrossRef]
380. Dainotti, M.G.; Omodei, N.; Srinivasaragavan, G.P.; Vianello, G.; Willingale, R.; O'Brien, P.; Nagataki, S.; Petrosian, V.; Nuygen, Z.; Hernandez, X.; et al. On the Existence of the Plateau Emission in High-energy Gamma-Ray Burst Light Curves Observed by Fermi-LAT. *Astrophys. J. Suppl.* **2021**, *255*, 13. [CrossRef]
381. Mania, D.; Ratra, B. Constraints on dark energy from H II starburst galaxy apparent magnitude versus redshift data. *Phys. Lett. B* **2012**, *715*, 9–14. [CrossRef]
382. Siegel, E.R.; Guzman, R.; Gallego, J.P.; Orduna Lopez, M.; Rodriguez Hidalgo, P. Towards a precision cosmology from starburst galaxies at z > 2. *Mon. Not. R. Astron. Soc.* **2005**, *356*, 1117–1122. [CrossRef]
383. Cao, S.; Ryan, J.; Ratra, B. Cosmological constraints from H ii starburst galaxy, quasar angular size, and other measurements. *Mon. Not. R. Astron. Soc.* **2022**, *509*, 4745–4757. [CrossRef]
384. González-Morán, A.L.; Chávez, R.; Terlevich, E.; Terlevich, R.; Fernández-Arenas, D.; Bresolin, F.; Plionis, M.; Melnick, J.; Basilakos, S.; Telles, E. Independent cosmological constraints from high-z H II galaxies: New results from VLT-KMOS data. *Mon. Not. R. Astron. Soc.* **2021**, *505*, 1441–1457. [CrossRef]
385. Allen, S.W.; Schmidt, R.W.; Ebeling, H.; Fabian, A.C.; van Speybroeck, L. Constraints on dark energy from Chandra observations of the largest relaxed galaxy clusters. *Mon. Not. R. Astron. Soc.* **2004**, *353*, 457. [CrossRef]
386. Waga, I.; Frieman, J.A. New constraints from high redshift supernovae and lensing statistics upon scalar field cosmologies. *Phys. Rev. D* **2000**, *62*, 043521. [CrossRef]
387. Chen, G.; Ratra, B. Cosmological constraints from compact radio source angular size versus redshift data. *Astrophys. J.* **2003**, *582*, 586–589. [CrossRef]
388. Podariu, S.; Daly, R.A.; Mory, M.P.; Ratra, B. Radio galaxy redshift-angular size data constraints on dark energy. *Astrophys. J.* **2003**, *584*, 577–579. [CrossRef]
389. Chae, K.H.; Chen, G.; Ratra, B.; Lee, D.W. Constraints on scalar—Field dark energy from the Cosmic Lens All—Sky Survey gravitational lens statistics. *Astrophys. J. Lett.* **2004**, *607*, L71–L74. [CrossRef]
390. Campanelli, L.; Fogli, G.L.; Kahniashvili, T.; Marrone, A.; Ratra, B. Galaxy cluster number count data constraints on cosmological parameters. *Eur. Phys. J. C* **2012**, *72*, 2218. [CrossRef]
391. Chen, Y.; Ratra, B. Galaxy cluster angular size data constraints on dark energy. *Astron. Astrophys.* **2012**, *543*, A104. [CrossRef]
392. Bonamente, M.; Joy, M.K.; La Roque, S.J.; Carlstrom, J.E.; Reese, E.D.; Dawson, K.S. Determination of the Cosmic Distance Scale from Sunyaev-Zel'dovich Effect and Chandra X-ray Measurements of High Redshift Galaxy Clusters. *Astrophys. J.* **2006**, *647*, 25–54. [CrossRef]
393. Maoz, D.; Bahcall, J.N.; Schneider, D.P.; Bahcall, N.A.; Djorgovski, S.; Doxsey, R.; Gould, A.; Kirhakos, S.; Meylan, G.; Yanny, B. The Hubble space telescope snapshot survey. 4. A Summary of the search for gravitationally lensed quasars. *Astrophys. J.* **1993**, *409*, 28–41. [CrossRef]
394. Falco, E.E. HST observations of gravitational lens systems. In *Gravitational Lenses in the Universe, Proceedings of the 31st Liège International Astrophysical Colloquium, Liège, Belgium, 21–25 June 1993*; Surdej, J., Fraipont-Caro, D., Gosset, E., Refsdal, S., Remy, M., Eds.; Universite de Liege, Institut d'Astrophysique: Liège, Belgium, 1993.
395. Crampton, D.; Morbey, C.L.; Le Fevre, O.; Hammer, F.; Tresse, L.; Lilly, S.J.; Schade, D.J. Surveys for Z > 1 field galaxies. *arXiv* **1994**, arXiv:astro-ph/9410041.
396. Kochanek, C.S.; Falco, E.E.; Schild, R. The FKS gravitational lens survey. *Astrophys. J.* **1995**, *452*, 109–139. [CrossRef]
397. Jaunsen, A.O.; Jablonski, M.; Pettersen, B.R.; Stabell, R. The not gl survey for multiply imaged quasars. *Astron. Astrophys.* **1995**, *300*, 323.
398. Riess, A.G.; Press, W.H.; Kirshner, R.P. A Precise distance indicator: Type Ia supernova multicolor light curve shapes. *Astrophys. J.* **1996**, *473*, 88. [CrossRef]
399. Garnavich, P.M.; Jha, S.; Challis, P.; Clocchiatti, A.; Diercks, A.; Filippenko, A.V.; Gilliland, R.L.; Hogan, C.J.; Kirshner, R.P.; Leibundgut, B.; et al. Supernova Limits on the Cosmic Equation of State. *Astron. Astrophys.* **1998**, *509*, 74–79. [CrossRef]
400. Myers, S.T.; Jackson, N.J.; Browne, I.W.A.; de Bruyn, A.G.; Pearson, T.J.; Readhead, A.C.S.; Wilkinson, P.N.; Biggs, A.D.; Blandford, R.D.; Fassnacht, C.D.; et al. The Cosmic Lens All-Sky Survey—I. Source selection and observations. *Mon. Not. R. Astron. Soc.* **2003**, *341*, 1–12. [CrossRef]
401. Browne, I.W.A.; Wilkinson, P.N.; Jackson, N.J.F.; Myers, S.T.; Fassnacht, C.D.; Koopmans, L.V.E.; Marlow, D.R.; Norbury, M.; Rusin, D.; Sykes, C.M.; et al. The Cosmic Lens All-Sky Survey—II. Gravitational lens candidate selection and follow-up. *Mon. Not. R. Astron. Soc.* **2003**, *341*, 13–32. [CrossRef]

402. Gurvits, L.I.; Kellermann, K.I.; Frey, S. The "angular size-redshift" relation for compact radio structures in quasars and radio galaxies. *Astron. Astrophys.* **1999**, *342*, 378.

403. Atteia, J.L.; Cordier, B.; Wei, J. The SVOM mission. *Int. J. Mod. Phys. D* **2022**, *31*, 2230008. [CrossRef]

404. Amati, L.; O'Brien, P.T.; Götz, D.; Bozzo, E.; Santangelo, A.; Tanvir, N.; Frontera, F.; Mereghetti, S.; Osborne, J.P.; Blain, A.; et al. The THESEUS space mission: Science goals, requirements and mission concept. *Exp. Astron.* **2021**, *52*, 183–218. [CrossRef]

405. Fardon, R.; Nelson, A.E.; Weiner, N. Dark energy from mass varying neutrinos. *JCAP* **2004**, *10*, 5. [CrossRef]

406. Peccei, R.D. Neutrino models of dark energy. *Phys. Rev.* **2005**, *D71*. [CrossRef]

407. Chitov, G.Y. Quintessence, Neutrino Masses and Unification of the Dark Sector. *arXiv* **2013**, arXiv:1112.4798. [CrossRef]

408. Asaka, T.; Shaposhnikov, M. The νMSM, dark matter and baryon asymmetry of the universe. *Phys. Lett. B* **2005**, *620*, 17–26. [CrossRef]

409. Asaka, T.; Blanchet, S.; Shaposhnikov, M. The nuMSM, dark matter and neutrino masses. *Phys. Lett. B* **2005**, *631*, 151–156. [CrossRef]

410. Bezrukov, F.; Shaposhnikov, M. The Standard Model Higgs boson as the inflaton. *Phys. Lett. B* **2008**, *659*, 703–706. [CrossRef]

411. Shaposhnikov, M.; Tkachev, I. The νMSM, inflation, and dark matter. *Phys. Lett. B* **2006**, *639*, 414–417. [CrossRef]

412. Dimopoulos, K.; Valle, J.W.F. Modeling quintessential inflation. *Astropart. Phys.* **2002**, *18*, 287–306. [CrossRef]

MDPI

St. Alban-Anlage 66

4052 Basel

Switzerland

www.mdpi.com

Universe Editorial Office

E-mail: universe@mdpi.com

www.mdpi.com/journal/universe